T0329331

MARINE ECOTOXICOLOGY

MARINE ECOTOXICOLOGY

CURRENT KNOWLEDGE AND FUTURE ISSUES

Edited by

Julián Blasco

Peter M. Chapman

Olivia Campana

Miriam Hampel

AMSTERDAM • BOSTON • HEIDELBERG • LONDON
NEW YORK • OXFORD • PARIS • SAN DIEGO
SAN FRANCISCO • SINGAPORE • SYDNEY • TOKYO
Academic Press is an imprint of Elsevier

Academic Press is an imprint of Elsevier
125 London Wall, London EC2Y 5AS, United Kingdom
525 B Street, Suite 1800, San Diego, CA 92101-4495, United States
50 Hampshire Street, 5th Floor, Cambridge, MA 02139, United States
The Boulevard, Langford Lane, Kidlington, Oxford OX5 1GB, United Kingdom

Library of Congress Cataloging-in-Publication Data
A catalog record for this book is available from the Library of Congress

British Library Cataloguing-in-Publication Data
A catalogue record for this book is available from the British Library

ISBN: 978-0-12-803371-5

Publisher: Candice Janco
Acquisition Editor: Laura Kelleher
Editorial Project Manager: Emily Thomson
Production Project Manager: Edward Taylor
Designer: Greg Harris

Typeset by TNQ Books and Journals

Contents

7. Sediment Toxicity Testing

S.L. SIMPSON, O. CAMPANA, K.T. HO

8. Mesocosm and Field Toxicity Testing in the Marine Context

A.C. ALEXANDER, E. LUIKER, M. FINLEY, J.M. CULP

9. Ecological Risk and Weight of Evidence Assessments

P.M. CHAPMAN

10. Global Change

J.L. STAUBER, A. CHARITON, S. APTE

Contributors

A.C. Alexander University of New Brunswick, Fredericton, NB, Canada

D. Álvarez-Muñoz Catalan Institute for Water Research, Girona, Spain

B. Anderson University of California, Davis, CA, United States

S. Apte CSIRO Land and Water, Kirrawee, NSW, Australia

D. Barceló Catalan Institute for Water Research, Girona, Spain; IDAEA-CSIC, Barcelona, Spain

J. Blasco Institute for Marine Sciences of Andalusia (CSIC), Puerto Real, Cádiz, Spain

O. Campana University of York, York, United Kingdom

P.M. Chapman Chapema Environmental Strategies Ltd, North Vancouver, BC, Canada

A. Chariton CSIRO Oceans and Atmosphere, Kirrawee, NSW, Australia

J.M. Culp University of New Brunswick, Fredericton, NB, Canada

M. Finley Government of Wisconsin, Madison, WI, United States

D.R. Fox Environmetrics Australia, Beaumaris, VIC, Australia; University of Melbourne, Parkville, VIC, Australia

M. Hampel University of Cadiz, Puerto Real, Cádiz, Spain

K.T. Ho U.S. Environmental Protection Agency, Narragansett, RI, United States

T. Jager DEBtox Research, De Bilt, The Netherlands

M. Llorca IDAEA-CSIC, Barcelona, Spain

E. Luiker University of New Brunswick, Fredericton, NB, Canada

M.L. Martín Díaz University of Cadiz, Puerto Real, Cádiz, Spain

B. Phillips University of California, Davis, CA, United States

S.L. Simpson CSIRO Land and Water, Sydney, NSW, Australia

J.L. Stauber CSIRO Land and Water, Kirrawee, NSW, Australia

W.-X. Wang Hong Kong University of Science and Technology, Kowloon, Hong Kong

Preface

Oceans, coastal zones, and transitional waters such as estuaries and coastal lagoons are zones of high productivity, which are critical to the overall function of our Earth's ecosystem. They can and are being adversely affected by a variety of stressors ranging from climate and habitat change, invasive/introduced species, eutrophication including harmful algal blooms, to chemical contaminants.

Ecotoxicology, which studies the effects of the above stressors, is a relatively new discipline. The term "ecotoxicology" was initially coined by Truhaut in 1969 (Truhaut, 1977). It is defined as the science that predicts effects of potentially toxic agents or other stressors to natural ecosystems and to target species (Hoffman et al., 2003).

Previous books on this topic have focused on freshwater rather than marine ecosystems. Thus, the Editors of this book saw the need for a book dedicated to Marine Ecotoxicology for undergraduate and postgraduate students, which provides reference information that should also be of use to senior researchers, particularly related to future research needs.

This book, which is divided into 10 chapters, was developed with the cooperation and input of some of the most senior researchers in the field. The chapters in this book proceed sequentially as follows: contaminants, statistics, contaminant modeling, bioaccumulation and biomonitoring, biomarkers, water column toxicity tests, sediment toxicity tests, mesocosm and field tests, ecological risk and weight of evidence, and global change.

We, the Editors, learned a lot from the authors of the 10 chapters. We hope that you, the readers, also learn a lot and use the knowledge provided herein to further advance the science of marine ecotoxicology.

Julián Blasco
Peter M. Chapman
Olivia Campana
Miriam Hampel

References

Truhaut, R., 1977. Ecotoxicology: objectives, principles and perspectives. Ecotoxicol. Environ. Saf. 1, 151–173.

Hoffman, D.J., Rattner, B.A., Burton Jr., G.A., Cairns Jr., J., 2003. Introduction. In: Hoffman, D.J., Rattner, B.A., Burton Jr., G.A., Cairns Jr., J. (Eds.), Handbook of Ecotoxicology, second ed. Lewis Publishers, Boca Raton, FL, USA, pp. 1–15.

Acknowledgments

We thank all of the contributors for their excellent and hard work. This book would not have been possible without them. We also thank Elsevier and especially Emily Thomson, our Editorial Project Manager, for continuing positive support. Finally, we thank our respective families for their support and understanding of the time we spent on this project.

CHAPTER

1

Contaminants in the Marine Environment

D. Álvarez-Muñoz[1], *M. Llorca*[2], *J. Blasco*[3], *D. Barceló*[1,2]

[1]Catalan Institute for Water Research, Girona, Spain; [2]IDAEA-CSIC, Barcelona, Spain; [3]Institute for Marine Sciences of Andalusia (CSIC), Puerto Real, Cádiz, Spain

List of Abbreviations

AE Atomic emission
AEO Alcohol polyethoxylates
AES Alkyl ether sulfates
APCI Atmospheric pressure chemical ionization
APEO Alkylphenol polyethoxylates
APPI Atmospheric pressure photoionization
AS Alkyl sulfates
ASP Amnesic shellfish poisoning
AZP Azaspiracid shellfish poisoning
BFRs Brominated flame retardants
BPS Brominated polystyrene
CFP Ciguatera fish poisoning
CNMS Carbon-based nanomaterials
CPE Cloud point extraction
D3 Hexamethylcyclotrisiloxane
D4 Octamethylcyclotetrasiloxane
D5 Decamethylcyclopentasiloxane
D6 Dodecamethylcyclohexasiloxane
DAD Diode array detector
deca-BDE Decabromodiphenylether
dl Dioxin-like
DLLME Dispersive liquid–liquid microextraction
DSP Diarrheic selfish poisoning
DW Dry weight
ECD Electron capture detector
EDCs Endocrine-disrupting compounds
EFSA European food safety agency
ELISA Enzyme-linked immunosorbent assay
ESI Electrospray ionization
FLD Fluorescence detector
FOSA C8 sulfonamide
GBC Graphitized black carbon
GC Gas chromatography
GF-AAS Graphite furnace atomic absorption spectrometry
GPC Gel permeation chromatography
HBCD Hexabromocyclododecane
HRMS High-resolution mass spectrometer
HSSE High-speed solvent extraction
HTpSPE High through planar SPE
ICP Inductively coupled plasma
ICP-MS ICP-mass spectrometry
IS Internal standards
K_d Water–sediment partition coefficient
K_{ow} Octanol–water partition coefficient
LAS Linear alkylbenzene sulfonates
LC Liquid chromatography
LLE Liquid–liquid extraction
MAE Microwave-assisted extraction
MASE Microwave-assisted solvent extraction
MP Microplastics
MS Mass spectrometer
MSPD Matrix solid phase dispersion
NMs Nanomaterials
NPs Nanoparticles
NSP Neurologic shellfish poisoning
NTA Nanoparticle tracking analysis

Marine Ecotoxicology
http://dx.doi.org/10.1016/B978-0-12-803371-5.00001-1

PAHs Polycyclic aromatic hydrocarbons
PBBs Polybrominated biphenyls
PBDEs Polybrominated diphenylethers
PCBs Polychlorinated biphenyls
PCDDs Polychlorinated dibenzo-*p*-dioxins
PCDFs Polychlorinated dibenzofurans
PDMS Polydimethylsiloxanes
PFASs Per- and polyfluorinated alkyl substances
PFOS Perfluorooctane sulfonate
PLE Pressurized liquid extraction
PPCPs Pharmaceuticals and personal care products
PSP Paralytic shellfish poisoning
QuEChERS Quick, easy, cheap, effective, rugged, and safe
SAX Strong anionic exchange
SBSE Stir bar sorptive extraction
SCX Strong cationic exchange
SFE Supercritical fluid extraction
SPE Solid phase extraction
SPME Solid phase microextraction
SWE Subcritical water extraction
TBBPA Tetrabromobisphenol A
TCDD 2,3,7,8-Tetrachlorodibenzo-*p*-dioxin
TEF Toxic effect
TEM Transmission electron microscopy
TR-FIA Time-resolved fluoroimmunoassay
USASE Ultrasound-assisted solvent extraction
USE Ultrasonic-assisted extraction
UV Ultraviolet
WW Wet weight
WWTPs Wastewater treatment plants

1.1 INTRODUCTION

There are thousands of chemicals present in the marine environment due to human activities. In any industrialized society, a wide variety of contaminants are released to the environment every day from residential, commercial, and industrial uses. All of them make a complex mixture of hazardous chemicals that poses a potential risk not only to wildlife but also to human health through the possible ingestion of contaminated seafood. In this chapter a revision of the chemical substances that pollute the marine environment is presented. Metals, persistent organic contaminants, and emerging organic contaminants are included. Their main sources of contamination and physicochemical properties are described, as well as the most used analytical techniques for their detection and quantification in the marine environment. Their usual levels found in seawater, sediments, and marine organisms are reported, providing an exposure scenario to wildlife that will be further studied along the book. Besides, there is a special mention to microplastics, a group of contaminants that has generated an increasing concern in the scientific and general community. Regarding analytical approaches applied for measuring these contaminants, the use of nontarget techniques is pointed out as the way forward for future environmental monitoring studies.

1.2 SOURCES AND PROPERTIES

1.2.1 Metals

The term "heavy metals" in ecotoxicology is related to environmental pollutants, whereas "trace metals" correspond to metals only at trace concentrations (0.01%). In many occasions, both terms have been employed for referring to the same metals. Nieboer (1980) proposed a chemical classification based on Lewis acid properties on metal ions, separating in Class A, B, and Borderline according to their degree of "hardness" or "softness" as acids or bases. Although this classification is well supported by chemical characteristics, the mentioned terms "heavy metals" and "trace metals" are frequently employed in the scientific literature. We are going to use the term "metals" for fitting metals with ecotoxicological relevance (Ag, As, Au, Bi, Cd, Co, Cr, Cu, Fe, Ga, Hg, In, Ir, Mn, Mo, Ni, Pb, Pd, Pt, Rh, Sb, Se, Sn, Ti, V, and Zn).

Metals occur in the rocks of the Earth and soil resulting from erosion processes. They can be transported by stream and rivers in dissolved form or associated to particulate matter. By atmosphere, they are transported on particulate matter, aerosols, and for some metals (eg, Hg) as vapor. Besides the natural source of metals, anthropogenic activities are responsible of metal inputs in the environment. The metals are introduced into seawater by river runoff, atmospheric transport, hydrothermal venting, groundwater seeps, and diffusion from sediment and transport for outer space: although only the first three fluxes are considered the main inputs. The relationship between fluxes and mass balances permit to establish the relationship between both metal sources. The major pathway by which anthropogenic metals enter in the ocean is via aeolian input. Libes (2009) summarized the natural and anthropogenic fluxes of metals showing the ratio between atmospheric/riverine inputs to the oceans, these ratios ranged between close 300 for Pb and 3.6 for As. In contrast to other pollutants metals are nonbiodegradables. Their fate, behavior, and toxicity are controlled by the chemical and physicochemical characteristics of marine environment. Metals can be adsorbed (adsorption process is defined as the reversible binding of a solute on a solid matrix in seawater or sediment) and partially scavenged by water column, and they settle down on the sediment. Colloid particles in the water column (clay, Fe and Mn oxides or hydroxides, calcium carbonate, and organic matter) act as reactive surfaces for binding metals. Nevertheless, metal speciation is the main process that controls its behavior, reactivity, toxicity, and ecological risk. Many elements dissolved in seawater can be present as various inorganic complexes; thus mercury is present mostly as $HgCl_4^{2-}$ and other elements are as chloride complex mostly ($AgCl_3^{2-}$, $AuCl_2^-$) and others are as carbonate complexes or hydroxyl ions. To resolve the speciation in complex systems, several computer programs are available (eg, MINEQL, MINTEQ, MINEQL+, WHAM, and CHESS). These programs are useful to examine different processes and for calculation of chemical equilibrium including dissolved species, precipitation, dissolution, redox transformations, and adsorption process [for more detail read Chapter 4 in Mason (2013)].

1.2.2 Persistent Organic Contaminants

1.2.2.1 Polycyclic Aromatic Hydrocarbons

Polycyclic aromatic hydrocarbons (PAHs) are one of the most important classes of environmental pollutants; they are common by-products of combustion processes. Although they are produced naturally by forest fires and volcanoes, most of PAHs in the environment are the result of industrial and other human activities such as processing of coal, crude oil, combustion of natural gas, including for heating, combustion of refuse, vehicle traffic, cooking, and tobacco smoking (WHO, 2000). Their molecular structure is formed exclusively by carbon and hydrogen atoms, consisting in two or more fused benzene ring arrangements. They have a relatively low solubility in water and it decreases approximately logarithmically when molecular mass increases (Johnsen et al., 2005). By contrast, they are highly lipophilic and many of them adsorb to suspended particulate matter, which then settles through the water column, and accumulate in aquatic sediments (Karickhoff et al., 1979). Besides, most PAHs have low vapor pressure and in the air heavy PAHs (more than four rings) tend to adsorb to particulate matter, while lighter PAHs (less than four rings) tend to remain gaseous until they are removed via precipitation (Skupinska et al., 2004). Therefore, PAHs may enter the environment either by adsorbing particulate matter through atmosphere and aquatic environments, reaching long distances, or by dissolution, resulting from more local sources (Douben, 2003). This is the case of two-ring PAHs, and to a lesser extent three-ring PAHs, that dissolve in water making them more available for aquatic biota (Mackay and Callcott, 1998).

1.2.2.2 Surfactants

Surfactants are one of the most ubiquitous families of organic compounds. Although their main application is the formulation of household and industrial detergents, they are also used in the manufacturing of other products such as cosmetics, paints, textile, dyes, polymers, agrochemicals, petroleum. Their molecular structure is characterized by consisting of two parts: a hydrophilic and a hydrophobic one. The presence of polar and nonpolar groups in each molecule is essential for detergency and solubilization of surfactants and allows them to form micelles in solution. The dirt is solubilized inside the micelles avoiding re-deposition and enabling the washing process. Surfactants are classified in four groups according to the nature of its ionic charge in solution: anionic, cationic, nonionic, and amphoteric. This classification is based on the hydrophilic part of the molecule because the hydrophobic part is always nonionic and generally it is constituted by an alkyl chain. The term anionic refers to the negatively charged surfactants, those with positive charge are called cationic, nonionic are the uncharged surfactants, and amphoteric are those ones that may be positively or negatively charged depending on the pH of the medium. The most important are the anionic and nonionic ones for their versatility and high consumption. Common anionic surfactants are linear alkylbenzene sulfonates (LAS), alkyl sulfates (AS), alkyl ether sulfates (AES), and soap. The major groups of nonionic surfactants are alcohol polyethoxylates (AEO) and alkylphenol polyethoxylates (APEO) among others.

Surfactants are soluble compounds and their main sources of contamination are effluents of wastewater treatment plants (WWTPs). However, their solubility decreases with the length of the alkyl chain and, depending on the surfactant type, it also varies with the hydrophilic group (position or number of units). They are low volatile compounds with the exception of nonylphenol that have been detected in the atmosphere (Van Ry et al., 2000). The octanol–water partition coefficient (K_{ow}) depends on the surfactant type and it increases with the hydrophobicity of the molecule. It has been long recognized that compounds accumulate in organisms to an extent that is directly related to the magnitude of their K_{ow}. In the case of surfactants, the water–sediment partition coefficient (K_d) values are generally higher than the K_{ow} which indicates a higher affinity of surfactants for sediments.

1.2.2.3 Polychlorinated Biphenyl

Polychlorinated biphenyls (PCBs) are synthetic substances manufactured and used worldwide from the 1930s to mid-1970s (Solaun et al., 2015). They are characterized by the attachment of chlorine atoms (between 1 and 10) to a biphenyl structure, with a common chemical formula $C_{12}H_{10-x}Cl_x$. This group of compounds comprises 209 structurally related congeners (Solaun et al., 2015) and 130 of the different PCB arrangements and orientations are commercially used (United_Nations, 1999). The main properties of PCBs include high chemical stability, high heat capacity, low flammability, and insulating properties, which allow their use in a wide variety of industrial and commercial applications. For example, they are used as dielectric fluids in transformers and capacitors, in printing ink, paints, dusting agents, pesticides, hydraulic fluids, plasticizers, adhesives, fire retardants, and lubricants (Erickson and Robert, 2011; Solaun et al., 2015). Although manufacture of PCBs is forbidden under the Stockholm Convention on Persistent Organic Pollutants, they are still released to the environment by disposal of large-scale electrical equipment and waste (WHO, 2010). They are also included in the Water Framework Directive (European_Commission, 2002) for their monitoring. In the case of marine environment, the major routes of PCB inputs are coastal discharges from rivers, urban and industrial outfalls, and atmospheric deposition (Dörr and Liebezeit, 2009; Solaun et al., 2015; UNEP, 2002). However, due to their lipophilic nature, PCBs are not expected to remain in water compartments. This group of compounds tends to accumulate within the lipid-rich tissue of biota and sediments (Erickson and Robert, 2011; Solaun et al., 2015). In addition, they can adversely affect reproduction and may affect immune systems (OSPAR_Commission, 2010). The most toxic PCBs are those with a planar structure which have similar chemical properties to dioxins and furans (OSPAR_Commission, 2010). Twelve of these compounds are considered as dioxin-like (dl)-PCBs because, under certain conditions, they may form dibenzofurans through partial oxidation (Tuomisto, 2011).

1.2.2.4 *Pesticides*

Pesticides are designed to kill, reduce, or repel insects, weeds, rodents, fungi, or other organisms that can threaten public health or economy. Their mode of action is by targeting systems or enzymes in the pests which may be identical or very similar in human beings; therefore, they pose risks to human health and the environment. They can be classified by target pest (ie, insecticides, fungicides, rodenticides, pediculicides) or by chemical identity (ie, pyrethroids, organophosphates, organochlorines, carbamates, glyphosate, triazoles). For example, all pesticides in the class organophosphate are derivates of phosphoric acid, and all organochlorines are composed of carbon, hydrogen, and chlorine. The pesticides formulation includes the main molecule of pesticide (like the ones previously mentioned) called active ingredient, and other ingredients that can be inert or may enhance the pesticidal properties. Typical formulations of pesticides include liquids, dusts, wettable powders, and emulsifiable concentrates. They are used to protect plants within agriculture, forestry, and horticulture, and as biocidal products (eg, timber impregnation, boat keel paints, and slime control products). They are usually spread on arable land and enter the aquatic environment through surface water runoff or as a result of "spray drift." Percolation of water to the drainage system can carry pesticides out into rivers and afterward to the sea. Leaching of water through the soil can also carry pesticides down to the groundwater. Therefore, two chemical characteristics of high interest are water solubility and volatility. The more water soluble a pesticide is, the greater is the potential for runoff and leaching. The more volatile a pesticide is, the greater is the potential for drift. The water solubility value is given for the active ingredient at room temperature, either 20 or 25°C, and is usually presented as milligrams of solute per liter of water (mg/L). The vapor pressure is normally given in millimeters of mercury (mm Hg). A pesticide with low vapor pressure does not escape into air so there is a potential to accumulate in water if it is soluble. If not water soluble, the pesticide may accumulate in soil or biota depending on its octanol–water partition coefficient (K_{ow}).

1.2.2.5 *Dioxins*

This group includes polychlorinated dibenzo-*p*-dioxins (PCDDs), dl compounds such as polychlorinated dibenzofurans (PCDFs; 10 of them), and some PCBs (specifically 12 of them which form dibezofurans under partial oxidation). Although the term "dioxin" implies other conformations, PCDDs/PCDFs and dl-PCBs are referred to as dioxins because their molecules contain a dibenzo-1,4-dioxin skeletal structure with 1,4-dioxin as the central ring. PCDDs and PCDFs are the by-products of industrial processes including the manufacture of chlorophenols and phenoxy herbicides, chlorine bleaching of paper pulp, and smelting. They can also be generated by natural events such as volcanic eruptions and forest fires (WHO, 2010). In addition, PCDFs were common contaminants of commercial PCB mixtures (WHO, 2010). The main sources of contamination are their releases to air from inadequate incineration and from contaminated soils and aquatic sediments (WHO, 2010). Once there, dioxin and dl compounds are stable against chemical and microbiological degradation. They tend to bioaccumulate in tissues rich in lipids and bioconcentrate through food chains due to their high fat solubility (WHO, 2010). In addition, because of their high number of chlorinated components, they persist longer in the environment and show greater bioaccumulation leading to biomagnifications in marine food chain (WHO, 2010). These factors increase their potential hazards to humans and animals. The most toxic dioxin is 2,3,7,8-tetrachlorodibenzo-*p*-dioxin (TCDD) which, per definition, has a toxic effect (TEF) of one (Van den Berg et al., 2006).

1.2.3 Emerging Organic Contaminants

1.2.3.1 Per- and Polyfluorinated Alkyl Substances

Per- and polyfluorinated alkyl substances (PFASs) are manmade compounds synthesized for more than 60 years. The main molecular structure contains the moiety "$C_nF_{2n+1}^-$" (Buck et al., 2011). PFASs are a wide group of compounds varying in their structure and, thus, exhibit different properties, environmental fate, and toxicity, but their common trend is a general high stability given by the carbon-chain bond (one of the strongest in nature) (Llorca, 2012). Perfluoroalkanes present a double hydrophobic and oleophobic character, and when they are mixed with hydrocarbons and water, form three immiscible phases. These compounds are employed as fire-resistant additives and oil, stain, grease, and water repellents (Llorca, 2012). They are used to provide nonstick surfaces on cookware and waterproof, breathable membranes for clothing, and in many industry segments including the aerospace, automotive, building/construction, chemical processing, electronics, semiconductors, and textile industries (EPA, 2009; Llorca, 2012). Their main environmental sources have been identified from direct (discharges or emissions at production sites, aqueous fire-fighting foams, and consumer and industrial products) and from indirect (degradation of perfluoroalkyl-based products, paper industry, WWTPs, or cooking) sources (Llorca, 2012). Most of the PFASs are physical, chemical, and biologically stable and have been found to be widely spread in the environment (Llorca, 2012), and therefore they have been considered as POPs. For example, the salt perfluorooctane sulfonate (PFOS) has been included under the Stockholm Convention (USEPA, 2006).

1.2.3.2 Brominated Flame Retardants

Flame retardants are chemicals that are added to or reacted with combustible materials to increase their fire resistance (WHO, 1997). They are applied to combustibles materials like plastics, woods, paper, and textiles. Flame-retarded polymers are widely used in cars, consumer electronics, computers, electrical equipment, and building materials. They can find their way into the environment as wastewaters of industrial facilities that produce FRs and manufacturing facilities that incorporate such compounds into products, through volatilization and leaching from products during manufacturing or usage, upon breakdown of foam products, or by disposal of products (eg, electronic equipments), through leaching from landfills, combustion and recycling of waste products, or adsorption onto dust particles (Alaee et al., 2003; Murphy, 2001). They are classified into four major chemical groups: inorganic, organophosphorous, halogenated organic, and nitrogen-based compounds (Alaee and Wenning, 2002). Halogenated organic flame retardants are further classified as containing either chlorine or bromine. Approximately 25% of all FRs (volume basis) contain bromine (Andersson et al., 2006). They can be divided into different subgroups depending on their mode of incorporation into the polymers. Additive brominated flame retardants (BFRs) are mixed together with the other components of the polymers; they comprise compounds such as polybrominated biphenyls (PBBs), polybrominated diphenylethers (PBDEs), and hexabromocyclododecane (HBCD). Reactive BFRs are a group of compounds such as tetrabromobisphenol A (TBBPA) that are chemically bonded to the plastics. Polymeric BFRs include compounds like brominated polystyrene (BPS), where bromine atoms are incorporated in the backbone of the polymer resulting in a more stable chemical structure with very high molecular weight, low volatility, bioavailability, and toxicity (Guerra et al., 2011). PBBs, PBDEs, TBBPA, and HBCD are also characterized by low vapor pressures and low solubility in water. In general, volatility decreases with rising bromine content; therefore highly brominated compounds are estimated to be nonvolatile, while low-brominated compounds will be more mobile in water and have a greater tendency to evaporate from surface water. The values of log K_{ow} range from 4.5 for low-brominated compounds (TBBPA) to 10 for decabromodiphenylether (deca-BDE) showing high bioaccumulation in aquatic biota for low-brominated compounds.

1.2.3.3 *Endocrine-Disrupting Compounds*

Endocrine disruptors are compounds with the ability to interfere with the endocrine system of different organisms causing important alterations in their normal development. They can produce adverse reproductive, neurological, and immune effects in both humans and wildlife. The mechanisms of endocrine-disrupting compounds (EDCs) involve divergent pathways including (but not limited to) estrogenic, antiandrogenic, thyroid, peroxisome proliferator-activated receptor γ, retinoid, and actions through other nuclear receptors; steroidogenic enzymes; neurotransmitter receptors and systems; and many other pathways that are highly conserved in wildlife and humans (Diamanti-Kandarakis et al., 2009). A wide range of chemicals can cause endocrine disruption; so this group encompasses a heterogeneous class of molecules with different structures and physicochemical properties depending on type of contaminant. The main sources of contamination will also vary with this. EDCs include insecticides, fungicides, herbicides, pharmaceuticals, industrial contaminants (Hotchkiss et al., 2008). Concretely bisphenols, alkylphenols, estrogens, and perfluorinated compounds are prevalent over other EDCs in the aquatic ecosystem (Vandermeersch et al., 2015).

1.2.3.4 *Pharmaceuticals and Personal Care Products*

Pharmaceutically active compounds play an important role in assuring both humans and animals health. They are complex molecules with different functionalities and physicochemical and biological properties. Under environmental conditions, they can be neutral, cationic, anionic, or zwitterionic and they also often have basic or acidic functionalities. Normally pharmaceuticals are classified according to their therapeutical purpose in different groups or families such as antibiotics, psychiatric drugs, analgesics, anti-inflammatories, tranquilizers, hormones, β-blockers, diuretics. Depending on their solubility and intestinal permeability, pharmaceuticals can be divided into four classes: class I is high soluble and high permeable, class II low soluble and high permeable, class III high soluble and low permeable, and class IV low soluble and low permeable. Any pharmaceutical compound to be absorbed by the organism must be present in the form of solution at the site of absorption and therefore different techniques like particle size reduction, crystal engineering, salt formation, solid dispersion, use of surfactant and complexation are used for the enhancement of the solubility of poorly soluble drugs. Once pharmaceuticals are administrated, they can be biotransformed by the organism into more polar molecules; however, metabolism is frequently incomplete and excretion rates range from 0% to 100%. Consequently one of the main sources of environmental contamination is sewage effluents. Other important sources are waste disposal, aquaculture, animal husbandry, and horticulture (Gaw et al., 2014). They provide a continuous input of these compounds into the environment and although pharmaceuticals have high transformation/removal rates they are compensated by their continuous introduction into the aquatic systems.

Personal care products include a wide range of compounds such as disinfectants (eg, triclosan), fragrances (eg, musks), insect repellants (eg, DEET), preservatives (eg, parabens), and ultraviolet (UV) filters (eg, methylbenzylidene camphor). They are products destined to external use on the human body including cosmetics, gels, and soaps among others (Brausch and Rand, 2011). The external body applications limit any metabolic change and therefore they enter into the environment unaltered through regular usage (Ternes et al., 2004). Some studies revealed that many of these compounds are environmentally persistent, are bioactive, and have the potential for bioaccumulation (Brausch and Rand, 2011).

1.2.3.5 Marine Biotoxins

Marine biotoxins are naturally occurring chemicals synthesized by marine microalgae (phytoplankton or benthic microalgae). Under specific climatic and hydrographic conditions, phytoplankton species can have a high proliferation rate resulting in high-density algae clouds (blooms) (Gerssen et al., 2010). These blooms are sometimes beneficial to aquaculture and marine biology. To date, it is known that 40 species belonging to the classes of dinoflagellates and diatoms algae produce marine toxins (named phycotoxins) during algae blooms (Gerssen et al., 2010). The structure of these molecules differs in size and conformation although have cyclic and aliphatic chains with C, H, O and, sometimes, N atoms in common. The structure of the most studied toxins can be seen elsewhere (Gerssen et al., 2010). The toxins can be accumulated in fish, shellfish, and other marine organisms without representing a toxicological problem for them. However, when contaminated shellfish are consumed by humans, marine toxins can cause diseases that range from gastrointestinal to cardiologic and to neurological problems (EURLMB, 2016; Hallegraeff et al., 1995). Different groups are of major human concern including hydrophilic toxins: (1) domoic acid which causes amnesic shellfish poisoning (ASP) and (2) saxitoxins causing paralytic shellfish poisoning (PSP) (Gerssen et al., 2010). Among lipophilic toxins, it is noteworthy to mention (1) brevetoxins causing neurologic shellfish poisoning (NSP), (2) okadaic acid, dinophysistoxins, and pecetenotoxins that cause diarrheic selfish poisoning (DSP), (3) azaspiracids, responsible of azaspiracid shellfish poisoning (AZP) (Gerssen et al., 2010), and (4) ciguatoxins, which belongs to a family of toxins that causes ciguatera fish poisoning (CFP) which is linked to a wide variety of gastrointestinal, neurological, or cardiovascular symptoms (Vandermeersch et al., 2015). The expansion of harmful and toxic algal blooms is a topic of major concern nowadays (Van Dolah, 2000). Because of the expansion and the effects in humans, the European Food Safety Agency (EFSA) has recommended different allowable limits in selfish for some of them (Alexander et al., 2008a,b,c, 2009a,b,c).

1.2.3.6 Nanomaterials

Nanomaterials (NMs) are commonly classified into carbon-based NMs, metal oxide nanoparticles (NPs), metal NPs, polymeric NMs, and quantum dots (Sanchís, 2015). This section will be focused on carbon-based nanomaterials (CNMs). NMs are emitted into the environment because of natural and anthropogenic sources (either incidental or nanotechnological sources) (Sanchís, 2015). Natural sources of CNM include combustion and other highly energetic processes such as volcanic eruptions (Buseck and Adachi, 2008; Jehlička et al., 2003), lightning (Daly et al., 1993), sandstorms (Gu et al., 2003), and wildfires (Sanchís, 2015). In the case of anthropogenic sources, it should be mentioned that combustion is the main source of CNMs, followed by friction processes and mining and, finally, nanotechnology that intentionally manufactures NMs (Sanchís, 2015). The group of CNMs include amorphous carbon NPs and different carbon allotropes (Hirsch, 2010), with the distinctive properties of sp^2- and sp^3-hybridized carbon bonds with characteristics of their physics and chemistry at the nanoscale (Sanchís, 2015). Within this group, the most studied compounds are pristine fullerenes: unsaturated unfunctionalized carbon molecules with 20 to several hundreds of carbon atoms arranged as a hollow polyhedron (Astefanei et al., 2015; Maruyama et al., 1991; Prinzbach et al., 2000; Sanchís, 2015).

1.2.3.7 *Polydimethylsiloxanes*

Polydimethylsiloxanes (PDMS) are mainly used as precursors in the production of silicones although they also have applications as fuel additives in cleaners, car waxes, and polishes. They are also used directly in hygiene and personal care products (cosmetics, deodorants, hair conditioners, skin creams), in biomedical applications (implants for cosmetic surgery), and in household products. They have a polymeric molecular structure that contains a backbone of alternating atoms of silicon and oxygen with organic side groups such as methyl, phenyl, or vinyl, attached to silicon. Therefore, the basic unit is $[SiO(CH_3)_2]_n$ where n, the number of repeating monomers, can range from zero to several thousand. By adjusting the –SiO– chain lengths, the functionality of the side groups and the cross-linking between molecular chains, silicones can be synthesized into an almost infinite variety of materials, each with unique chemical properties and performance characteristics. Besides, they can be linear or cyclic, the second ones being more difficult to degrade in the environment. In general PDMS are poorly soluble compounds with high Henry's law constant (Hamelink et al., 1996) which make them very volatile. Ninety percent of the production evaporates into the atmosphere where it is mainly decomposed into hydroxyl radicals (Dewil et al., 2007) and only a 10% reach WWTPs (Allen et al., 1997). Once they are in the WWTPs, they can also volatilize but due to their water–sediment partition coefficient ($K_d = 2.2–5.0$) they also adsorb strongly to sludge which makes their volatilization difficult (David et al., 2000; Whelan et al., 2009). The use of the sludge as a fertilizer is a major route of entry of PDMS into the environment, especially in soils. Although trace concentrations of PDMS are detected in effluents, they represent a chronic source of contamination to the aquatic environment. Once PDMS are released into the water, they tend to associate with or dissolve particulate organic matter and settle down; therefore, the highest concentrations are usually found in sediments near wastewater discharge areas (Sparham et al., 2011).

1.3 ANALYTICAL APPROACHES

1.3.1 Metals

Many of metals are found in seawater at very low concentrations and in complex matrix formed by salts and many other substances (organic and inorganic). Moreover, contamination during sampling, storage, and processing of any element that you are interested in measuring can make it really difficult to measure. These considerations have provoked the idea that data on metal concentrations before 1975 cannot be considered trustworthy. The use of ultraclean techniques has allowed collecting valuable data for many metals around the world ocean. Specifically, the program Marine Biogeochemical Cycles of Trace Elements and Isotopes (GEOTRACES) has been successful for getting information about biogeochemical cycles of metals. Currently, any marine study with the objective to obtain information about metals at trace levels need to prevent the contamination problems. To avoid contamination by trace elements during the sampling, specific samplers have been designed, and specific plastics should be employed (Kremling, 2002). Specific information about materials, treatment, and other steps of ultraclean process is out of this chapter and can be found elsewhere (Tovar-Sánchez, 2012). Several analytical techniques have been employed for metal determination in seawater such as graphite furnace atomic absorption spectrometry (GF-AAS), inductively coupled plasma (ICP)-atomic emission (AE), ICP-mass spectrometry (ICP-MS), flowing stream systems (eg, flow injection analysis, FIA; sequential injection analysis, SIA) with sensitive detection techniques (spectrophotometer or fluorometer), voltammetry (anodic or cationic stripping voltammetry ASV or CSV). Both ICP-AE and ICP-MS are the most employed techniques as a result of their multielemental characteristic, but its direct use for seawater analysis is only possible with very diluted samples which give unsatisfactory detection limits. Although the use of gas dilution is a promising alternative in new ICP-MS equipment, its usefulness at ultratrace levels has not been satisfactory until now. To resolve these problems, preconcentration steps are necessary and a wide variety of selective extraction techniques can be employed (eg, liquid–liquid extraction with chelating agents such as dithiocarbamates or chelating cation resins-Chelex 100) (Kremling, 2002).

1.3.2 Persistent Organic Contaminants

1.3.2.1 Polycyclic Aromatic Hydrocarbons

As a consequence of their high hydrophobicity in marine environments, PAHs are mainly accumulated in organisms, particulate matter, and sediments. The low level of PAHs present in water requires concentrating the samples before instrumental analysis (Li et al., 2013). Different extraction and preconcentration techniques have been used such as liquid–liquid extraction (LLE), stir bar sorptive extraction (SBSE), solid phase extraction (SPE), and solid phase microextraction (SPME) (Kruger et al., 2011; Li et al., 2015; Robles-Molina et al., 2013). However, the volume of grab samples is often a limitation factor that has been overcome by the use of passive samplers. They have been employed for assessing temporal and spatial trends of dissolved PAHs in coastal areas (Alvarez et al., 2014). Regarding the extraction of sediments and biota it has been done with nonpolar solvents and different techniques like Soxhlet (De Boer and Law, 2003), supercritical fluid extraction (SFE) (Berg et al., 1999), pressurized liquid extraction (PLE) (Burkhardt et al., 2005), subcritical water extraction (SWE) (Ramos et al., 2002), microwave-assisted extraction (MAE) (Banjoo and Nelson, 2005; Pena et al., 2006), ultrasonic-assisted extraction (USE) (Banjoo and Nelson, 2005), and QuEChERS (Quick, Easy, Cheap, Effective, Rugged, and Safe) (Johnson, 2012) among others. However, the literature indicates a predominant use of ultrasonic bath according to Navarro et al. (2009). Among the several clean-up procedures developed for PAH extracts, those based on SPE are the most convenient with satisfactory results reported for silica (Pena et al., 2006) and florisil (Burkhardt et al., 2005; Navarro et al., 2009) sorbents. Regarding instrumental analysis gas chromatography (GC), rather than liquid chromatography (LC), is often the preferred approach for separation, identification, and quantification of PAHs, largely because GC generally affords greater selectivity, resolution, and sensitivity than LC (Poster et al., 2006).

1.3.2.2 Surfactants

Surfactants are usually commercialized as complex mixtures which can comprise hundreds of different isomers, homologues, and/or ethoxymers with different physicochemical properties. Therefore the analytical approach is different depending on whether the objective is to determine the total content of surfactants of a certain group (cationic, anionic, nonionic), or of analytes belonging to different classes of chemical compounds. The first case is easier to perform and only solid or liquid extraction (depending on the matrix type) must be carried out (Roslan et al., 2010), while in the second case the analytes also need to be isolated and/or preconcentrated after the appropriate extraction. Water samples are immediately preserved by addition of a biocide (usually formaldehyde) to minimize biodegradation (Lara-Martin et al., 2006a) and they are kept in the freezer and analyzed within a short period of time (48 h). Although LLE was previously used, nowadays, SPE on octadecylsilica (C18) is the most used technique for the extraction and purification of surfactants from both liquid and solid samples. It can be used in combination with strong anionic exchange (SAX) (Leon et al., 2000; Matthijs et al., 1999) for anionic surfactants. For nonionic surfactants, graphitized black carbon (GBC) (Houde et al., 2002) and silica (C2–C18) cartridges (Petrovic et al., 2001) combined with strong cationic exchange (SCX) and SAX (Dunphy et al., 2001) have been also used. In general, most authors use C18 and GBC cartridges because they are suitable for simultaneous extraction and isolation of a wide range of surfactants including anionic, nonionic, and their metabolites in a single stage. Solid samples like sediments and biota are usually dried in a heater and/or lyophilized (Alvarez-Munoz et al., 2004; Lara-Martin et al., 2006a). After this, the extraction can be performed by using different techniques such as Soxhlet (Lara-Martin et al., 2006b; Saez et al., 2000), sonication (Lara-Martin et al., 2011; Versteeg and Rawlings, 2003), PLE (Alvarez-Munoz et al., 2007; Lara-Martin et al., 2006a), microwave-assisted solvent extraction (MASE) (Croce et al., 2003), and matrix solid phase dispersion (MSPD), specially used for the extraction of surfactants from organisms (Tolls et al., 1999). Regarding solvents used, methanol is usually preferred for the extraction of LAS, AEO, AES, and their main degradation products, while for APEO and degradation products methanol is substituted by other less polar solvents like hexane or dichloromethane (Shang et al., 1999). The clean-up stage is performed similar to that previously reported for water samples depending on the target group of surfactants. The instrumental analysis is carried out mainly by using liquid or GC (after derivatizing) coupled to mass spectrometry (Lara-Martin et al., 2006a; Reiser et al., 1997), although other detectors like ultraviolet (UV) or fluorescence detector (FLD) have been also employed (Alvarez-Munoz et al., 2004; Croce et al., 2003).

1.3.2.3 Polychlorinated Biphenyl

The analysis of PCBs in the marine environments is devoted to the study of sediments and biota because due to their high hydrophobicity they are not present in water columns. The analytical process involves an extraction of lyophilized matrices followed by a strong clean-up of the matrix. In addition, it is important to remark on the use of internal standards (IS) to monitor the correct extraction of selected compounds. IS are usually isotopically labeled compounds belonging to the group of contaminants of interest, but in the case of PCBs a congener which is not detected in the nature can be used. Sediments are commonly extracted by solid–liquid extraction. The first methodologies were carried out with Soxhlet extraction for 16 h with n-hexane and dichloromethane (1:1, v/v) (Castells et al., 2008). Although Soxhlet extraction is still in use, faster techniques such as PLE (also known as accelerated solvent extraction, ASE) have been applied too. The process is based on the extraction of the analytes by using high pressure (1000–2000 psi) and temperature (in general higher than solvent vapor temperature). This process has been used with dichloromethane as solvent extractor for the analysis of PCBs in sediments (Barakat et al., 2013). A clean-up with alumina:silica or florisil chromatographic columns was applied followed by elution of the analytes with pentane:dichloromethane (1:1, v/v) (Barakat et al., 2013). Finally the eluates were reduced under nitrogen and heated temperature before instrumental analysis.

Biota extraction is usually carried out with similar procedures than described for sediments. For example, Solaun et al. (2015) used PLE with pentane:dichloromethane (1:1, v/v) at 100°C and 1750 psi. The extraction cell was filled with florisil and Na_2SO_4 to perform a preclean-up during the extraction and remove water drops. The authors preconcentrated the sample under a gentle stream of nitrogen. Then it was reconstituted with dichloromethane and filtered through 0.45 μm before the purification step that was carried out with gel permeation chromatography (GPC). The eluate was then evaporated and reconstituted in isooctane, cleaned with Na_2SO_4, centrifuged, and the organic phase was analyzed.

The instrumental analysis for PCBs is based on GC separation coupled to electron capture detector (ECD) or mass spectrometer analyzer (MS).

1.3.2.4 Pesticides

Although multiresidue analysis of pesticides has been carried out since the 1970s (Farre et al., 2014), it still remains a challenge because different chemical classes (with a wide range of physicochemical properties) are present at low concentrations in complex matrices. The sample preparation is a step necessary to isolate and concentrate these pesticide residues from the matrix, and it usually takes most of the analysis time. SPE is normally the preferred technique for water samples preparation (Hernandez et al., 2012; Masia et al., 2013a,b), although other techniques such as LLE (Jiang et al., 2013) or passive sampling techniques (Martinez Bueno et al., 2009) have been also applied. Several types of SPE sorbents have been found to be suitable for extracting pesticides from water samples such as GCB, Oasis HLB, Strata-X, Strata C18. For the extraction of pesticides belonging to a high range of polarities, C18-bonded silicas and styrene/divinyl benzene copolymers are the most commonly used (Primel et al., 2012).

For the extraction of pesticides from solid samples such as biota or sediments, QuEChERS is the most applied technique (Andreu and Pico, 2012; Bruzzoniti et al., 2014). It provides a versatile platform of many different protocols depending on the type of pesticide (influence of pH, degradability), the purpose of the analysis, etc. (Masia et al., 2014). Other studies on matrices different to water are based on array techniques that can be used alone or combined (Masia et al., 2014). These alternative techniques include LLE, membrane-assisted MSPD, turbulent flow chromatography, MAE, PLE, SPME, SBSE, SPE, and high through planar SPE (HTpSPE) extractions or just dilution of the sample and direct injection (see Masia et al., 2014 for details).

Nowadays SPE and/or QuEChERS in combination with LC-MS are the most applied techniques for the analysis of pesticides from water or solid samples. LC is preferred over GC because currently pesticides are quite polar, thermally labile, or not easily vaporized, and consequently, worst detected by GC (Masia et al., 2014). Online SPE is considered an elegant alternative to traditional offline SPE plus LC-MS detection. It provides an automated way of sample pretreatment with automatic direct analysis of a high number of samples.

1.3.2.5 Dioxins

Dioxins are only analyzed in sediments and biota since they are not expected to be present in water because of their high hydrophobicity. Dioxins and dl-PCBs have been extracted with Soxhlet for 24 h using toluene or other optimum solvents (Castro-Jiménez et al., 2013). Then a clean-up step was mandatory. This was carried out with acid digestion followed by chromatographic purification with silica gel. A subsequent clean-up with automated system Power-Prep based on sequential multilayer silica, basic alumina, and PX-21 carbon adsorbents (Castro-Jiménez et al., 2013) was also performed. The extraction of these compounds from biota was performed by using Soxhlet extraction for 24 h with toluene:cyclohexane (1:1, v/v) (Castro-Jiménez et al., 2013; Parera et al., 2013). The same clean-up previously described for sediments can be followed (Castro-Jiménez et al., 2013; Cloutier et al., 2014) or alternatively a more efficient one as described in Parera et al. (2013). The authors rotary evaporated the Soxhlet solvents and kept the residue in an oven overnight (105°C) to eliminate the solvents for gravimetrical fat determination. The fat residues were dissolved in n-hexane and removed using a silica gel column modified with H_2SO_4 (44%, w/w). Finally the clean-up was performed by using an automated system Power Prep. The fractions eluted were rotary concentrated and reduced to dryness under a gentle stream of nitrogen. Final extracts were reconstituted with nonane (Parera et al., 2013).

The instrumental analysis for dioxins is based on GC coupled to high-resolution mass spectrometer (HRMS) (Castro-Jiménez et al., 2013; Cloutier et al., 2014; Parera et al., 2013).

1.3.3 Emerging Organic Contaminants

1.3.3.1 *Per- and Polyfluorinated Alkyl Substances*

The analysis of PFASs entails the difficulty of the cross-contamination from lab material as well as from instrumental analytic system. Due to this issue, a blank extraction is always performed and analyzed in parallel to the samples. Water samples are usually filtered with glass fiber filters (Ø 47 mm) for removing particulate matter (Ahrens et al., 2009, 2010a,b). After addition of IS, they are extracted on SPE. Cartridges equipped with anionic interchange stationary phase are normally used due to the nature of PFASs. The eluate is reduced until dryness under nitrogen stream and reconstituted in methanol: water at the initial chromatographic conditions.

Sediment analyses have been performed by solid–liquid extraction with diluted acid in water first and then with ammonium hydroxide in methanol (Long et al., 2013) using an ultrasonic bath. Other authors have proved an efficient extraction using only methanol in ultrasound-assisted extraction for 1 h (Llorca et al., 2014). The supernatant from the centrifuged sample was purified with anionic SPE (Llorca et al., 2014) or cleaned with ultra-pure carbon powder followed by SPE (Long et al., 2013). Afterward the eluate was treated similar to water.

For biota samples a first step including lipid precipitation is necessary. For example, Llorca (2012) used methanol (NaOH 10 mM). The extraction was carried out for 1 h in an orbital digester and then centrifuged. The supernatant was purified using anionic SPE and the eluate was treated as reported earlier for water and sediment. The instrumental analysis was based on LC coupled to mass spectrometry in tandem (LC-MS/MS). However, volatile PFASs were also analyzed with GC coupled to mass spectrometry in tandem GC-MS/MS.

1.3.3.2 *Brominated Flame Retardants*

Due to the hydrophobic character of PBBs, PBDEs, TBBPA, and HBCD, and the low concentrations expected in water, the analysis of large volumes (up to 1000 mL) are typically required to ensure positive detection of these compounds. Nonpolar solvents are usually used during the extraction process. It can be carried out by LLE but this technique requires big amount of solvent and therefore it is often replaced by SPE (Covaci et al., 2007). C18 is the sorbent with the widest range of applications, especially for PBDEs (Fulara and Czaplicka, 2012); it retains both nonpolar and moderate polar compounds. Other phases like poly(styrenedivinylbenzene) copolymer has been successfully applied for the extraction of moderately polar and water-soluble new BFRs (Lopez et al., 2009). Other techniques such as SPME (Polo et al., 2006), SBSE (Quintana et al., 2007), dispersive liquid–liquid microextraction (DLLME) (Li et al., 2007), and cloud point extraction (CPE) (Fontana et al., 2009) have been also used for the determination of BFRs in water samples. For the analysis of solid samples like sediments or biota, water depletion by mixing the sample with sodium sulfate or by freeze drying is usually carried out before the extraction process takes place. This can be successfully performed by Soxhlet extraction (Losada et al., 2009), PLE (Lacorte et al., 2010), SPME (Montes et al., 2010), and microwave solid phase dispersion (Covaci et al., 2007) among others. Due to the relatively low selectivity of extraction techniques and complexity of the samples, a purification step is mandatory. For sediment samples, the clean-up process consists of removal of sulfur while for biota fats need to be eliminated. The best approach is fractionation with gel chromatography and adsorption chromatography using silica gel, alumina, or florisil columns with different activation degrees (Fulara and Czaplicka, 2012). Regarding instrumental techniques GC-MS or GC-ECD (Fontana et al., 2009; Rezaee et al., 2010) are the most commonly used techniques although LC-MS has been also applied (Quintana et al., 2007).

1.3.3.3 *Endocrine-Disrupting Compounds*

As previously reported, the molecular heterogeneity of this group of contaminants makes rather difficult the development of analytical methods including compounds with different physicochemical properties. As a consequence the majority of the published studies usually target a low number of compounds although efforts have been made to achieve multiresidue analytical methods (Jakimska et al., 2013). Normally the addition of surrogate IS to the sample (water, sediments, or biota) is the first step in the analytical protocol. Regarding water extraction, the most used techniques have been SPE with C18, $-NH_2$, or $-CN$ modified silica (Petrovic et al., 2002a). The eluates were evaporated near to dryness under a gentle stream of nitrogen and reconstituted in the adequate solvent for the chromatographic analysis either with LC or GC (Petrovic et al., 2002a). For solid samples like sediments and biota, a pretreatment consisting of freeze drying the sample is often required before the extraction. After this a certain amount of lyophilized or freeze-dried sample is analyzed. Sediments were extracted using solid–liquid extraction by sonication in an ultrasonic bath, or by means of PLE. The extract was then cleaned-up with solid–liquid adsorption chromatography in open columns (florisil, alumina, different type of carbon, etc.), or SPE with C18, −NH2, or −CN-modified silica (Petrovic et al., 2002a). Freeze-dried biota samples can be extracted using different techniques such as PLE (Al-Ansari et al., 2010; Rudel et al., 2013; Schmitz-Afonso et al., 2003), sonication (Pojana et al., 2007), High Speed Solvent Extraction (HSSE) (Kim et al., 2011), or MAE (Liu et al., 2011). These are time- and/or solvent-consuming techniques. However, other faster techniques such as QuEChERS have been also successfully applied to biota samples (Jakimska et al., 2013). The most often used purification technique includes SPE (like in sediments) using florisil adsorbent (Pojana et al., 2007), C18 cartridges (Schmitz-Afonso et al., 2003), or GPC (Navarro et al., 2010). The samples were then treated like in the case of water, and the instrumental analysis was based either on LC-MS/MS or on GC-MS depending on the nature of selected EDCs (Navarro et al., 2010; Petrovic et al., 2002a; Schmitz-Afonso et al., 2003).

1.3.3.4 *Pharmaceuticals and Personal Care Products*

Multiresidue analytical methods are the most commonly applied methodology for the analysis of pharmaceuticals and personal care products (PPCPs) in the marine environment (Rodríguez-Mozaz et al., 2015). In this kind of methodology compounds belonging to different therapeutic families (in the case of pharmaceuticals) such as antibiotics, psychiatric drugs, analgesics/anti-inflammatories, tranquilizers, hormones, β-blockers, diuretics are simultaneously analyzed. Due to the different physicochemical properties of the therapeutic groups included, a compromise in the selection of the experimental conditions is required to achieve acceptable recoveries for all target compounds. Besides, isotope-labeled standards are normally used to avoid inaccurate quantification and to compensate matrix effects. For sea water samples SPE followed by LC in combination with mass spectrometry has been the methodology of choice in the majority of analytical methods (Borecka et al., 2015; Gros et al., 2009, 2013; Jiang et al., 2014; Loos et al., 2013; Yang et al., 2011), although passive samplers as a simultaneous sampling and extraction methodology have been also used (Martinez Bueno et al., 2009; Munaron et al., 2012; Tertuliani et al., 2008). Regarding sediment and biota samples, the target analytes must be first extracted from the matrix, and a surrogate internal standard is usually added. Different extraction techniques can be used, among them organic solvent extraction combined with either shaking, vortexing, or ultrasonication has been the most recurrent one, probably because it is easy to be performed (Klosterhaus et al., 2013; Kwon et al., 2009; Na et al., 2013). PLE and MASE have been also used both in sediments and organisms (Alvarez-Muñoz et al., 2015a,b; Berrada et al., 2008; Hibberd et al., 2009; Jelic et al., 2009; McEneff et al., 2013; Wille et al., 2011). QuEChERS has been applied as extraction and purification technique in biota samples by some authors (Martinez Bueno et al., 2013, 2014; Villar-Pulido et al., 2011), although SPE on Oasis HLB has been the selected clean-up technique by excellence (Azzouz et al., 2011; Dodder et al., 2014; Kwon et al., 2009; Samanidou and Evaggelopoulou, 2007). Sometimes a deeper clean-up step was required, especially in those cases of fatty organisms, and GPC was applied (Huerta et al., 2013; Tanoue et al., 2014). The available literature for sediments is based on LC in combination with mass spectrometry as instrumental technique preferred (Klosterhaus et al., 2013; Kwon et al., 2009). For organisms it is also the most used technique although other kinds of detectors like FLD, UV, or diode array detector (DAD) have been also used (Cueva-Mestanza et al., 2008; Fernandez-Torres et al., 2011; He et al., 2012). Others techniques such as enzyme-linked immunosorbent assay (ELISA), and time-resolved fluoroimmunoassay (TR-FIA) (Chafer-Pericas et al., 2010a,b) have been also applied in biota samples.

1.3.3.5 Marine Biotoxins

The analysis of marine toxins in waters, sediments, and biota is hampered by the lack of pure standards, especially for lipophilic toxins, since it is difficult to synthesize these molecules because of its structure. In this case, the purification of the molecules generated by algae is mandatory. The current official methods for biota (the most interesting matrix for the study of marine biotoxins since seafood is consumed by humans) have been based on biological in vivo tests with mouse or rats (Gerssen et al., 2010). However, nowadays the development of new in vitro assays and biochemical (immunochemical) and chemical methods is growing up (FAO, 2004; Gerssen et al., 2010). In the case of chemical methods, these have been based on LC-MS/MS or LC-fluorescence analyzer depending on the molecule.

1.3.3.6 Nanomaterials

For nanomaterials, two different types of analysis should be differentiated named as quantitative (for concentration determinations) and qualitative analysis (eg, particle size). This section is focused on quantitative analysis in water, sediments, and biota. As it has been described for the other contaminants, the first step is the addition of surrogate IS.

Although it is not expected to detect free CNM in waters, it is known that they form aggregates suspended in waters. Water samples have been extracted following different strategies: (1) LLE with toluene, (2) SPE with C18 cartridges using toluene as eluent (Xiao et al., 2011), and (3) filtration and particulate extraction with toluene although it has not been tested in sea water yet (Sanchís et al., 2012). Then, toluene suspensions were rotator reduced and reconstituted in toluene or methanol:toluene depending on the initial conditions of mobile phase for the analysis (Sanchís, 2015).

Sediment samples were extracted with toluene using either Soxhlet extraction overnight, ultrasound-assisted extraction, or PLE (Sanchís, 2015). The resulting suspension was rotator reduced and reconstitute as it has been described for water samples. Regarding biota samples their extraction is currently under development although the methodologies applied so far are similar to the ones for sediment samples (Sanchís, 2015).

The quantitative analysis of CNMs has been based on LC-UV/Visible and LC-MS/MS (nowadays researchers are starting to use HRMS) but it also could be done by transmission electron microscopy (TEM) and nanoparticle tracking analysis (NTA). In the case of LC, the most common columns used for chromatographic separation have been C18-bonded silica columns and pyrenylpropyl-bonded silica columns. As previously mentioned, the analysis has been performed either by UV–Visible detector or MS. This last approach has been proved to be more efficient if it is equipped with atmospheric pressure chemical ionization (APCI) and atmospheric pressure photoionization (APPI) than electrospray ionization (ESI) source (Astefanei et al., 2015; Núñez et al., 2012; Sanchís, 2015).

1.3.3.7 *Polydimethylsiloxanes*

The number of analytical methods developed for the analysis of siloxanes in environmental matrices is still very limited. The low concentrations present in the aquatic environment together with the high risk of contamination during sample treatment (due to the volatility of the compounds) make this a very challenging task. The analysis of water samples has been mainly carried out in surface water, wastewater, and river water. Different extraction methods have been used like purge and trap (Kaj et al., 2005), headspace extraction (Sparham et al., 2008), headspace SPME (Companioni-Damas et al., 2012), LLE (Sanchis et al., 2013), membrane-assisted solvent extraction, and ultrasound-assisted dispersive liquid–liquid microextraction (Cortada et al., 2014). Regarding sediments most of the studies have been carried out by using solid–liquid extraction (Sparham et al., 2011; Warner et al., 2010; Zhang et al., 2011), although other techniques such as PLE and ultrasound-assisted solvent extraction (USASE) have been also applied (Sanchis et al., 2013; Sparham et al., 2011). Fish tissues have been extracted using SLE (Kaj et al., 2005; Warner et al., 2010), USASE (Sanchis et al., 2016), and purge and trap (Kierkegaard et al., 2010) methods. Most of the works have been performed using hexane as extraction solvent although other solvents like pentane (Wang et al., 2013) and ethyl acetate (Sparham et al., 2011) have been also used. Usually a clean-up step was not carried out to avoid analytes evaporation and extract manipulation, but in the case of complex matrices, such as fish, a simple centrifugation can clarify a cloudy extract of hexane (Sanchis et al., 2016). GC-MS analysis has been the technique of choice by the vast majority of methods because of the low molecular weight and low polarity of siloxanes. A small number of studies have been performed using GC coupled to other detectors as flame ionization (Huppmann et al., 1996; Dewil et al., 2007).

1.4 OCCURRENCE IN THE MARINE ENVIRONMENT

1.4.1 Metals

The use of clean techniques has allowed the improvement of knowledge about the distribution and behavior of metals in seawater. The concentration of trace elements in ocean waters are ranging between nM (10^{-9} mol/L) and pM (10^{-12} mol/L), although for Au this is at fM (10^{-15} mol/L) level (Donat and Bruland, 1995). GEOTRACES—international study of the marine biogeochemical cycles of trace elements and their isotopes—released the GEOTRACES Intermediate Data Product 2014, which corresponds to a compilation of digital data for 796 stations, including hydrographical parameters and dissolved and particulate trace elements such as Al, Ba, Cd, Cu, Fe, Mn, Mo, Ni, Pb, and rare earth elements (REEs), stable isotopes, and radioactive isotopes (The GEOTRACES group, 2015). The horizontal and vertical distribution of metals in seawaters is consequence of the balance between supply and removal. Vertical profiles of trace metals can be classified as: (1) nutrient type, (2) conservative, and (3) scavenged, although mixture behaviors are possible. The nutrient profile is showed by biolimiting elements, having low surface water and high at deep-water concentrations, and these elements are controlled by biological process. Elements with conservative behavior show a profile related to the salinity with lack of gradient and homogenous with depth. Scavenged distribution shows vertical profiles in which dissolved concentration decreases with increasing depth (eg, Mn, Pb, and Co). For these elements, the removal is mediated by precipitation or adsorption processes. A summary of vertical profiles for different elements have been collected by Nozaki (1997). Metal distribution is not uniform and coastal areas and estuaries are subjected to higher concentrations than oceanic waters. Historically, places where municipal, industrial, and sewage discharges have been carried out show metal pollution; some examples are New York Bight, Boston Harbor, and Thames Estuary among others (Kennish, 1997). Luoma and Rainbow (2008) summarized dissolved concentrations for UK estuaries, semi-enclosed bays, UK coastal waters, and undisturbed coastal waters showing that concentrations ranged for Cu: 0.1–4.64 μg/L, Zn: 0.27–5.1 μg/L, Cd: 0.003–0.19 μg/L, and Pb: 0.004–0.269 μg/L, with differences among estuaries and coastal data that does not exceed 10–20-fold. Information available at global scale about metal dissolved concentration is limited and assessing strategies to mitigate disturbances and to improve water quality need long-term studies of the biogeochemical cycling and fate of aquatic contaminants in the water column (Sañudo-Wihelmy et al., 2004; Tovar-Sánchez, 2012). Metals in the water column are removed in coastal zones in combination with particulate matter removal as consequence of mixing process between fresh and seawater in estuaries; these zones represent an important place for removal and trapping of metal in sediments. Dissolved metals (pore water, overlying water) have got a high affinity for particulate matter (sediment or suspended matter) and this relationship can be assessed using a partition coefficient (K_d). Metal concentrations in particulate matter are higher by several orders of magnitude than in dissolved form; this fact provokes that sediments act as a sink of metals, although a change in the physicochemical conditions can release them to water column increasing their bioavailability. To compare metal levels among sites, normalization criteria should be taken into account, because grain size or particulate size that affect metal concentrations is a confound factor; finer grain particles have got higher surface area per unit of mass and increases binding sites. The methodologies for normalization include chemical extraction, physical separation of fine-grained sediment, normalization to particle size-sensitive natural components, and statistical techniques (Luoma and Rainbow, 2008). Kennish (2000) summarized trace metals concentration—expressed as ppm on dry weight—from selected estuaries and coastal marine systems in USA and 19 UK estuaries. Some ecosystems showed high levels: San Francisco Bay (USA) (Cr: 1466 ppm; Cu: 160 ppm; Pb: 67 ppm; Cd: 0.51 ppm) and Restronguet Creek (UK) (Cu: 2398 ppm; Pb: 341 ppm, Cd: 1.5 ppm). However, sediment total metal concentration does not give information about bioavailability and different approaches can be considered, but this one is out of the scope of this chapter [more detail in Parsons et al. (2007)].

Organisms in water column, on or in sediment can accumulate metals from different sources (water, suspended matter). This mechanism is known as bioaccumulation, and physiology, life styles, etc. can modulate the process. In Chapter 4 of this book, this topic is treated in depth.

1.4.2 Persistent Organic Contaminants

1.4.2.1 Polycyclic Aromatic Hydrocarbons

As reported in previous sections PAHs solubility in water is very poor, therefore the concentration of the dissolved fraction is very low ranging from pg/L to ng/L. The lower molecular weight compounds are more water soluble and PAHs such as acenaphthylene, naphthalene, and benz(a)anthracene, containing among two and four aromatic rings, are the most usually detected in water. For example, Ren et al. (2010) measured total concentrations of PAHs ranging between 30.40 and 120.29 ng/L in East and South China seas and the cited compounds were the most commonly found. Sediments, however, act as a sink for these contaminants due to their hydrophobicity, and the concentrations found are much higher reaching up to several thousands of ng/g. For instance, Baumard et al. in 1998 found up to 8400 ng/g dry weight (dw) of total PAHs in sediment collected from Barcelona Harbor, in the Mediterranean Sea. Leon et al. (2014) have reported total concentrations of PAHs of 1006 ng/g dw in the same area. Due to their lipophilic character, PAHs also tend to accumulate in aquatic organisms especially in mollusks since vertebrates have a higher capacity to metabolize and excrete them (Meador et al., 1995). For example, up to 80 ng/g dw of PAHs was measured in mussels from the western Mediterranean (Baumard et al., 1998), and up to 40 ng/g dw in mullet fish from the same region (Leon et al., 2014).

1.4.2.2 Surfactants

Surfactants and their degradation metabolites can be found at any environmental compartment. Their presence in surface waters is widely documented and the levels usually range from less than 1 ng/L to several hundreds of μg/L. The most reported compounds are LAS and NPEOs as anionic and nonionic surfactants, respectively. For instance, between 1 and 296 μg/L of LAS were measured in Venice Lagoon (Italy) (Stalmans et al., 1991) and between 1 and 37 μg/L of NPEOs were found in Tarragona (Spain) (Petrovic et al., 2002b). Levels of surfactants in sediments are generally several orders of magnitude higher than those measured in water due to their moderate to high sorption capacity. LAS and NPEOs have been detected in sediments with concentrations ranging from low mg/kg to a couple of hundreds (Traverso-Soto et al., 2012). The occurrence of surfactants in marine organisms has been also reported especially for those compounds showing endocrine disruptor activity such as nonylphenol. For example, alkylphenols were quantified in oysters and snails from coastal areas of Taiwan with concentrations ranging from 20 to 5190 ng/g (Chin-Yuan et al., 2006).

1.4.2.3 Polychlorinated Biphenyls

The occurrence of PCBs in marine environment is characterized by being accumulated in biota, especially fish. This bioaccumulation induces the biomagnifications through marine food chain of some congener. This has been described within the work developed by Lu et al. (2014) regarding the study of chiral PCBs in Greenland sharks from Cumberland Sound and its Arctic marine food web. For example, zooplankton presented values of 153 ng/g lipid weight (lw) (McKinney et al., 2012), between 109 and 561 ng/g lw for herrings (McKinney et al., 2012), 438 ng/g lw for capelin (Lu et al., 2014), 867 ng/g lw sculpins (McKinney et al., 2012), 221 ng/g lw Greenland halibut (McKinney et al., 2012), between 76 and 4810 ng/g lw for char (McKinney et al., 2012), and up to 4600 ng/g lw in liver shark from the same region (Lu et al., 2014). In general, values detected in liver or plasma were higher than the ones observed in muscle because of the high lipophilic character of PCBs congeners (Lu et al., 2014).

For sediments, in a study carried out in Egyptian coast, levels of 2.29–377 ng/g dw were detected (Barakat et al., 2013). In another study in the Mediterranean coast, the researchers detected from 2.33 to 44.00 ng/g dw of PCBs in coastal sediments, and from 22.34 to 37.74 ng/g in an area near a marine emissary (Besòs River, Spain) (Castells et al., 2008).

1.4.2.4 Pesticides

The presence of pesticides in coastal environments is not unexpected since major agricultural areas are located in coastal plains and river valleys. The levels found in seawater usually range from the low ng/L up to μg/L. Even pesticides that have been banned worldwide such as DDT are still present in the marine environment around the world due to their resistance to degradation. For example, in a study carried out collecting sea surface slicks on a global circumnavigation (Menzies et al., 2013) measured up to 96.6 ng/L of DDT and its metabolites in Gatun Locks, Panama Canal, up to 285.1 ng/L of chlorodane and related compounds in Pago Pago, American Samoa, and up to 1213.1 ng/L of chlorinated benzene in Cooks Bay, Moorea, French Society Islands. Pesticides have been also found at ng/g levels in sediments from estuarine areas. For instance, Zheng et al. (2016) determined that organochlorines such as DDT, DDD, and DDE were the main contaminants in sediments from the Jiulong Estuary in China, reaching up to 311 ng/g dw of DDE. Regarding biota, although the levels of pesticides measured in superficial waters generally range below lethal exposure concentrations for wildlife, pesticides can be accumulated by aquatic organisms and provoke sublethal adverse effects. For example, in the same study previously cited by Zheng et al. detected up to 3094 ng/L of procymidone in seawater, and further experiments demonstrated that this compound could disrupt the expression of vitellogenin in estuarine fish. Concentrations of pesticides in organisms range in the ng/g levels. For example, Zhou et al. (2014) collected 11 mollusks species from coastal areas along the Yangtze River Delta (China) showing concentrations of DDTs, hexachlorocyclohexane, and chlordanes from 6.22 to 398.19, 0.66–7.11, and 0.14–4.08 ng/g wet weight (ww), respectively.

1.4.2.5 Dioxins

The occurrence of dioxins in marine environments is characterized by human impact. For example, in a study carried out in biota from Ebro River Delta (Spanish Mediterranean coast) including mussels, carpet shells, murexes, congers, gilthead seabreams, flounders, and sardines, the authors detected levels of dioxins between 0.29 and 2.29 pg/g (ww; expressed as the ΣPCDD/F) and from 24.6 to 5503 pg/g ww for Σdl-PCBs (Parera et al., 2013). The study of crustaceans and fish from Blanes submarine canyon (North-Western Mediterranean Sea), a more pristine site in Mediterranean Sea, the authors detected from 110 to 795 pg/g lipid weight (lw) of ΣPCDD/F (Castro-Jiménez et al., 2013). The studies characterizing temporal trends of dioxins showed a slight decrease of the concentration in biota (Parera et al., 2013).

In the case of sediments, Castro-Jiménez et al. (2013) quantified dioxins in the deep sea from Blanes submarine canyon (North-Western Mediterranean Sea) as a sum of $\sum$PCDD/F between 102 and 680 pg/g dw.

1.4.3 Emerging Organic Contaminants

1.4.3.1 Per- and Polyfluorinated Alkyl Substances

PFASs are widespread in marine environment having also a long-range transport since they have been detected in remote areas such as

Arctic and Antarctic continents. In water, Ahrens et al. (2010b) quantified the most recalcitrant PFASs (PFOA and PFOS) as well as C6 carboxylic acid and C8 sulfonamide (FOSA) in surface water from North Sea, Baltic Sea, and Norwegian Sea, ranging between 0.02 and 6.16 ng/L. In another study, PFOS were detected at concentrations ranging from below 11 up to 51 pg/L in the Antarctic Circumpolar current zone (Ahrens et al., 2010a).

Regarding the analysis of marine sediments, in a study characterizing sediments from Pacific Northwest (Puget Sound, Washington, USA) C4 carboxylic acid, PFOS and FOSA between 0.13 and 1.50 ng/g were detected (Long et al., 2013). In the Greek zone of the Mediterranean Sea, higher concentrations of PFOA, PFOS, C10 carboxylic acid, and sulfonate compounds were observed (levels from 8.2 up to 146 ng/kg) (Llorca et al., 2014). In another study carried out in Cap de Creus Canyon (deep Western Mediterranean Sea) concentrations ranging from 120 ng/kg to 11,650 ng/kg were quantified for PFOS, PFOA, C4, C6, and C9 carboxylic acids (Sanchez-Vidal et al., 2015).

Concerning biota from the marine environment, a study undertaken with bluegill sunfish (*Lepomis macrochirus*) fillets showed high concentrations of PFOS (2.08–275 ng/g dw) in locations near historical PFASs sources from North Carolina (Delinsky et al., 2009). In another publication the presence of PFASs in seafood from Spanish market was studied. Hake roe, swordfish, striped mullet, young hake, and anchovy were the species analyzed. The results showed the presence of C5, C9, and C10 carboxylic acids, C4 and C10 sulfonates, as well as PFOA and PFOS at concentrations ranging from 0.09 to 50 ng/g dw (Llorca et al., 2009). A similar study took place in European markets where bivalves, herring, whiting, pangasius, hake, salmon, cod, and tuna were studied with concentrations of C4, C5, C6, and C9 carboxylic acids, PFOA, PFOS, and FOSA from 0.09 to 54 ng/g dw (Llorca, 2012).

1.4.3.2 Brominated Flame Retardants

BFRs have the potential for long-range atmospheric transport; as a result they have been found in sediment and fish in areas remote from known sources indicating they are widespread environmental contaminants. For example, PBDEs were found at high concentrations in fish, crabs, Arctic ringed seals, and other marine mammals in northern Canada (Ikonomou et al., 2002), and also in fish and mussels from Greenland (Christensen et al., 2002). Sediment samples from a Norwegian fjord were also analyzed and HBCD concentrations between 35 and 9000 μg/kg organic carbon were determined along a transect away from a known point source (Haukas et al., 2009). BFRs have been also measured around known contamination sources: for example in sediment cores from Tokyo Bay (42), the Clyde Estuary in Scotland (47), Xiamen offshore areas (48), etc., the levels reported usually range from the low μg/kg up to several thousands.

1.4.3.3 Endocrine-Disrupting Compounds

EDCs in seawater have been detected in Venice Lagoon (bay of the Adriatic Sea, Italy) at concentrations of 2.8–211 ng/L (Pojana et al., 2007). In the same study the authors detected a similar profile of compounds in sediments at concentrations ranging from 3.1 to 289 ng/g dw (Pojana et al., 2007). These levels were slightly lower than the ones reported later on by Salgueiro-González et al. (2014) in sediments from Galicia coast, with concentrations between 20.1 and 1409 ng/g dw, although these levels were just for nonylphenol and octylphenol. Regarding EDCs in marine organisms, the concentrations found ranged from a few ng/g to μg/g when they are expressed as a summatory of different homologues. For example, mean concentrations of parabens ranging from 0.005 ng/g to 1.45 ng/g ww were found in fish from China (Liao et al., 2013), up to 98.4 ng/g dw in mullet fish from the Tagus Estuary, Portugal (Alvarez-Muñoz et al., 2015b), and up to a maximum of

1255 ng/g ww of total akylphenolic compounds were measured in shrimps from Fiumicino (Italy) (Ferrara et al., 2008).

1.4.3.4 *Pharmaceuticals and Personal Care Products*

Their presence in coastal areas has been studied all over the world [for details see Rodríguez-Mozaz et al. (2015)] mainly in estuaries, harbors, lagoons, and enclosed or semi-enclosed areas. More than hundred pharmaceuticals have been detected at least once in marine waters, among them antibiotics is the most researched group followed by psychiatric drugs, analgesics, and anti-inflammatories. The most recurrent compounds determined in marine waters are acetaminophen, ibuprofen, diclofenac, erythromycin, clarithromycin, sulfamethoxazole, trimethoprim, carbamazepine, gemfibrozil, and atenolol. In general, the levels of pharmaceuticals measured in the aquatic environment range from the low ng/L up to a few μg/L. For example, Moreno-Gonzalez et al. (2015) have detected 0.5 ng/L of propranolol, 11.8 ng/L of salicylic acid, 40.7 ng/L of erythromycin, and 64.8 ng/L of sulfamethoxazole (among others) in Mediterranean seawater during summer time. In a study carried out in Bohai Bay (China), Zhou et al. (2011) detected concentrations of antibiotics in seawater ranging from 2.3 ng/L of sulfamethoxazole to 6800 ng/L of norfloxacine. Sediments are natural repositories of many contaminants present in the water column and in the case of pharmaceuticals they have been identified as a major sink for antibiotics (Kim and Carlson, 2007). The most recurring compounds identified in sediments are antibiotics like tetracycline, oxytetracycline, norfloxacin, ofloxacin, enrofloxacin, and ciprofloxacin (Rodríguez-Mozaz et al., 2015) with concentrations reaching up to 458 ng/g of ofloxacin in the Yangtze estuary (Shi et al., 2014) in China. Regarding organisms, the most detected pharmaceuticals are carbamazepine and oxytetracycline. The concentrations of pharmaceuticals reported usually ranged in the low nanogram per gram levels with few exceptions. For example, Alvarez-Muñoz et al. (2015b) have reported up to 36.1 ng/g of venlafaxine and 13.3 ng/g of azithromycin in bivalves from Po delta, Italy. Li et al. (2012) found up to 370 ng/g dw of norfloxacin, 242 ng/g dw of ofloxacin, 208 ng/g dw of ciprofloxacin, and 147 ng/g dw of enrofloxacine in wild mollusk collected from the Bohai Sea, China.

As regards to PCPs, few studies about the occurrence of these compounds in marine environments have been published. For example, musk fragrances have been detected in aquatic biota (Kannan et al., 2005) as well as UV filters in dolphins from different regions (Gago-Ferrero et al., 2013). In the case of sea water some authors have reported the presence of insect repellent in the North Sea (Weigel et al., 2002).

1.4.3.5 *Marine Biotoxins*

Concentrations ranging from 6 to 900 ng/g of azaspiracids in mussels from Mediterranean Sea and Black Sea have been reported by different authors (Elgarch et al., 2008; Taleb et al., 2006). The levels reported in Atlantic (Northeast) for mussels and oysters are slightly higher (1.4–4200 ng/g) (Furey et al., 2002; James et al., 2002; Magdalena et al., 2003; Vandermeersch et al., 2015). In the case of ciguatoxins, this family of toxins has been detected in fish, coral cod, and kingfish among others, in the Pacific (Eastern Central) and Atlantic (Western Central) at concentrations ranging from 0.1 to 52.9 ng/g (Dickey, 2008; Stewart et al., 2010; Vandermeersch et al., 2015).

1.4.3.6 *Nanomaterials*

To the best of our knowledge, there are no studies reporting the occurrence of CNM in the marine environment (considering water, sediments, and biota). However, the presence of aerosol-bound fullerenes in the Mediterranean Sea atmosphere by Sanchís et al. (2011) has been confirmed.

1.4.3.7 Siloxanes

As previously reported due to their low solubility and high vapor pressure, siloxanes are present in water only at trace concentrations (Sparham et al., 2008). However, they can be found at higher levels in sediments and biota due to their high hydrophobicity. For instance, up to 920 ng/g dw were determined in sediments from the Inner Oslofjord, with decamethylcyclopentasiloxane (D5) being the most abundant compound with the highest concentrations detected near the WWTP of Bekkelaget (Schlabach et al., 2007), and between 60 and 260 ng/g dw of the same compound were measured in the Humber Estuary (Kierkegaard et al., 2011). In less populated areas such as the Arctic, the values decreased and concentrations around 2 ng/g dw were detected in sediments (Sparham et al., 2011). Regarding aquatic organisms PDMS have been studied in seafood samples collected in markets from Barcelona (Spain), the levels found ranged from low pg/g to 30 ng/g ww with octamethylcyclotetrasiloxane (D4) and D5 being the most frequently detected compounds (Sanchís et al., 2015). Another study also reported concentrations in wild mussel, turbot, and cod ranging from 50 to 321.3 ng/g ww of hexamethylcyclotrisiloxane (D3), 1.3–134.4 ng/g ww of D4, 3.3–2200 ng/g of D5 and 0.9–151.5 ng/g of dodecamethylcyclohexasiloxane (D6) (Schlabach et al., 2007).

1.5 MICROPLASTICS IN THE MARINE ENVIRONMENT

Besides the multiple compounds present in the marine environment that have been reviewed along this chapter, microplastics (MP) constitute another group of contaminants of emerging concern. Although plastic litter and pellets are not a new problem, the awareness of microplastics in the environment and their propensity to cause environmental damage is a relatively recent development which is now attracting an increasing amount of attention (Hartl et al., 2015). Under environmental conditions, larger plastics items degrade to so-called microplastics, fragments typically smaller than 5 mm in diameter (Wagner et al., 2014). Due to their high mobility they are widespread through the world's seas; concretely large oceanic gyres are "hot spots" of plastics pollution. Most studies investigate neustonic and pelagic MP (Wagner et al., 2014); however, MP are also present in sediments with concentrations ranging from 1 to 100 items/kg (Hidalgo-Ruz et al., 2012). MP can be ingested by aquatic organisms and accumulated throughout the food web (Wright et al., 2013). Marine organisms such as mussel (Browne et al., 2008), turtles (Stamper et al., 2009), and fish (Boerger et al., 2010), among others, have been shown to ingest microplastics. They accumulate in the digestive tract of the organism leading to nutritional impairment and general decline condition (Browne et al., 2008). MP can also act as vectors for other organic pollutants such as PCBs, DDT, and PAHs (Mato et al., 2001; Rios et al., 2007) acting therefore as a source of wildlife exposure to these chemicals (Oehlmann et al., 2009; Teuten et al., 2009).

Regarding monitoring methodologies, there are no well-established guidelines for sampling MP yet, but efforts are being done by international initiatives and workshops toward this aim. Once sampled the most common separation technique is by using a high density flotation technique followed by examination of the fragments under a microscope, although some high-density polymers, such as PVC or polyester, may be underestimated using this technique.

So far, scientific efforts focus on studying MP abundance and effects in the marine environment. More research is required to understand their potential harm, including source and fate in the marine environment, as well as the development of appropriate biomarkers of exposure in marine organisms, and implications for human health (Hartl et al., 2015). It is also important to investigate the interaction of MP with other contaminants, their chemical burden, absorption/desorption kinetics, and the transfer of chemicals from plastic to biota.

1.6 FUTURE TRENDS

An important issue that deserves to be highlighted is the fact that all contaminants reviewed in this chapter are simultaneously present in many marine environments, especially coastal and estuarine areas. They make a "cocktail" of hazardous chemicals that poses a potential risk to wildlife and human health. Even very low concentrations (ng/L or pg/L) can be toxicologically relevant. Besides, the sum of concentrations of different contaminants simultaneously present can result in a high total level of pollution. There is scientific evidence that when organisms are exposed to a number of different chemical substances, these may act jointly, affecting the overall level of toxicity by addition, antagonism, potentiation, synergies, etc.

Traditionally target analysis of a specific group of contaminants has been the approach followed, as reported along the chapter for the different groups of contaminants. However, nowadays the major challenge is to develop systematic ways of addressing chemical mixtures in environmental assessment (European_Commission, 2012) and to identify priority mixtures of contaminants of potential concern. Target analysis offers good sensitivity and reliable identification of the compounds, but it has a significant disadvantage as it always misses all compounds not included in the method. There are good reasons to believe that the concentration of unidentified compounds, named "unknowns," is much higher than the concentration of the knowns and they could better explain the toxicity of a certain sample. The application of nontarget analysis techniques seems to be the way forward to fill this knowledge gap. It is a powerful tool for the identification of environmental pollutants without a preceding selection of the compounds of interest. Besides, the improvement of trace analytical methods allows the identification of less abundant components (Chetwynd et al., 2014; David et al., 2014), and the use of high-resolution mass analyzers permit a retrospective analysis to look for pollutants which earlier were unknown (Acuña et al., 2015). Nontarget analysis is therefore the future trend in environmental chemistry. It has been mainly applied so far in water samples including wastewater (Gomez-Ramos et al., 2011), river water (Ruff et al., 2015), surface water (Ibañez et al., 2008), ground water (Ibañez et al., 2011), and drinking water (Muller et al., 2011) and to a lesser extension in sediments and organism (Alvarez-Munoz et al., 2015; Rostkowski et al., 2011) due to the complexity of these matrices, especially in the case of biological samples (Simon et al., 2015).

1.7 CONCLUSIONS

Chemical contamination of the marine environment (both estuarine and coastal areas) is a highly complex issue. Up to 15 different groups of pollutants including metals, persistent organic contaminants, and emerging organic contaminants can be simultaneously present at different levels in marine ecosystems. All of them make a "cocktail" of hazardous substances that may pose negative implications for the aquatic environment, human health (through the possible ingestion of contaminated seafood), and related coastal activities such as fishing, aquaculture, or recreational activities.

Contaminant groups such as metals, PAHs, PCBs, surfactants, pesticides, dioxins, PPCPs, EDCs, PFASs, BFRs, NMs, marine biotoxins, PDMS, and MP have been reviewed along the chapter. Their main physicochemical properties have been reported and it has been shown that they are directly related to the molecular structure of a certain contaminant. Their main sources of contamination have been identified and in the majority of the cases their introduction in the aquatic environment is due to human activities. The most used analytical techniques for their detection and quantification in the marine environment have been also presented with SPE being the most extended technique for the analysis of liquid samples. Their occurrence and usual levels found in seawater, sediments, and marine organisms have been also reported showing a wide geographical distribution in most cases.

Acknowledgments

This work was partly supported by the Generalitat de Catalunya (Consolidated Research Group: Catalan Institute for Water Research 2014SGR291) from the European Union Seventh Framework Programme (FP7/2007–2013) under the ECsafeSEAFOOD project (grant agreement no. 311820) and SEA-on-a-CHIP project (grant agreement no. 614168), and the project CTM2012-3872-C03-03 funded by MINECO and ERDF.

References

Acuña, J., Stampachiacchiere, S., Pérez, S., Barceló, D., 2015. Advances in liquid chromatography-high resolution mass spectrometry for quantitative and qualitative environmental analysis. Anal. Bioanal. Chem. 407, 6289–6299.

Ahrens, L., Barber, J.L., Xie, Z., Ebinghaus, R., 2009. Longitudinal and latitudinal distribution of perfluoroalkyl compounds in the surface water of the Atlantic Ocean. Environ. Sci. Technol. 43, 3122–3127.

Ahrens, L., Xie, Z., Ebinghaus, R., 2010a. Distribution of perfluoroalkyl compounds in seawater from Northern Europe, Atlantic Ocean, and Southern Ocean. Chemosphere 78, 1011–1016.

Ahrens, L., Gerwinski, W., Theobald, N., Ebinghaus, R., 2010b. Sources of polyfluoroalkyl compounds in the North Sea, Baltic Sea and Norwegian Sea: evidence from their spatial distribution in surface water. Mar. Pollut. Bull. 60, 255–260.

Alaee, M., Wenning, R.J., 2002. The significance of brominated flame retardants in the environment: current understanding, issues and challenges. Chemosphere 46, 579–582.

Alaee, M., Arias, P., Sjodin, A., Bergman, A., 2003. An overview of commercially used brominated flame retardants, their applications, their use patterns in different countries/regions and possible modes of release. Environ. Int. 29, 683–689.

Al-Ansari, A.M., Saleem, A., Kimpe, L.E., Sherry, J.P., McMaster, M.E., Trudeau, V.L., Blais, J.M., 2010. Bioaccumulation of the pharmaceutical 17alpha-ethinylestradiol in shorthead redhorse suckers (*Moxostoma macrolepidotum*) from the St. Clair River, Canada. Environ. Pollut. 158, 2566–2571.

Alexander, J., Benford, D., Cockburn, A., Cradevi, J.P., Dogliotti, E., Domenico, A.D., Fernandez-Cruz, M.L., Fink-Gremmels, J., Furst, P., Galli, C., Grandjean, P., Gzyl, J., Heinemeyer, G., Johansson, N., Mutti, A., Schlatter, J., Van Leeuwen, R., Van Peteghem, C., Verger, P., 2008a. Marine biotoxins in shellfish – yessotoxin group. EFSA J. 907, 1–62.

Alexander, J., Benford, D., Cockburn, A., Cradevi, J.P., Dogliotti, E., Domenico, A.D., Fernandez-Cruz, M.L., Fink-Gremmels, J., Furst, P., Galli, C., Grandjean, P., Gzyl, J., Heinemeyer, G., Johansson, N., Mutti, A., Schlatter, J., Van Leeuwen, R., Van Peteghem, C., Verger, P., 2008b. Marine biotoxins in shellfish – azaspiracid group. EFSA J. 723, 1–52.

Alexander, J., Audunsson, G.A., Benford, D., Cockburn, A., Cradevi, J.P., Dogliotti, E., Domenico, A.D., Fernandez-Cruz, M.L., Fink-Gremmels, J., Furst, P., Galli, C., Grandjean, P., Gzyl, J., Heinemeyer, G., Johansson, N., Mutti, A., Schlatter, J., Van Leeuwen, R., Van Peteghem, C., Verger, P., 2008c. Marine biotoxins in shellfish – okadaic acid and analogues. EFSA J. 589, 1–62.

Alexander, J., Benford, D., Cockburn, A., Cradevi, J.P., Dogliotti, E., Domenico, A.D., Fernandez-Cruz, M.L., Fink-Gremmels, J., Furst, P., Galli, C., Grandjean, P., Gzyl, J., Heinemeyer, G., Johansson, N., Mutti, A., Schlatter, J., Van Leeuwen, R., Van Peteghem, C., Verger, P., 2009a. Marine biotoxins in shellfish – saxitoxin group. EFSA J. 1019.

Alexander, J., Benford, D., Cockburn, A., Cradevi, J.P., Dogliotti, E., Domenico, A.D., Fernandez-Cruz, M.L., Fink-Gremmels, J., Furst, P., Galli, C., Grandjean, P., Gzyl, J., Heinemeyer, G., Johansson, N., Mutti, A., Schlatter, J., Van Leeuwen, R., Van Peteghem, C., Verger, P., 2009b. Marine biotoxins in shellfish – domoic acid. EFSA J. 1181.

Alexander, J., Benford, D., Cockburn, A., Cradevi, J.P., Dogliotti, E., Domenico, A.D., Fernandez-Cruz, M.L., Fink-Gremmels, J., Furst, P., Galli, C., Grandjean, P., Gzyl, J., Heinemeyer, G., Johansson, N., Mutti, A., Schlatter, J., Van Leeuwen, R., Van Peteghem, C., Verger, P., 2009c. Marine biotoxins in shellfish – pectenotoxin group. EFSA J. 1109, 1–47.

Allen, R.B., Kochs, P., Chandra, G., 1997. Organosilicon Materials, Handbook of Environmental Chemistry. Springer-Verlag.

Alvarez, D.A., Maruya, K.A., Dodder, N.G., Lao, W., Furlong, E.T., Smalling, K.L., 2014. Occurrence of contaminants of emerging concern along the California coast (2009–10) using passive sampling devices. Mar. Pollut. Bull. 81, 347–354.

Alvarez-Munoz, D., Saez, M., Lara-Martin, P.A., Gomez-Parra, A., Gonzalez-Mazo, E., 2004. New extraction method for the analysis of linear alkylbenzene sulfonates in marine organisms. Pressurized liquid extraction versus Soxhlet extraction. J. Chromatogr. A 1052, 33–38.

Alvarez-Munoz, D., Gomez-Parra, A., Gonzalez-Mazo, E., 2007. Testing organic solvents for the extraction from fish of sulfophenylcarboxylic acids, prior to determination by liquid chromatography-mass spectrometry. Anal. Bioanal. Chem. 388, 1013–1019.

Alvarez-Munoz, D., Indiveri, P., Rostkowski, P., Horwood, J., Greer, E., Minier, C., Pope, N., Langston, W.J., Hill, E.M., 2015. Widespread contamination of coastal sediments in the Transmanche Channel with anti-androgenic compounds. Mar. Pollut. Bull. 95, 590–597.

Alvarez-Muñoz, D., Huerta, B., Fernandez-Tejedor, M., Rodríguez-Mozaz, S., Barceló, D., 2015a. Multi-residue method for the analysis of pharmaceuticals and some of their metabolites in bivalves. Talanta 136, 174–182.

Alvarez-Muñoz, D., Rodríguez-Mozaz, S., Maulvault, A.L., Tediosic, A., Fernández-Tejedor, M., Van den Heuvele, F., Kottermanf, M., Marques, A., Barceló, D., 2015b. Occurrence of pharmaceuticals and endocrine disrupting compounds in macroalgaes, bivalves, and fish from coastal areas in Europe. Environ. Res. 143, 56–64.

Andersson, P.L., Oberg, K., Orn, U., 2006. Chemical characterization of brominated flame retardants and identification of structurally representative compounds. Environ. Toxicol. Chem. 25, 1275–1282.

Andreu, V., Pico, Y., 2012. Determination of currently used pesticides in biota. Anal. Bioanal. Chem. 404, 2659–2681.

Astefanei, A., Núñez, O., Galceran, M.T., 2015. Characterisation and determination of fullerenes: a critical review. Anal. Chim. Acta 882, 1–21.

Azzouz, A., Souhail, B., Ballesteros, E., 2011. Determination of residual pharmaceuticals in edible animal tissues by continuous solid-phase extraction and gas chromatography-mass spectrometry. Talanta 84, 820–828.

Banjoo, D.R., Nelson, P.K., 2005. Improved ultrasonic extraction procedure for the determination of polycyclic aromatic hydrocarbons in sediments. J. Chromatogr. A 1066, 9–18.

Barakat, A.O., Mostafa, A., Wade, T.L., Sweet, S.T., El Sayed, N.B., 2013. Distribution and ecological risk of organochlorine pesticides and polychlorinated biphenyls in sediments from the Mediterranean coastal environment of Egypt. Chemosphere 93, 545–554.

Baumard, P., Budzinski, H., Michon, Q., Garrigues, P., Burgeot, T., Bellocq, J., 1998. Origin and bioavailability of PAHs in the Mediterranean Sea from mussel and sediment records. Estuar. Coast. Shelf Sci. 47, 77–90.

Berg, B.E., Lund, H.S., Kringstad, A., Kvernheim, A.L., 1999. Routine analysis of hydrocarbons, PCB and PAH in marine sediments using supercritical CO_2 extraction. Chemosphere 38, 587–599.

Berrada, H., Borrull, F., Font, G., Marce, R.M., 2008. Determination of macrolide antibiotics in meat and fish using pressurized liquid extraction and liquid chromatography-mass spectrometry. J. Chromatogr. A 1208, 83–89.

Boerger, C.M., Lattin, G.L., Moore, S.L., Moore, C.J., 2010. Plastic ingestion by planktivorous fishes in the North Pacific Central Gyre. Mar. Pollut. Bull. 60, 2275–2278.

Borecka, M., Siedlewicz, G., Halinski, L.P., Sikora, K., Pazdro, K., Stepnowski, P., Bialk-Bielinska, A., 2015. Contamination of the southern Baltic Sea waters by the residues of selected pharmaceuticals: method development and field studies. Mar. Pollut. Bull. 94, 62–71.

Brausch, J.M., Rand, G.M., 2011. A review of personal care products in the aquatic environment: environmental concentrations and toxicity. Chemosphere 82, 1518–1532.

Browne, M.A., Dissanayake, A., Galloway, T.S., Lowe, D.M., Thompson, R.C., 2008. Ingested microscopic plastic translocates to the circulatory system of the mussel, *Mytilus edulis* (L). Environ. Sci. Technol. 42, 5026–5031.

Bruzzoniti, M.C., Checchini, L., De Carlo, R.M., Orlandini, S., Rivoira, L., Del Bubba, M., 2014. QuEChERS sample preparation for the determination of pesticides and other organic residues in environmental matrices: a critical review. Anal. Bioanal. Chem. 406, 4089–4116.

Buck, R.C., Franklin, J., Berger, U., Conder, J.M., Cousins, I.T., de Voogt, P., Jensen, A.A., Kannan, K., Mabury, S.A., van Leeuwen, S.P.J., 2011. Perfluoroalkyl and polyfluoroalkyl substances in the environment: terminology, classification, and origins. Integr. Environ. Assess. Manag. 7, 513–541.

Burkhardt, M.R., Zaugg, S.D., Burbank, T.L., Olson, M.C., Iverson, J.L., 2005. Pressurized liquid extraction using water/isopropanol coupled with solid phase extraction cleanup for semivolatile organic compounds, polycyclic aromatic hydrocarbons (PAH) and alkylated PAH homolog groups in sediment. Anal. Chim. Acta 549, 104–116.

Buseck, P.R., Adachi, K., 2008. Nanoparticles in the atmosphere. Elements 4, 389–394.

Castells, P., Parera, J., Santos, F.J., Galceran, M.T., 2008. Occurrence of polychlorinated naphthalenes, polychlorinated biphenyls and short-chain chlorinated paraffins in marine sediments from Barcelona (Spain). Chemosphere 70, 1552–1562.

Castro-Jiménez, J., Rotllant, G., Ábalos, M., Parera, J., Dachs, J., Company, J.B., Calafat, A., Abad, E., 2013. Accumulation of dioxins in deep-sea crustaceans, fish and sediments from a submarine canyon (NW Mediterranean). Prog. Oceanogr. 118, 260–272.

Chafer-Pericas, C., Maquieira, A., Puchades, R., Miralles, J., Moreno, A., 2010a. Fast screening immunoassay of sulfonamides in commercial fish samples. Anal. Bioanal. Chem. 396, 911–921.

Chafer-Pericas, C., Maqueira, A., Puchades, R., Miralles, J., Moreno, A., Pastor-Navarro, N., Espinos, F., 2010b. Immunochemical determination of oxytetracycline in fish: comparison between enzymatic and time-resolved fluorometric assays. Anal. Chim. Acta 662, 177–185.

Chetwynd, A.J., David, A., Hill, E.M., Abdul-Sada, A., 2014. Evaluation of analytical performance and reliability of direct nanoLC-nanoESI-high resolution mass spectrometry for profiling the (xeno)metabolome. J. Mass Spectrom. 49, 1063–1069.

Chin-Yuan, C., Li-Lian, L., Wang-Hsien, D., 2006. Occurrence and seasonal variation of alkylphenols in marine organisms from the coast of Taiwan. Chemosphere 65, 2152–2159.

Christensen, J.H., Glasius, M., Pecseli, M., Platz, J., Pritzl, G., 2002. Polybrominated diphenyl ethers (PBDEs) in marine fish and blue mussels from southern Greenland. Chemosphere 47, 631–638.

Cloutier, P., Fortin, F., Fournier, M., Brousseau, P., Groleau, P.-A., Desrosiers, M., 2014. Development of an analytical method for the determination of low-level of dioxin and furans in marine and freshwater species. J. Xenobiot. 4, 73–75.

Companioni-Damas, E.Y., Santos, F.J., Galceran, M.T., 2012. Analysis of linear and cyclic methylsiloxanes in water by headspace-solid phase microextraction and gas chromatography-mass spectrometry. Talanta 89, 63–69.

Cortada, C., dos Reis, L.C., Vidal, L., Llorca, J., Canals, A., 2014. Determination of cyclic and linear siloxanes in wastewater samples by ultrasound-assisted dispersive liquid-liquid microextraction followed by gas chromatography-mass spectrometry. Talanta 120, 191–197.

Covaci, A., Voorspoels, S., Ramos, L., Neels, H., Blust, R., 2007. Recent developments in the analysis of brominated flame retardants and brominated natural compounds. J. Chromatogr. A 1153, 145–171.

Croce, V., Paggio, S., Pagnoni, A., Polesello, S., Valsecchi, S., 2003. Determination of 4-nonylphenol and 4-nonylphenol ethoxylates in river sediments by microwave assisted solvent extraction. Ann. Chim. 93, 297–304.

Cueva-Mestanza, R., Torres-Padron, M.E., Sosa-Ferrera, Z., Santana-Rodriguez, J.J., 2008. Microwave-assisted micellar extraction coupled with solid-phase extraction for preconcentration of pharmaceuticals in molluscs prior to determination by HPLC. Biomed. Chromatogr. 22, 1115–1122.

Daly, T.K., Buseck, P.R., Williams, P., Lewis, C.F., 1993. Fullerenes from a fulgurite. Science 259, 1599–1601.

David, M.D., Fendinger, N.J., Hand, V.C., 2000. Determination of Henry's law constants for organosilicones in actual and simulated wastewater. Environ. Sci. Technol. 34, 4554–4559.

David, A., Abdul-Sada, A., Lange, A., Tyler, C.R., Hill, E.M., 2014. A new approach for plasma (xeno)metabolomics based on solid-phase extraction and nanoflow liquid chromatography-nanoelectrospray ionisation mass spectrometry. J. Chromatogr. A 1365, 72–85.

De Boer, J., Law, R.J., 2003. Developments in the use of chromatographic techniques in marine laboratories for the determination of halogenated contaminants and polycyclic aromatic hydrocarbons. J. Chromatogr. A 1000, 223–251.

Delinsky, A.D., Strynar, M.J., Nakayama, S.F., Varns, J.L., Ye, X., McCann, P.J., Lindstrom, A.B., 2009. Determination of ten perfluorinated compounds in bluegill sunfish (*Lepomis macrochirus*) fillets. Environ. Res. 109, 975–984.

Dewil, R., Appels, L., Baeyens, J., Buczynska, A., Van Vaeck, L., 2007. The analysis of volatile siloxanes in waste activated sludge. Talanta 74 (1), 14–19.

Diamanti-Kandarakis, E., Bourguignon, J.P., Giudice, L.C., Hauser, R., Prins, G.S., Soto, A.M., Zoeller, R.T., Gore, A.C., 2009. Endocrine-disrupting chemicals: an Endocrine Society scientific statement. Endocr. Rev. 30, 293–342.

Dickey, R.W., 2008. Ciguatera toxins: chemistry, toxicology, and detection. Food Sci. Technol. 173, 479. New York, Marcel Dekker.

Dodder, N.G., Maruya, K.A., Lee Ferguson, P., Grace, R., Klosterhaus, S., La Guardia, M.J., Lauenstein, G.G., Ramirez, J., 2014. Occurrence of contaminants of emerging concern in mussels (*Mytilus* spp.) along the California coast and the influence of land use, storm water discharge, and treated wastewater effluent. Mar. Pollut. Bull. 81, 340–346.

Donat, J.R., Bruland, K.W., 1995. In: Salbu, B., Steiness, E. (Eds.), Trace Elements in Natural Waters. CRC-Press, Boca Raton, FL, pp. 247–281.

Dörr, B., Liebezeit, G., 2009. Organochlorine compounds in blue mussels, *Mytilus edulis*, and Pacific oysters, *Crassostrea gigas*, from seven sites in the Lower Saxonian Wadden Sea, Southern North Sea. Bull. Environ. Contam. Toxicol. 83, 874–879.

Douben, P.E.T., 2003. PAHs: An Ecotoxicological Perspective. John Wiley & Sons.

Dunphy, J.C., Pessler, D.G., Morrall, S.W., Evans, K.A., Robaugh, D.A., Fujimoto, G., Negahban, A., 2001. Derivatization LC/MS for the simultaneous determination of fatty alcohol and alcohol ethoxylate surfactants in water and wastewater samples. Environ. Sci. Technol. 35, 1223–1230.

Elgarch, A., Vale, P., Rifai, S., Fassouane, A., 2008. Detection of diarrheic shellfish poisoning and azaspiracids toxins in Moroccan mussels: comparison of LC-MS method with the commercial immunoassay kit. Mar. Drugs 6, 587–594.

EPA, 2009. Provisional Health Advisories for Perfluorooctanoic Acid (PFOA) and Perfluorooctane Sulfonate (PFOS).

Erickson, M.D., Robert, G.K.I., 2011. Applications of polychlorinated biphenyls. Environ. Sci. Pollut. Res. 18, 135–151.

EURLMB, 2016. European Union Reference Laboratory for Marine Biotoxins. http://aesan.msssi.gob.es/en/CRLMB/web/faqs/biotoxinas.shtml.

European_Commission, 2002. Water Framework Directive: Directive 2000/60/EC of the European Parliament and of the Council Establishing a Framework for the Community Action in the Field of Water Policy. http://ec.europa.eu/environment/water/water-framework/index_en.html.

European_Commission, 2012. The Combination Effects of Chemicals. Chemical Mixtures. Communication from the Commission to the Council 252.

FAO, 2004. Food and Agriculture Organization of the United Nations: Marine Biotoxins.

Farre, M., Pico, Y., Barcelo, D., 2014. Application of ultra-high pressure liquid chromatography linear ion-trap orbitrap to qualitative and quantitative assessment of pesticide residues. J. Chromatogr. A 1328, 66–79.

Fernandez-Torres, R., Bello Lopez, M.A., Olias Consentino, M., Callejon Mochon, M., 2011. Simultaneous determination of selected veterinary antibiotics and their main metabolites in fish and mussel samples by high-performance liquid chromatography with diode array-fluorescence (HPLC-DAD-FLD) detection. Anal. Lett. 44, 2357–2372.

Ferrara, F., Ademollo, N., Delise, M., Fabietti, F., Funari, E., 2008. Alkylphenols and their ethoxylates in seafood from the Tyrrhenian Sea. Chemosphere 72, 1279–1285.

Fontana, A.R., Silva, M.F., Martinez, L.D., Wuilloud, R.G., Altamirano, J.C., 2009. Determination of polybrominated diphenyl ethers in water and soil samples by cloud point extraction-ultrasound-assisted back-extraction-gas chromatography-mass spectrometry. J. Chromatogr. A 1216, 4339–4346.

Fulara, I., Czaplicka, M., 2012. Methods for determination of polybrominated diphenyl ethers in environmental samples–review. J. Sep. Sci. 35, 2075–2087.

Furey, A., Braña-Magdalena, A., Lehane, M., Moroney, C., James, K.J., Satake, M., Yasumoto, T., 2002. Determination of azaspiracids in shellfish using liquid chromatography/tandem electrospray mass spectrometry. Rapid Commun. Mass Spectrom. 16, 238–242.

Gago-Ferrero, P., Alonso, M.B., Bertozzi, C.P., Marigo, J., Barbosa, L., Cremer, M., Secchi, E.R., Azevedo, A., Lailson-Brito Jr., J., Torres, J.P., 2013. First determination of UV filters in marine mammals. Octocrylene levels in Franciscana dolphins. Environ. Sci. Technol. 47, 5619–5625.

Gaw, S., Thomas, K.V., Hutchinson, T.H., 2014. Sources, impacts and trends of pharmaceuticals in the marine and coastal environment. Philos. Trans. R. Soc. Lond. B Biol. Sci. 369.

Gerssen, A., Pol-Hofstad, I.E., Poelman, M., Mulder, P.P., Van den Top, H.J., De Boer, J., 2010. Marine toxins: chemistry, toxicity, occurrence and detection, with special reference to the Dutch situation. Toxins (Basel) 2, 878–904.

Gomez-Ramos, M.D.M., Pérez-Parada, A., García-Reyes, J.F., Fernández-Alba, A.R., Agüera, A., 2011. Use of an accurate-mass database for the systematic identification of transformation products of organic contaminants in wastewater effluents. J. Chromatogr. A 1218.

Gros, M., Petrovic, M., Barcelo, D., 2009. Tracing pharmaceutical residues of different therapeutic classes in environmental waters by using liquid chromatography/quadrupole-linear ion trap mass spectrometry and automated library searching. Anal. Chem. 81, 898–912.

Gros, M., Rodriguez-Mozaz, S., Barcelo, D., 2013. Rapid analysis of multiclass antibiotic residues and some of their metabolites in hospital, urban wastewater and river water by ultra-high-performance liquid chromatography coupled to quadrupole-linear ion trap tandem mass spectrometry. J. Chromatogr. A 1292, 173–188.

Gu, Y., Rose, W.I., Bluth, G.J.S., 2003. Retrieval of mass and sizes of particles in sandstorms using two MODIS IR bands: a case study of April 7, 2001 sandstorm in China. Geophys. Res. Lett. 30.

Guerra, P., Alaee, M., Eljarrat, E., Barceló, D., 2011. Brominated Flame Retardants. Chapter 1: "Introduction to Brominated Flame Retardants: Commercially Products, Applications, and Physicochemical Properties". Springer.

Hallegraeff, G., McCausland, M., Brown, R., 1995. Early warning of toxic dinoflagellate blooms of *Gymnodinium catenatum* in southern Tasmanian waters. J. Plankton Res. 17, 1163–1176.

Hamelink, J.L., Simon, P.B., Silberhorn, E.M., 1996. Henry's law constant, volatilization rate and aquatic half-life of octamethylcylotetrasiloxanes. Environ. Sci. Technol. 30, 1946–1952.

Hartl, M.G.J., Gubbins, E., Gutierrez, T., Fernandes, T.F., 2015. Review of Existing Knowledge – Emerging Contaminants: Focus on Nanomaterials and Microplastics in Waters. CSsCoEf, ed. Edinburgh. www.crew.ac.uk/pulications.

Haukas, M., Hylland, K., Berge, J.A., Nygard, T., Mariussen, E., 2009. Spatial diastereomer patterns of hexabromocyclododecane (HBCD) in a Norwegian fjord. Sci. Total Environ. 407, 5907–5913.

He, X., Wang, Z., Nie, X., Yang, Y., Pan, D., Leung, A.O.W., Cheng, Z., Yang, Y., Li, K., Chen, K., 2012. Residues of fluoroquinolonas in marine aquaculture environment of the Pearl River Delta, South China. Environ. Geochem. Health 34, 323–335.

Hernandez, F., Portoles, T., Ibanez, M., Bustos-Lopez, M.C., Diaz, R., Botero-Coy, A.M., Fuentes, C.L., Penuela, G., 2012. Use of time-of-flight mass spectrometry for large screening of organic pollutants in surface waters and soils from a rice production area in Colombia. Sci. Total Environ. 439, 249–259.

Hibberd, A., Maskaoui, K., Zhang, Z., Zhou, J.L., 2009. An improved method for the simultaneous analysis of phenolic and steroidal estrogens in water and sediment. Talanta 77, 1315–1321.

Hidalgo-Ruz, V., Gutow, L., Thompson, R.C., Thiel, M., 2012. Microplastics in the marine environment: a review of the methods used for identification and quantification. Environ. Sci. Technol. 46, 3060–3075.

Hirsch, A., 2010. The era of carbon allotropes. Nat. Mater. 9, 868–871.

Hotchkiss, A.K., Rider, C.V., Blystone, C.R., Wilson, V.S., Hartig, P.C., Ankley, G.T., Foster, P.M., Gray, C.L., Gray, L.E., 2008. Fifteen years after "Wingspread"—environmental endocrine disrupters and human and wildlife health: where we are today and where we need to go. Toxicol. Sci. 105, 235–259.

Houde, F., DeBlois, C., Berryman, D., 2002. Liquid chromatographic-tandem mass spectrometric determination of nonylphenol polyethoxylates and nonylphenol carboxylic acids in surface water. J. Chromatogr. A 961, 245–256.

Huerta, B., Jakimska, A., Gros, M., Rodriguez-Mozaz, S., Barcelo, D., 2013. Analysis of multi-class pharmaceuticals in fish tissues by ultra-high-performance liquid chromatography tandem mass spectrometry. J. Chromatogr. A 1288, 63–72.

Huppmann, R., Lohoff, H.W., Schröder, H.F., 1996. Cyclic siloxanes in the biological waste water treatment process — Determination, quantification and possibilities of elimination. Fres. J. Anal. Chem. 354 (1), 66–71.

Ibañez, M., Sancho, J.V., Hernández, F., McMillan, D., Rao, R., 2008. Rapid non-target screening of organic pollutants in water by ultraperformance liquid chromatography coupled to time-of-light mass spectrometry. Trends Anal. Chem. 27, 481–489.

Ibañez, M., Sancho, J.V., Pozo, J.O., Hernández, F., 2011. Use of quadrupole time-of-flight mass spectrometry to determine proposed structures of transformation products of the herbicide bromacil after water chlorination. Rapid Commun. Mass Spectrom. 25, 3103–3113.

Ikonomou, M.G., Rayne, S., Fischer, M., Fernandez, M.P., Cretney, W., 2002. Occurrence and congener profiles of polybrominated diphenyl ethers (PBDEs) in environmental samples from coastal British Columbia, Canada. Chemosphere 46, 649–663.

Jakimska, A., Huerta, B., Barganska, Z., Kot-Wasik, A., Rodriguez-Mozaz, S., Barcelo, D., 2013. Development of a liquid chromatography-tandem mass spectrometry procedure for determination of endocrine disrupting compounds in fish from Mediterranean rivers. J. Chromatogr. A 1306, 44–58.

James, K.J., Furey, A., Lehane, M., Ramstad, H., Aune, T., Hovgaard, P., Morris, S., Higman, W., Satake, M., Yasumoto, T., 2002. First evidence of an extensive northern European distribution of azaspiracid poisoning (AZP) toxins in shellfish. Toxicon 40, 909–915.

Jehlička, J., Svatoš, A., Frank, O., Uhlík, F., 2003. Evidence for fullerenes in solid bitumen from pillow lavas of Proterozoic age from Mítov (Bohemian Massif, Czech Republic). Geochim. Cosmochim. Acta 67, 1495–1506.

Jelic, A., Petrovic, M., Barcelo, D., 2009. Multi-residue method for trace level determination of pharmaceuticals in solid samples using pressurized liquid extraction followed by liquid chromatography/quadrupole-linear ion trap mass spectrometry. Talanta 80, 363–371.

Jiang, H., Zhang, Y., Chen, X., Lvb, J., Zou, J., 2013. Simultaneous determination of pentachlorophenol, niclosamide and fenpropathrin in fishpond water using an LC-MS/MS method for forensic investigation. Anal. Methods 5, 111–115.

Jiang, J.J., Lee, C.L., Fang, M.D., 2014. Emerging organic contaminants in coastal waters: anthropogenic impact, environmental release and ecological risk. Mar. Pollut. Bull. 85, 391–399.

Johnsen, A.R., Wick, L.Y., Harms, H., 2005. Principles of microbial PAH-degradation in soil. Environ. Pollut. 133, 71–84.

Johnson, Y.S., 2012. Determination of polycyclic aromatic hydrocarbons in edible seafood by QuEChERS-based extraction and gas chromatography-tandem mass spectrometry. J. Food Sci. 77, T131–T137.

Kaj, L., Schlabach, M., Andersson, J., Cousins, A.P., Schmidbauer, N., Brorstrom-Lunden, E., 2005. Siloxanes in the Nordic Environment.

Kannan, K., Reiner, J.L., Yun, S.H., Perrotta, E.E., Tao, L., Johnson-Restrepo, B., Rodan, B.D., 2005. Polycyclic musk compounds in higher trophic level aquatic organisms and humans from the United States. Chemosphere 61, 693–700.

Karickhoff, C.W., Brown, D.S., Scott, T.A., 1979. Sorption of hydrophobic pollutants on natural sediments. Water Res. 13, 241–248.

Kennish, M.J., 1997. Practical Handbook of Estuarine and Marine Pollution. CRC Press, Boca Raton, FL.

Kennish, M.J., 2000. Practical Handbook of Marine Science (Boca Raton, FL).

Kierkegaard, A., Adolfsson-Erici, M., McLachlan, M.S., 2010. Determination of cyclic volatile methylsiloxanes in biota with a purge and trap method. Anal. Chem. 82, 9573–9578.

Kierkegaard, A., Van Egmond, R., McLachlan, M.S., 2011. Cyclic volatile methylsiloxane bioaccumulation in flounder and ragworm in the Humber estuary. Environ. Sci. Technol. 45, 5936–5942.

Kim, S.C., Carlson, K., 2007. Temporal and spatial trends in the occurrence of human and veterinary antibiotics in aqueous and river sediment matrices. Environ. Sci. Technol. 41, 50–57.

Kim, J.W., Ramaswamy, B.R., Chang, K.H., Isobe, T., Tanabe, S., 2011. Multiresidue analytical method for the determination of antimicrobials, preservatives, benzotriazole UV stabilizers, flame retardants and plasticizers in fish using ultra high performance liquid chromatography coupled with tandem mass spectrometry. J. Chromatogr. A 1218, 3511–3520.

Klosterhaus, S.L., Grace, R., Hamilton, M.C., Yee, D., 2013. Method validation and reconnaissance of pharmaceuticals, personal care products, and alkylphenols in surface waters, sediments, and mussels in an urban estuary. Environ. Int. 54, 92–99.

Kremling, K., 2002. Determination of trace elements. In: Grasshoff, K., Kremling, K., Ehrhardt, M. (Eds.), Methods of Seawater Analysis. Wiley-VCH, Weinheim, pp. 253–273.

Kruger, O., Christoph, G., Kalbe, U., Berger, W., 2011. Comparison of stir bar sorptive extraction (SBSE) and liquid-liquid extraction (LLE) for the analysis of polycyclic aromatic hydrocarbons (PAH) in complex aqueous matrices. Talanta 85, 1428–1434.

Kwon, J.W., Armbrust, K.L., Vidal-Dorsch, D., Bay, S.M., 2009. Determination of 17alpha-ethynylestradiol, carbamazepine, diazepam, simvastatin, and oxybenzone in fish livers. J. AOAC Int. 92, 359–369.

Lacorte, S., Ikonomou, M.G., Fischer, M., 2010. A comprehensive gas chromatography coupled to high resolution mass spectrometry based method for the determination of polybrominated diphenyl ethers and their hydroxylated and methoxylated metabolites in environmental samples. J. Chromatogr. A 1217, 337–347.

Lara-Martin, P.A., Gomez-Parra, A., Gonzalez-Mazo, E., 2006a. Development of a method for the simultaneous analysis of anionic and non-ionic surfactants and their carboxylated metabolites in environmental samples by mixed-mode liquid chromatography-mass spectrometry. J. Chromatogr. A 1137, 188–197.

Lara-Martin, P.A., Gomez-Parra, A., Gonzalez-Mazo, E., 2006b. Simultaneous extraction and determination of anionic surfactants in waters and sediments. J. Chromatogr. A 1114, 205–210.

Lara-Martin, P.A., Gonzalez-Mazo, E., Brownawell, B.J., 2011. Multi-residue method for the analysis of synthetic surfactants and their degradation metabolites in aquatic systems by liquid chromatography-time-of-flight-mass spectrometry. J. Chromatogr. A 1218, 4799–4807.

Leon, V.M., Gonzalez-Mazo, E., Gomez-Parra, A., 2000. Handling of marine and estuarine samples for the determination of linear alkylbenzene sulfonates and sulfophenylcarboxylic acids. J. Chromatogr. A 889, 211–219.

Leon, V.M., Garcia, I., Martinez-Gomez, C., Campillo, J.A., Benedicto, J., 2014. Heterogeneous distribution of polycyclic aromatic hydrocarbons in surface sediments and red mullet along the Spanish Mediterranean coast. Mar. Pollut. Bull. 87, 352–363.

Li, Y., Wei, G., Wang, X., 2007. Determination of decabromodiphenyl ether in water samples by single-drop microextraction and RP-HPLC. J. Sep. Sci. 30, 2698–2702.

Li, W., Shi, Y., Gao, L., Liu, J., Cai, Y., 2012. Investigation of antibiotics in mollusks from coastal waters in the Bohai Sea of China. Environ. Pollut. 162, 56–62.

Li, N., Qi, L., Shen, Y., Li, Y., Chen, Y., 2013. Amphiphilic block copolymer modified magnetic nanoparticles for microwave-assisted extraction of polycyclic aromatic hydrocarbons in environmental water. J. Chromatogr. A 1316, 1–7.

Li, J.Y., Cui, Y., Su, L., Chen, Y., Jin, L., 2015. Polycyclic aromatic hydrocarbons in the largest deepwater port of East China Sea: impact of port construction and operation. Environ. Sci. Pollut. Res. 22, 12355–12365.

Liao, C., Chen, L., Kannan, K., 2013. Occurrence of parabens in foodstuffs from China and its implications for human dietary exposure. Environ. Int. 57–58, 68–74.

Libes, S.M., 2009. Introduction to Marine Biogeochemistry. Academic Press, Burlington.

Liu, J., Wang, R., Huang, B., Lin, C., Wang, Y., Pan, X., 2011. Distribution and bioaccumulation of steroidal and phenolic endocrine disrupting chemicals in wild fish species from Dianchi Lake, China. Environ. Pollut. 159, 2815–2822.

Llorca, M., Farré, M., Picó, Y., Barceló, D., 2009. Development and validation of a pressurized liquid extraction liquid chromatography-tandem mass spectrometry method for

perfluorinated compounds determination in fish. J. Chromatogr. A 1216, 7195–7204.
Llorca, M., Farré, M., Karapanagioti, H.K., Barceló, D., 2014. Levels and fate of perfluoroalkyl substances in beached plastic pellets and sediments collected from Greece. Mar. Pollut. Bull. 87, 286–291.
Llorca, M., 2012. Analysis of Perfluoroalkyl Substances in Food and Environmental Matrices (Bachelor thesis). University of Barcelona, Department of Environmental Chemistry, IDAEA-CSIC.
Long, E.R., Dutch, M., Weakland, S., Chandramouli, B., Benskin, J.P., 2013. Quantification of pharmaceuticals, personal care products, and perfluoroalkyl substances in the marine sediments of Puget Sound, Washington, USA. Environ. Toxicol. Chem. 32, 1701–1710.
Loos, R., Tavazzi, S., Paracchini, B., Canuti, E., Weissteiner, W., 2013. Analysis of polar organic contaminants in surface water of the northern Adriatic Sea by solid-phase extraction followed by ultrahigh-pressure liquid chromatography-QTRAP® MS using a hybrid triple-quadrupole linear ion trap instrument. Anal. Bioanal. Chem. 405, 5875–5885.
Lopez, P., Brandsma, S.A., Leonards, P.E., De Boer, J., 2009. Methods for the determination of phenolic brominated flame retardants, and by-products, formulation intermediates and decomposition products of brominated flame retardants in water. J. Chromatogr. A 1216, 334–345.
Losada, S., Roach, A., Roosens, L., Santos, F.J., Galceran, M.T., Vetter, W., Neels, H., Covaci, A., 2009. Biomagnification of anthropogenic and naturally-produced organobrominated compounds in a marine food web from Sydney Harbour, Australia. Environ. Int. 35, 1142–1149.
Lu, Z., Fisk, A.T., Kovacs, K.M., Lydersen, C., McKinney, M.A., Tomy, G.T., Rosenburg, B., McMeans, B.C., Muir, D.C., Wong, C.S., 2014. Temporal and spatial variation in polychlorinated biphenyl chiral signatures of the Greenland shark (*Somniosus microcephalus*) and its arctic marine food web. Environ. Pollut. 186, 216–225.
Luoma, S.N., Rainbow, P.S., 2008. Metal Contamination in Aquatic Environments. Science and Lateral Management. Cambridge University Press, Cambridge.
Mackay, D., Callcott, D., 1998. Partitioning and Physical Chemical Properties of PAHs. Springer, Berlin Heidelberg.
Magdalena, A.B., Lehane, M., Krys, S., Fernandez, M.L., Furey, A., James, K.J., 2003. The first identification of azaspiracids in shellfish from France and Spain. Toxicon 42, 105–108.
Martinez Bueno, M.J., Hernando, M.D., Agüera, A., Fernandez-Alba, A.R., 2009. Application of passive sampling devices for screening of micro-pollutants in marine aquaculture using LC-MS/MS. Talanta 77, 1518–1527.
Martinez Bueno, M.J., Boillot, C., Fenet, H., Chiron, S., Casellas, C., Gomez, E., 2013. Fast and easy extraction combined with high resolution-mass spectrometry for residue analysis of two anticonvulsants and their transformation products in marine mussels. J. Chromatogr. A 1305, 27–34.
Martinez Bueno, M.J., Boillot, C., Munaron, D., Fenet, H., Casellas, C., Gomez, E., 2014. Occurrence of venlafaxine residues and its metabolites in marine mussels at trace levels: development of analytical method and a monitoring program. Anal. Bioanal. Chem. 406, 601–610.
Maruyama, S., Lee, M.Y., Haufler, R.E., Chai, Y., Smalley, R.E., 1991. Thermionic emission from giant fullerenes. Z. Phys. D Atoms Mol. Clusters 19, 409–412.
Masia, A., Campo, J., Vazquez-Roig, P., Blasco, C., Pico, Y., 2013a. Screening of currently used pesticides in water, sediments and biota of the Guadalquivir River Basin (Spain). J. Hazard. Mater. 263 (Pt 1), 95–104.
Masia, A., Ibanez, M., Blasco, C., Sancho, J.V., Pico, Y., Hernandez, F., 2013b. Combined use of liquid chromatography triple quadrupole mass spectrometry and liquid chromatography quadrupole time-of-flight mass spectrometry in systematic screening of pesticides and other contaminants in water samples. Anal. Chim. Acta 761, 117–127.
Masia, A., Blasco, C., Picó, Y., 2014. Last trends in pesticide residue determination by liquid chromatography-mass spectrometry. Trends Anal. Chem. 2, 11–24.
Mason, R.P., 2013. Trace Metals in Aquatic Systems. Wiley Blackwell, Chichester.
Mato, Y., Isobe, T., Takada, H., Kanehiro, H., Ohtake, C., Kaminuma, T., 2001. Plastic resin pellets as a transport medium for toxic chemicals in the marine environment. Environ. Sci. Technol. 35, 318–324.
Matthijs, E., H, M.S., Kiewiet, A., Rijs, G.B.J., 1999. Environmental monitoring of linear alkylbenzene sulfonates, alcohol ethoxylate, alcohol ethoxy sulfate, alcohol sulfate and soap. Environ. Toxicol. Chem. 18, 2634–2644.
McEneff, G., Barron, L., Kelleher, B., Paull, B., Quinn, B., 2013. The determination of pharmaceutical residues in cooked and uncooked marine bivalves using pressurised liquid extraction, solid-phase extraction and liquid chromatography-tandem mass spectrometry. Anal. Bioanal. Chem. 405, 9509–9521.
McKinney, M.A., McMeans, B.C., Tomy, G.T., Rosenberg, B., Ferguson, S.H., Morris, A., Muir, D.C.G., Fisk, A.T., 2012. Trophic transfer of contaminants in a changing Arctic marine food web: Cumberland Sound, Nunavut, Canada. Environ. Sci. Technol. 46, 9914–9922.
Meador, J.P., Stein, J.E., Reichert, W.L., Varanasi, U., 1995. Bioaccumulation of polycyclic aromatic hydrocarbons by marine organisms. Rev. Environ. Contam. Toxicol. 143, 79–165.

Menzies, R., Soares Quinete, N., Gardinali, P., Seba, D., 2013. Baseline occurrence of organochlorine pesticides and other xenobiotics in the marine environment: Caribbean and Pacific collections. Mar. Pollut. Bull. 70, 289–295.

Montes, R., Rodriguez, I., Cela, R., 2010. Solid-phase microextraction with simultaneous oxidative sample treatment for the sensitive determination of tetra- to hexa-brominated diphenyl ethers in sediments. J. Chromatogr. A 1217, 14–21.

Moreno-Gonzalez, R., Rodriguez-Mozaz, S., Gros, M., Barcelo, D., Leon, V.M., 2015. Seasonal distribution of pharmaceuticals in marine water and sediment from a Mediterranean coastal lagoon (SE Spain). Environ. Res. 138, 326–344.

Muller, A., Schulz, W., Ruck, W.K., Weber, W.H., 2011. A new approach to data evaluation in the non-target screening of organic trace substances in water analysis. Chemosphere 85, 1211–1219.

Munaron, D., Tapie, N., Budzinski, H., Andral, B., Gonzalez, J.-L., 2012. Pharmaceuticals, alkylphenols and pesticides in Mediterranean coastal waters: results from a pilot survey using passive samplers. Estuar. Coast. Shelf Sci. 114, 82–92.

Murphy, J., 2001. Modifying Specific Properties: Flammability-Flame Retardants.

Na, G., Fang, X., Cai, Y., Ge, L., Zong, H., Yuan, X., Yao, Z., Zhang, Z., 2013. Occurrence, distribution, and bioaccumulation of antibiotics in coastal environment of Dalian, China. Mar. Pollut. Bull. 69, 233–237.

Navarro, P., Etxebarria, N., Arana, G., 2009. Development of a focused ultrasonic-assisted extraction of polycyclic aromatic hydrocarbons in marine sediment and mussel samples. Anal. Chim. Acta 648, 178–182.

Navarro, P., Bustamante, J., Vallejo, A., Prieto, A., Usobiaga, A., Arrasate, S., Anakabe, E., Puy-Azurmendi, E., Zuloaga, O., 2010. Determination of alkylphenols and 17beta-estradiol in fish homogenate. Extraction and clean-up strategies. J. Chromatogr. A 1217, 5890–5895.

Nieboer, E.R.D., 1980. The replacement of the nondescript term "heavy metal" by abiologically and chemically significant classification of metal ions. Environ. Pollut. 1, 3–26.

Nozaki, Y., 1997. A fresh look at element distribution in the North Pacific Ocean. Eos 78, 221–223.

Núñez, Ó., Gallart-Ayala, H., Martins, C.P.B., Moyano, E., Galceran, M.T., 2012. Atmospheric pressure photoionization mass spectrometry of fullerenes. Anal. Chem. 84, 5316–5326.

Oehlmann, J., Schulte-Oehlmann, U., Kloas, W., Jagnytsch, O., Lutz, I., Kusk, K.O., Wollenberger, L., Santos, E.M., Paull, G.C., Van Look, K.J., Tyler, C.R., 2009. A critical analysis of the biological impacts of plasticizers on wildlife. Philos. Trans. R. Soc. Lond. B Biol. Sci. 364, 2047–2062.

OSPAR_Commission, 2010. Polychlorinated Byphenyls. Status and Trend in Marine Chemical Pollution.

Parera, J., Ábalos, M., Santos, F.J., Galceran, M.T., Abad, E., 2013. Polychlorinated dibenzo-p-dioxins, dibenzofurans, biphenyls, paraffins and polybrominated diphenyl ethers in marine fish species from Ebro River Delta (Spain). Chemosphere 93, 499–505.

Parsons, J., Belzunce-Segarra, M.J., Cornelissen, G., Gustafsson, Ö., Grontenhuis, T., Harms, H., Janssen, C.R., Kukkonen, J., Van Noort, P., Ortega-Calvo, J.J., Etexeberria, S., 2007. Characterisation of contaminants in sediments – effects of bioavailability on impact. In: Barceló, D., Petrovic, M. (Eds.), Sustainable Management of Sediment Resources: Sediment Quality and Impact Assessment of Pollutants. Elsevier, Amsterdam, pp. 35–60.

Pena, T., Pensado, L., Casais, C., Mejuto, C., Phan-Tan-Luu, R., Cela, R., 2006. Optimization of a microwave-assisted extraction method for the analysis of polycyclic aromatic hydrocarbons from fish samples. J. Chromatogr. A 1121, 163–169.

Petrovic, M., Diaz, A., Ventura, F., Barcelo, D., 2001. Simultaneous determination of halogenated derivatives of alkylphenol ethoxylates and their metabolites in sludges, river sediments, and surface, drinking, and wastewaters by liquid chromatography-mass spectrometry. Anal. Chem. 73, 5886–5895.

Petrovic, M., Eljarrat, E., López de Alda, M.J., Barceló, D., 2002a. Recent advances in the mass spectrometric analysis related to endocrine disrupting compounds in aquatic environmental samples. J. Chromatogr. A 974, 23–51.

Petrovic, M., Fernandez-Alba, A.R., Borrull, F., Marce, R.M., Gonzalez, M.E., Barcelo, D., 2002b. Occurrence and distribution of nonionic surfactants, their degradation products, and linear alkylbenzene sulfonates in coastal waters and sediments in Spain. Environ. Toxicol. Chem. 21, 37–46.

Pojana, G., Gomiero, A., Jonkers, N., Marcomini, A., 2007. Natural and synthetic endocrine disrupting compounds (EDCs) in water, sediment and biota of a coastal lagoon. Environ. Int. 33, 929–936.

Polo, M., Llompart, M., Garcia-Jares, C., Gomez-Noya, G., Bollain, M.H., Cela, R., 2006. Development of a solid-phase microextraction method for the analysis of phenolic flame retardants in water samples. J. Chromatogr. A 1124, 11–21.

Poster, D.L., Schantz, M.M., Sander, L.C., Wise, S.A., 2006. Analysis of polycyclic aromatic hydrocarbons (PAHs) in environmental samples: a critical review of gas chromatographic (GC) methods. Anal. Bioanal. Chem. 386, 859–881.

Primel, E.G., Caldas, S.S., Escarrone, A.L.V., 2012. Multiresidue analytical methods for the determination of pesticides and PPCPs in water by LC-MS/MS: a review. Cent. Eur. J. Chem. 10, 876–899.

Prinzbach, H., Weiler, A., Landenberger, P., Wahl, F., Worth, J., Scott, L.T., Gelmont, M., Olevano, D., v. Issendorff, B., 2000. Gas-phase production and photoelectron spectroscopy of the smallest fullerene, C20. Nature 407, 60–63.

Quintana, J.B., Rodil, R., Muniategui-Lorenzo, S., Lopez-Mahia, P., Prada-Rodriguez, D., 2007. Multiresidue analysis of acidic and polar organic contaminants in water samples by stir-bar sorptive extraction-liquid desorption-gas chromatography-mass spectrometry. J. Chromatogr. A 1174, 27–39.

Ramos, L., Kristenson, E.M., Brinkman, U.A., 2002. Current use of pressurised liquid extraction and subcritical water extraction in environmental analysis. J. Chromatogr. A 975, 3–29.

Reiser, R., Toljander, H.O., Giger, W., 1997. Determination of alkylbenzenesulfonates in recent sediments by gas chromatography/mass spectrometry. Anal. Chem. 69, 4923–4930.

Ren, H., Kawagoe, T., Jia, H., Endo, H., Kitazawa, A., Goto, S., Hayashi, T., 2010. Continuous surface seawater surveillance on poly aromatic hydrocarbons (PAHs) and mutagenicity of East and South China Seas. Estuar. Coast. Shelf Sci. 86, 395–400.

Rezaee, M., Yamini, Y., Faraji, M., 2010. Evolution of dispersive liquid-liquid microextraction method. J. Chromatogr. A 1217, 2342–2357.

Rios, L.M., Moore, C., Jones, P.R., 2007. Persistent organic pollutants carried by synthetic polymers in the ocean environment. Mar. Pollut. Bull. 54, 1230–1237.

Robles-Molina, J., Gilbert-Lopez, B., Garcia-Reyes, J.F., Molina-Diaz, A., 2013. Comparative evaluation of liquid-liquid extraction, solid-phase extraction and solid-phase microextraction for the gas chromatography-mass spectrometry determination of multiclass priority organic contaminants in wastewater. Talanta 117, 382–391.

Rodríguez-Mozaz, S., Álvarez-Muñoz, D., Barceló, D., 2015. Environmental Problems in Marine Biology: Methodological Aspects and Applications. Chapter: Pharmaceuticals in Marine Environment: Analytical Techniques and Applications. CRC Press, Taylor & Francis.

Roslan, R.N., Hanif, N.M., Othman, M.R., Azmi, W.N., Yan, X.X., Ali, M.M., Mohamed, C.A., Latif, M.T., 2010. Surfactants in the sea-surface microlayer and their contribution to atmospheric aerosols around coastal areas of the Malaysian peninsula. Mar. Pollut. Bull. 60, 1584–1590.

Rostkowski, P., Horwood, J., Shears, J.A., Lange, A., Oladapo, F.O., Besselink, H.T., Tyler, C.R., Hill, E.M., 2011. Bioassay-directed identification of novel antiandrogenic compounds in bile of fish exposed to wastewater effluents. Environ. Sci. Technol. 45, 10660–10667.

Rudel, H., Bohmer, W., Muller, M., Fliedner, A., Ricking, M., Teubner, D., Schroter-Kermani, C., 2013. Retrospective study of triclosan and methyl-triclosan residues in fish and suspended particulate matter: results from the German Environmental Specimen Bank. Chemosphere 91, 1517–1524.

Ruff, M., Mueller, M.S., Loos, M., Singer, H.P., 2015. Quantitative target and systematic non-target analysis of polar organic micro-pollutants along the River Rhine using high-resolution mass-spectrometry identification of unknown sources and compounds. Water Res. 87, 145–154.

Saez, M., Leon, V.M., Gomez-Parra, A., Gonzalez-Mazo, E., 2000. Extraction and isolation of linear alkylbenzene sulfonates and their intermediate metabolites from various marine organisms. J. Chromatogr. A 889, 99–104.

Salgueiro-González, N., Turnes-Carou, I., Muniategui-Lorenzo, S., López-Mahía, P., Prada-Rodríguez, D., 2014. Analysis of endocrine disruptor compounds in marine sediments by in cell clean up-pressurized liquid extraction-liquid chromatography tandem mass spectrometry determination. Anal. Chim. Acta 852, 112–120.

Samanidou, V.F., Evaggelopoulou, E.N., 2007. Analytical strategies to determine antibiotic residues in fish. J. Sep. Sci. 30, 2549–2569.

Sanchez-Vidal, A., Llorca, M., Farré, M., Canals, M., Barceló, D., Puig, P., Calafat, A., 2015. Delivery of unprecedented amounts of perfluoroalkyl substances towards the deep-sea. Sci. Total Environ. 526, 41–48.

Sanchís, J., Berrojalbiz, N., Caballero, G., Dachs, J., Farré, M., Barceló, D., 2011. Occurrence of aerosol-bound fullerenes in the Mediterranean Sea atmosphere. Environ. Sci. Technol. 46, 1335–1343.

Sanchís, J., Farré, M., Barceló, D., 2012. Analysis and Fate of Organic Nanomaterials in Environmental Samples. Elsevier.

Sanchis, J., Martinez, E., Ginebreda, A., Farre, M., Barcelo, D., 2013. Occurrence of linear and cyclic volatile methylsiloxanes in wastewater, surface water and sediments from Catalonia. Sci. Total Environ. 443, 530–538.

Sanchís, P., Llorca, M., Farré, M., Barceló, D., 2015. Volatile methyl siloxanes in market seafood and fish from the Xuquer River, Spain. Sci. Total Environ. 443, 530–538.

Sanchis, J., Llorca, M., Pico, Y., Farre, M., Barcelo, D., 2016. Volatile dimethylsiloxanes in market seafood and freshwater fish from the Xuquer River, Spain. Sci. Total Environ. 545–546, 236–243.

Sanchís, J., 2015. Analysis of Nanomaterials and Nanostructures in the Environment (Bachelor thesis). Universitat de Barcelona, Department of Environmental Chemistry, IDAEA-CSIC.

Sañudo-Wihelmy, S.A., Tovar-Sánchez, A., Fisher, N.S., Flegal, A.R., 2004. Examining dissolved toxic metals in U.S. estuaries. Environ. Sci. Technol. 38, 34A–38A.

Schlabach, M., Andersen, M.S., Green, N., Schoyen, M., Kaj, L., 2007. Siloxanes in the Environment of the Inner Oslofjord. NILU OR 27/2007.

Schmitz-Afonso, I., Loyo-Rosales, J.E., de la Paz Aviles, M., Rattner, B.A., Rice, C.P., 2003. Determination of alkylphenol and alkylphenolethoxylates in biota by liquid chromatography with detection by tandem mass spectrometry and fluorescence spectroscopy. J. Chromatogr. A 1010, 25–35.

Shang, D.Y., Ikonomou, M.G., Macdonald, R.W., 1999. Quantitative determination of nonylphenol polyethoxylate surfactants in marine sediment using normal-phase liquid chromatography-electrospray mass spectrometry. J. Chromatogr. A 849, 467–482.

Shi, H., Yang, Y., Liu, M., Yan, C., Yue, H., Zhou, J., 2014. Occurrence and distribution of antibiotics in the surface sediments of the Yangtze Estuary and nearby coastal areas. Mar. Pollut. Bull. 83, 317–323.

Simon, E., Lamoree, M., Hamers, T., Boer, J., 2015. Challenges in effect-directed analysis with a focus on biological samples. Trends Anal. Chem. 67, 179–191.

Skupinska, K., Misiewicz, I., Kasprzycka-Guttman, T., 2004. Polycyclic aromatic hydrocarbons: physiochemical properties, environmental appearance and impact on living organisms. Acta Pol. Pharm. 61, 233–240.

Solaun, O., Rodríguez, J., Borja, A., Larreta, J., Valencia, V., 2015. Relationships between polychlorinated biphenyls in molluscs, hydrological characteristics and human pressures, within Basque estuaries (northern Spain). Chemosphere 118, 130–135.

Sparham, C., Van Egmond, R., O'Connor, S., Hastie, C., Whelan, M., Kanda, R., Franklin, O., 2008. Determination of decamethylcyclopentasiloxane in river water and final effluent by headspace gas chromatography/mass spectrometry. J. Chromatogr. A 1212, 124–129.

Sparham, C., Van Egmond, R., Hastie, C., O'Connor, S., Gore, D., Chowdhury, N., 2011. Determination of decamethylcyclopentasiloxane in river and estuarine sediments in the UK. J. Chromatogr. A 1218, 817–823.

Stalmans, M., Matthijs, E., De Oude, N.T., 1991. Fate and effects of detergent chemicals in the marine and estuarine environment. Water Sci. Technol. 24, 115–126.

Stamper, M.A., Spicer, C.W., Neiffer, D.L., Mathews, K.S., Fleming, G.J., 2009. Morbidity in a juvenile green sea turtle (*Chelonia mydas*) due to ocean-borne plastic. J. Zoo Wildl. Med. 40, 196–198.

Stewart, I., Eaglesham, G.K., Poole, S., Graham, G., Paulo, C., Wickramasinghe, W., Sadler, R., Shaw, G.R., 2010. Establishing a public health analytical service based on chemical methods for detecting and quantifying Pacific ciguatoxin in fish samples. Toxicon 56, 804–812.

Taleb, H., Vale, P., Amanhir, R., Benhadouch, A., Sagou, R., Chafik, A., 2006. First detection of azaspiracids in mussels in north west Africa. J. Shellfish Res. 25, 1067–1070.

Tanoue, R., Nomiyama, K., Nakamura, H., Hayashi, T., Kim, J.W., Isobe, T., Shinohara, R., Tanabe, S., 2014. Simultaneous determination of polar pharmaceuticals and personal care products in biological organs and tissues. J. Chromatogr. A 1355, 193–205.

Ternes, T.A., Joss, A., Siegrist, H., 2004. Peer reviewed: scrutinizing pharmaceuticals and personal care products in wastewater treatment. Environ. Sci. Technol. 38, 392A–399A.

Tertuliani, J.S., Alvarez, D.A., Furlong, E.T., Meyer, M.T., Zaugg, S.D., Koltun, G.F., 2008. Occurrence of organic wastewater compounds in the Tinkers Creek watershed and two other tributaries to the Cuyahoga River, Northeast Ohio. In: U.S. Geological Survey Scientific Investigations Report, vol. 5173, p. 60.

Teuten, E.L., Saquing, J.M., Knappe, D.R., Barlaz, M.A., Jonsson, S., Bjorn, A., Rowland, S.J., Thompson, R.C., Galloway, T.S., Yamashita, R., Ochi, D., Watanuki, Y., Moore, C., Viet, P.H., Tana, T.S., Prudente, M., Boonyatumanond, R., Zakaria, M.P., Akkhavong, K., Ogata, Y., Hirai, H., Iwasa, S., Mizukawa, K., Hagino, Y., Imamura, A., Saha, M., Takada, H., 2009. Transport and release of chemicals from plastics to the environment and to wildlife. Philos. Trans. R. Soc. Lond. B Biol. Sci. 364, 2027–2045.

Tolls, J., Haller, M., Sijm, D.T., 1999. Extraction and isolation of linear alkylbenzenesulfonate and its sulfophenylcarboxylic acid metabolites from fish samples. Anal. Chem. 71, 5242–5247.

Tovar-Sánchez, A., 2012. Sampling approaches for trace element determination in seawater. In: P, J. (Ed.), Comprehensive Sampling and Sample Preparation. Analytical Techniques for Scientists, vol. 1. Elsevier-AP, Amsterdam, pp. 318–334.

Traverso-Soto, J.M., González-Mazo, E., Lara-Martín, P.A., 2012. Analysis of Surfactants in Environmental Samples by Chromatographic Techniques.

Tuomisto, J., 2011. Synopsis on Dioxins and PCBs. Report/National Institute for Health and Welfare (THL)= Raportti/Terveyden ja hyvinvoinnin laitos: 14/2011.

UNEP, 2002. PCB Transformers and Capacitors: From Management to Reclassification and Disposal. Geneva, United Nations Environment Programme, UNEP Chemicals.

United_Nations, 1999. Guidelines for the Identification of PCBs and Materials Containing PCBs. United Nations Environment Program.

USEPA, 2006. 2010/15 Stewardship Program. http://www.epa.gov/oppt/pfoa/pubs/stewardship/index.html.

Van den Berg, M., Birnbaum, L.S., Denison, M., De Vito, M., Farland, W., Feeley, M., Fiedler, H., Hakansson, H., Hanberg, A., Haws, L., Rose, M., Safe, S., Schrenk, D., Tohyama, C., Tritscher, A., Tuomisto, J., Tysklind, M., Walker, N., Peterson, R.E., 2006. The 2005 World Health Organization re-evaluation of human and mammalian toxic equivalency factors for dioxins and dioxin-like compounds. Toxicol. Sci. 93, 223–241.

Van Dolah, F.M., 2000. Marine algal toxins: origins, health effects, and their increased occurrence. Environ. Health Perspect. 108, 133.

Van Ry, D.A., Dachs, J., Gigliotti, C.L., Brunciak, P.A., Nelson, E.D., Eisenreich, S.J., 2000. Atmospheric seasonal trends and environmental fate of alkylphenols in the Lower Hudson River Estuary. Environ. Sci. Technol. 34, 2410–2417.

Vandermeersch, G., Lourenço, H.M., Alvarez-Muñoz, D., Cunha, S., Diogène, J., Cano-Sancho, G., Sloth, J.J., Kwadijk, C., Barceló, D., Allegaert, W., Bekaert, K., Fernandes, J.O., Marques, A., Robbens, J., 2015. Environmental contaminants of emerging concern in seafood – European database on contaminants levels. Environ. Res. 143, 29–45.

Versteeg, D.J., Rawlings, J.M., 2003. Bioconcentration and toxicity of dodecylbenzene sulfonate (C12LAS) to aquatic organisms exposed in experimental streams. Arch. Environ. Contam. Toxicol. 44, 237–246.

Villar-Pulido, M., Gilbert-Lopez, B., Garcia-Reyes, J.F., Martos, N.R., Molina-Diaz, A., 2011. Multiclass detection and quantitation of antibiotics and veterinary drugs in shrimps by fast liquid chromatography time-of-flight mass spectrometry. Talanta 85, 1419–1427.

Wagner, M., Scherer, C., Alvarez-Muñoz, D., Brennholt, N., Bourrain, X., Buchinger, S., Fries, E., Grosbois, C., Klasmeier, J., Marti, T., Rodriguez-Mozaz, S., Urbatzka, R., Vethaak, A.D., Winther-Nielsen, M., Reifferscheid, G., 2014. Microplastics in freshwater ecosystems: what we know and what we need to know. Environ. Sci. Eur. 26, 12.

Wang, D.G., Steer, H., Tait, T., Williams, Z., Pacepavicius, G., Young, T., Ng, T., Smyth, S.A., Kinsman, L., Alaee, M., 2013. Concentrations of cyclic volatile methylsiloxanes in biosolid amended soil, influent, effluent, receiving water, and sediment of wastewater treatment plants in Canada. Chemosphere 93, 766–773.

Warner, N.A., Evenset, A., Christensen, G., Gabrielsen, G.W., Borga, K., Leknes, H., 2010. Volatile siloxanes in the European Arctic: assessment of sources and spatial distribution. Environ. Sci. Technol. 44, 7705–7710.

Weigel, S., Kuhlmann, J., Hühnerfuss, H., 2002. Drugs and personal care products as ubiquitous pollutants: occurrence and distribution of clofibric acid, caffeine and DEET in the North Sea. Sci. Total Environ. 295, 131–141.

Whelan, M.J., Sanders, D., Van Egmond, R., 2009. Effect of Aldrich humic acid on water–atmosphere transfer of decamethylcyclopentasiloxane. Chemosphere 74, 1111–1116.

WHO, 1997. Flame Retardants: A General Introduction. Environmental Health Criteria Geneva. World Health Organization.

WHO, 2000. Air Quality Guidelines for Europe, second ed. WHO Regional Publications (European series).

WHO, 2010. Exposure to Dioxins and Dioxin-Like Substances: A Major Public Health Concern Preventing Disease through Healthy Environments.

Wille, K., Kiebooms, J.A.L., Claessens, M., Rappé, K., Vanden Bussche, J., Noppe, H., Van Praet, M., De Wulf, E., Van Caerter, P., Janssen, R.C., De Brabander, H.F., Vanhaecke, L., 2011. Development of analytical strategies using U-HPLC-MS/MS and LC-ToF-MS for the quantification of micropollutants in marine organisms. Anal. Bioanal. Chem. 400, 1459–1472.

Wright, S.L., Thompson, R.C., Galloway, T.S., 2013. The physical impacts of microplastics on marine organisms: a review. Environ. Pollut. 178, 483–492.

Xiao, Y., Chae, S.-R., Wiesner, M.R., 2011. Quantification of fullerene (C60) in aqueous samples and use of C70 as surrogate standard. Chem. Eng. J. 170, 555–561.

Yang, Y., Fu, J., Peng, H., Hou, L., Liu, M., Zhou, J.L., 2011. Occurrence and phase distribution of selected pharmaceuticals in the Yangtze Estuary and its coastal zone. J. Hazard. Mater. 190, 588–596.

Zhang, Z., Qi, H., Ren, N., Li, Y., Gao, D., Kannan, K., 2011. Survey of cyclic and linear siloxanes in sediment from the Songhua River and in sewage sludge from wastewater treatment plants, Northeastern China. Arch. Environ. Contam. Toxicol. 60, 204–211.

Zheng, S., Chen, B., Qiu, X., Chen, M., Ma, Z., Yu, X., 2016. Distribution and risk assessment of 82 pesticides in Jiulong River and estuary in South China. Chemosphere 144, 1177–1192.

Zhou, S., Tang, Q., Jin, M., Liu, W., Niu, L., Ye, H., 2014. Residues and chiral signatures of organochlorine pesticides in mollusks from the coastal regions of the Yangtze River Delta: source and health risk implication. Chemosphere 114, 40–50.

Zhou, L.J., Ying, G.G., Zhao, J.L., Yang, J.F., Wang, L., Yang, B., Liu, S., 2011. Trends in the occurrence of human and veterinary antibiotics in the sediments of the Yellow River, Hai River and Liao River in northern China. Environ. Pollut. 159, 1877–1885.

CHAPTER

2

Contemporary Methods for Statistical Design and Analysis

D.R. Fox[1,2]

[1]Environmetrics Australia, Beaumaris, VIC, Australia; [2]University of Melbourne, Parkville, VIC, Australia

2.1 INTRODUCTION

The present chapter deliberately avoids rehashing the familiar statistical concepts covered in any university "STAT101" course. Not only are there numerous texts on all aspects of "classical" statistical theory and practice (eg, Cox and Snell, 2000; Bickel and Doksum, 2016; Kutner et al., 2016), but there now exists a substantial body of "guideline" material, including statistical guidance documents, that has been amassed by environmental jurisdictions around the world including Australia, New Zealand, the European Union, Canada, and the United States. Accordingly, this chapter attempts to shift the focus away from where we have been and more toward where we need to go. For example, redeploying the scarce experimental resources that are liberated by abandoning wasteful analysis of variance (ANOVA) procedures for computing no observed effect concentrations (NOECs) and actually *modelling* the concentration–response (C-R) phenomenon rather than reducing it to a trite statement of "no observed effect at concentration *x*."

This chapter deals with a number of facets associated with the statistical aspects of modern-day ecotoxicology. It is neither complete nor comprehensive and invariably reflects the author's biases and interests. For example, models of contaminant uptake, transport, and fate have not been discussed nor is there any discussion of stochastic models used to describe population dynamics. However, the topics that are covered are hopefully sufficiently representative of the main statistical issues and challenges facing ecotoxicologists in their pursuit of sustainable ecosystem protection.

2.2 TOXICITY MEASURES

To monitor and manage contaminants (ie, potential toxicants) in a marine environment, it is necessary to measure both their concentration in receiving waters and their "effects" on prescribed components of the ecosystem. *Effects* can be either acute or chronic—the difference primarily a function of *dose* which is the product of *concentration* and *time*. Acute effects are those

Marine Ecotoxicology
http://dx.doi.org/10.1016/B978-0-12-803371-5.00002-3

that result from short-term exposure (relative to the organism's life span) to high concentrations; chronic effects result from long-term exposure (relative to the organism's life span) to relatively low concentrations.

The ability of a substance to cause acute and/or chronic effects in an organism is referred to as its *toxicity*. What constitutes an "effect" is a critical question and one not always easily answered; however, ecotoxicologists use a variety of endpoints ranging from sublethal (for example, compromised reproductive capability) to lethal (death), which define the *dependent variables* in C-R experiments (discussed in greater detail in Section 2.6).

The *quantification* of toxicity is most commonly assessed from an examination of an organism's *response* (with respect to a particular endpoint) to changes in toxicant concentration giving rise to the so-called C-R curve.

The development of toxicity metrics is rooted in human toxicity studies that date back to the turn of the 20th century. In 1908 Theodore Cash published a paper in the British Medical Journal, in which he described early dose–response experiments (Cash, 1908). In his paper he talks of "the nearest fulfillment of a mathematical relationship seemed to be achieved by working upwards from that amount of any drug which produced the minimum of appreciable action." This dose, he notes, was variously referred to as the "*Grenzdose*" or "*limit dose*." Cash (1908) also introduced the additional terms "*minimal effective dose*" and "*maximal ineffective dose*," although it was later suggested these terms be replaced with the *median lethal dose* or LD_{50} (Trevan, 1927). However, note that an LC refers to a concentration outside of biological tissues, while an LD refers to a concentration inside biological tissues.

What is interesting is the plethora of toxicity metrics that have since emerged. Those commonly used in ecotoxicology include:

- LD_x or LC_x—the dose/concentration that is lethal to x% of organisms;
- EC_x (effect concentration)—the concentration at which x% of organisms will exhibit an "effect";
- IC_x (inhibitory concentration)—the concentration at which x% "impairment" occurs;
- NEC (no effect concentration)—the highest concentration below which no effect occurs;
- NOEC/NOAEL [no observed (adverse) effect concentration]—the largest of a small number of discrete test concentrations for which the mean response is not statistically different from the control response;
- LOEC/LOAEL [lowest observed (adverse) effect concentration]—the smallest of a small number of discrete test concentrations for which the mean response *is* statistically different from the control response;
- MATC (maximum acceptable toxicant concentration)—defined as the geometric mean of the NOEC and LOEC;
- BEC_x (bounded effect concentration)—the highest concentration for which one may claim (with 95% confidence) that its effect does not exceed x% (Hoekstra and van Ewijk, 1993)

Not surprisingly, this array of metrics has motivated numerous studies to compare their efficacy as regulatory instruments as well as to reconcile the implications of their use for ecosystem protection (Crane and Newman, 2000; Hose and Van den Brink, 2004; Payet, 2004; Shieh et al., 2001; van der Hoeven et al., 1997). This is an important issue that is far from resolved. The choice of *which* particular toxicity measure to use in practice is largely dictated by the requirements of standard testing protocols. Where flexibility of choice exists, it is unfortunately true that practitioners often treat these measures as either exchangeable or related via a simple scaling. The most common example of the latter is the arbitrary scaling of measures of acute toxicity by some order of magnitude in an attempt to harmonize the results with measures of chronic toxicity. Statistically, there is no justification for this although Fox (2006) did attempt to at least remove the arbitrariness of the scaling factor—otherwise known as the *acute-to-chronic ratio* or ACR—by allowing the data to determine an "optimal" value for the ACR.

The regulatory use of toxicity measures is confused and confusing. For many years NOECs and LOECs were the preferred measures of toxicity although more recently their use has attracted strong criticism (Chapman et al., 1996; Jager, 2012; Fox, 2008; Fox et al., 2012) and even calls to ban their use altogether (van Dam et al., 2012; Warne and van Dam, 2008; Landis and Chapman, 2011). However, such views have themselves been criticized with others arguing that calls to ban the NOEC are misinformed and the result of simplistic statistical thinking (Green et al., 2013).

Jurisdictions around the world have published various "Guideline" documents and while most articulate preferred general strategies, many leave the detail up to the analyst. For example, the Organization for Economic Cooperation and Development (OECD) Guidelines recommend the use of model-based toxicity measures although "the choice is left to the reader" as far as the statistical detail is concerned (OECD, 2014). At the other end of the spectrum, the recently revised Australian and New Zealand Guidelines for toxicants in fresh and marine waters takes a more prescriptive approach (Warne et al., 2013). With respect to chronic toxicity measures, these guidelines give preference in the following order:

1. NEC;
2. $EC_x/IC_x/LC_x$ where $x \leq 10$ (*NB: all three equally ranked)*;
3. BEC_{10};
4. EC_x or LC_x where $15 \leq x \leq 20$;
5. NOEC or NOEC estimated from MATC, LOEC, or LC_{50} values.

In the event there are insufficient chronic toxicity data the Guidelines (OECD, 2014) recommend that acute EC_{50} and/or IC_{50} and/or LC_{50} be converted to "chronic equivalent data" for the derivation of a protective concentration.

While the science of toxicity metrics continues to evolve, for example, the use of Bayesian methods (Link and Albers, 2007; Fox, 2010; Cliffroy et al., 2013; Zhang et al., 2012; Grist et al., 2006; Jaworska et al., 2010), the challenge for regulators and practitioners is to ensure that toxicity data and the statistical methods used to generate them are (1) ecologically relevant, (2) fit-for-purpose, (3) robust to moderate violations of assumptions, (4) scientifically defensible, and (5) statistically credible. We shall return to these themes in Section 2.7 when we consider statistical issues associated with Species Sensitivity Distributions (SSDs). At this stage, a number of important statistical considerations arise:

1. The nature of the endpoint determines the type of random variable and associated statistical treatment of toxicity data. Random variables are either *discrete* or *continuous*. As a rule-of-thumb, if the endpoint is the result of observation (eg, mortality status) rather than measurement, then the random variable is discrete and the data are typically obtained as a result of *counting* (number of dead organisms). Alternatively, if the endpoint requires physical measurement of some property (eg, length of fish), then the random variable is continuous (eg, growth rate). This distinction is not simply academic—it is a necessary (but not sufficient) requirement to satisfy (2) above. Unfortunately insufficient attention is paid to this point in many ecotoxicological studies with the result that inappropriate statistical tests and/or models are used with incompatible data constructs, for example,

t-tests and ANOVA used to analyze binary (eg, dead/alive) data obtained from small (eg, <20) samples.

2. The arbitrary scaling and pooling of toxicity metrics to generate "sufficient" data from an assumed common distribution of toxicity (as is done in SSD modeling) is difficult to justify ecologically and impossible to justify statistically. The (implicit) assumption that the sample percentiles of toxicity data corresponding to either a single endpoint across multiple species or multiple endpoints across a single species have the same underlying statistical distribution is surely false. One only needs to consider the dichotomy of discrete/continuous distributions arising from different endpoints to make this assertion.
3. The *statistical properties* of a toxicity metric need careful elicitation and evaluation before contemplation for use. It is insufficient to appeal to heuristics or intuition when arguing the case for the adoption of a "new" metric, procedure, or estimation strategy. It may well be that the use of an arbitrary ACR of 100 is expected to yield a conservative estimate of a "protective concentration," but such an approach completely ignores the competing risks and objectives in any environmental assessment. There are at least two ways we "can get it wrong" in setting environmental standards (cf. Type I and Type II errors in statistical hypothesis testing). Using a limit which fails to adequately protect the environment is one way. Using a limit which inappropriately denies human activity is another.

2.3 DESIGN CONSIDERATIONS

Much has been written about the design and analysis of environmental data in general (USEPA, 2002, 2006) and ecotoxicological data more specifically (Environment Canada, 2005; CCME, 2007; European Commission, 2011; Newman, 2012; ANZECC/ARMCANZ, 2000a,b; OECD, 2012, 2014) although most of this is based on and repeats standard frequentist principles that are taught in all introductory statistics courses (Sparks, 2000). The OECD (2012) and Environment Canada (2005) documents exemplify this point. Topics covered in both documents include hypothesis testing, Type I/II errors, statistical power, randomization, replication, outliers, and data transformations. While these statistical concepts are important, it could be argued that the emphasis placed on classical/frequentist statistics has stifled the development of strategies and procedures that are better equipped to handle the many and varied perturbations of assumed conditions encountered in ecotoxicology. The assumption of *randomness* is a case in point. It is safe to say that the majority of statistical methods are predicated on the joint notions of randomness and independence. Indeed statistical theory demands that toxicity data used in SSD modeling be obtained from a *randomly* selected sample of species. The well-accepted reality is that this is *never* the case (Fox, 2015 and references therein). The irony is that while guideline documents stress the importance of randomness (eg, "randomization should prevail in all aspects of the design and procedures for a toxicity test," Environment Canada, 2005), they simultaneously recommend procedures that ensure samples are biased. For example, the revised Australian and New Zealand Water Quality Guidelines recommend using toxicity data from at least eight species from at least four taxonomic groups (Batley et al., 2014). Such *purposive* sampling is the antithesis of randomness and while the distinction between *probability sampling* and *judgmental sampling* for ecotoxicological studies is not always made clear, good advice does exist albeit in the broader context of environmental monitoring (USEPA, 2002, 2006).

So while not diminishing the importance of sound statistical design to guide all aspects of the data collection and analysis process, the reality is that ecotoxicological studies tend to be severely constrained by (1) high cost of data acquisition; (2) inability (or compromised ability) to invoke the core statistical principles of *randomness*, *replication*, and *blocking*; and (3) nonconformity. The high cost of data acquisition is a function of the logistics of field sampling coupled with the expensive laboratory analyses required to generate toxicity data. The strict definition of randomness means that every "unit" in the target "population" under investigation has an equal chance of being included in the sample and we have already seen that protocols exist (eg, Australian Water Quality Guidelines, ANZECC/ARMCANZ, 2000a,b) which ensure this cannot happen. In addition, standardized protocols for laboratory-based toxicity tests tend to be available for only a relatively small number of animals or organisms thus ensuring another layer of nonrandom selection. Replication improves the quality of estimation and inference but is a casualty of (1), while control through the use of "blocking" where experimental units are organized according to some other exogenous variable(s) is only an option if the major sources of extraneous variation are known in advance. "Nonconformity" refers to the propensity of ecotoxicological data to violate many of the prerequisites or assumptions required by most statistical tests and procedures described in the various guideline documents. These include, but are not limited to, violations of assumptions concerning: independence; distributional form; variance structures; sample size; outliers; censoring; and response-generating mechanism.

Rather than summarizing standard statistical design theory, which is readily available in textbooks (eg, Hinkelmann and Kempthorne, 2008; Gad, 2006) and the aforementioned guideline documents, the remainder of this section is devoted to the exploration of some more "contemporary" aspects of experimental design in ecotoxicology.

ANOVA techniques have (and continue) to play a significant role in the analysis of ecotoxicological data. Even though the use of this technique is expected to diminish as scientists move away from generating NOEC data, ANOVA methods still have an important role to play in testing hypotheses concerning the toxic effects of chemicals in the environment. The identification of an appropriate experimental design is a critical first step in the use of ANOVA and related tools of statistical inference. At the very least the experimental design should be such that it:

- allows for the *unbiased* and *efficient* estimation of all effects of interest;
- controls (to the extent possible) sources of extraneous variation likely to affect the measured response (for example, a temperature or salinity effect in a C-R experiment); and
- makes minimal use of limited resources.

Reducing bias, improving precision, and controlling extraneous variation tend to result in a greater number of treatment combinations and/or increased replication—both of which increase the cost of the experiment. It is therefore surprising that *orthogonal fractional factorial designs* have not been used more widely in ecotoxicology (Dey, 1985). While it is not possible to provide a comprehensive treatment of this important topic in this book, we illustrate the potential benefits with the use of a simple example.

In a recent paper, Webb et al. (2014) described a toxicology experiment that "did not succumb to standard experimental design." The challenge was to satisfy the requirements of the three dot-points above in a way that accommodated unique physical and logistical constraints. Their solution relied upon advanced mathematical and computational skills—the detail of which is beyond the scope of this book. The situation described in Webb et al. (2014) motivates the following hypothetical example illustrating the use of a fractional factorial design.

2.3.1 Example

A study into the potential impacts associated with the discharge of waste water from a proposed desalination plant relied on toxicity testing. For one of these tests, researchers were interested in whether or not the hypersaline effluent was toxic to marine organisms. Other factors thought to be important were the time of day (TOD) when exposure to the toxicant commenced as well as the temperature (temp) and salinity (salin) of the waste stream. As in the Webb et al. (2014) study, the manner in which test samples were stored was potentially another source of variation that needed to be controlled for. In this case, beakers could be placed on shelves that were arranged in three racks each having four shelves. The positioning of beakers (representing different combinations of dose, TOD, temp, and salin) on shelves was important due to the potential influences of light levels, proximity to the door, and thermal stratification. The basic experiment involved the direct manipulation of the four factors: dose (present/absent); TOD (am/pm); temp (15°C/25°C); and salin (ambient/elevated) coupled with the two factors determining beaker position (racks and shelves). A "full factorial" experiment would require all 384 combinations of these factors to be tested at least once. Not only is such an experiment time-consuming and expensive, it may be unnecessary. *If* information is only sought on the "main effects" (ie, the effect of each factor separately) *and* higher-order interaction effects are (or can be assumed to be) negligible, then significant savings in experimental effort can be realized with the use of a fraction of the treatments in the full design—hence the name *fractional factorial design*. In addition, if the treatments comprising this fraction are carefully selected, it is possible to estimate the main effects independently of each other—a property that is clearly desirable but not guaranteed by either a random or subjective selection. In statistics, the *independence* of two random variables has the geometrical interpretation of *orthogonality* (ie, being at right angles to each other). Hence, fractional factorial designs which permit the independent estimation of effects are referred to as *orthogonal fractional factorial designs*. These are not new and date back to the work of Adelman (1961), Bose and Bush (1952), Rao (1950), Kempthorne (1947), and others. As one might expect, the method of identifying *how many* and *which* treatments to include in the fractional design such that the orthogonality requirement is met is far from simple and requires a good understanding of advanced mathematical concepts such as linear algebra, Hadamard matrices, and Galois Field theory. Thankfully, statistical software tools make this task easier although more complex designs tend not to be included. R (R Development Core Team, 2004) is the only free package of such tools; it has a large and rapidly increasing library of user-contributed functions including packages for creating and analyzing fractional factorial designs. Readers interested in learning more should consult the R website (CRAN, 2015) and the texts by Lawson (2015) and Gad (2006).

Returning to the present example, an illustration of the savings in experimental effort is indicated by the experimental design represented by the allocations in Table 2.1 which has reduced the number of treatments from 384 to a mere 25. This design is what is referred to as *Resolution III*, meaning that main effects are estimated independently of each other but not independently of interactions

TABLE 2.1 Orthogonal Fractional Factorial Design for Effluent Toxicity Study

Run	Shelf	Rack	Dose	TOD	Temp	Salin
1	1	3	Absent	am	15	Ambient
2	2	3	Absent	am	15	Ambient
3	3	3	Absent	am	15	Ambient
4	3	3	Present	pm	25	Elevated
5	4	3	Present	pm	25	Elevated
6	1	1	Absent	am	25	Elevated
7	2	1	Absent	pm	25	Ambient
8	3	1	Present	pm	15	Ambient
9	3	1	Present	am	15	Ambient
10	4	1	Absent	am	15	Elevated
11	1	1	Absent	pm	15	Elevated
12	2	1	Present	am	15	Elevated
13	3	1	Present	am	25	Ambient
14	3	1	Absent	am	25	Ambient
15	4	1	Absent	pm	15	Ambient
16	1	2	Present	am	25	Ambient
17	2	2	Present	am	15	Elevated
18	3	2	Absent	pm	15	Elevated
19	3	2	Absent	pm	15	Ambient
20	4	2	Absent	am	25	Ambient
21	1	2	Present	pm	15	Ambient
22	2	2	Absent	pm	25	Ambient
23	3	2	Absent	am	25	Elevated
24	3	2	Absent	am	15	Elevated
25	4	2	Present	am	15	Ambient

(hence the need to know or assume that interaction effects are negligible). Other resolution designs may be available which are less restrictive but require greater experimental resources (typically in the form of more treatment combinations).

More will be said about "optimal" designs in the context of planning a C-R experiment in Section 2.6.

2.4 DATA PROCESSING AND HANDLING

The rapid rise of the phenomenon referred to as "big data" has rejuvenated interest in what scientists regard as the necessary, but mundane, task of data storage, handling, and manipulation (referred to by some as "statistical janitorial work," New York Times, 2014).

According to one definition "big data" is characterized by "data sets with sizes beyond the ability of commonly used software tools to capture, curate, manage, and process data within a tolerable elapsed time" (Wikipedia, 2015). Areas of science generating big data include astronomy, telecommunications, genomics, and natural resource management. While these are worthy of investments in R&D effort, concern has been expressed that the rush to "get on board" with the big data push may be diverting attention away from the equally important issue of "little data" (Environmetrics Australia, 2014a).

While ecotoxicological data sets may never fall within the realm of big data, they can span orders of magnitude in size. For example users of the USEPA ECOTOX database (USEPA, 2015) can download up to 10,000 records on single chemical toxicity data, whereas SSDs are usually modeled using no more than 20 observations and often as few as 5 or 6.

As with design considerations (Section 2.3), the various guideline documents (or accompanying documents) published by jurisdictions around the world generally provide comprehensive treatment of commonly used data analysis techniques. While most of this is useful, some is outdated. For example Appendix I of the Canadian Guidelines (Environment Canada, 2005) is a blank sheet of logarithmic paper for plotting results of C-R experiments *by hand*.

Another feature of all of these guideline documents is their strong reliance on flowcharts for the statistical treatment of data. While the statistically naïve will find these both useful and comforting, rigid adherence to a highly structured approach to preliminary data analysis runs counter to the objectives of *exploratory data analysis* (EDA). Indeed the essence of EDA is to uncover patterns, trends, anomalies, correlations, outliers, and other important features of a data set via a fairly *unstructured* approach using powerful computer graphics and specialized software tools designed to tease out hidden structure. There is no roadmap for this process.

The task of preparing, organizing, and manipulating data is both necessary and time-consuming, with some claiming these activities constitute up to 80% of the data analysis effort (Dasu and Johnson, 2003). The lack of consistent advice or a standard code of practice is largely responsible for this situation. Wickham (2014) has identified the following common problems with digital data sets:

- Column headers are values, not variable names;
- Multiple variables are stored in one column;
- Variables are stored in both rows and columns;
- Multiple types of observational units are stored in the same table; and
- A single observational unit is stored in multiple tables.

To address the lack of guidance, Wickham (2009) has developed the R packages `tidyr`, `dplyr`, and `ggplot2`. The first two are used for "data wrangling"—that is the often tedious procedures associated with getting data into a format that is amenable for further analysis and interrogation. `ggplot2` is a powerful graphics package that allows users to interactively explore and display data by modifying a base plot by adding and removing layers. Space restrictions preclude a more comprehensive discussion of these packages, however some of their capabilities are briefly examined next.

2.4.1 Manipulating Data: The R Package `tidyr`

The `tidyr` and `dplyr` packages work hand-in-hand. Both `dplyr` and `tidyr` are designed to clean up or "mung" messy data (Quora, 2014). `tidyr` works on the simple and consistent philosophy that columns of a dataframe (R's terminology for a rectangular array of data) represent variables and rows represent observations. To see how this works, consider the data in Table 2.2, which give the number of surviving marine organisms (out of 100) for each of four replicates as a function of toxicant concentration. Although this tabular presentation of results is compact, it is not well suited to further statistical analysis.

The function `gather()` fixes this. The R code below illustrates how the tabular data stored in a CSV file called `example_1.csv` is read into

TABLE 2.2 Number of Surviving Organisms (out of 100) for Each of Four Replicates as a Function of Toxicant Concentration (%)

CONCEN (%)	1	2	3	4
0	89	92	88	93
3.1	87	93	97	91
6.3	96	91	90	94
12.5	93	89	95	95
25	76	67	73	85
50	0	0	0	0
100	0	0	0	0

R and then converted into the "standard" format:

```
dat<-read.csv("example_1.csv") # read in tabular data
names(dat)<-c("concen","1","2","3","4") # assign column names
dat1<-gather(dat,rep,surv,2:5) # convert to standard format
head(dat1) # display first 6 rows of converted data

   concen  rep  surv
1     0.0    1    89
2     3.1    1    87
3     6.3    1    96
4    12.5    1    93
5    25.0    1    76
6    50.0    1     0
```

An example of a data set in which variables appear in both rows and columns is shown in Table 2.3. This is a portion of data from a larger experiment to assess the toxicity of effluent from a desalination plant by examining the effect of different effluent concentrations (as a percentage of undiluted effluent) on macroalgal growth (as measured by gametophyte length). Information on the covariates pH, salinity, and dissolved oxygen was also recorded.

There are a total of six variables in Table 2.3 (concentration, length, pH, salinity, dissolved oxygen, and replicate); these appear in the rows and columns of the table. The function `gather()` is first used to separate out the variables followed by function `spread()` which creates multiple columns according to levels of other factor(s).

Again, the R code below illustrates how the tabular data stored in a CSV file called `example_2.csv` are read into R and then converted into the "standard" format.

```
Dat<-read.csv("example_2.csv") # read in tabular data
dat2<-gather(dat,type,rep,A:D) # stacks all reps into single col
names(dat2)[3:4]<-c("rep","value") # assign names to cols
dat3<-spread(dat2,type) # creates separate cols from 'type'
str(dat3) # get details of structure of new dataframe

'data.frame':   28 obs. Of   6 variables:
 $ concen: num   0 0 0 0 3.1 3.1 3.1 3.1 6.3 6.3 ...
 $ rep  : Factor w/ 4 leve"s"""""""""""""""D": 1 2 3 4 1 2 3 4 1 2 ...
 $ DO  : num  98 98 98 98 103 ...
 $ length: num   19 23.5 20.9 20.6 21 ...
 $ pH  : num  8 8 8 8 7.98 7.98 7.98 7.98 8.01 8.01 ...
 $ salin : num  37.3 37.3 37.3 37.3 37.3 37.3 37.3 37.3 36.7 36.7 ...
```

TABLE 2.3 Portion of Gametophyte Length Data for a Macroalgal Growth Test. Columns are Effluent Concentration, Measurement Type, Replicates A–D.

Concentration (%)	Type	A	B	C	D
0	Length	19.027	23.476	20.908	20.559
0	pH	8	8	8	8
0	Salin	37.3	37.3	37.3	37.3
0	DO	98	98	98	98
3.1	Length	20.98	20.581	21.867	19.663
3.1	pH	7.98	7.98	7.98	7.98
3.1	Salin	37.3	37.3	37.3	37.3
3.1	DO	103.2	103.2	103.2	103.2

The result is dataframe `dat3`, which is now in "standard" format with six columns (variables) and 28 rows (observations).

2.4.2 Visualizing Data: The R Package `ggplot2`

According to its developer `ggplot2` "is unlike most other graphics packages because it has a deep underlying grammar" (Wickham, 2009). The main strength of `ggplot2` is its ability to create publication-quality graphics by interactively adding "layers." These might consist of a plot of the same variables but from a different data set or "objects" such as a smooth fit generated by one of R's packages.

The following example from Eduard Szöcs' website (Szöcs, 2015a) shows how `ggplot2` can be used to prepare an annotated SSD for chlorpyrifos data taken from the USEPA ECOTOX database (USEPA, 2015). The data are stored in an R dataframe called `df` (Fig. 2.1A and B).

```
df <- df[order(df$val), ]          # rearrange toxicity data in ascending order
df$frac <- ppoints(df$val, 0.5)    # use intrinsic function ppoints to compute
                                   # empirical estimates of cumulative
                                   # probabilities
#
# Next use ggplot2 to build up SSD
#
require(ggplot2)                   # load the ggplot2 package
p<-ggplot(data = df)               # sets up base layer of plot and stores as
                                   # an R object

  p<-p+geom_point(aes(x = val, y = frac), size = 5)
                                   # adds a layer of points to base layer and
                                   # stores back into object p

  p                                # plot the object simply by naming it
                                   # see Figure 2.1A
```

```
# Next use log-scale for concen, add species and axis labels,
# and change background theme.

  p<-p+geom_text(aes(x = val, y = frac, label = species), hjust = 1.1,
    size = 4) + theme_bw() + scale_x_log10(limits=c(0.0075, max(df$val))) +
    labs(x = expression(pas‘e('Concentration of Chlorpyrifos‘[ ', m‘, ‘g
    ', L^-‘, ' ]')), y‘= 'Fraction of species affec'ed')

  p                    # plot the object simply by naming it
                       # see Figure 2.1B
```

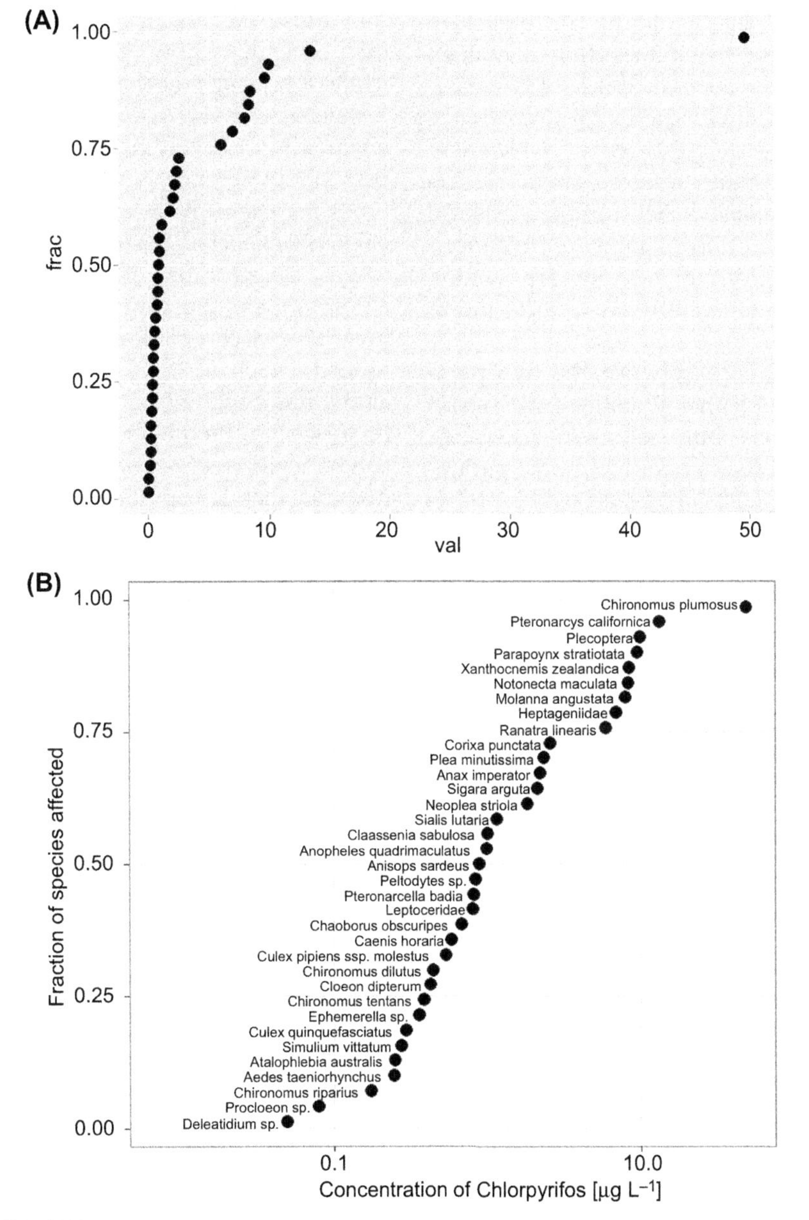

FIGURE 2.1 Using ggplot2 to create empirical SSD: (A) base layer; (B) annotated plot.

2.5 ESTIMATION AND INFERENCE

Statistical inference is concerned with the dual problems of *estimation and hypothesis testing*. While these are related, there are nevertheless subtle differences and objectives (Fig. 2.2).

Estimation is concerned with quantifying unknown model parameters in a manner that is in some way "optimal." Hypothesis testing procedures, on the other hand, commence with a statement concerning the value of an unknown parameter and use the information contained in a sample to assess the plausibility of this statement. Two statistical paradigms for problems of inference have emerged: *Frequentist statistics* (ie, the "classical" statistics referred to in Section 2.1) and *Bayesian statistics*. Most of the statistical procedures used in ecotoxicology are frequentist based although Bayesian methods are gaining popularity (Evans et al., 2010; Fox, 2010; Billoir et al., 2008).

The main difference between the two paradigms is that Frequentist statistics treat model parameters as *fixed* but unknown constants, while Bayesian statistics treat the model parameters as *random variables* characterized by probability distributions (referred to as a *prior* densities). Data are used in a Frequentist mode to find an "optimal" estimate of the unknown parameter(s), while in Bayesian statistics, data are used to update the prior probability distributions using Bayes' formula. The updated distributions are known as *posterior* densities and are the basis for subsequent parameter inference. The two modes of estimation and inference have proven highly divisive among the statistical community with much journal space and many conferences devoted to arguments about the relative merits of the two approaches. Thankfully those tensions have largely dissipated with many scientists now acknowledging the legitimacy of both approaches. The adoption of either a Bayesian or Frequentist framework for any given study is likely to reflect considerations of personal preference, ease-of-use, and efficacy.

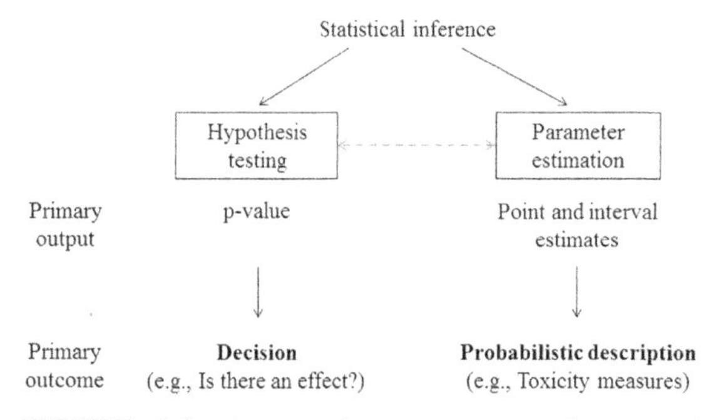

FIGURE 2.2 Conceptual representation of statistical inference in an ecotoxicological setting. *Taken from Fox, D.R., Billoir, E., Charles, S., Delignette-Muller, M.L., Lopes, C., 2012. What to do with NOECs/NOELs – prohibition or innovation? Integr. Environ. Assess. Manag. 8, 764–766.*

As the use of Bayesian techniques in ecotoxicology is expected to steadily increase, the challenge posed for regulators is how to accommodate the element of subjectivity that invariably accompanies this mode of inference. Indeed, much of the "controversy" referred to above has centered on this issue. It is perhaps because the Frequentist framework is "data-driven" and treats *subjective probability* as inadmissible that it has enjoyed such a prominent place in setting environmental standards. On the other hand, it is precisely the element of subjectivity in a Bayesian approach that many argue provides for a richer, more informative analysis. An attractive feature of the Bayesian framework is that the prior density can be used as the vehicle by which expert opinion is introduced into the analysis. However, as noted by Barnett and O'Hagan (1997), the process of eliciting information from one or more experts "is fraught with technical difficulties." More recent studies using structured elicitation techniques have sought to reduce the sources of expert bias and error in the presence of epistemic uncertainty (McBride et al., 2012).

However, concerns remain about the use of subjective information in the derivation of "safe" concentrations of toxicants in marine and fresh waters although, as noted by Fox

(2015), approaches using Frequentist statistical methods are not as objective as many would think. It is also difficult to see how a "protective" concentration derived using Bayesian methods would stand up to scrutiny in a court of law—even if the underlying methodology is sound. One can foresee the credibility of a Bayesian-derived figure being challenged on a number of fronts. For example, it might be argued that the selection of "experts" responsible for eliciting priors in the Bayesian analysis was biased or that the inevitable comparison with the result from a more traditional Frequentist analysis yields inconsistent and/or irreconcilable differences. An example of the latter is provided in Cliffroy et al. (2013) who compared HC_5 estimates (the 5% hazardous concentrations derived from SSDs) using both Frequentist and Bayesian approaches. While they found that their Bayesian method tended to underestimate the reference value, discrepancies between the two ranged from a factor of 0.2 to a factor of nearly 6. While the propensity for underestimation was regarded by the authors as a positive feature of the approach due to the increased level of protection it afforded, it is a view unlikely to be shared by all stakeholders in an environmental planning and assessment process.

Although it is not possible to give a comprehensive account of statistical estimation procedures here, it is relevant to mention the *general linear model* as it is the foundation of many common methods such as regression, ANOVA, and analysis of covariance (ANCOVA) (Graybill, 1976). The general model common to all of these techniques is given by Eq. (2.1).

$$Y = X\beta + \varepsilon \tag{2.1}$$

where Y is a $(n \times 1)$ vector of responses; X is a $(n \times p)$ "design" matrix that contains coding to describe the experimental factors and/or values of covariates $\{x_1, x_2, \ldots\}$; β is a $(p \times 1)$ vector of parameters to be estimated; and ε is a $(n \times 1)$ vector of random error terms assumed to be independently normally distributed with mean zero and variance σ_ε^2. The simplest version of Eq. (2.1) corresponds to the simple regression of Y on x, that is:

$$Y_i = a + bx_i + e_i \tag{2.2}$$

In this case $Y^T = [y_1, y_2, \ldots, y_n]$; $X = \begin{bmatrix} \underline{1}^T & \underline{x}^T \end{bmatrix}$; $\beta^T = [a \quad b]$ where $\underline{1}$ is a $(1 \times n)$ vector of ones and $\underline{x}$ is a $(1 \times n)$ vector containing the values of the covariate x. When some or all of the covariates are *factors*, then the corresponding $\underline{x}$ is a vector of "dummy" codes. For example, $\underline{x}$ might simply indicate the presence or absence of a toxicant in which case the values assigned to $\underline{x}$ could be $\{0, 1\}$.

Irrespective of the whether X contains all measured values, all dummy codes, or a combination of both, the parameters of the model are estimated using the same equation. That is:

$$\hat{\beta} = \left(X^T X\right)^{-1} X^T Y \tag{2.3}$$

Furthermore, the variance—covariance matrix of the estimated parameters is:

$$\mathrm{Cov}\left[\hat{\beta}\right] = \sigma_\varepsilon^2 \left(X^T X\right)^{-1} \tag{2.4}$$

A key understanding is how the statistical *design* influences the *quality* of inference. In the case of the simple linear regression above, it is readily verified that $(X^T X)$ is given by Eq. (2.5) and its inverse given by Eq. (2.6).

$$X^T X = \begin{bmatrix} n & \sum_{i=1}^{n} x_i \\ \sum_{i=1}^{n} x_i & \sum_{i=1}^{n} x_i^2 \end{bmatrix} \tag{2.5}$$

$$\left(X^T X\right)^{-1} = \frac{1}{n \sum_{i=1}^{n} x_i^2 - \left(\sum_{i=1}^{n} x_i\right)^2} \times \begin{bmatrix} \sum_{i=1}^{n} x_i^2 & -\sum_{i=1}^{n} x_i \\ -\sum_{i=1}^{n} x_i & n \end{bmatrix} \tag{2.6}$$

Thus we see from Eqs. (2.4) and (2.6) that the variances and covariances of the estimated parameters are entirely a function of the x_i's (for a given σ_ε^2) and not of the responses, y_i. Furthermore, that the off-diagonal entry in Eq. (2.6) is nonzero tells us that the parameters $\{a, b\}$ are not estimated independently (unless $\sum_{i=1}^{n} x_i = 0$).

Returning to the fractional factorial design given by Table 2.1 in Section 2.3 the factors and number of levels are *shelf* (4), *rack* (3), *dose* (2), *TOD* (2), *temp* (2), and *salin* (2). Without going into the details, the number of parameters p required in the design matrix X in Eq. (2.1) is

$p = \sum_j^k L_j - k + 1$ where k is the number of factors and L_j is the number of levels for factor j. So for the design of Table 2.1, $p = 15 - 6 + 1 = 10$ resulting in a (25×10) design matrix X. For the coding of X (not given here) corresponding to the data in Table 2.1, the following X^TX matrix was obtained:

$$X^TX = \begin{bmatrix} 25 & 0 & 0 & 0 & 0 & 0 & 0 & 0 & 0 & 0 \\ 0 & 15 & 0 & 0 & 0 & 0 & 0 & 0 & 0 & 0 \\ 0 & 0 & 15 & 0 & 0 & 0 & 0 & 0 & 0 & 0 \\ 0 & 0 & 0 & 15 & 0 & 0 & 0 & 0 & 0 & 0 \\ 0 & 0 & 0 & 0 & 10 & 0 & 0 & 0 & 0 & 0 \\ 0 & 0 & 0 & 0 & 0 & 10 & 0 & 0 & 0 & 0 \\ 0 & 0 & 0 & 0 & 0 & 0 & 25 & 0 & 0 & 0 \\ 0 & 0 & 0 & 0 & 0 & 0 & 0 & 25 & 0 & 0 \\ 0 & 0 & 0 & 0 & 0 & 0 & 0 & 0 & 25 & 0 \\ 0 & 0 & 0 & 0 & 0 & 0 & 0 & 0 & 0 & 25 \end{bmatrix}$$

The orthogonality of the fractional factorial design is immediately apparent from the fact that all off-diagonal entries in X^TX are zero. As mentioned earlier, this is desirable since all main effects will be estimated independently of each other. Furthermore, the diagonal structure of X^TX means that the inverse is readily obtained by replacing the diagonal elements by their reciprocals. Given that the diagonal entries are not all identical means that the model parameters will be estimated with varying precision. Ideally our coding scheme would have resulted in the diagonal entries all being equal to 25. In deciding how good our design is, it would be useful to have a measure of the overall efficiency of the proposed design with the ideal design. One way of doing this is to examine the *determinant* (Wikipedia, 2016) of the matrix $(X^TX)^{-1}$. Thus, one criterion for selecting a "good" fractional factorial design is one which minimizes this determinant. This is the basis of *D-optimality*, which is discussed further in Section 2.6. For the present example, it can be verified that the design given in Table 2.1 is approximately 70% efficient when compared to a full factorial design.

As mentioned at the beginning of this section, estimation and hypothesis testing procedures overlap, although the emphasis is different. Estimating parameters of SSDs or complex C-R models often involves the use of sophisticated mathematical and statistical tools. Even more challenging is the determination of standard errors of quantities derived from the fitted model. As an example Eq. (2.7) is a four-parameter logistic function used to model C-R data.

$$y = \beta_0 + \frac{\beta_1 - \beta_0}{1 + \exp\{\beta_2[\ln x - \ln \beta_3]\}} \tag{2.7}$$

This is a complex, nonlinear model. Not only it is nonlinear in concentration (x), but it is also nonlinear in the parameters $\{\beta_0, \beta_1, \beta_2, \beta_3\}$. There is no simple way of estimating the parameters; this will invariably require the use of specialized computer software such as the R package `drc` discussed in Section 2.9. Once fitted, the model is generally used to estimate the response $\widehat{y}_0$ corresponding to a concentration x_0 by replacing parameters in Eq. (2.1) by their estimates $\left\{\widehat{\beta}_0, \widehat{\beta}_1, \widehat{\beta}_2, \widehat{\beta}_3\right\}$ and/or estimate the concentration $\widehat{x}_0$ for a given response y_0 using Eq. (2.8).

$$\widehat{x}_0 = \exp\left\{\frac{\ln\left(\frac{\widehat{\beta}_1 - y_0}{\widehat{\beta}_0 - y_0}\right) + \widehat{\beta}_2 \ln\widehat{\beta}_3}{\widehat{\beta}_2}\right\} \tag{2.8}$$

Given values for the parameter estimates, use of Eqs. (2.7) and (2.8) could be performed on a calculator. However, the computation of the standard error of the resulting estimate has no "closed-form" expression and more sophisticated methods are required. Statistical *resampling* techniques such as the *bootstrap* or *jack-knife* are often used for this purpose. These involve repeatedly sampling (with replacement) from the sample data. Parameter estimates are obtained for each sample and the quantity of interest is computed using Eq. (2.7) or Eq. (2.8). The variation among the collection of estimates thus obtained provides an estimate of the standard error for that quantity. Clearly, this is a complicated and repetitive task that is ideally suited to computer implementation.

2.6 CONCENTRATION–RESPONSE MODELING

This section provides an overview of some of the statistical issues associated with C-R modeling as well as indicating opportunities for future development. It is not intended to provide an exhaustive treatment of all aspects of C-R methodologies—for example, the use of toxicokinetic–toxicodynamic (TKTD) models to evaluate toxicity at the level of individual organisms.

In the present context, C-R experiments yield the data that are used to set "safe" exposure to concentrations of a toxicant, which is the subject of Section 2.7 on SSDs. We note that the terms "C-R" and "dose–response" tend to be used interchangeably, although strictly speaking these are not the same since *dose* is a function of *concentration*, *frequency*, and *duration* (of exposure).

C-R experiments and the mathematical and statistical modeling tools developed to analyze the data they generate have a long and distinguished history which we briefly trace next.

A plot of an animal or organism's response to increasing levels of a toxicant can assume a wide variety of shapes, ranging from a simple linear relationship to more complex curves displaying *hormesis* (a stimulatory effect at low concentrations) and *hysteresis* (where the trace of the response as a function of increasing concentration is not the reverse of the trace of the response as a function of decreasing concentration). Two examples of idealized C-R curves are shown in Fig. 2.3.

Also indicated in Fig. 2.3 is the NEC and two estimates of an EC_5 (ie, the concentration at which a 5% effect is expected. While hypothetical, Fig. 2.3 does highlight a real and commonly occurring issue in C-R modeling, which is the sensitivity of the derived metric(s) to model choice and parameterization.

If the response in Fig. 2.3 represents the fraction of *unaffected* organisms, then a plot of the *affected* fraction assumes the shape shown in Fig. 2.4. This elongated "S" shape arises not only in C-R modeling but also in the context of population growth models.

Early investigations into population dynamics by the British scholar, the Reverend Thomas Malthus (Malthus, 1978) used simple exponential models to describe population growth. However, these lacked realism because biological populations do not grow indefinitely. More realistic growth models displayed asymptotic behavior; one of the most important of these is the logistic function which was published in 1838 by Francois Verhulst (1838). The logistic equation remained a relatively obscure mathematical result until it was rediscovered in 1920 by Raymond Pearl and Lowell Reed at John Hopkins University (Pearl and Reed, 1920). In the 1930s prominent statisticians Chester Bliss and Ronald Fisher took a slightly different approach to bioassay modeling by treating the stimulus (dose) as the covariate and, because of variability in individual tolerance levels, treated the response as a *random variable* (Bliss, 1935). In this formulation a probability model is associated with the response, which is often a normal distribution, but this is not a requirement. Bliss

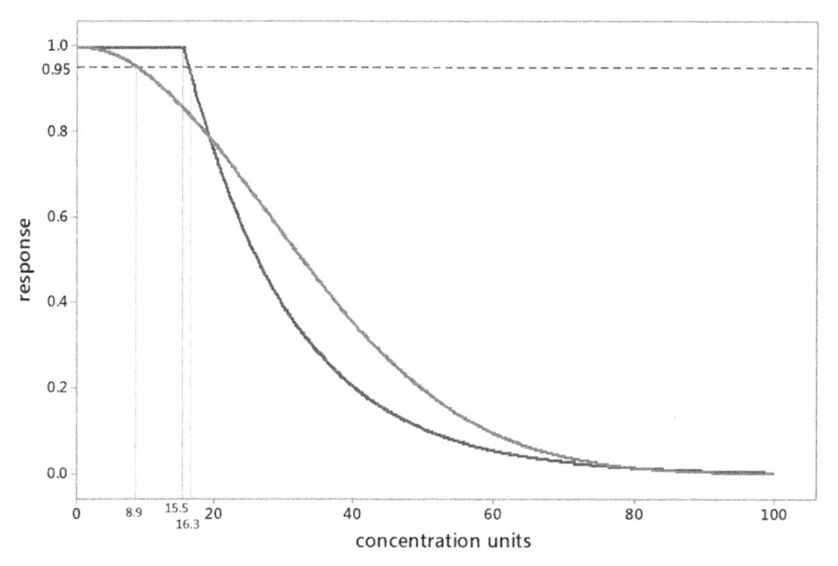

FIGURE 2.3 Idealized C-R curves with threshold effect [*blue* (dark gray in print versions)] having $EC_5 = 16.3$; and without [*red* (gray in print versions)] having $EC_5 = 8.9$. The no effect concentration (NEC) is 15.5.

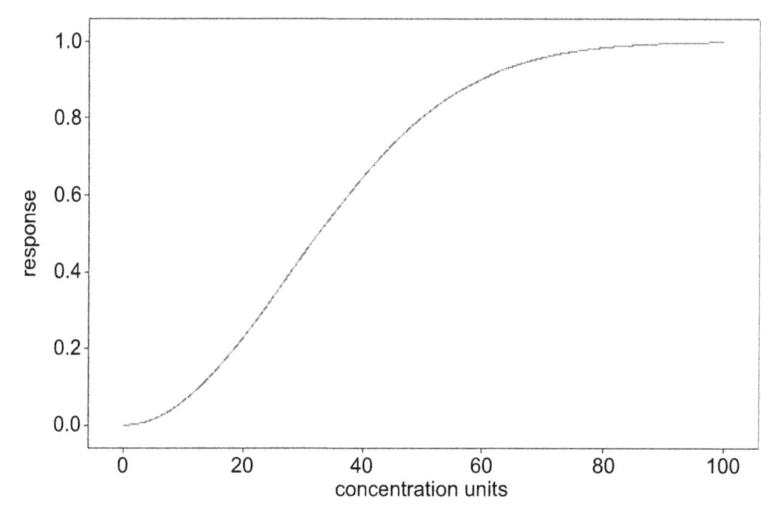

FIGURE 2.4 Typical elongated S-shape curve for fraction of affected organisms as a function of toxicant concentration.

was responsible for coining the term *probit* as shorthand for probability unit (Bliss, 1935).

Probit analysis quickly established itself as the de facto modeling approach to describe any relation of a discrete binary outcome to one or more explanatory variables. In 1944 US statistician Joseph Berkson advocated the use of the logistic function as an alternative to probit analysis (Berkson, 1944) and introduced the term *logit* as shorthand for the mathematical transformation of proportions using the logarithm of the odds ratio (the odds ratio is loosely defined as $\left(\frac{\text{probability ("success")}}{\text{probability ("failure")}}\right)$.

As mentioned in Section 2.2, ANOVA techniques have been used for many years in conjunction with C-R experiments to derive the now widely discredited NOEC/NOEL toxicity metrics. While not wishing to revisit the long list of objections here, it is nonetheless constructive to briefly explore the critical flaw with these measures in our quest for better designed C-R experiments.

In the context of C-R experiments, hypothesis tests should be used to assess the validity of statements about a toxic effect while estimation techniques should be used to derive a measure of a toxic effect (Fox et al., 2012). Indeed, it is a gross waste of experimental resources to derive a toxicity measure using ANOVA methods since it makes no use whatsoever of the store of information contained in the *response-generating mechanism* (ie, the C-R *model*). Additionally, ANOVA requires replication at each concentration, whereas this is not a requirement for C-R modeling. It has been argued that hypothesis testing is preferable to modeling if the relationship between the effect and the toxicant concentration is unknown (Newman and Clements, 2008) or poorly defined (Green, 2016). However, this is precisely the role of EDA techniques discussed in Section 2.4. Ultimately, the problem with ANOVA methods and the associated multiple comparison procedures used to derive toxicity measures is that concentration is treated as a *factor* not a *covariate*. This means that concentrations, which are carefully measured in a laboratory, are stripped of quantitative information and are simply treated as *labels* for which estimates of uncertainty and statements of precision are inadmissible (Fox et al., 2012; Fox and Landis, 2016a).

Although recommendations to move to model-based estimation procedures date back almost 20 years (for example, OECD, 1998), change has been slow and hindered by pockets of inertia (Green et al., 2013; Green, 2016) even in the face of extensive advocacy (Landis and Chapman, 2011; Warne and van Dam, 2008; van Dam et al., 2012; Fox and Landis, 2016b).

To highlight the fundamental difficulty, Fig. 2.5 shows the results of an experiment to investigate the toxicity of a herbicide to cucumber growth. When plotted on a log-scale using the measured concentration data, the nature of the relationship between response and concentration is revealed (Fig. 2.5A); however, ANOVA only uses it as a label to distinguish different dose groups (Fig. 2.5B). Furthermore, while Fig. 2.5A indicates an "effect" in the region between 10^{-4} and 10^{-3}, use of ANOVA methods identifies dose groups 1–7 in Fig. 2.5B as not being statistically significant from the control group. Taking the "largest" of these results in dose group 7 being the estimate of the effect level. Dose group 7 corresponds to an original concentration of 0.0055 units or $10^{-2.26}$ which is well into the effect range of Fig. 2.5A.

If we suspect (or better still, *know*) that the responses for dose group 7 had been corrupted, they could be legitimately removed and the analysis re-performed. Doing so results in the NOEC *increasing* to 0.0065 units, while a model-based analysis (not shown here) showed the estimate of the NEC remained relatively unaffected by the removal of the offending data and actually *decreased* slightly. Finally, while a statement of precision or confidence for model-based estimates can be readily obtained, such measures are not only meaningless for the ANOVA-based analysis but simply *cannot* be obtained (Fox et al., 2012).

Clearly, the use of models to generate toxicity estimates provides for a richer and more informative mode of inference. However the "cost" of these gains is usually an increase in computational complexity; although this may have posed problems in the past, the availability of sophisticated tools such as R and special purpose software such as the `drc`

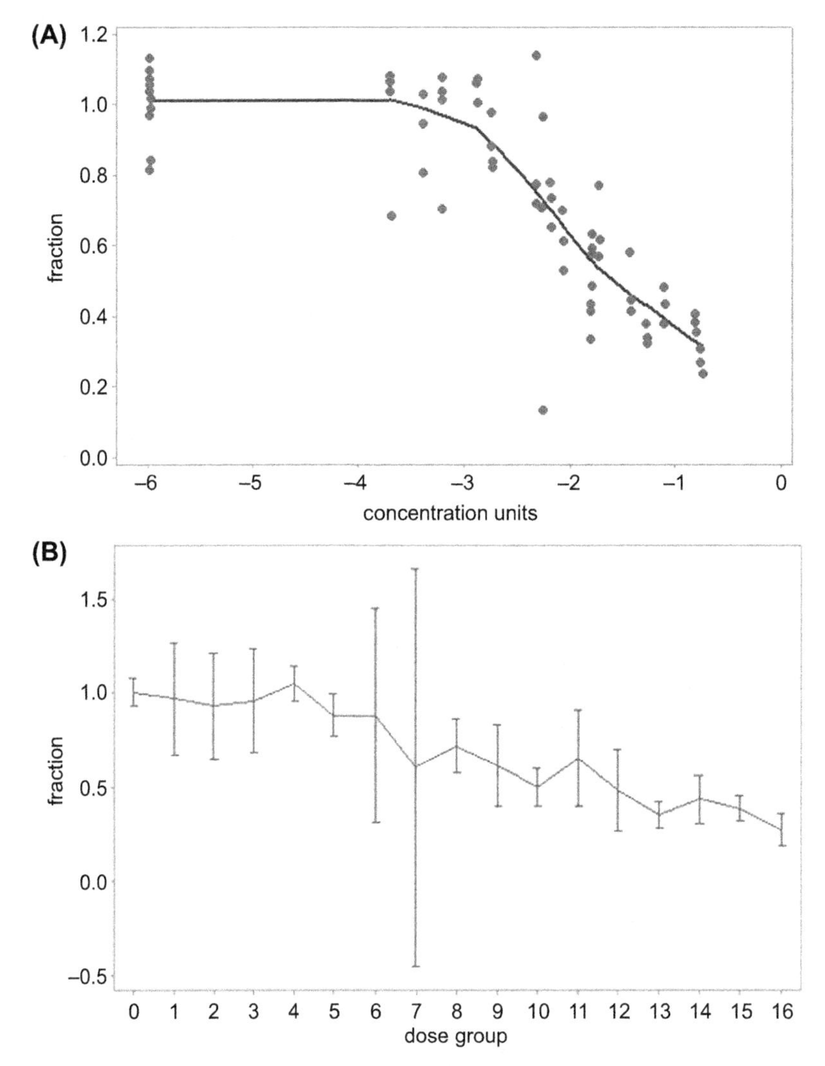

FIGURE 2.5 Cucumber shoot weight (as a fraction of control weight) versus: (A) herbicide concentration on log-scale (−6 representing control group), *solid line* is smooth and (B) as a nominal dose group (0 representing control group), *vertical lines* show 95% confidence intervals for mean response for each group. *Adapted from Moore, D.R.J., Warren-Hicks, W.J., Qian, S., Fairbrother, A., Aldenberg, T., Barry, T., Luttik, R., Ratte, H.-T., 2010. Uncertainty analysis using classical and Bayesian hierarchical models. In: Warren-Hicks, W.J., Hart, A. (Eds), Application of Uncertainty Analysis to Ecological Risks of Pesticides, CRC Press, Boca Raton FL, USA.*

package means that estimation and inference for complex C-R models can be undertaken with relative ease.

As mentioned in Section 2.5, the use of Bayesian methods in ecotoxicology is becoming more prevalent; while the issue of how to formally accommodate subjective assessment into a regulatory framework remains an open question, there is little doubt that this mode of statistical thinking and analysis will only continue to increase. By way of example, and following the procedure in Fox (2010), Bayesian methods may be used to fit the exponential-threshold model given by Eqs. (2.9a) and (2.9b) to the data appearing in Fig. 2.5A.

$$Y_i \overset{d}{\sim} g_Y(\bullet) \tag{2.9a}$$

$$E[Y_i|x_i] = \mu_i = \alpha \exp[-\beta(x_i - \gamma)I(x_i - \gamma)] \tag{2.9b}$$

where Y_i denotes the response at concentration x_i and $I(z) = \begin{cases} 1 & z > 0 \\ 0 & z \leq 0 \end{cases}$.

$E[Y_i|x_i]$ denotes the mathematical expectation of Y_i conditional on x_i; the notation $\overset{d}{\sim}$ is read as "is distributed as." The parameters α, β, γ have the following interpretations: α is the mean response for concentrations between 0 and γ; β controls the rate of response for concentrations greater than γ; and γ is the NEC.

Although seemingly complex, the Bayesian analysis can be programmed with about 10 lines of code using freely available software such as OpenBUGS (Openbugs, 2009), JAGS (Sourceforge, 2015), or Stan (Stan Development Team, 2015).

A point of divergence between Frequentist and Bayesian methods of inference is that Frequentists use point/interval estimates and/or hypothesis testing, while Bayesians base all inference on the *posterior density*. Hypothesis testing is a Frequentist concept that has no Bayesian analogue. As an example, Fig. 2.6 summarizes the results of sampling from relevant posterior distributions of α, β, γ as well as that of the EC_5. Conventional summary statistics such as means and standard deviations can be obtained for these distributions; however, the preferred Bayesian point estimate is either the median or mode of the relevant posterior distribution. Instead of a confidence interval, Bayesians determine the interval of highest posterior density or *HPD interval*. This is obtained by determining the shortest interval for which the area under the posterior density is some nominal value—for example 0.95. A 95% credibility interval for the parameter γ is found to be [0, 0.0014] consistent with a subjective assessment based on the data in Fig. 2.5A. Note that the estimated posterior probability of obtaining a value of γ at least as large as the previously determined NOEC for this data is only 0.003 suggesting an

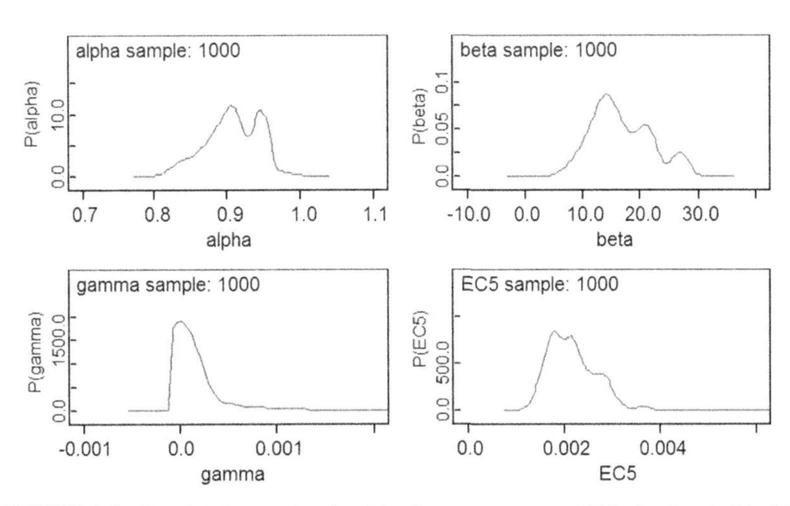

FIGURE 2.6 Empirical posterior densities for parameters and EC_5 for data in Fig. 2.5.

irreconcilable difference between the Bayesian model and the Frequentist estimate.

Another facet of C-R modeling that is attracting increasing attention is the *design* aspect. The high cost of obtaining and analyzing data to populate C-R models brings into focus the need for an experimental design that yields the most precise estimates of critical parameters that can be obtained from a fixed sample size. A long-held view by practicing ecotoxicologists is that a minimum of three replicates is necessary for each concentration used in a C-R experiment. This convention is no doubt driven by the replication requirements of the one-way ANOVA methods used to generate NOECs. Without replication, the mean-square error term in the ANOVA model cannot be estimated and thus there is no basis for inference—including the determination of a NOEC. Even though there is a growing trend to use modeling approaches, it is not uncommon to see experimental designs employing a minimum of three replicates at each concentration. Not only this is unnecessary, it is potentially wasteful of precious experimental resources. To be clear, there is a simple dictum in statistics that states:

$$\text{data} = \text{model} + \text{error}$$

The "error" term in the above equation is a "repository" for all variation in the data that are not accounted for by the model—which is why it is often referred to as the "residual." There are many and varied sources of "error," but two important ones are *stochastic error* and *lack-of-fit*. Stochastic error is what some software packages refer to as "*pure error*"; it represents the unpredictable, randomly occurring variations from "expectation" and, as such, can only be described *probabilistically*. Lack-of-fit, on the other hand, arises from model misspecification and represents the discrepancies that arise because the chosen model is simply incapable of describing the response exactly. The situation is shown in Fig. 2.7 which is a plot of leaf length

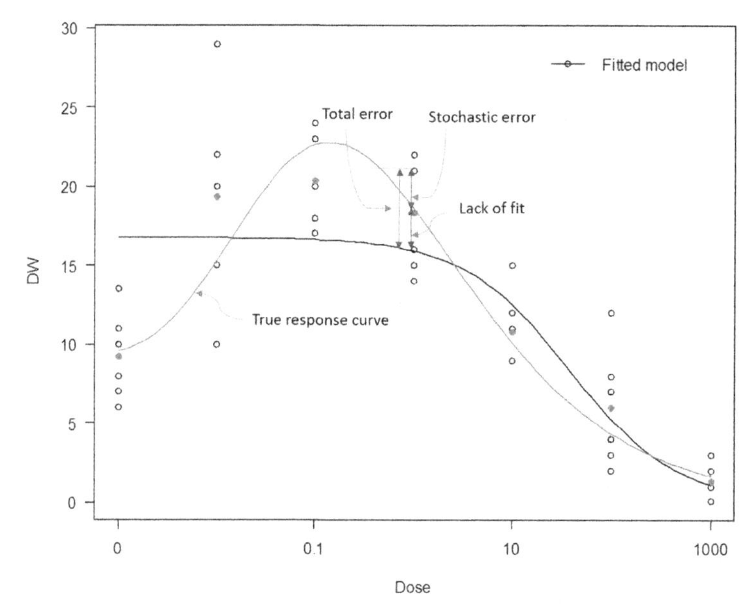

FIGURE 2.7 Relationship between total error, stochastic error, and lack-of-fit. *Open circles* are data and *solid red circles* (gray in print versions) the mean response at each concentration. Vertical axis is measured leaf length (cm); horizontal axis is concentration of metsulfuron-methyl (mg/L).

data provided with the R package `drc`. Thus, provided there is replication, an independent estimate of the pure error can be obtained and this can be subtracted from what many packages report as mean square of residuals to obtain an estimate of the lack-of-fit.

Returning to the issue of design, there is no need to replicate unless the significance of the lack-of-fit term is required to be tested and, even so, there are other "less costly" ways of assessing the adequacy of alternative model forms. For example, the `drc` package has a function to compare the utility of a list of models using either the akaike information or AIC (a measure of the utility of a model fitted to data) criterion or the Bayesian information criterion (similar to the AIC whose computation does not require specification of prior distributions in a Bayesian analysis).

As noted by the OECD (2012) *the choice of the number of dose levels and the dose spacing is crucial to achieving the objectives of the study (e.g., hazard identification or dose-response/risk assessment) and is important for subsequent statistical analysis.* With the requirement to replicate removed, experimental resources can now be re-directed to increasing the density of information extracted around critical areas of the C-R curve—for example, the NEC. While the design of the C-R experiment (ie, the choice of number of doses and dose spacing) can be done subjectively or based on heuristic rules, more formal procedures are available which identify the "best" design by optimizing some criterion such as the D-optimality criterion introduced in Section 2.5.

2.6.1 Locally D-Optimal Designs for C-R Experiments

Currently, the use of mathematical optimization methods to design C-R experiments is not widespread, although this is expected to change as ecotoxicologists move away from ANOVA-based to model-based estimation and inference. The D-optimality criterion was introduced in Section 2.5 in the context of fractional factorial designs, which are usually analyzed using ANOVA techniques. As was the case in the example in Section 2.5, the optimality criterion was a function of the *design matrix* X only. This means that it is possible to compute the efficiency of any given design by constructing X—something which can be done in advance of any data collection. Unfortunately, this is not possible when designing a C-R experiment for which the statistical model is nonlinear because evaluation of the optimality criterion requires knowledge of the model parameters. Thus, we have a "circular" situation of wanting to design an experiment to efficiently estimate model parameters, but identification of the "best" design requires us to first gather data to estimate the parameters. Although a "best guess" for the parameters could be used (for example, Chèvre and Brazzale, 2008 used what they called "plausible values"), this is not entirely satisfactory as there is no way of assessing the quality of the guess and hence of any design constructed based on it.

More sophisticated solutions to this problem exist. For example, Li and Fu (2013) couple Bayesian methods with an adaptive approach to the design problem. The procedure is complex and requires an understanding of advanced mathematical and statistical concepts. A simplified outline of a modified version of their method is as follows:

1. Allocate a small fraction of the total sample size to a "pilot experiment," which uses a subjective design (possibly informed by the results of similar studies). For example, in a "conventional" ANOVA-type C-R experiment assume the total sample size n is given as $n = k \cdot r$ where k is the number of concentrations to be used and r is the number of replicates at each concentration. The pilot experiment could be based on a sample size $n_1 = k$ for example.

2. Using the results of the pilot experiment, obtain the empirical posterior distribution, $p(\Theta|data)$ for the vector of parameters of interest, Θ (how this is done will depend on the particular software package used). Identify a list of candidate designs, $\{D_i\}$ $i = 1,\ldots,m$ where D_i is an assignment of concentrations to the remaining sample of size $n_2 = n - n_1 = k(r - 1)$.
3. For a given design, D_i:
   a. sample from the posterior distribution $p(\Theta|data)$ to obtain a realization of the predicted responses $\underline{Y} = \{Y_1, \ldots, Y_{n_2}\}$. Repeat this L times;
   b. For each of the samples in (a), evaluate the *utility function* $g(\underline{Y})$ (eg, $g(\cdot)$ might be the D-optimality criterion); and
   c. Estimate $E[g(\underline{Y})|D_i]$ by taking the average of the L values obtained in (a).
4. The "best" design among the set of candidate designs $\{D_i\}$ is that which maximizes the average in (c).

2.7 SPECIES SENSITIVITY DISTRIBUTION MODELING

This section presents a brief overview of the current status of SSD modeling. Discussion of the computational and inferential aspects of the SSD methodology is not covered here as these have been well documented elsewhere—for example, Posthuma et al. (2002) and Duboudin et al. (2004).

In a sense, the SSD is the inferential tool that allows us to move beyond individual toxicity metrics to saying something about the toxicity of a chemical for an entire "population"; however, that population is defined (although generally taken to mean all species comprising an ecosystem).

The SSD is a statistical device and as such there are important statistical considerations that accompany its use and interpretation. As with all inferential statistical techniques, an important distinction is made between the *population* and the *sample*. For statisticians, a population is simply the largest collection of "things" that (s)he is interested in; this may be biological, animate, or inanimate. For example, an electronics engineer might be interested in making inference about the failure rate for a population of circuit boards, while a plant physiologist might be interested in estimating the stomatal response for all plants of a certain species to elevated CO_2 levels. In ecotoxicology the SSD allows us to make inference about the toxic effect of a substance on *all* species in a defined population when only a small fraction of responses have been observed.

While the SSD has been an important development in ecotoxicology, its use has also been plagued with problems and controversy (Forbes and Forbes, 1993; Forbes and Calow, 2002; Hickey and Craig, 2012; Wheeler et al., 2002a,b; Zajdlik, 2006, 2015). At least conceptually, the borrowing of ideas from statistical inference makes sense. If we take a random sample of species from our defined population we can fit a theoretical probability distribution to the collection of toxicity measures obtained from the sample. The advantage of working with this theoretical construct is that it (presumably) encapsulates our understanding of toxicity based on a small sample and extends it to the entire population—which is why the use of the SSD in this manner is sometimes referred to as an "extrapolation" technique. But there are some severe shortcomings which, despite more than 30 years of use and refinement, remain problematic and potentially undermine the credibility of the whole approach. Numerous papers have been written which have documented the many shortcomings of the SSD methodology (and the companion issue of flawed metrics such as NOECs) (Newman et al., 2000; Okkerman et al., 1991; Wang et al., 2014). While some of these have obvious if not always practical solutions (for example, increasing the sample size to improve the quality of inference), other issues endure without resolution. Most significant among these are (1) unlike other areas of science (for example, physics, chemistry, and thermodynamics) where there are theoretical justifications for why a particular functional form should apply, there is no such basis for SSDs; and (2) despite being a core statistical assumption and requirement, the selection of species used to obtain the sample statistics referred to above is *never* random.

With respect to (1), current practice has enshrined the use of a small number of probability models (notably the log-normal, logistic, and log-logistic distributions) as suitable descriptors of the SSD. However, there is no guiding theory in ecotoxicology to justify one distributional form over another. While for many applications this would not be a problem because, overall, the fits afforded by all choices are reasonable, in ecotoxicology this "degree of freedom" is particularly problematic. Standard goodness-of-fit tests do not help in the elicitation of the "best" distribution as the invariably small sample size results in low-powered goodness-of-fit tests. This means that *any* plausible candidate model is unlikely to be rejected and, for this reason, the vexatious issue of choice of functional form is unlikely to ever be resolved. The other reason why the choice of functional form is so problematic in SSD modeling is because our ultimate interest lies in the most ill-defined portion of the fitted curve—namely the extreme left tail (assuming larger values have a greater adverse outcome). The ultimate objective of fitting the SSD is to estimate either the fraction of species adversely affected by a prescribed concentration of toxicant or the concentration that is hazardous to no more than $x\%$ of all species. This latter quantity is identified as the HC_x and is numerically equivalent to the xth percentile of the SSD, typically the HC_5.

On the issue of species selection, Posthuma et al. (2002) note "one of the serious failings of SSDs is the assumption that a non-random sample of test species represents the receiving communities." Given the importance of this key assumption, it is surprising that so little has been done to overcome its violation in ecotoxicology. Indeed, various guideline documents and regulatory requirements around the world actually embed nonrandomness into SSDs by mandating the use of purposive sampling. For example, the revised Australian and New Zealand Water Quality Guidelines recommend using toxicity data from at least eight species from at least four taxonomic groups (Batley et al., 2014; Warne et al., 2014). Random sampling will likely remain an unattainable ideal in SSD modeling for at least two reasons: (i) it is impossible to identify all species in an ecosystem; and (ii) testing protocols only exist for a handful of (non-randomly selected) species. Interestingly, although this has been a long-standing and widely acknowledged problem, it is only recently that any serious attempt has been made to both quantify and ameliorate the effects of nonrandom species selection in SSD fitting and HC_x estimation (Fox, 2015).

Although more research is needed into this crucial topic, Fox (2015) has shown that under the very mild assumption that specifies a beta probability density function for the species selection function, the actual SSD is *not* the assumed SSD (unless of course the selection process is truly random). In particular Fox (2015) notes that the *actual* distribution, when an *assumed* log-logistic SSD having parameters (α, β) is used to describe toxicity data (X) that have been selected according to a beta distribution with parameters (a, b), is a modified F distribution having the following probability density function *(pdf)*:

$$g_X(x; a, b, \alpha, \beta) = \frac{b\,\beta}{a\,\alpha}\left(\frac{x}{\alpha}\right)^{\beta-1} dF\left[\frac{b}{a}\left(\frac{x}{\alpha}\right)^{\beta}; 2a, 2b\right] \tag{2.10}$$

where the notation $dF(\cdot\,; \nu_1, \nu_2)$ denotes a standard F distribution having ν_1 and ν_2 degrees of freedom. A requirement satisfied by this *pdf* is that when species selection is truly random (corresponding to the special case of the beta distribution with $a = b = 1$), the actual distribution of the sample data is the assumed log-logistic. The impact of nonrandom species selection can be profound—resulting in HC_x estimation errors of a factor of 20 or more (Fox, 2015). Fortunately, the impact of selection bias on an estimated HC_x can be ameliorated through the use of a *bias correction factor* or *bcf*. For the assumed log-logistic SSD with a beta selection function with parameters (a, b), the *bcf* to apply to an HC_x estimated from the SSD fitted to the (biased) data is:

$$bcf(x) = \left[\frac{b}{a}\frac{x}{\xi_x\,100 - x}\right]^{1/\beta} \tag{2.11}$$

where ξ_x is the xth percentile from the standard F distribution having $2a$ and $2b$ degrees of freedom (Fox, 2015). The only difficulty with Eq. (2.11) is that it requires knowledge of the true value of the shape parameter β of the assumed log-logistic SSD. As a workaround Fox (2015) suggests replacing β in Eq. (2.11) with its sample estimate $\widehat{\beta}$. By way of example, suppose the fitted log-logistic SSD had parameters $\widehat{\alpha} = 5.46$ and $\widehat{\beta} = 1.76$. Furthermore, the species selection function was known to preferentially select the more sensitive species, which was adequately described by the beta density with $a = 0.5$ and $b = 2.0$. Using either published tables of the F distribution having 1 and 4 degrees of freedom or, more conveniently, EXCEL's intrinsic `F.INV()` function, it is readily determined that for $x = 5$, $\xi_5 = 0.004453$ and thus $bcf(5) = 8.94$. In other words, the HC_5 estimated from the SSD fitted to the sample data needs to be increased by almost an order of magnitude to compensate for the biased selection process.

Another area of active research, which aims to improve the effectiveness and applicability of SSD modeling, is the explicit incorporation of time, which has always been the "missing dimension" in the whole approach. As we have seen, a small, biased sample of toxicity data is used to estimate the parameters of the theoretical SSD. Whether regression or ANOVA-based, these metrics are invariably derived from C-R experiments for which standard laboratory protocols have been developed. Part of that standardization includes the fixing of the duration over which to run the experiment—typically 24, 48, or 96 h. Thus, the toxicity measures derived from these experiments are, in reality, only relevant for one particular period of exposure. As noted by Fox and Billoir (2013), there have been few, if any, attempts to fully integrate the temporal component of SSDs. Using only simple assumptions about the manner in which time was introduced into the SSD model, Fox and Billoir (2013) were able to quantify the impacts of this extra dimension on important quantities such as the HC_x, which of course itself becomes a function of the duration of the C-R experiment. More recently, Kon Kam King et al. (2015) used a "toxicodynamic" (TD) model to add a temporal dimension to the SSD. The TD model is a "lumped" version of the more familiar TKTD models used to describe various biological processes related to the uptake, processing, and toxic effects within an individual organism (eg, Jager et al., 2011). In this formulation, the number of individuals N_{ijk} of species j surviving until time t_k after having lived until time t_{k-1} when exposed to the ith concentration level is described by a binomial distribution where the probability of survival between t_k and t_{k-1} is a four-parameter function of time and exposure concentration, one of which is the *NEC*. A Bayesian hierarchical model was used to estimate parameters and make inference about the time-varying nature of the HC_5. Using published data on the salinity tolerance of 217 macroinvertebrates from the Murray Darling Basin region in Australia, Kon Kam King et al. (2015) demonstrated a strong HC_5–time dependency and concluded that an HC_5 using data from a 72-h C-R experiment was likely to overestimate the HC_5 obtained from a longer running experiment. From an environmental perspective, this is a disturbing outcome given that the potentially affected fraction of species having long-term exposure to a salinity not exceeding a conventionally derived HC_5 is likely to be much higher than the assumed 5%.

2.8 STATISTICAL SOFTWARE TOOLS FOR ECOTOXICOLOGY

This section provides an overview of some common software tools used in ecotoxicology. It is not intended to be exhaustive and focuses mainly on the analysis of C-R data and SSD fitting. TKTD models were briefly mentioned in the previous section and, while these methods can be used to obtain toxicity estimates, their focus is to predict toxic effects over time by modeling the relationship between an individual's exposure to a contaminant and internal concentrations in the body of the organism. These models therefore need to explicitly describe processes controlling uptake, elimination, internal distribution, and metabolism. Closely related to TKTD models are those developed around dynamic energy budget (DEB) theory. The most common of these is DEBtox which first appeared in 1996 as an accompaniment to the book by Kooijman and Bedaux (1996). DEB theory "is all about mechanistically linking (time-varying) external concentrations of a toxicant to the effects on life-history traits such as survival, growth and reproduction, over time" (Debtox Information Site, 2011). It does this by constructing rules that govern how organisms partition energy to control all life stages.

The information and communications technology "revolution" and the aforementioned rise of "big data" have led to rapid and unparalleled developments in computing hardware and software. In ecotoxicology, this has resulted in the deployment of more sophisticated models whose parameters can be estimated using computationally intensive techniques such as Markov Chain Monte Carlo (MCMC) and other resampling strategies (Gamerman, 2006; Gilks et al., 1998). To some extent, ecotoxicological guideline documents and textbooks have not caught up with these advances.

More contemporary approaches to the manipulation, presentation, visualization, and statistical analyses of data take advantage of purpose-built software tools. While stand-alone, computer programs, designed to achieve a single outcome such as the estimation of toxicity values and/or protective concentrations, have been in widespread use for many years—for example, the ToxCalc program (Tidepool Scientific Software, 2016), there has been a seismic shift in academia and science more generally to the R statistical computing environment which commenced around 2012. There are two good reasons for this: (1) the R system is open source and therefore completely free of charge; and (2) the range of intrinsic and user-contributed packages is unrivalled.

Given that the number of user-contributed R packages is predicted to reach 10,000 in 2016 (Environmetrics Australia, 2014b) and in keeping with a forward-looking assessment of trends in ecotoxicology, the remainder of this section focuses on ecotoxicological uses of R.

2.8.1 R Package `webchem`

`webchem` is an R package to retrieve chemical information from the web (Szöcs, 2015b). It can interrogate and retrieve information from a number of internet sources including:

- National Cancer Institute's Chemical Identifier Resolver;
- Royal Society of Chemistry's ChemSpider;
- National Center for Biotechnology Information's PubChem BioAssay Database;
- University of California's Chemical Translation Service;
- Pesticide Action Network's Pesticide Database;
- Effects information from aquatic and terrestrial ecotoxicology from the Federal Environment Agency of Germany's ETOX database;

- University of Hertfordshire's pesticide chemical identity, physicochemical, human health, and ecotoxicological database; and
- United States National Library of Medicine's TOXNET—a group of databases covering chemicals and drugs, diseases and the environment, environmental health, occupational safety and health, poisoning, risk assessment and regulations, and toxicology.

For example, entering the command get_cid ("Triclosan") produces the following list of chemical IDs for triclosan:

```
#>  [1] "5564"     "131203"   "627458"   "15942656" "16220126" "16220128"
#>  [7] "16220129" "16220130" "18413505" "22947105" "23656593" "24848164"
#> [13] "25023954" "25023955" "25023956" "25023957" "25023958" "25023959"
#> [19] "25023960" "25023961" "25023962" "25023963" "25023964" "25023965"
#> [25] "25023966" "25023967" "25023968" "25023969" "25023970" "25023971"
#> [31] "25023972" "25023973" "45040608" "45040609" "67606151" "71752714"
#> [37] "92024355" "92043149" "92043150"
```

while the following examples illustrate how to convert between IDs:

```
cts_convert(query = '3380-34-5', from = 'CAS', to = 'PubChem CID')
#> [1] "5564"   "34140"
cts_convert(query = '3380-34-5', from = 'CAS', to = 'ChemSpider')
#> [1] "31465"
(inchk <- cts_convert(query = '50-00-0', from = 'CAS', to = 'inchikey'))
#> [1] "WSFSSNUMVMOOMR-UHFFFAOYSA-N"
```

2.8.2 R Package `ggplot2`

`ggplot2` is a feature-rich, general purpose graphics package and, although not specifically developed for ecotoxicologists, it can be used to great effect to produce highly customizable SSDs (Figs. 2.8 and 2.9).

2.8.3 R Package `drc`

The package `drc` provides a comprehensive suite of modeling and analyses tools for fitting complex models to data generated by C-R experiments. An excellent overview describing the features and use of `drc` can be found in Ritz and Streibig (2005). As an example, the R code below fits a four-parameter and a five-parameter log-logistic function to C-R data. Using the `anova()` function with the fitted models as arguments provides a quick check on the value of the extra parameter. In this case, the reported *p-value* of 0.3742 suggests no significant improvement in overall fit using the 5-parameter model. Plotting the data and fitted models is simple and elegant as shown by the brevity of code required and the resulting output (Fig. 2.10).

```
> head(df)   #   inspect first 6 rows of dataframe df

   concen    response
1       0   0.8457565
2       0   0.9741697
3       0   0.9874539
4       0   1.0228782
5       0   1.0405904
6       0   1.0583026

> fit1<-drm(df$response ~ df$concen,fct=LL.4())   # fit 4-parameter log-logistic
> fit2<-drm(df$response ~ df$concen,fct=LL.5())   # fit 5-parameter log-logistic
>
> anova(fit1,fit2)   #   compare models
1st model
  fct:   LL.4()
2nd model
  fct:   LL.5()

ANOVA table

          ModelDf     RSS   Df   F value   p value
1st model      59  1.2922
2nd model      58  1.2746    1    0.8019    0.3742

> # plot results
>
> plot(fit1,broken=TRUE,legend=TRUE,legendText="4-param log-logistic",
+ col="red",type="confidence",ylab="fraction",xlab="concen")
>
> plot(fit2,add=TRUE,legend=TRUE,legendText="5-param log-
logistic",legendPos=c(0.19,1.2),
+ col="blue",type="confidence",ylab="fraction",xlab="concen")
>
> plot(fit2,add=TRUE,type="obs",pch=16,cex=0.6)
```

2.8.3 R Function `fitdistr()` and Package `fitdistrplus`

These are general-purpose distribution-fitting programs. `fitdistr()` is part of the `MASS` package and fits univariate probability distributions using maximum-likelihood estimation. The `fitdistrplus` package has more features and is capable of fitting a variety of univariate probability models to both censored and noncensored data using a variety of algorithms including maximum likelihood estimation, method of moments, quantile matching, and maximum goodness of fit. `fitdistrplus` is at the heart of the online SSD tool `MOSAIC` (see below).

Parameter estimates for a lognormal distribution fitted to the data of Fig. 2.8 are readily obtained using either `fitdistr()` or `fitdistrplus`:

```
> fitdistr(data$EC50,"lognormal")
   meanlog       sdlog
   1.9084029     1.8440993
   (0.4024155)  (0.2845507)

> fit<-fitdist(data$EC50, "lnorm",method="mme")
> fit
Fitting of the distribution ' lnorm ' by matching moments
Parameters:
          estimate
meanlog   3.825651
sdlog     1.716488
```

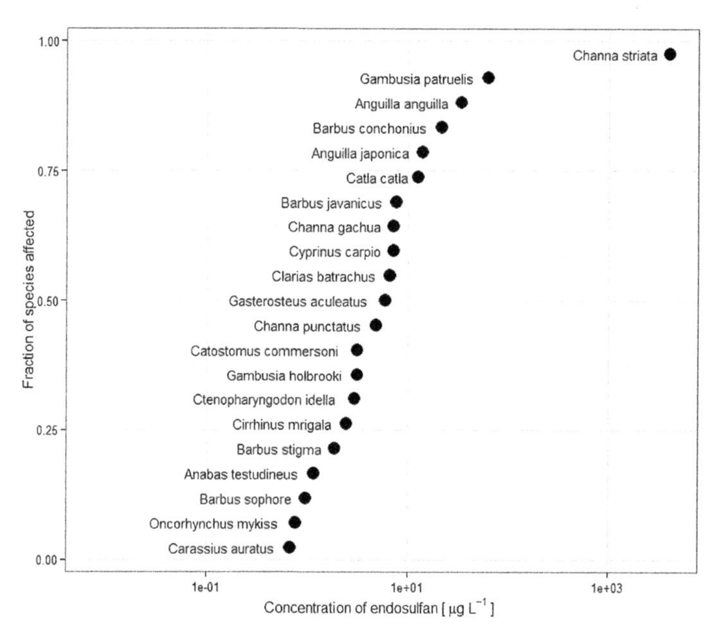

FIGURE 2.8 Empirical SSD for endosulfan EC_{50} data for non-Australian fish. *From Hose, G.C., Van den Brink, P.J., 2004. Confirming the species-sensitivity distribution concept for endosulfan using laboratory, mesocosm, and field data. Environ. Contam. Toxicol. 47, 511–520.*

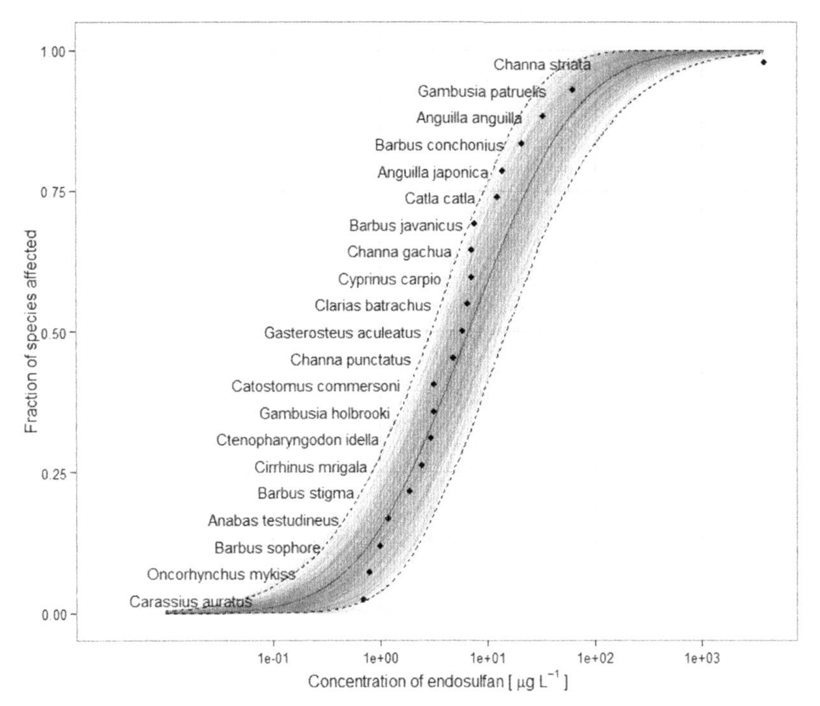

FIGURE 2.9 Theoretical lognormal SSD [*red curve* (dark gray in print versions)] for endosulfan data of Fig. 2.8 (*solid black circles*) together with lognormal distributions fitted to bootstrap samples [*blue curves* (gray in print versions)] and 95% confidence limits (*black dashed curves*). *R source code adapted from http://bit.ly/1OrFC5n.*

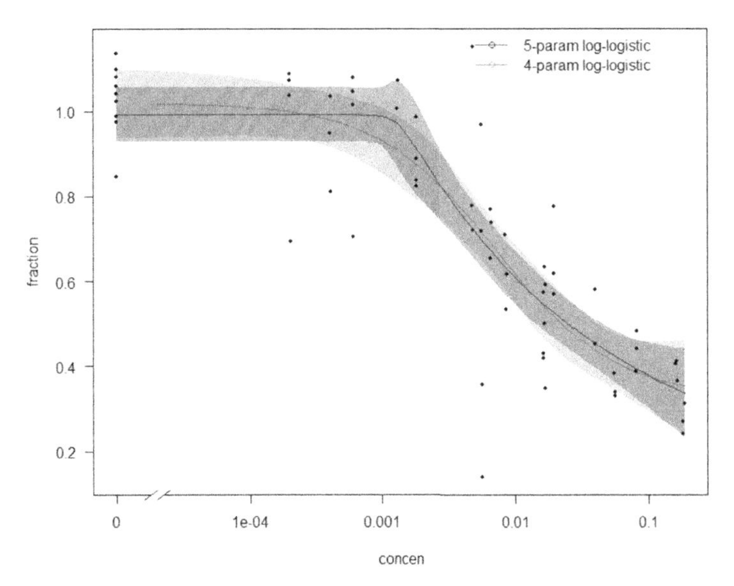

FIGURE 2.10 Plot of data and fitted models using `drc` package. Colored bands are 95% confidence intervals for mean response.

2.8.4 MOSAIC

MOSAIC is an online tool that interfaces with R to fit one of a number of predefined probability models to empirical toxicity data. It also uses bootstrapping to provide estimates of the standard error of parameter estimates as well as for derived HC_x values. Developed by researchers at the Biometry and Evolutionary Biology Laboratory at the University of Lyon, the tool can be accessed online (Biometry and Evolutionary Biology Laboratory, 2015). A screenshot of the output for the fish data of Fig. 2.8 is shown in Fig. 2.11.

2.8.5 BurrliOZ

BurrliOZ (CSIRO, 2015) is similar in concept to MOSAIC in that it is an interface to R and is used to fit either a log-logistic or Burr-type distribution to toxicity data. However, unlike MOSAIC, BurrliOz is downloaded and run locally on a desktop computer. It was originally released as a standalone program developed by statisticians at Australia's Commonwealth Scientific and Industrial Research Organisation (CSIRO) to support the 2000 revision of the Australian and New Zealand Guidelines for fresh and marine water quality (ANZECC/ARMCANZ, 2000a). While offering

FIGURE 2.11 Screenshot of MOSAIC output for lognormal SSD fitted to data of Fig. 2.8.

similar capabilities to MOSAIC, BurrliOz is more cumbersome and, although it also uses R to perform the necessary calculations, unlike MOSAIC the R source code cannot be accessed. A screenshot from BurrliOz applied to the data of Fig. 2.8 is shown in Fig. 2.12.

2.9 FUTURE POSSIBILITIES

It is true to say that advances in statistical methodology over the past 40 years have to a large extent been driven by advances in computing power and software developments. Witness, for example the explosive growth of R, the pervasiveness of computationally intensive techniques such as MCMC, and the ability to produce publication-quality graphics with the click of a mouse. It therefore seems reasonable to suggest that opportunities and developments in *statistical ecotoxicology* will exhibit a similar dependency. Already, stand-alone desktop applications for undertaking many of the analyses presented in this chapter are gradually being replaced by online tools such as MOSAIC (Biometry and Evolutionary Biology Laboratory, 2015) and BurrliOz (CSIRO, 2015). The increasing number of customized ecotoxicology packages written in R coupled with the development of "wrapper" programs and languages such as R-Studio's Shiny software (RStudio, 2016) will improve user-friendliness thereby encouraging greater uptake and use of these tools. This convergence of technologies will, we predict, result in the development of ecotoxicological "apps" for mobile computing devices such as tablets, smart phones, and iPads. So, at one level, a new frontier in statistical ecotoxicology will be entirely digital.

In terms of theoretical challenges and opportunities, those identified by Eggen et al. (2004) remain relevant and are associated with:

- Low concentrations of pollutants and chronic effects;
- Multiple effects by single pollutants;
- Complex mixtures of pollutants;
- Multiple stressors; and
- Ecosystem complexity.

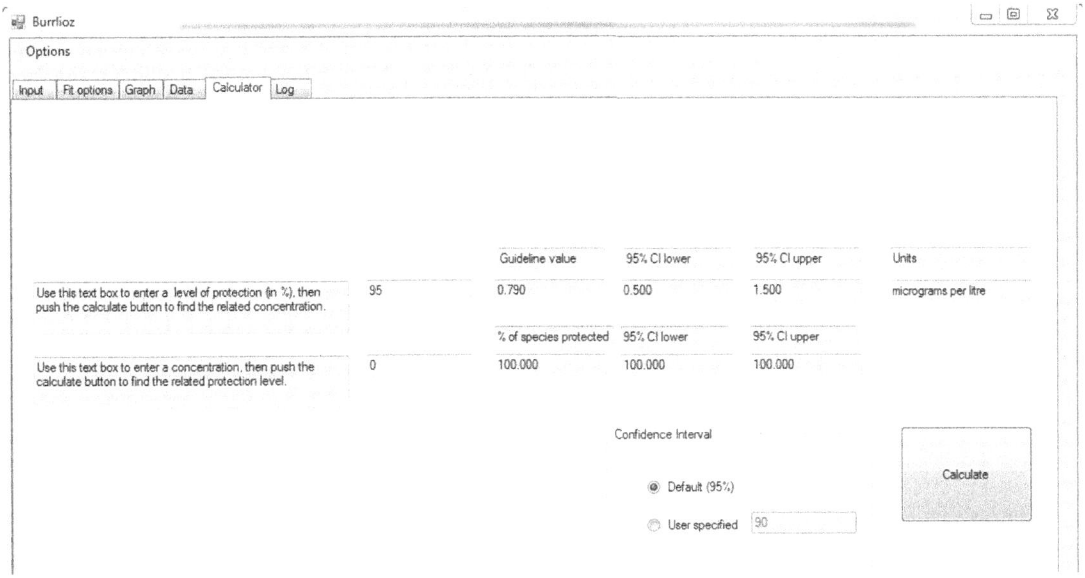

FIGURE 2.12 Screenshot of BurrliOz HC_x calculator screen for log-logistic SSD fitted to data of Fig. 2.8.

Statistical tools such as maximum likelihood estimation, probit analysis, and logistic regression and to a lesser extent generalized linear models and Bayesian methods now underpin contemporary methods in ecotoxicological estimation and inference. The challenge as we see it, is to more tightly integrate this important facet of ecotoxicology into university courses and professional training programs. Much of the statistical framework in ecotoxicology has been the result of organic growth rather than planned development. Regrettably, biometricians are a vanishing species in natural resource management agencies around the world. An unintended consequence of this decline is a tendency of researchers in ecotoxicology to avoid new and potentially challenging statistical concepts thereby increasing the propensity to maintain the status quo in terms of the statistical analysis of ecotoxicological data. It is, we suspect, also partly responsible for on-going debates that have outlived their usefulness about indeterminate questions of the sort *is the log-normal distribution better than the log-logistic distribution as an SSD?* or *what value of x should be used to reconcile ECx and NOECs?* More productive lines of investigation will be associated with:

- Addressing and compensating for small, biased samples in SSD and HC_x estimation;
- Development of statistical methods to validate predictions from TKTD models and SSD-based approaches;
- How to design a C-R experiment that maximizes information content for minimum cost;
- How, or establish if it is possible, to set an HC_x for a mixture of chemicals;
- How to seamlessly integrate the temporal dimension into SSD modeling and HC_x estimation rather than marginalizing it;
- Strategies for error propagation to incorporate uncertainties arising from data collection process, imprecise model specification, and statistical treatment of data;
- Refinement of modeling capabilities to undertake external and internal exposure assessments;
- Elicitation of "expert" opinion in the setting of Bayesian priors and protocols for reaching consensus when these are used in a regulatory context.

Because of its importance and widespread use, the SSD methodology and companion techniques have been a focus of this chapter. Most of the dot-points above relate to ways of improving both the theoretical and applied aspects of the approach. It is likely that the SSD will retain its place as the preferred (and often mandated) method for establishing protective concentrations although one group of experts recently identified a need to "move away from using SSDs as a purely statistical construct applied to poorly understood species sensitivity data to one in which SSDs provide the framework for a more process-based approach" (ECETOC, 2014).

In its defense the SSD has provided a consistent and reproducible way of establishing water quality criteria and while not totally free of subjective choices concerning choice of distribution, parameterization, and estimation techniques, it has removed the arbitrariness of the assessment factor method it replaced. Problems that have been identified with this "statistical construct" are not the result of flawed statistical science, but invariably arise from violations of critical assumptions and/or inappropriate application. How the research community deals with these issues remains to be seen.

References

Adelmann, S., 1961. Irregular fractions of the 2^n factorial experiments. Technometrics 3, 479–496.

ANZECC/ARMCANZ, 2000a. Australian and New Zealand Guidelines for Fresh and Marine Water Quality. Australian and New Zealand Environment and Conservation

Council/Agricultural and Resource Management Council of Australia and New Zealand, Canberra, ACT, Australia.

ANZECC/ARMCANZ, 2000b. Australian Guidelines for Water Quality Monitoring and Reporting. Australian and New Zealand Environment and Conservation Council, Agriculture and Resource Management Council of Australia and New Zealand (National Water Quality Management Strategy number 7).

Barnett, V., O'Hagan, A., 1997. Setting Environmental Standards: The Statistical Approach to Handling Uncertainty and Variation. Chapman and Hall, London, UK.

Batley, G.E., Braga, O., Van Dam, R., Warne, M.S.J., Chapman, J.C., Fox, D.R., Hickey, C., Stauber, J.L., 2014. Technical Rationale for Changes to the Method for Deriving Australian and New Zealand Water Quality Guideline Values for Toxicants. Council of Australian Government's Standing Council on Environment and Water, Sydney, Australia.

Berkson, J., 1944. Application of the logistic function to bioassay. J. Am. Stat. Assoc. 39, 357–365.

Bickel, P.J., Doksum, K.A., 2016. Mathematical Statistics: Basic Ideas and Selected Topics. CRC Press, Boca Raton, FL, USA.

Billoir, E., Delignette-Muller, M.L., Péry, A.R.R., Charles, S., 2008. A Bayesian approach to analyzing ecotoxicological data. Environ. Sci. Technol. 42, 8978–8984.

Biometry and Evolutionary Biology Laboratory, 2015. MOSAIC: Modelling and Statistical Tools for Ecotoxicology. University of Lyon, France. Available at: http://pbil.univ-lyon1.fr/software/mosaic/ssd/.

Bliss, C.I., 1935. The calculation of the dosage-mortality curve. Ann. Appl. Biol. 22, 134–167.

Bose, R.C., Bush, K.A., 1952. Orthogonal arrays of strength two and three. Ann. Math. Stat. 23, 508–524.

Cash, J.T., 1908. The relationship of action to dose especially with reference to repeated administration of indaconitine. Br. Med. J. 1908, 1213–1218.

CCME, 2007. A Protocol for the Derivation of Water quality Guidelines for the Protection of Aquatic Life. Canadian Council of Ministers of the Environment, Ottawa, ON, Canada, 37 pp.

Chapman, P.M., Cardwell, R.S., Chapman, P.F., 1996. A warning: NOECs are inappropriate for regulatory use. Environ. Toxicol. Chem. 15, 77–79.

Chèvre, N., Brazzale, A.R., 2008. Cost-effective experimental design to support modelling of concentration-response functions. Chemosphere 72, 803–810.

Cliffroy, P., Keller, M., Pasanisi, A., 2013. Estimating hazardous concentrations by an informative Bayesian approach. Environ. Toxicol. Chem. 32, 602–611.

Cox, D.R., Snell, E.J., 2000. Statistics: Principles and Examples. Chapman and Hall/CRC, Boca Raton, FL, USA.

CRAN, 2015. Design of Experiments (DoE) and Analysis of Experimental Data. Available at: https://cran.r-project.org/web/views/ExperimentalDesign.html.

Crane, M., Newman, M.C., 2000. What level of effect is a no observed effect? Environ. Toxicol. Chem. 19, 516–519.

CSIRO, 2015. BurrliOz 2.0. Available at: https://research.csiro.au/software/burrlioz/.

Dasu, T., Johnson, T., 2003. Exploratory Data Mining and Data Cleaning. John Wiley & Sons. http://ca.wiley.com/WileyCDA/WileyTitle/productCd-0471268518,subjectCd-CSB0.html.

Debtox Information Site, 2011. Debtox Information: Making Sense of Ecotoxicity Test Results. Available at: http://www.debtox.info/about_debtox.html.

Dey, A., 1985. Orthogonal Fractional Factorial Designs. John Wiley and Sons, Haslted Press, New York.

Duboudin, C., Ciffroy, P., Magaud, H., 2004. Effects of data manipulation and statistical methods on species sensitivity distributions. Environ. Toxicol. Chem. 23, 489–499.

ECETOC, 2014. Estimating Toxicity Thresholds for Aquatic Ecological Communities From Sensitivity Distributions 11–13 February, 2014. Amsterdam Workshop Report No. 28. European Centre for Ecotoxicology and Toxicology of Chemicals.

Eggen, R.I.L., Behra, R., Burkhardt-Holm, P., Esche, B.I., Schweigert, N., 2004. Challenges in ecotoxicology. Environ. Sci Technol. 38, 58A–64A.

Environment Canada, 2005. Guidance Document on Statistical Methods for Environmental Toxicity Tests. EPS 1/RM/46, Ottawa, ON, Canada.

Environmetrics Australia, 2014a. Big Data Is Watching You. Available at: http://www.environmetrics.net.au/index.php?news&nid=81.

Environmetrics Australia, 2014b. The Explosive Growth of R. Available at: http://environmetrics.net.au/index.php?news&nid=79.

European Commission, 2011. Technical guidance for deriving environmental quality standards. In: Guidance Document No. 27, Common Implementation Strategy for the Water Framework Directive, Brussels, Belgium, 204 pp.

Evans, D.A., Newman, M.C., Lavine, M., Jaworska, J.S., Toll, J., Brooks, B.W., Brock, T.C.M., 2010. The Bayesian vantage for dealing with uncertainty. In: Warren-Hicks, J., Hart, A. (Eds.), Application of Uncertainty Analysis to Ecological Risks of Pesticides. CRC Press, Boca Raton, FL, USA.

Forbes, T.L., Forbes, V.E., 1993. A critique of the use of distribution-based extrapolation models in ecotoxicology. Funct. Ecol. 7, 249–254.

Forbes, V.E., Calow, P., 2002. Species sensitivity distributions: a critical appraisal. Hum. Ecol. Risk Assess. 8, 473–492.

Fox, D.R., 2006. Statistical issues in ecological risk assessment. Hum. Ecol. Risk Assess. 12, 120–129.

Fox, D.R., 2008. NECs, NOECs, and the ECx. Australas. J. Ecotoxicol. 14, 7–9.

Fox, D.R., 2010. A Bayesian approach for determining the no effect concentration and hazardous concentration in ecotoxicology. Ecotoxicol. Environ. Saf. 73, 123–131.

Fox, D.R., 2015. Selection bias correction for species sensitivity distribution modelling and hazardous concentration estimation. Environ. Toxicol. Chem. 34, 2555–2563.

Fox, D.R., Landis, W.G., April 18, 2016a. Don't be fooled - A NOEC is no substitute for a poor concentration-response experiment. Environ. Toxicol. Chem. http://dx.doi.org/10.1002/etc.3459.

Fox, D.R., Landis, W.G., 2016b. Comment on ET&C perspectives November 2015 - A Holistic View. Environ. Toxicol. Chem. 35, 1337–1339.

Fox, D.R., Billoir, E., 2013. Time dependent species sensitivity distributions. Environ. Toxicol. Chem. 32, 378–383.

Fox, D.R., Landis, W.G., April 18, 2016a. Don't be fooled – A NOEC is no substitute for a poor concentration-response experiment. Environ. Toxicol. Chem. http://dx.doi.org/10.1002/etc.3459.

Fox, D.R., Landis, W.G., 2016b. Comment on ET&C perspectives November 2015 – A Holistic View. Environ. Toxicol. Chem. 35, 1337–1339.

Fox, D.R., Billoir, E., Charles, S., Delignette-Muller, M.L., Lopes, C., 2012. What to do with NOECs/NOELs – prohibition or innovation? Integr. Environ. Assess. Manag. 8, 764–766.

Gad, S.C., 2006. Statistics and Experimental Design for Toxicologists and Pharmacologists, 4th ed. Taylor and Francis, Boca Raton, FL, USA.

Gamerman, D., 2006. Markov Chain Monte Carlo: Stochastic Simulation for Bayesian Inference. Chapman and Hall/CRC, Boca Raton, FL, USA.

Gilks, W.R., Richardson, S., Spiegelhalter, D.J., 1998. Markov Chain Monte Carlo in Practice. Chapman and Hall/CRC, Boca Raton, FL, USA.

Graybill, F.A., 1976. Theory and Application of the Linear Model. Duxbury Press, North Scituate.

Green, J.W., 2016. Issues with using only regression models for ecotoxicology studies. Integr. Environ. Assess. Manag. 12, 198–199.

Green, J.W., Springer, T.A., Staveley, J.P., 2013. The drive to ban the NOEC/LOEC in favor of ECx is misguided and misinformed. Integr. Environ. Assess. Manag. 9, 12–16.

Grist, E.P.M., O'Hagan, A., Crane, M., Sorokin, N., Sims, I., Whitehouse, P., 2006. Comparison of frequentist and Bayesian freshwater species sensitivity distributions for chlorpyrifos using time-to-event analysis and expert elicitation. Environ. Sci. Technol. 40, 295–301.

Hickey, G.L., Craig, P.S., 2012. Competing statistical methods for the fitting of normal species sensitivity distributions: recommendations for practitioners. Risk Anal. 32, 1232–1243.

Hinkelmann, K., Kempthorne, O., 2008. Design and Analysis of Experiments. In: Introduction to Experimental Design, vol. 1. Wiley, NJ, USA.

Hoekstra, J.A., Van Ewijk, P.H., 1993. The bounded effect concentration as an alternative to the NOEC. Sci. Tot. Environ. 134, 705–711.

Hose, G.C., Van den Brink, P.J., 2004. Confirming the species-sensitivity distribution concept for endosulfan using laboratory, mesocosm, and field data. Environ. Contam. Toxicol. 47, 511–520.

Jager, T., 2012. Bad habits die hard: the NOECs persistence reflects poorly on ecotoxicology. Environ. Toxicol. Chem. 31, 228–229.

Jager, T., Albert, C., Preuss, T.G., Ashauer, R., 2011. General unified threshold model of survival – a toxicokinetic-toxicodynamic framework for ecotoxicology. Environ. Sci. Technol. 45, 2529–2540.

Jaworska, J., Gabbert, S., Aldenberg, T., 2010. Towards optimization of chemical testing under REACH: a Bayesian network approach to integrated testing strategies. Regul. Toxicol. Pharmacol. 57, 157–167.

Kempthorne, O., 1947. A simple approach to confounding and fractional replication in factorial experiments. Biometrika 34, 255–272.

Kon Kam King, G., Delignette-Muller, M.L., Kefford, B.J., Piscart, C., Charles, S., 2015. Constructing time-resolved species sensitivity distributions using hierarchical toxicodynamic model. Environ. Sci. Technol. 49, 12465–12473.

Kooijman, S., Bedaux, J.J.M., 1996. The Analysis of Aquatic Toxicity Data. VU University Press, Amsterdam, Netherlands.

Kutner, M.H., Nachtsheim, C.J., Neter, J., Li, W., 2016. Applied Linear Statistical Models, 5th ed. Irwin/McGraw-Hill, New York, USA.

Landis, W.G., Chapman, P.M., 2011. Well past time to stop using NOELs and LOELs. Integr. Environ. Assess. Manag. 7, vi–viii.

Lawson, J., 2015. Design and Analysis of Experiments with R. CRC Press, Boca Raton, FL, USA.

Li, J., Fu, H., 2013. Bayesian adaptive D-optimal design with delayed responses. J. Biopharm. Stat. 23, 559–568.

Link, W.A., Albers, P.H., 2007. Bayesian multimodel inference for dose-response studies. Environ. Toxicol. Chem. 26, 1867–1872.

Malthus, T.R., 1798. An Essay on the Principle of Population. J. Johnson, London, UK.

McBride, M.F., Garnett, S.T., Szabo, J.K., Burbidge, A.H., Butchart, S.H.M., Christidis, L., Dutson, G., Ford, H.A., Loyn, R.H., Watson, D.M., Burgman, M.A., 2012.

Structured elicitation of expert judgments for threatened species assessment: a case study on a continental scale using email. Methods Ecol. Evol. 3, 906–920.

Moore, D.R.J., Warren-Hicks, W.J., Qian, S., Fairbrother, A., Aldenberg, T., Barry, T., Luttik, R., Ratte, H.-T., 2010. Uncertainty analysis using classical and Bayesian hierarchical models. In: Warren-Hicks, W.J., Hart, A. (Eds.), Application of Uncertainty Analysis to Ecological Risks of Pesticides. CRC Press, Boca Raton FL, USA.

Newman, M.C., 2012. Quantitative Ecotoxicology. CRC Press, Boca Raton, FL, USA.

Newman, M.C., Clements, W.H., 2008. Ecotoxicology: A Comprehensive Treatment. CRC Press, Boca Raton, FL, USA.

Newman, M.C., Ownby, D.R., Mézin, L.C.A., Powell, D.C., Christensen, T.R.L., Leerberg, S.B., Anderson, B.-A., 2000. Applying species sensitivity distributions in ecological risk assessment: assumptions of distribution type and sufficient numbers of species. Environ. Toxicol. Chem. 19, 508–515.

New York Times, 2014. For Big-data Scientists, 'Janitor Work' is Key Hurdle to Insights. Available at: http://www.nytimes.com/2014/08/18/technology/for-big-data-scientists-hurdle-to-insights-is-janitor-work.html?_r=2.

OECD (Organization for European Cooperation and Development), 1998. Report of the OECD workshop on statistical analysis of aquatic toxicity data. In: OECD Environmental Health and Safety Publications Series on Testing and Assessment No. 10, Paris, France.

OECD, 2012. Guidance document 116 on the conduct and design of chronic toxicity and carcinogenicity studies, supporting test guidelines 451, 452, and 453. In: Series on Testing and Assessment No. 116, Paris, France, 2nd ed.

OECD, 2014. Current approaches in the statistical analysis of ecotoxicity data: a guidance to application (annexes to this publication exist as a separate document). In: OECD Series on Testing and Assessment, No. 54, Paris, France.

Okkerman, P.C., Plassche, E.J.V.D., Slooff, W., Van Leeuwen, C.J., Canton, J.H., 1991. Ecotoxicological effects assessment: a comparison of several extrapolation procedures. Ecotoxicol. Environ. Saf. 21, 182–193.

Openbugs, 2009. Bayesian Inference Using Gibbs Sampling. Available at: http://www.openbugs.net/w/FrontPage.

Payet, J., 2004. Assessing Toxic Impacts on Aquatic Ecosystems in Life Cycle Assessment (LCA) (Ph.D. thesis). École Polytechnique Fédérale de Lausanne, Lausanne, Switzerland.

Pearl, R., Reed, L.J., 1920. On the rate of growth of the population of the United States since 1790 and its mathematical representation. Proc. Natl. Acad. Sci. USA 6, 275–288.

Posthuma, L., Suter, G.W., Traas, T.P., 2002. Species Sensitivity Distributions in Ecotoxicology. Lewis Publishers, Boca Raton, FL, USA.

Quora, 2014. What Is Data Munging? Available at: https://www.quora.com/What-is-data-munging.

R Development Core Team, 2004. R: A Language and Environment for Statistical Computing. R Foundation for Statistical Computing, Vienna, Austria. ISBN: 3-900051-00-3, URL http://www.R-project.org.

Rao, C.R., 1950. The theory of fractional replication in factorial experiments. Sankhya 10, 81–86.

Ritz, C., Streibig, J.C., 2005. Bioassay analysis using R. J. Stat. Softw. 12, 1–22.

RStudio, 2016. Shiny Software. Available at: https://www.rstudio.com/products/shiny/.

Shieh, J.N., Chao, M.R., Chen, C.Y., 2001. Statistical comparisons of the no-observed-effect concentration and the effective concentration at 10% inhibition (EC10) in algal toxicity tests. Water Sci. Technol. 43, 141–146.

Sourceforge, 2015. JAGS: Just Another Gibbs Sampler. Available at: http://mcmc-jags.sourceforge.net/.

Sparks, T., 2000. Statistics in Ecotoxicology. Wiley and Sons, Chichester, UK.

Stan Development Team, 2015. Stan Modeling Language Users Guide and Reference Manual, Version 2.8.0. Available at: http://mc-stan.org/documentation/.

Szöcs, E., 2015a. Species Sensitivity Distributions (SSD) with R. Available at: http://edild.github.io/ssd/.

Szöcs, E., 2015b. Introducing the Webchem Package. Available at: http://edild.github.io/webchem/.

Tidepool Scientific Software, 2016. ToxCalc Software. Available at: https://tidepool-scientific.com/ToxCalc/ToxCalc.html.

Trevan, J.W., 1927. The error of determination of toxicity. Proc. R. Soc. B 101 (712), 483–514.

USEPA (U.S. Environmental Protection Agency), 2002. Guidance on Choosing a Sampling Design for Environmental Data Collection for Use in Developing a Quality Assurance Project Plan. EPA QA/G-5S, Washington, DC, USA.

USEPA, 2006. Data Quality Assessment: Statistical Methods for Practitioners. EPA QA/G-9S, Washington, DC, USA.

USEPA, 2015. ECOTOX User Guide: ECOTOXicology Database System. Version 4.0. Available at: http://www.epa.gov/ecotox/.

van Dam, R.A., Harford, A.J., Warne, M.S., 2012. Time to get off the fence: the need for definitive international guidance on statistical analysis of ecotoxicity data. Integr. Environ. Assess. Manag. 8, 242–245.

van der Hoeven, N., Noppert, F., Leopold, A., 1997. How to measure no effect. Part I: towards a new measure of chronic toxicity in ecotoxicology. Introduction and workshop results. Environmetrics 8, 241–248.

Verhulst, P.F., 1838. Notice sur la loi que la population poursuit dans son accroissement. Corresp. Math. Phys. 10, 113–121.

Wang, Y., Zhang, L., Meng, F., Zhou, Y., Jin, X., Giesy, J.P., Liu, F., 2014. Improvement on species sensitivity distribution methods for deriving site-specific water quality criteria. Environ. Sci. Pollut. Res. 22, 5271–5282.

Warne, MStJ., van Dam, R., 2008. NOEC and LOEC data should no longer be generated or used. Australas. J. Ecotox. 14, 1–5.

Warne, MStJ., Batley, G.E., Braga, O., Chapman, J.C., Fox, D.R., Hickey, C.W., Stauber, J.L., van Dam, R.A., 2013. Revisions to the derivation of the Australian and New Zealand guidelines for toxicants in fresh and marine waters. Environ. Sci. Pollut. Res. 21, 51–60.

Warne, MStJ., Batley, G.E., Van Dam, R.A., Chapman, J., Fox, D., Hickey, C., Stauber, J., 2014. Revised method for deriving Australian and New Zealand water quality guideline values for toxicants. In: Prepared for the Council of Australian Government's Standing Council on Environment and Water, Sydney, Australia.

Webb, J.M., Smucker, B.J., Bailer, A.J., 2014. Selecting the best design for nonstandard toxicology experiments. Environ. Toxicol. Chem. 33, 2399–2406.

Wheeler, J.P., Grist, E.P.M., Leung, K.M.Y., Morritt, D., Crane, M., 2002a. Species sensitivity distributions: data and model choice. Mar. Pollut. Bull. 45, 192–202.

Wheeler, J.R., Leung, K.M.Y., Morritt, D., Whitehouse, P., Sorokin, N., Toy, R., Holt, M., Crane, M., 2002b. Freshwater to saltwater toxicity extrapolation using species sensitivity distributions. Environ. Toxicol. Chem. 21, 2459–2467.

Wickham, H., 2009. ggplot2: Elegant Graphics for Data Analysis. Springer, New York, NY, USA.

Wickham, H., 2014. Tidy data. J. Stat. Softw. 59, 1–23.

Wikipedia, 2015. Big Data. Available at: https://en.wikipedia.org/wiki/Big_data.

Wikipedia, 2016. Determinant. Available at: https://en.wikipedia.org/wiki/Determinant.

Zajdlik, B.A., 2006. Potential statistical models for describing species sensitivity distributions. In: Prepared for Canadian Council of Ministers for the Environment CCME Project # 382–2006. Available at: http://www.ccme.ca/assets/pdf/pn_1415_e.pdf.

Zajdlik, B.A., 2015. A statistical evaluation of the safety factor and species sensitivity distribution approach to deriving environmental quality guidelines. Integr. Environ. Assess. Manag.. http://dx.doi.org/10.1002/ieam.1694.

Zhang, J., Bailer, A.J., Oris, J.T., 2012. Bayesian approach to estimating reproductive inhibition potency in aquatic toxicity testing. Environ. Toxicol. Chem. 31, 916–927.

CHAPTER

3

Dynamic Modeling for Uptake and Effects of Chemicals

T. Jager

DEBtox Research, De Bilt, The Netherlands

3.1 INTRODUCTION

A model is a simplified representation of a part of the real world. In the natural sciences, these models are generally presented in a mathematical form so that their performance can be evaluated quantitatively. Building models is not an aim in itself, but rather a means to obtain a deeper understanding of the mechanisms underlying processes observed in the real world. Without models, it would be impossible to interpret the integrated (often nonlinear) effect of multiple factors that act on our system of interest. Furthermore, models are essential to make predictions for untested situations. In ecotoxicology, this can be useful for the optimal design of experimental work, but model predictions are particularly beneficial for a science-based risk assessment; for example, to evaluate the impacts of releasing a new chemical into the environment or to evaluate the effectiveness of different mitigation strategies. The usefulness of models for regulatory purposes is demonstrated in environmental chemistry, as fate models are an integral part of virtually all frameworks for environmental risk assessment around the world. These models integrate quantitative knowledge about the transport and transformation of chemicals. Subsequently, they are used to understand chemical fate in the environment and to predict the environmental concentrations (generally over time and space) from chemical properties and emission scenarios. In this way, fate models support risk assessments for new chemicals (before they are emitted into the environment) and new situations (eg, an oil spill in a particular location). Even though a model is always a simplification of reality, and therefore always "wrong," modeling should be an integral part of our quest to mechanistically understand and effectively manage the world around us.

In ecotoxicology, the models that are most commonly applied are those for the uptake of chemicals in individual organisms over time (toxicokinetic or TK models, discussed in Section 3.3). For the interpretation of toxic effects on individuals, hypothesis testing and dose–response curves are traditionally used (Chapter 2, Statistics). Even though these approaches can be viewed as crude models, there is no attempt to explain the observed effects from underlying principles (the purpose is to test for significant

Marine Ecotoxicology
http://dx.doi.org/10.1016/B978-0-12-803371-5.00003-5

effects, or to interpolate, in a given set of data). Therefore, the two most important aims of modeling (understanding and prediction) are not served, and hence such descriptive approaches are not addressed in this chapter. More mechanistic models for explaining the effects on individuals over time (toxicodynamic or TD models, discussed in Section 3.4) are gaining interest but are not as commonly applied yet in ecotoxicology as TK models. Ecotoxicological modeling also increasingly takes place at lower levels of biological organization (the molecular and cellular levels) as well as higher levels (populations, food chains, and ecosystems), but these models are not discussed in detail in this chapter. The individual level is a central level of biological organization for several reasons but, for ecotoxicologists, the most important ones are twofold: effects at lower levels of organization are hardly ecologically relevant unless they affect the life-history traits of individuals (ie, growth, reproduction, survival), and effects at higher levels of organization ultimately follow from changes in individual life-history traits by toxicant stress. When learning about modeling in ecotoxicology, the level of the individual is a good place to start, and it is therefore the focus of this chapter.

For marine ecotoxicology, we can in general apply the same models as those used for freshwater and terrestrial organisms. The modeling principles are the same, and the similarities between species are more important than the differences. There are, however, several biological aspects that feature more prominently in marine organisms than in their freshwater and terrestrial counterparts. The life cycle of most marine invertebrates (as well as most ray-finned fish) includes larval stages with very different morphology, feeding habits, and lifestyle from the adult stage (Pechenik, 1999). Another typical aspect of many marine invertebrate life cycles is a prominent storage of energy to deal with seasonal variations in food availability and to fuel periodic spawning events (Giese, 1959; Lee et al., 2006). These major changes in morphology, ecology, and composition over the life cycle require careful consideration in our attempts to model the uptake and effects of toxicants.

This chapter first presents the basic principles of model construction, the evaluation of their usefulness, and the confrontation to data. Subsequently, it discusses the most important models for the field of ecotoxicology, at the level of the individual organism (higher levels of organization are briefly touched upon in Section 3.7). The focus is on concepts rather than on mathematical details; a firm grasp of the concepts is extremely important for biologists to be able to read and interpret modeling studies, and also essential before diving into the mathematics and coding. Furthermore, understanding the concepts is needed to design experimental tests in such a way that they can contribute to mechanistic modeling work. In this chapter, the amount of mathematics is therefore restricted to a minimum and mainly used to illustrate the general principles. This chapter does not attempt to provide a full review of all the models (or model types) that are, or have been, used in ecotoxicology; instead, it focuses on a few simple examples to illustrate the approach taken, the assumptions made, and their usefulness. A supporting website is set up for this chapter (http://www.debtox.info/marecotox.html) with more information and presenting links to further texts and software (including Matlab files to reproduce the case studies).

3.2 GENERAL MODELING PRINCIPLES

Before we can go into specific models for (marine) ecotoxicology, it is of utmost importance to consider some general modeling principles. This section presents these principles; in the next section (toxicokinetics), they are demonstrated using a simple toxicokinetic model.

3.2.1 Systems and States

In modeling terms, a "system" is basically a set of interacting components forming a unity, with boundaries separating it from the rest of the world. In biology, systems can be individual organisms, but also organs or cells within individuals, or populations of individuals, or entire ecosystems. For this chapter, the individual organism is the system of choice, because the focus is on the uptake and effects of chemicals on individuals.

The state of a system is specified by its state variables, a relevant property of the system that can change (generally over time). For example, the internal concentration is a state variable in TK models. To predict the future development of a system, we need to know the current value of the state variables, but not their history. These variables thus fully capture the current state of the system, at least, as far as deemed necessary given the purpose of the model. The selection of appropriate state variables is thus a critical step in model design; for the same system, different research questions may well lead to different sets of state variables.

In general, the change in the state variables over time depends on the current value of the states. Therefore, dynamic models are usually formulated in the form of Ordinary Differential Equations (ODEs); equations where the derivative (the change in a state) is a function of the value of the state itself.

3.2.2 The Role of Assumptions

Modeling does not start with mathematics, but with the identification of a scientific problem and the formulation of a research aim. Based on observations of the system to be modeled (eg, from the literature), a set of simplifying assumptions is formulated. To be useful in a scientific setting, these assumptions should represent simplifications of the mechanisms that we assume to underlie the behavior of the system. Clearly, biological systems are complex, and simplifying this complexity in a useful manner is a considerable intellectual task.

Models thus follow from simplifying assumptions about (biological) reality; useful and well-described models follow *uniquely* from a clear and consistent set of assumptions. Unfortunately, models are often presented in the form of equations only, and it is left to the reader to reconstruct the underlying assumptions that are made or implied. This practice unfortunately hampers the acceptability of models (and their results) in scientific and regulatory settings.

3.2.3 Model Complexity and the Need for Generality

Models are simplifications of complex systems. However, complex systems do not necessarily require complex models. Model complexity should be closely linked to the purpose of a model and the information available to parameterize it, and only to a much lesser extent by the (perceived) complexity of the system itself. Complex models are difficult to test (errors might easily go undetected) and require a lot of information to parameterize. And, more importantly, they will teach us very little about the system that we are modeling. The general strategy should thus be to start as simple as possible and only include more detail if absolutely necessary for the purpose at hand.

Another aspect that should drive model design and model complexity is the degree of generality that is to be achieved. A model for the effects of chemical A on species B under the set of environmental conditions C could include a lot of detail on A, B, and C and might thereby easily become very complex. However, developing a new model from scratch for each permutation of A, B, and C would be an inefficient use of time and resources; furthermore, the models (and their results) would be impossible to compare. Therefore, there is a lot to be gained by looking at what species and chemicals have in common, rather than focusing on their unique details.

An important guiding principle in model design is the selection of appropriate scales, both in terms of space and time. Combining very different scales into one model is inefficient and bound to lead to problems. For example, there is little to be gained by modeling processes at the molecular scale, which play out at the scale of milliseconds and nanometres, to explain effects on the life history of a multicellular animal (playing out at a time scale of days to years, and a spatial scale of millimeters to meters).

3.2.4 Mechanistic vs Descriptive Models

The distinction between mechanistic and descriptive models is not a straightforward one. In ecotoxicology, dose–response curves and statistical testing (see Chapter 2, Statistics) are clearly descriptive. Such approaches are used to *describe* the data as they are, and we therefore do not learn much from their application about the underlying mechanisms. Furthermore, they do not allow for useful extrapolations beyond the conditions of the experimental test. For these reasons, summary statistics resulting from descriptive methods (such as the EC*x* and NOEC, see Chapter 2, Statistics) have limited usefulness for science and risk assessment (Jager, 2011). A mechanistic model should be able to provide an *explanation* for the patterns, which should (at least in principle) provide a platform for extrapolation beyond the test conditions (eg, from constant to time-varying exposure, and from ad libitum to limiting food availability). However, if we go deep enough, all mechanistic models will include descriptive elements, and the powers of extrapolation will have limitations in practice.

3.2.5 Compartment Modeling and Building Blocks

An important modeling principle that we can use to simplify biological complexity is the compartment approach. A compartment is a model component, generally with well-defined boundaries, that can be assumed homogeneous or "well mixed." This approach has a long history in modeling the environmental fate of chemicals (see MacLeod et al., 2010) and in toxicokinetics (see Barron et al., 1990). In the toxicokinetics example that will follow in Section 3.3, we treat the organism (or more precisely: its internal concentration) as a single compartment. Compartments can have clear physical boundaries (such as an individual organism or an organ) but may also represent different pools within the same physical boundaries (eg, a parent compound and its metabolite can form two compartments within one organism).

Models can be made up of a single compartment or of multiple compartments that are connected in some way. In connecting models, it is essential to distinguish between extensive and intensive quantities. If you divide a system into two, its extensive quantities (such as its mass or the moles of a chemical it contains) will also be divided by two. Its intensive quantities (such as the concentration of a chemical it contains) will stay the same (assuming that the system is homogeneous). When extensive quantities are transported from one compartment to another, we need to consider the conservation laws: when 1 mg of chemical is transported out of compartment A to compartment B, 1 mg of chemical should arrive in compartment B. This may sound trivial, but this is a point where mistakes are often made. In ecotoxicology, we are generally more interested in chemical concentrations in organisms than in chemical masses. However, the conservation laws do not hold for intensive quantities such as concentrations: when compartment A decreases by 1 mg/L as a result from transport to compartment B, it is not necessary that the concentration in compartment B increases by 1 mg/L (this would only hold when the volumes of A and B are equal). Therefore, it is often a sensible strategy to start building models on mass basis, and move to concentrations later (by dividing masses by the appropriate volume or mass of the compartment).

There is a wealth of mathematical functions to choose from for describing the processes affecting state variables. However, most ecotoxicological models are constructed from a small set of building blocks, as summarized in Table 3.1. Borrowing the terminology from chemistry, these are called "kinetics" of a certain "order" (the power to which A is raised on the right side of the equations in Table 3.1). These simple kinetics form the building blocks of more complex models and can be combined by adding and subtracting them in an ODE (as will be demonstrated in the next section). It is good to realize that, in this context, the terms "kinetics" and "dynamics" can be used as synonyms. However, in a tradition that comes from pharmacology, the terms "toxicokinetics" and "toxicodynamics" are reserved for different processes (explained in Sections 3.3 and 3.4).

TABLE 3.1 The Four Most Common Types of Kinetics in Ecotoxicological Modeling

Kinetics	Explanation	Example ODE
Zero-order kinetics	The change in the state variable is independent of the state variable itself	$\frac{d}{dt}A = k$
First-order kinetics	The change in the state variable linearly depends on the value of one state variable	$\frac{d}{dt}A = kA$
Second-order kinetics (law of mass action)	The change in the state variable linearly depends on the product of two state variables	$\frac{d}{dt}A = kA^2$
Hyperbolic kinetics (Michaelis–Menten kinetics)	The change in the state variable is a hyperbolic function of the value of one state variable	$\frac{d}{dt}A = \frac{kA}{K+A}$

In the examples, A is the state variable, and k and K are parameters that do not depend on the state variable. Note that the interpretation (and the units) of k will not be the same in each case.

3.2.6 Translation into Mathematics and Model Testing

For a useful application of models, the assumptions have to be translated into mathematics, and often subsequently into computer code. These two steps are more technical and are not dealt with here in any detail. However, the model can still be put to the test without diving into the mathematics or coding the equations in a software. It is generally useful to start with the model assumptions (if they are explicitly provided): if the assumptions are unrealistic, there is no need to go into the equations and code anyway. Assumptions that violate the laws of physics or are inconsistent with observations are unlikely to produce useful models. Conversely, if a model produces unrealistic outputs, it makes sense to go back to the underlying assumptions (after ensuring the correctness of the translation into equations and computer code).

A quick and useful check on the correctness of a model equation is to perform a dimension analysis. We can only add or subtract quantities that have the same dimension (and units), and if we multiply or divide quantities, we also divide their dimensions (and units). Both sides of an equal sign should lead to the same dimensions (and units), and arguments of transcendental functions (such as exponentials and logarithms) should be dimensionless. A model that fails a dimension analysis is not a useful mechanistic representation of reality.

In many cases, obtaining a better understanding of model behavior is possible by looking at the differential equations themselves. Differential equations specify how a state changes as a function of the value of the state variable and the model parameters (and additional forcings). If the change in a state can be zero, this implies that there is an equilibrium for this state, and we can analyze whether this equilibrium is stable or not (whether the system will return to the equilibrium state after a small perturbation).

3.2.7 Confronting Models With Data

In ecotoxicology, models will often be confronted with experimental or monitoring data. This can be done in a calibration process (to fit model parameters to observations) or in a validation attempt (comparing model predictions to independent observations). First, it is important to distinguish between quantal (discrete) and graded (continuous) responses (Chapter 2, Statistics). In a quantal response, only the presence or absence of a response is determined on each individual (eg, whether it is dead or alive). As a consequence, the LC50 thus represents the concentration at which half of the test population is expected to die (after a specified exposure duration). In contrast, a graded response implies that the degree of response in each individual is determined (eg, the number of eggs produced per female). An EC50 for reproduction thus represents the predicted concentration at which the average female shows a 50% reduction in egg production. These types of responses require different models and different statistical procedures in their confrontation with data.

In model calibration, we need to judge the deviations between model and observations, such that we can prefer one fit over another and find the parameters that yield the "best possible fit" to the data. Deviations are inevitable as observations contain error and the model is always "wrong" (as it is a simplification of reality). We thus also need a model for these deviations (or residuals) and that is the area of statistics. The most-commonly applied statistical model is to assume that the deviations are random samples from independent normal distributions with mean zero and a constant variance. This set of assumptions underlies the popular least-squares analysis. However, in many practical cases, these assumptions will be violated. Quantal responses, such as survival, are discrete responses; assuming that the residuals follow a normal distribution is therefore a questionable practice (especially with small numbers of individuals). Body weight of an organism is a graded response, and a normal distribution may be more appropriate. However, a constant variance is questionable (large body weights tend to show more variation), and if we measure the same individuals over time, independence is violated. To what extent such violations affect the interpretation of a data set is not always clear. In any case, it is good to explicitly mention the assumptions made for the statistical model, just as for the dynamic model that describes the process.

When we have selected a statistical model for the residuals, we can fit the model to the data and calculate confidence intervals on the parameter estimates. It is essential to realize that a confidence interval is only meaningful when both the process model and the statistical model are meaningful (the interval is "true" given that the models are "true"). In practice, no model is ever "true," which implies that all confidence intervals should be regarded as approximations. A wide confidence interval can mean several things: the parameter is not (substantially) influencing the model output, or the data provide no (or insufficient) information on this parameter, or the parameter is heavily correlated to another one (fixing the other parameter could yield a tight interval for the first parameter). A tight confidence interval implies that none of these issues are at play. Several techniques are available to construct confidence intervals (see Chapter 2, Statistics), which may yield different intervals.

Modelers will often be confronted by the question of whether their model is "validated." But what is meant by a validated model? Model validation is the process by which model outputs are (systematically) compared to independent real-world observations to judge the quantitative and qualitative correspondence with reality. A validated model is therefore a model for which this comparison has been performed, but it says nothing about the outcome. There is no strict distinction between models that are, and those that are not, "valid"; much will depend on the purpose for which the model is to be used.

3.3 TOXICOKINETICS

The term toxicokinetics (TK) is used to capture the collection of processes by which organisms take up chemicals, distribute them throughout their bodies, biotransform them to other compounds, and ultimately remove them by excretion from the body. TK models attempt to explain/describe internal concentrations in an organism as a function of exposure level and exposure time. This form of modeling is well established in ecotoxicology, largely influenced by the models used in environmental chemistry (ie, fate modeling). The connection to fate models is a natural one because TK also deals with transport and distribution processes. Even though organisms are modeled instead of environmental compartments, most of the processes involved are conceptually similar. For example, diffusive processes play an important role in both model types, and biotransformation in organisms is (from a modeling perspective) equivalent to biodegradation in an environmental compartment.

TK models generally consist of one or more compartments that exchange chemical with the environment and with each other. The transport mechanisms that we should consider include diffusion (eg, uptake and elimination of dissolved chemicals through the skin or gills), advection (eg, uptake of chemicals into the gut with food), and active transport (eg, using transporter enzymes to actively excrete chemicals). Additionally, chemicals may be biotransformed by organisms, yielding metabolites with different properties from the parent compound (which might necessitate more compartments; see Chapter 4, Bioaccumulation).

3.3.1 The One-Compartment Model With First-Order Kinetics

The simplest useful TK model is the one-compartment model with first-order kinetics. This model will be used here to introduce TK methods but also to make the abstract modeling principles from the previous section more concrete. This is the only model that will be dealt with in this chapter in this level of detail.

Consider an individual organism that exchanges a chemical with its aqueous environment. The system of interest is now the individual organism, and the only state variable that we follow for this system is the internal chemical concentration. By making these choices, we have already made a crucial simplifying assumption, namely that the organism can be treated as a single well-mixed compartment. In reality, an organism is obviously more complex, but this simplification often works surprisingly well in practice. Furthermore, we assume for now that the volume of the organism is constant (ie, no growth or shrinking). For the exchange of the chemical, the simplest useful set of assumptions is that the uptake flux of chemical from the environment into the organism is proportional to the external concentration, and the elimination flux of chemical from the

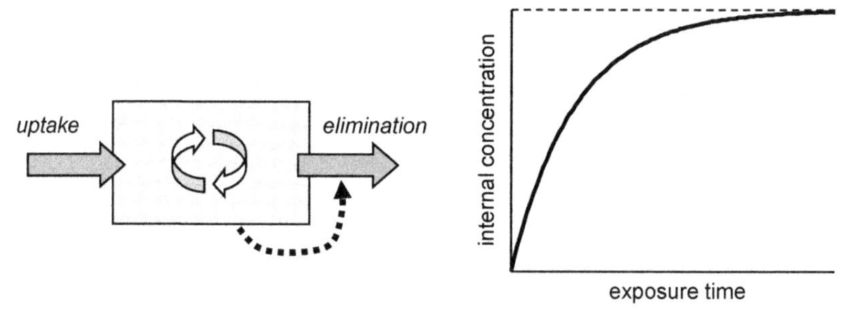

FIGURE 3.1 Schematic representation of the one-compartment model. *Thick arrows* embody chemical transport, whereas the *broken arrow* indicates the influence of the internal concentration on the elimination flux (the essence of first-order kinetics, see Table 3.1). The plot shows the general time course for the internal concentration, given first-order kinetics and constant exposure. The *broken line* indicates the equilibrium level.

organism to the environment is proportional to the internal concentration in the organism (first-order kinetics, see Table 3.1 and Fig. 3.1). Another common simplification is to assume that the outside world is infinitely large and also well mixed. As a result, the uptake by, and elimination from, the organism has no effect on the external concentration. For an organism in the sea, this assumption is safe to make, but for an experimental test, its applicability needs to be carefully considered. Alternatively, the external medium should have its own compartment: the system of choice is then the test container including the animal(s), and the system now has (at least) two state variables.

The assumptions for the simple one-compartment model can be translated into a differential equation for the internal concentration C_i:

$$\frac{dC_i}{dt} = k_u C_w - k_e C_i \quad \text{with } C_i(0) = 0 \qquad (3.1)$$

Here, C_w is the external concentration (the forcing of the system), and the model parameters are the uptake rate constant k_u and the elimination rate constant k_e. The change in internal concentration (the derivative dC_i/dt) depends on the current value for the internal concentration (the state C_i). This ODE is thus a combination of a zero-order term ($k_u \cdot C_w$) and a first-order term ($k_e \cdot C_i$), as explained in Table 3.1. The differential equation itself only specifies how the internal concentration *changes* over time; it says nothing about the actual internal concentration at a given time point. For that, we need to specify the value of the state variable at one time point, for example $t = 0$, as was done here (the additional assumption is now that the organism starts at $t = 0$ with $C_i = 0$).

From this ODE, we can already determine a few properties of our system. Initially, C_i is zero, so the positive part of the ODE ($k_u \cdot C_w$) will be larger than the negative part ($k_e \cdot C_i$), thus the derivative will be positive (ie, the internal concentration will start to increase). Next, we can ask whether there is an equilibrium situation in this system. An equilibrium implies that there is no change in the state variable C_i, and hence that $dC_i/dt = 0$. We can easily see from Eq. (3.1) that there is an equilibrium for:

$$C_i = \frac{k_u}{k_e} C_w \qquad (3.2)$$

The ratio k_u/k_e is also known as the bioconcentration factor; it specifies the concentration ratio between inside and outside in equilibrium. By looking at the sign of dC_i/dt, we can see that this equilibrium is stable: if C_i is below the equilibrium value, the derivative is positive (C_i will increase over time), and when C_i is above the equilibrium value, the derivative will be negative (C_i will decrease). Thus, small perturbations from equilibrium will yield a derivative that will bring the system back to equilibrium. In chemistry, a distinction is usually made between equilibrium (ie, thermodynamic equilibrium) and a steady state (no change in a state variable despite various processes continuing to affect it). In mathematics, this distinction is not made. This chapter uses the term "equilibrium," even though in some cases, "steady state" may be more appropriate.

Instead of using two rate constants, we can rewrite the ODE of Eq. (3.1) using the bioconcentration factor (K in L/kg):

$$\frac{dC_i}{dt} = k_e(KC_W - C_i) \quad \text{with } C_i(0) = 0 \tag{3.3}$$

This specification is mathematically fully equivalent to Eq. (3.1); which one to choose is largely arbitrary. If the chemical exchange is thought to be driven by passive diffusion, the second formulation is more natural (chemical exchange is proportional to a concentration difference). However, when the uptake and elimination fluxes are separate processes (eg, mediated through active processes), the first formulation is preferred.

Next, we need to perform a dimension analysis to see if there are no obvious consistency problems. We can do this by checking the units of all elements in the ODE. Suppose that C_i has the unit mol/kg, then dC_i/dt has the unit mol/kg/day (if we use days as the unit for time). To match the units on both sides of the equality in Eq. (3.1), k_e needs to have the simple unit of day^{-1}. If we take mol/L as unit for C_w, the unit of k_u should be L/kg/day. Clearly, the uptake rate constant is of a different nature than the elimination rate constant. One might suggest that 1 L of water weighs 1 kg, and hence simplify the units of k_u to day^{-1}. However, 1 L of water is not comparable to 1 L of organism; these two properties do not cancel out in a division. Even if we use mol L^{-1} as unit for C_i, it is important to specify the units of k_u as $L_{water}/L_{organism}$/day.

Once values for the model parameters have been established, and forcing of the system specified (the external concentration in the water, which may be time variable), we can predict how the system moves from one state to the next. To do that, an implementation of the model into software is generally needed. If the external concentration is constant (and the parameters as well), this ODE can be solved analytically to:

$$C_i(t) = \frac{k_u}{k_e} C_W \left(1 - e^{-k_e t}\right) \tag{3.4}$$

How this equation follows from the ODE in Eq. (3.1) is outside of the scope of this chapter; more background can be obtained from textbooks on mathematical modeling (eg, Doucet and Sloep, 1993). However, it is easy to check whether this is indeed a solution to Eq. (3.1) by calculating the derivative dC_i/dt from Eq. (3.4) and confirm that the equality in Eq. (3.1) indeed holds.

From Eq. (3.4), several observations can be made. The entire dynamics of the system is dictated by the part between parentheses, and thus by k_e. The elimination rate constant thus determines how long it takes to reach a certain percentage of the equilibrium situation. The uptake rate constant (k_u) only determines the final internal concentration (together with k_e and C_w) through the part before parentheses. This finding underlines the different nature of both rate constants (as was already seen from their units), and the importance of k_e for the dynamics of the system (which is why it can easily be used as the sole rate constant in Eq. (3.3)).

In this explicit form of Eq. (3.4), the model can be easily implemented into a range of software programs (including spreadsheets) for analysis and fitting to data. The option to solve an ODE to an analytical solution is only available for the simplest of models. In practice, we are rapidly forced to turn to more specialist software that includes solvers to approximate the solution of the ODE numerically.

3.3.2 Beyond the Simple One-Compartment Model

For many applications, we need to consider extensions of the simple model of Eq. (3.1). These extensions may include more compartments or different types of kinetics of the compartment(s). An extension with more compartments is needed when we cannot view the organism as a single well-mixed compartment, which is generally judged by a failure of a one-compartment model to explain observed kinetics. We might use multiple compartments for the organism as rather abstract "central" and "peripheral" compartments, or to represent different tissues in the organism (see Barron et al., 1990). We might also want to consider various chemical pools (eg, an inert and active pool, or a parent compound and metabolites), which would also require additional compartments. The most elaborate multicompartment models take the form of "physiologically based" TK models (PBTK), which define compartments for specific organs (or groups of organs) and connect them with a blood flow (eg, Nichols et al., 1990). Such models have only been constructed for a few animal groups (mammals, birds, and fish), and improvement in predictive capabilities may not always outweigh the additional complexity. It is tempting to include more and more biological and toxicological realism into models, but it is a good strategy to include more compartments only if it is absolutely needed to capture the dynamics of the system. If the chemical exchange with the environment is slow, relative to the internal distribution processes, the entire organism will behave like one well-mixed compartment. In such a situation, using a PBTK model will only add more parameters without improving performance.

When building a multicompartment model, it is of utmost importance to ensure that there is a mass balance: the number of molecules of a chemical that leaves compartment A to compartment B, must also arrive there (see Section 3.2.5). Therefore, it is generally advisable to start building the model on a mass basis (ie, making chemical mass in each compartment the state variables), and only recalculate to concentrations at a later stage.

Many marine organisms build up a storage to survive seasonal periods with adverse conditions and/or to fuel periodic spawning events. The buildup and use of such a storage has consequences for TK as it influences body size as well as body composition (and thereby the affinity for chemicals). Whether or not this storage requires its own compartment depends on the exchange rates with the rest of the body: if this exchange is fast (relative to exchange with the environment), the whole organism still behaves like one compartment (albeit with parameters that vary with the amount of storage). An example of such a more complex one-compartment model was presented for mussels by Van Haren et al. (1994). At spawning, part of the stored mass will leave the body with the gametes, taking with it a part of the chemical body burden (maternal transfer). Thus, spawning can constitute an important elimination route for hydrophobic compounds. A conceptual diagram of a model including such a storage is provided in Fig. 3.2.

Other extensions incorporate a deviation from first-order kinetics. Consider, for example, the case where uptake or elimination of a chemical is taking place using active transport (ie, mediated by transport enzymes). For small chemical concentrations, the rate of transport is proportional to the concentration driving it: every molecule will find a free enzyme, so the absolute transport rate (in molecules per time) is limited by the availability of the chemical (and the handling time of the enzyme, which we assume is constant). At high concentrations, however,

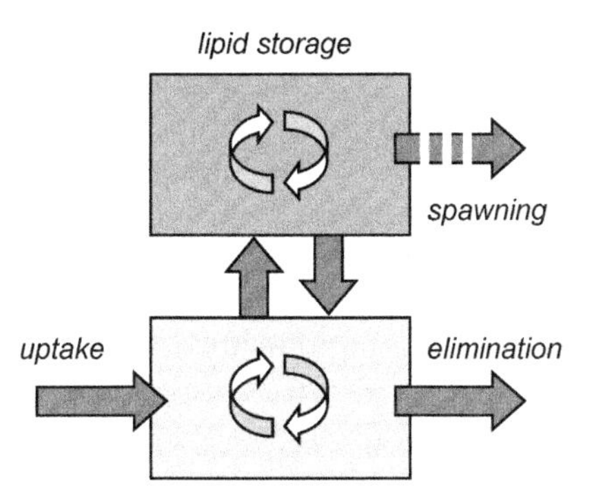

FIGURE 3.2 Schematic diagram for a TK model that includes a lipid storage and losses due to periodic reproduction. *Arrows* represent transport of a chemical.

the number of transport enzymes may not be enough to keep up with the supply of toxicant molecules, and the transport rate saturates: a twofold increase in chemical concentration will produce less than a twofold increase in transport rate as the number of enzymes is limiting the transport rate. In those cases, a hyperbolic relationship (Michaelis–Menten kinetics) may be more appropriate (Table 3.1).

A different type of extension considers changes in size of the compartment(s). Obviously, animals may grow or shrink in size, which has consequences for the toxicokinetics. The first consequence is that an increase is body size has a negative effect on the internal concentration (dilution by growth). Similarly, a decrease in body size will concentrate the chemical in the body. A change in size will, however, generally also affect the values of the uptake and elimination rate constants, as these parameters depend on the surface:volume ratio of the organism (Gobas et al., 1986). If the organism does not change in shape over its life cycle, the surface area will scale with volume to the power 2/3. Therefore, the surface:volume ratio decreases with volume to the power 1/3 and hence proportional to the length of the organism. This can be effectively demonstrated by assuming a cube-shaped organism. The volume is given by a^3 (where a is the length of an edge) and the surface area by $6a^2$. The surface:volume ratio is $6/a$, and thus inversely proportional to the length of an edge. The inverse proportionality to length also holds for other shapes, as long as the organism does not change shape with growth, but the proportionality constant obviously depends on the shape and on what length measure is used.

Taken together, the growth dilution and change of surface:volume ratio will lead to a general extension of the one-compartment model of Eq. (3.3) (Kooijman and Bedaux, 1996; Jager and Zimmer, 2012):

$$\frac{dC_i}{dt} = k_e \frac{L_m}{L}(KC_w - C_i) - \frac{C_i}{V}\frac{dV}{dt} \quad \text{with } C_i(0) = 0 \tag{3.5}$$

The last term accounts for growth dilution as body volume (V) changes; it can be derived from a model for chemical mass and the standard rules for differentiation. If there is no growth, dV/dt is zero, so this factor disappears. If growth is exponential, the relative growth rate (dV/dt divided by V) is constant, and growth dilution has a similar effect as elimination (and can also be represented by a rate constant, see Chapter 4, Bioaccumulation). The elimination rate is modified in Eq. (3.5) by a factor L_m/L. In this way, the effective k_e decreases with increasing body length L, and the numerical value of k_e can be interpreted as the elimination rate at maximum body size L_m.

3.3.3 Case Study

The one-compartment TK model with first-order kinetics is illustrated using data on accumulation and depuration of the pharmaceutical tetrazepam in the mussel *Mytilus galloprovincialis* (Gomez et al., 2012). The experimental design included an exposure phase of seven days, followed by a 7-day depuration in clean water. The fit in Fig. 3.3 (with parameter estimates in Table 3.2) was produced by simultaneous fitting

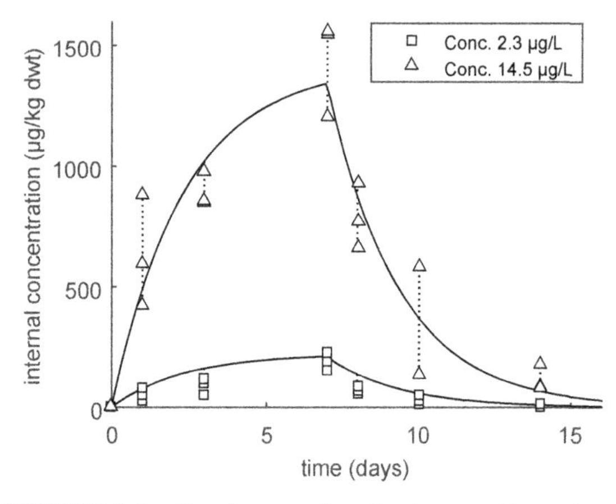

FIGURE 3.3 Simultaneous fit to the tissue residues of tetrazepam in *M. galloprovincialis* at two exposure levels. Exposure during the first seven days, followed by a depuration phase in clean water.

TABLE 3.2 Parameter Estimates From the Fit in Fig. 3.3, With 95% Likelihood-Based Confidence Intervals

Parameter	Estimated value (95% CI)
Uptake rate constant (k_u)	47 (38–56) L kg^{-1}dwt day^{-1}
Elimination rate constant (k_e)	0.48 (0.39–0.59) day^{-1}

of Eq. (3.1) on the tissue residues at both exposure concentrations. The fit can be improved by allowing a different value for the uptake rate constant k_u at high and low exposure (yielding a lower value at the low exposure). Depending on the purpose of the analysis and the confidence in the various measurements, one may decide to use the simultaneous fit or to fit both sets separately (as was done by the original authors).

For the statistical model, the fit was made assuming independent and normally distributed deviations between the model and data (with constant variance). Measuring body residues is a destructive procedure, so independence of the observation is a reasonable assumption. Internal concentration is a continuous variable, so a normal distribution is a good place to start. A constant variance is, however, questionable. The variance of concentration measurements will probably increase with increasing value of the mean. To accommodate this, one could apply a variance that varies with the mean (eg, by adding weights to the residuals) or apply a transformation (eg, log or square-root). It should be stressed that transformation also transforms the shape of the distribution; assuming a normal distribution for the residuals on a log scale implies a log-normal distribution on the normal scale.

3.4 GENERAL INTRODUCTION ON TOXICODYNAMICS

Toxicokinetic models can of course be used by themselves (eg, to study accumulation in a food chain), but they can also be paired with other models to interpret and predict toxic effects. Toxic effects are generally not caused by an external concentration directly, but only after the chemical has been taken up into the organism and transported to a target site. The term toxicodynamics (TD) is used for the collection of processes that link the internal concentration (at a target site) to the effects on endpoints over time. Hence, the combination of TK and TD models is termed TKTD modeling. In contrast to TK, TD modeling is less well established in ecotoxicology. Traditionally, external concentrations are linked directly to effects on life-history traits at a single time point (represented by the broken arrow in Fig. 3.4), for example using a dose–response curve. TK has been influenced by the approaches used in environmental chemistry, and similar types of simplification as applied so successfully in TK are also feasible for TD.

In TKTD modeling, a TK model is linked to a TD model to provide a dynamic picture of the

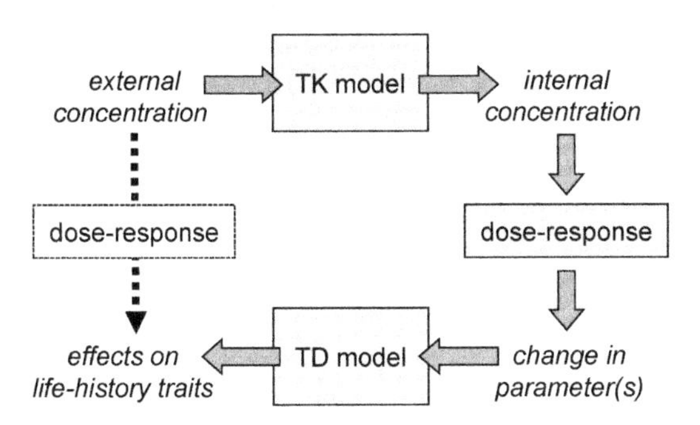

FIGURE 3.4 Instead of linking external concentration to effects directly (*broken arrow*), TKTD models link TK and TD models to provide a causal relationship between external concentrations (which might be time varying) and effects on life-history traits over time.

toxic effects (Fig. 3.4). TKTD modeling has advantages over descriptive dose–response models (Jager et al., 2006; Ashauer and Escher, 2010). Most importantly, they formalize our knowledge about the underlying processes leading to toxicity, and thereby allow us to test that knowledge quantitatively. This is a crucial step to understand, and ultimately predict, differences in sensitivity between species, toxicity between chemicals, and the interactions in mixtures of chemicals as well as between chemicals and environmental stress. TKTD models explicitly incorporate the temporal aspect of toxicity, which means they can be used to interpret and predict the evolution of toxicity over time, and effects due to time-varying exposure (eg, an oil spill). The most useful TD model depends on the questions asked, but also on the nature of the endpoint; a model for a behavioral endpoint will likely differ from a model for effects on reproduction.

3.4.1 Linking TK to TD Models

We assume that it is an internal chemical concentration at a target site inside the organism that causes the observed effects. But which concentration and where? The simplest starting point is to assume that the whole-body concentration of the parent compound is the best metric to link to a TD model. Even if this assumption is not valid, it might still provide a good explanation for the observed effects as long as the "real" dose metric is more-or-less proportional to the whole-body concentration of the parent compound. Using more elaborate TK models provides the possibility to link effects to concentrations in a specific tissue, or to the concentration of a specific metabolite.

Once we have established the most useful dose metric from our TK model, the next question is how to link it to a TD model. We can assume that an increase in the value of the dose metric (eg, the internal concentration) affects one or more processes (read: parameters) in the TD model, which leads to effects that can be observed on the endpoint (see Fig. 3.4). As an example (discussed in more detail in Section 3.6), the internal concentration may be linked to the feeding rate of the organism, and the decrease in energy intake can be translated into effects on growth and reproduction over the exposure duration.

What type of relationship should we take between the dose metric and the affected process? There is very little theoretical and empirical evidence to rely on. In practice, a linear-with-threshold relationship (Fig. 3.5) works well

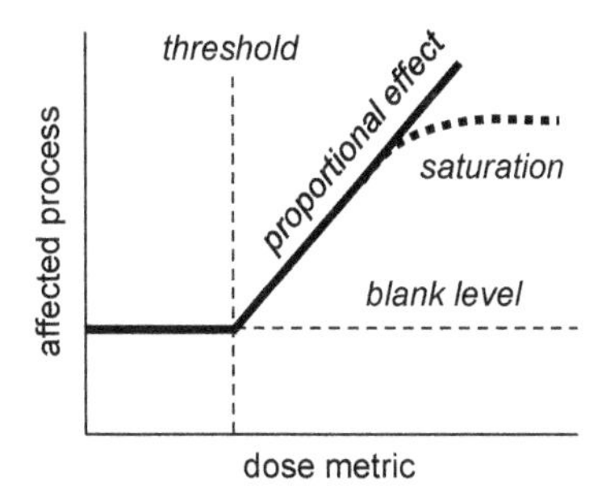

FIGURE 3.5 A linear-with-threshold relationship between the dose metric (eg, the internal concentration) and an affected process (ie, a parameter in a TD model that governs a physiological process). The *dotted curve* illustrates a potential saturating effect at higher doses.

(Kooijman, 1996; Jager et al., 2011) and has distinct advantages. Aquatic organisms are not living in pure H_2O; the medium around them contains a myriad of chemicals, all of them potentially toxic. Every toxicity test is therefore, in essence, a mixture toxicity experiment. To avoid having to deal with all of these chemicals in the "natural background," we can assume that the concentrations of these chemicals are below their respective thresholds, and thus pragmatically ignore their contribution to the toxic effects. A linear relationship between the concentration above the threshold and the affected process makes intuitive sense: every additional molecule of the chemical has the same contribution to the overall effect. At least for small effects, this is a reasonable assumption, but for larger effects, interactions between molecules and the target site may become more complex (eg, leading to a saturation of the effect with increasing dose, see the dotted line in Fig. 3.5).

3.4.2 Using TK Models in the Absence of Body-Residue Data

In many cases, we like to model the effects on an organism over time, but we do not have access to measured body residues. In those cases, we can still apply TKTD models; the development of the toxic effect over time contains information on the time needed to achieve equilibrium (and thus about the elimination rate constant, k_e). However, there is no information about the absolute value of the internal concentration, and hence no possibility to estimate k_u (Eq. (3.1)) or K (Eq. (3.3)) from the effects data alone. We could set K to an arbitrary value, but a more elegant approach is scaling: we can divide both sides of Eq. (3.3) by K, which yields a new state variable for the body residue, the scaled internal concentration C_i^*:

$$\frac{dC_i^*}{dt} = k_e\left(C_W - C_i^*\right) \quad \text{with } C_i^* = \frac{C_i}{K} \text{ and } C_i^*(0) = 0 \tag{3.6}$$

This new state is proportional to the real (but unknown) internal concentration and has the dimensions of the external concentration (in equilibrium, $C_i^* = C_W$). If C_i^* is used as the dose metric in Fig. 3.5, the threshold will also have the dimension of an external concentration. The C_i^* as function of time can subsequently be linked to a TD module. The scaled TK model now only has a single parameter (k_e), which can be estimated from data for effects over time.

3.4.3 Extension With a Damage Compartment or Receptor Kinetics

So far, we have assumed that some internal chemical concentration is responsible for the toxic effect. It may, however, also be the case that chemicals inside the body lead to some form of damage that is repaired at a certain rate. This damage may be the proximate cause for the toxic effect, where exposure to the chemical is the ultimate cause. Damage can be included into TKTD models as an additional compartment with first-order kinetics (Lee et al., 2002; Jager et al., 2011). Depending on the value of the rate constants, the damage compartment can show a different time course than the internal concentration. This may help to explain why the kinetics of the internal concentration does not always match the dynamic patterns in toxicity (Ashauer et al., 2010).

In TKTD models, damage is often treated as something rather abstract that cannot be measured. A more concrete extension is to implement explicit receptor dynamics. We might assume that the chemical, once taken up, binds to a receptor, and the complex is responsible for the toxic effect (Jager and Kooijman, 2005). Binding to the receptor depends on the meeting frequency between the chemical and the free receptors, and it therefore follows second-order kinetics (see Table 3.1). Bound receptors may be released at a rate that depends on the number of bound receptors (first-order kinetics). The behavior of such a receptor model will differ from the first-order damage extension in the fact that it will saturate at higher concentrations (at some point, the number of free receptors will be so small that it limits the binding rate of the chemical).

In general, these extensions are only useful when there are measured body residues available, such that the kinetics of the internal concentration can be fixed. Data on effects over time alone hardly ever allow for the distinction between toxicokinetics and damage/receptor kinetics.

3.5 EFFECTS ON SURVIVAL

3.5.1 Why Do Animals Die?

Any TD model for the endpoint survival should address the question of why animals die when exposed to a chemical stress, or better: why don't they all die at the same time at a given concentration? The most popular explanation for this phenomenon is the "individual tolerance" hypothesis. The assumptions underlying this hypothesis are that an individual dies immediately when its internal concentration exceeds a threshold value and that this threshold differs between individuals in a population according to a certain probability distribution (popular choices are the log-normal or log-logistic distribution). When 50% of the individuals in a toxicity test have died after a certain exposure time, we assume that the animals that are left are the more tolerant ones. This view of the death process is closely linked to the idea of a Critical Body Residue (CBR, see McCarty and Mackay, 1993) in which an adverse outcome (such as death) is associated with exceeding a critical internal concentration. A certain internal concentration is thus expected to yield the same percentage of mortality in a test population, irrespective of the exposure concentration or time. The toxic effect is therefore only time dependent as long as the internal concentration is increasing over time.

An alternative explanation is "stochastic death," which assumes that all individuals are identical but that death itself is best viewed as a stochastic process (see Bedaux and Kooijman, 1994). The internal concentration of a toxicant increases the probability of an individual to die. So when 50% of the animals die in a test, these were not more sensitive than the survivors but just unlucky. It is unlikely that death is inherently stochastic, but the number of processes and variables at play may be so huge that stochasticity can provide a good description (similarly, throwing dice is not an inherently stochastic process, but well described by it). In contrast to the individual tolerance approach, there is no immediate death above the threshold, but an increased probability to die, and hence there is no constant relationship between the actual internal concentration and the percentage of mortality.

It turns out to be surprisingly difficult to choose between these two hypotheses; both tend to describe the patterns in the data well (albeit with different parameter values). In reality, both mechanisms are likely to play a role, and mixed models have been proposed (Jager et al., 2011). Only the stochastic death model is discussed here in more detail. This is sufficient to demonstrate the construction and use of TKTD models for the endpoint survival. The individual-tolerance hypothesis is discussed elsewhere, including some of its unrealistic properties (Kooijman, 1996; Newman and McCloskey, 2000).

3.5.2 The Stochastic Death Model

Central to the stochastic death model is the concept of the "hazard rate"; a statistical principle that is used to describe stochastic "failure" of a system or product over time. The hazard rate, multiplied by a very short time interval, gives the probability of death in that interval, given that the organism is alive at the start of it. A constant hazard rate leads to an exponentially decreasing probability to survive over time. When exposed to a chemical stress, the hazard rate is usually not constant, but increases with increasing internal concentration (at the target site).

Departing from the scaled TK model of Eq. (3.6), we can now define a linear-with-threshold relationship (Fig. 3.5) between the scaled internal concentration (C_i^*) and the hazard rate (h):

$$h(t) = k_k \max\left(0, C_i^*(t) - c_0\right) + h_b \tag{3.7}$$

where c_0 is the threshold concentration and k_k the killing rate (the proportionality between the scaled internal concentration above the threshold and the hazard rate). Background mortality is included through h_b, which may be taken constant if deaths during the toxicity test can be treated as simply random (eg, due to handling or other accidents). For independent causes of death, the hazard rates may be added. Here, the hazard rate is thus treated as a physiological process in the individual organism.

Integrating the hazard rate over time provides the survival probability S over the time course of interest:

$$S(t) = \exp\left(-\int_0^t h(\tau)d\tau\right) \tag{3.8}$$

The full model linking external concentrations to survival probability over time thus consists of three equations (Eqs. (3.6)–(3.8)), with only four parameters to be estimated from the data. This shows that a meaningful TKTD model can be very simple and parameter sparse. Given the amount of detail that is present in routine survival data, fitting more parameters is unlikely to be useful, unless other sources of information are available (eg, body residues or information on other species or chemicals). One particularly powerful approach is to simultaneously fit multiple data sets, assuming some parameters are the same in each set while others are unique for a set (eg, Jager and Kooijman, 2005). This allows the use of more complex models with more parameters, by keeping the number of model parameters per *data set* low.

To fit this TKTD model to data, we need to consider the nature of the deviations (the residuals). Normal independent distributions for the residuals are not the first choice: survival data can only assume discrete values (as each animal is either dead or alive), and the survival probability is bounded between zero and one (while the tails of the normal distribution extend to plus and minus infinity). Fortunately, a better-suited alternative is available in the form of the multinomial distribution (Bedaux and Kooijman, 1994; Jager et al., 2011). This is an extension of the binomial distribution to more than two discrete outcomes of a random sample (here: the time interval in which an individual dies). A thorough treatment of the statistics for fitting survival models is outside of the scope of this chapter; more details can be found elsewhere (Bedaux and Kooijman, 1994; Jager et al., 2011).

3.5.3 Case Study

To demonstrate the stochastic death model defined above, an example is provided for the boreal marine copepod *Calanus finmarchicus* exposed to mercury (Øverjordet et al., 2014). Copepodite stage V was used for this experiment, which is the final developmental stage before adulthood. The data set comprises observations on survival over four days, with seven exposure treatments (30 animals each) and a control (60 animals). The TKTD model outlined in the previous section was fitted to this data set (Fig. 3.6); the resulting parameter estimates are provided in Table 3.3. The statistical model to judge the deviations between model and data is formed by the multinomial distribution (Jager et al., 2011). This distribution is an excellent match to deal with discrete random events over time. The resulting likelihood function was maximized to find the best-fitting parameter values. Further, the likelihood function can be profiled to construct robust confidence intervals on the estimates (Meeker and Escobar, 1995).

The model does a reasonable job in explaining the observed survival patterns over time, although the fit is certainly not perfect. Since the model is a very simple one, it is tempting

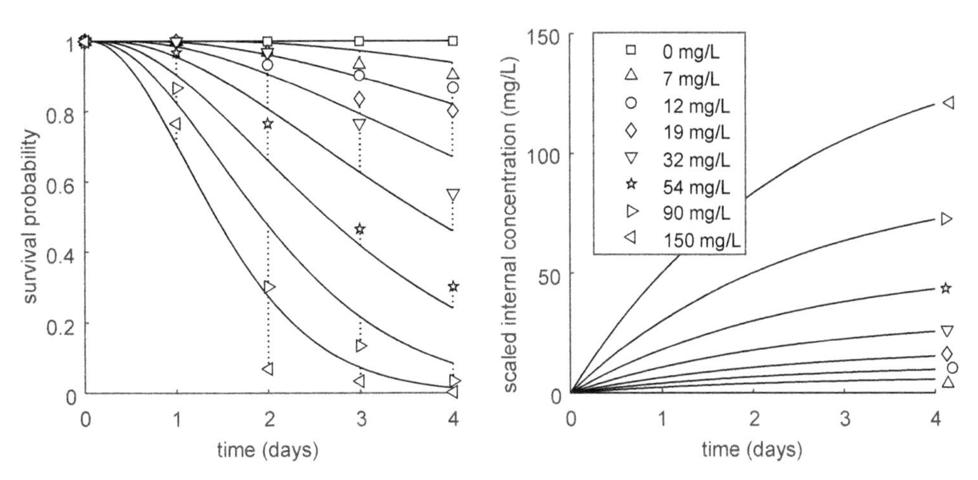

FIGURE 3.6 Fit of the TKTD model for survival to data for the marine copepod *C. finmarchicus* exposed to mercury (left panel). Right panel shows the predicted scaled internal concentration, as implied by the fit on the survival data. Parameter estimates are provided in Table 3.3.

to now go back to the underlying assumptions, modify them, and extend the model accordingly to provide a better fit. However, we need to realize the nature of the data. We basically need to estimate survival probabilities from the observed death frequencies in a test population. Repeating the experiment might lead to a very different outcome, simply because of the stochasticity in the death process. Therefore, model extensions should be considered very carefully and are most promising when the test comprises a large numbers of individuals. In any case, it makes sense to also try the alternative model (individual tolerance) on this data set; this stage of copepodites can be variable in lipid content, which may translate into interindividual differences in sensitivity (Hansen et al., 2011).

The toxicokinetics of mercury were derived from the survival pattern over time (no body residues were determined in the experiment). Only one single TK parameter was used, the elimination rate (k_e), which determines the time needed to reach equilibrium. Since the scaled TK model (Eq. (3.6)) was used, the scaled internal concentration has the dimensions of the external concentration, and in equilibrium, its value will equal the external concentration (see right panel of Fig. 3.6). For the same reason, the parameters for the threshold (c_0) and killing rate (k_k) also have external concentration in their units (see also Fig. 3.5).

TABLE 3.3 Parameter Estimates From the Fit in Fig. 3.6, With 95% Likelihood-Based Confidence Intervals

Parameter	Estimated value (95% CI)
Elimination rate constant (k_e)	0.41 (0.034–0.92) day^{-1}
Threshold for effects (c_0)	2.9 (0.35–4.9) mg/L
Killing rate proportionality (k_k)	0.015 (0.0087–0.11) L/mg/day
Background hazard rate (h_b)	0 (0–0.030) day^{-1}

3.6 EFFECTS ON SUBLETHAL ENDPOINTS

3.6.1 TD Models for Sublethal Effects

For sublethal effects, such as those on growth and reproduction, we cannot use the same TD models as for survival. For survival, we could simply describe death as a stochastic process and link the internal concentration to a hazard rate. Clearly, growth and reproduction cannot be treated as stochastic events, as they are graded responses: we observe the degree of response on each individual (see Section 3.2.7). Therefore, we need a TD approach that includes a representation of the mechanisms by which organisms grown and reproduce. A useful starting point for such models is provided from the perspective of an energy budget. Animals have to obey the conservation laws for mass and energy. When we observe a reduction in reproductive output as a consequence of chemical exposure, this implies that less energy is lost in the form of offspring compared to the control. Where did this energy go? It might be that this energy was never assimilated in the first place (ie, a reduction in the feeding rate or in the assimilation efficiency of energy from food) or that the energy was used by the stressed animal in a different manner (eg, to mediate the negative effects of chemical stress). Following the flows of energy is a natural approach when effects are observed on endpoints that constitute a sink for energy such as growth and reproduction. Clearly, there are also endpoints for which the energy budget might not be such a useful starting point, such as changes in behavior and developmental malformations.

Energy-budget models specify how energy and building blocks are taken up from the environment (through feeding, in animals), and how they are used to fuel the resource-demanding processes (Fig. 3.7). Central to these models is

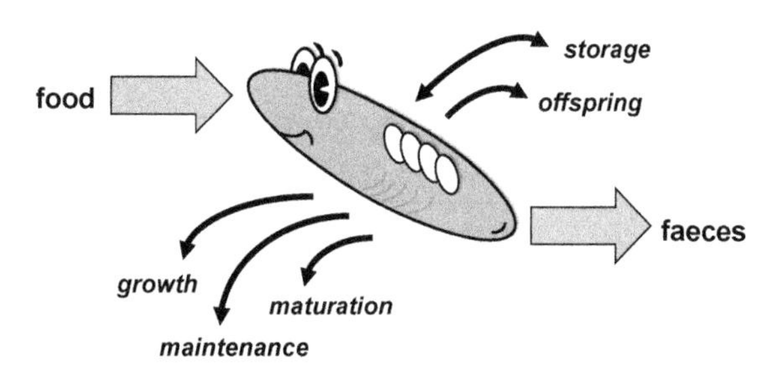

FIGURE 3.7 Principle of energy balance for an animal. Food is taken up; part of the energy in the food is assimilated and used for various processes as shown.

the conservation of mass and energy; we basically have to do the book-keeping so that all mass and energy is accounted for. Furthermore, we have to be consistent with the second law of thermodynamics, in the sense that no transformation of mass or energy can be 100% efficient; inevitably, losses occur in each transformation step.

In the simplest case of an energy budget, one can attempt to measure these fluxes to create a closed mass/energy balance. Often, ingestion, defecation, and respiration are measured and the difference between assimilated and dissipated energy is referred to as "scope for growth" (eg, Widdows and Johnson, 1988). In this context, "growth" thus refers to both body growth and reproduction. The limitation of this approach is that it relies heavily on measurements of respiration, feeding, and defecation (which are not routinely done in toxicity testing) and that it is descriptive (the fluxes need to be measured at each point in time). To use the energy perspective in a TD model, a dynamic and more mechanistic approach is useful.

3.6.2 Dynamic-Energy Budgets

Dynamic Energy Budget (DEB) theory is at the moment the most comprehensive and best-tested framework for energy-budget modeling (Nisbet et al., 2000; Kooijman, 2001). A gentle introduction into the concepts of DEB theory and its application in ecotoxicology can be found elsewhere (Jager, 2015b). A range of practical models has been derived from this theory, including dedicated models for ecotoxicological applications, often referred to as "DEBtox" (Kooijman and Bedaux, 1996; Billoir et al., 2008; Jager and Zimmer, 2012). These models use easy-to-interpret compound parameters (such as maximum reproduction rate), which are combinations of the underlying primary parameters that directly link to energetic processes (such as the feeding rate or the specific maintenance costs). The models are relatively simple to use and the parameters easy to interpret, but the use of compound parameters lacks transparency and easily leads to errors (Billoir et al., 2008; Jager and Zimmer, 2012). Furthermore, model extension (eg, to other endpoints such as respiration or inclusion of lipid storage) is hampered by the lack of an explicit mass balance. A detailed description of a simple DEB model with explicit mass balance can be found elsewhere (Jager, 2015a).

Energy-budget models represent an enormous simplification of an organism, and inevitably many biological "details" are lost. The level of simplification is, however, comparable to that so successfully applied in TK models. This implies that we have the possibility to parameterize and use this model on the basis of the (relatively) limited data sets that are available in ecotoxicology. Departing from DEB theory, there is no need to build a new model from scratch for each species. All organisms are related through evolution, and similar organisms share a similar structure for metabolic organization. Animals form, metabolically speaking, a rather homogeneous group, as they all obtain their resources from feeding on other organisms. That being said, for many species and many applications, we need to consider modifications of the standard DEB models. For example, development with larval stages is common in the marine environment. The larvae of many species are quite different from the adult form, require different food, and may also metabolically differ. For example, most larvae seem to have a lower metabolic rate compared to the later stages, and some form of acceleration takes place over the life cycle (Kooijman, 2014). Another modification to consider for marine organisms is the seasonal buildup of a storage buffer to survive periods of food scarcity and/or to fuel spawning events. Such a storage can likely be viewed as a "reproduction buffer" in the DEB context (Saraiva et al., 2012; Jager and Ravagnan, 2015), although the rules for its build-up and use are species specific and require further detailed study. In making modifications, we should always keep in mind that the rules for metabolic organization are subject to natural selection, and thus shaped by evolution. Therefore, we need to come up with similar adaptations for related species.

The advantage of using a DEB model as our TD model is that we can explain toxic effects over the life cycle on all endpoints in an integrated framework. It makes little sense to look at endpoints like growth and reproduction in isolation, as they are closely linked. Reproduction generally starts at a constant body size, and body size affects feeding rates and thereby reproduction rates. A toxicant effect on growth will thus indirectly affect reproduction as well. The interactions between these traits, and with others (eg, respiration and feeding), are an integral part of the energy budget. Another (although related) advantage is that we can use DEB models to make educated extrapolations to untested situations, such as food limitation, different temperatures, time-varying exposure, and mixture toxicity. Opinions can differ on which energy-budget model is most appropriate (for which question), but some form of energy balancing is unavoidable for a TD model dealing mechanistically with effects on energy-demanding traits such as growth and reproduction.

3.6.3 Linking TK to the Energy Budget

For ecotoxicological applications, we can use an energy-budget model as our TD model, and link it to a TK model of our choice. Before considering how TK links to TD, consider the reverse: how do life-history traits of the animal (as affected by toxicants) affect the toxicokinetics? As we are considering sublethal effects in this section, the organisms will be growing, so we should consider the extension for growing animals of Eq. (3.5). Obviously, there are more ways in which TK can be modified by the life-history traits (in particular reproduction and storage), but this is not dealt with here (see eg, Van Haren et al., 1994).

The TK model provides us with the dose metric (eg, the scaled internal concentration). We can link the dose metric to a metabolic process assuming a linear-with-threshold relationship (Fig. 3.5). The metabolic process, in this respect, is a primary model parameter of the energy budget. But which parameter is affected? An effect on each parameter has specific consequences for the patterns of growth and reproduction (and other energy-requiring processes) over the life cycle of the animal. Such a pattern can be called a metabolic or physiological mode of action (Alda Álvarez et al., 2006). For example, a toxic effect on the assimilation process implies less energy input for the organism, resulting in slower growth, smaller ultimate body size, delayed reproduction, and lower reproduction rate (Jager, 2015b). In contrast, a toxic effect on the costs for an egg will lead to a reduction in reproduction rate with no effects on growth or on the start of reproduction. We can thus analyze the effects of a chemical on the life history of the organism to deduce the most likely parameter affected.

3.6.4 Fitting the TKTD Energy-Budget Model to Data

In fitting a model to sublethal effects over time, we need to select an appropriate model for the deviations between model and observations (the residuals). A thorough treatment of the statistical aspects is outside the scope of this chapter, but some general remarks are in order. In contrast to survival, the sublethal endpoints growth and reproduction are graded: we can measure the degree of response in each individual (rather than just the distinction between alive and dead). For this reason, we can depart from the set of assumptions that underlie the least-squares analysis: independent normal distributions with constant variance. As already mentioned in Section 3.2.7, if we follow a group of animals over time, independence is violated, and a constant variance may also not hold. Furthermore, if reproduction is measured as number of eggs, it is a discrete variable, and a normal distribution is less appropriate when we observe small numbers of offspring. A more appropriate statistical treatment of the data requires that we dive deeper into the nature of the deviations between model and observation. Measurement error (which is what we usually think of when performing a least-squares analysis) is generally small or absent when determining body size and reproduction. However, the individuals in a test are not identical; they differ in their energy-budget parameters and in their response to the chemical. Finally, the model is a simplification of biological complexity and therefore "wrong" (which adds to the deviations between model and data).

In practice, these nuisances are usually ignored and least-squares optimization is used (eg, using transformation or weighing to accommodate changes in variance with the mean of the observations). The reason is that it is quite complicated to come up with better alternatives. As long as the model is capable of providing an adequate representation of the observations, it is unlikely that the details of the statistical treatment will substantially influence the values for the best-fitting parameters. The confidence intervals will, however, suffer more, and they therefore need to be interpreted very carefully indeed.

3.6.5 Case Study

To illustrate the use of DEB-based TKTD models, consider a data set for the effects of nonylphenol on growth and reproduction of the marine polychaete *Capitella teleta*. For the analysis, the "DEBtox" model (Jager and Zimmer, 2012) is used. The equations are not provided here but can be found in the original publication (including its derivation). The TK model comprises a scaled version of Eq. (3.5) (which includes the effects of growth), and the TD model is a simple energy budget. The data were taken from Hansen et al. (1999), and have been analyzed in a full DEB framework by Jager and Selck (2011).

Like many data sets, this set does not allow for a straightforward analysis, for two reasons: the first exposure treatment consistently performs better than the control ("hormesis"), and initial growth must have been slower than expected from the observations ("acceleration"). To start with the first point: the term hormesis is used to describe (but not to explain) the reversal of the direction of the effect with increasing exposure. Here, we see an increase in growth and reproduction at the lowest dose, which implies a larger output of mass and energy from the organism (assuming that there were no large differences in egg size). As the organism needs to obey the conservation laws, this increased output has to be fueled by an increase in feeding or in assimilation efficiency. Of course, there is always intraspecies variation in life history, so it is possible that the "stimulation" was simply accidental. If the stimulation is real, we are basically left with two options: nonylphenol increases food availability or quality (eg, by stimulating bacterial growth) or it acts as a medicine that counteracts an unidentified stress in the control. The hormetic response is discussed in more detail elsewhere (Jager and Selck, 2011); for the purpose of this chapter, we ignore it and assume that the lowest dose is representative for the "real" control behavior.

The other model deviation comprises the apparent "acceleration," as the initial part of the growth curve cannot match the von Bertalanffy growth curve that follows from the standard DEBtox model (see Fig. 3.8, panel bottom left). This is a quite common feature, especially for organisms featuring larval stages (Kooijman, 2014). The experiment started with newly released larvae of *C. teleta*, which are free swimming. After some time, the larvae settle in the sediment and metamorphose to their juvenile form. In the absence of any information in the initial part of the growth curve, any model adaptation will be little more than mere speculation. For this reason, we simply introduced a "lag time" before growth and toxicant uptake commence.

With these additional assumptions, the model can be fitted to the data set (Fig. 3.8, parameter estimates are provided in Table 3.4). In total, eight parameters needed to be estimated from the data. However, as there are six curves (each with 11 points), the total number of parameters is reasonable, given the available data. For the statistical model, it is difficult to provide a good match to the problem. Growth and reproduction were measured on the same group of individuals over time, so independence is violated. Furthermore, the two endpoints will not be independent as reproduction depends on body size and is measured on the same animals. Body volume is a continuous variable, but reproduction is a discrete one (although this is a minor problem given the large number of eggs produced). Another issue to consider is that reproduction in the model is a rate (eggs per day), but what we measure is the number of eggs produced in a certain observation interval. One way to solve this is to work with the cumulative number of eggs (as shown in Fig. 3.8). This is straightforward for the calculations and interpretations but adds to the problems for the statistical model by increasing the dependence between the observations. There is no simple solution to this problem; a pragmatic approach is followed by assuming independent normal distributions of the residuals with constant variance after square-root transformation (to decrease the weight of observations with a large value).

The mode of action that best described the data is an increase in the costs for growth and the costs for reproduction by the same factor. The body size data clearly indicate that growth costs are affected: this mode of action makes it more expensive for the animal to grow but does not affect ultimate size. Effects on growth also indirectly affect reproduction, delaying the start of reproduction and taking more time to reach the maximum reproduction rate (as smaller organisms eat less, and thus reproduce less). However, to fully capture the effects on

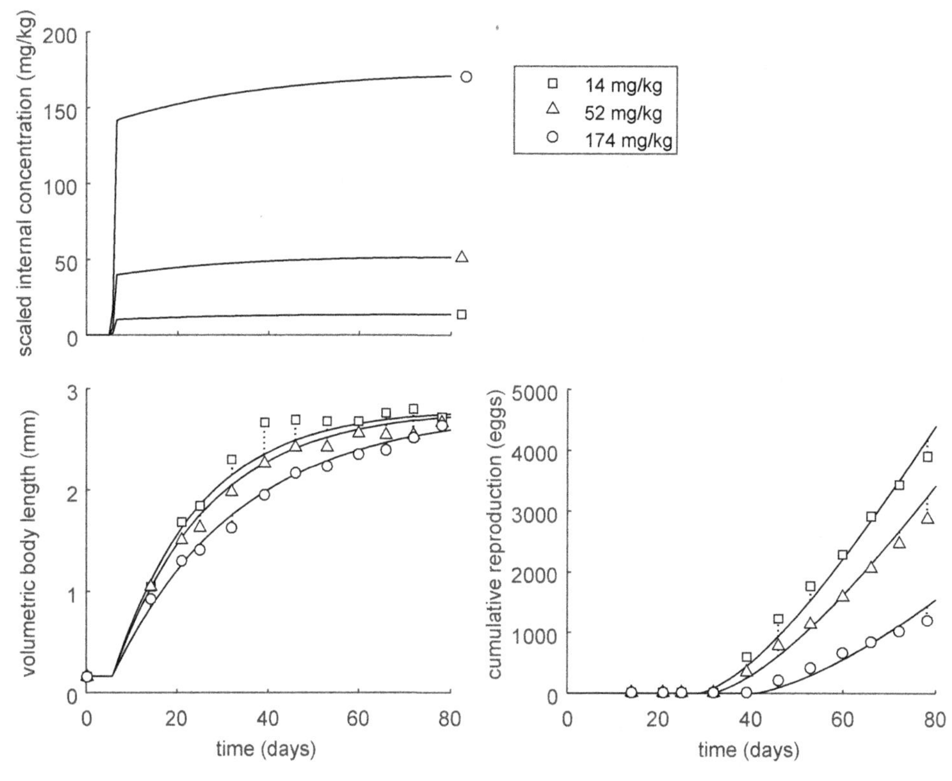

FIGURE 3.8 Simultaneous fit of DEBtox to growth and reproduction data for nonylphenol in *C. teleta*. Top panel shows the predicted scaled internal concentration over time. Bottom panel shows the exposure treatments for growth and reproduction over time. All chemical concentrations in mg/kg dry sediment.

growth in this data set, an additional effect on reproduction was needed.

The top panel in Fig. 3.8 shows the scaled internal concentration, associated with the fit on growth and reproduction in the lower panels. The only TK parameter that is estimated from the effects data is the elimination rate (k_e). The accumulation pattern is different from the general pattern in Fig. 3.1 as the TK model of Eq. (3.5) accounts for the effects of an increase in body size over time (diluting the internal concentration and changing the surface:volume ratio of the animal). As in the survival example, a scaled TK model is used, which is why the internal chemical concentration approaches the external concentration after some exposure time (the dimension of the scaled internal concentration is that of an external concentration, see Section 3.4.2). The confidence interval for the elimination rate extends to infinity, which implies that instantaneous equilibrium also provides a good fit to the data.

So, what can be learned from this analysis? First, this analysis provides new and more fundamental questions on this species and the test conditions: how is it possible that a low concentration of nonylphenol apparently increases the food availability and what happens in the early life stages? These questions require dedicated experimental work. Second, this analysis provides us with a physiological mode of action for nonylphenol that is fully consistent with all of

TABLE 3.4 Parameter Estimates From the Simultaneous Fit in Fig. 3.8, With 95% Likelihood-Based Confidence Intervals

Parameter	Estimated value (95% CI)
Initial volumetric length	0.16 (n.e.) mm
Volumetric length at puberty	2.1 (2.0–2.1) mm
Maximum volumetric length	2.8 (2.7–2.9) mm
Maximum reproduction rate	120 (110–137) eggs/day
Energy investment ratio	10 (n.e.) [–]
von Bertalanffy rate constant	0.053 (0.045–0.062) day^{-1}
Lag time for growth to start	5.6 (3.9–7.2) day
Elimination rate constant	0.44 (0.081–∞) day^{-1}
Threshold for effects	14 (14–21) mg/kg
Tolerance concentration	240 (160–310) mg/kg

n.e. means that the parameter is not estimated but fixed to a certain value.

the information present in this data set. To test this result, the next step would be to use the calibrated model to make predictions for other traits (eg, feeding and respiration rates) or under different circumstances (eg, under food limitation or pulsed exposure). These predictions should then be tested experimentally, leading to new/additional assumptions, which lead to model refinements, yielding new predictions, etc. Such an alternation of modeling and experiments is optimal to progress our understanding of the action of a chemical in a particular species but is rare in practice.

3.7 POPULATION LEVEL AND HIGHER

In environmental risk assessment, the concern is not with the well-being of individual animals but with the long-term health of populations, communities, and ecosystems. If there is no effect on individuals (under all realistic environmental conditions), there will be no effect on the ecosystem, but does the converse hold as well? How does an x% reduction of the reproductive output of individuals translate to changes at higher levels of organization? These higher levels are hardly amenable to systematic experimental testing, so models are essential tools to address these questions. The topic of population and food-chain modeling has a long tradition in ecology, and many good textbooks are available (see the supporting website http://www.debtox.info/marecotox.html). The focus of this chapter is on modeling at the individual level, so it only treats the higher levels of biological organization in a cursory manner.

The life-history traits of individuals (ie, feeding, survival, growth, reproduction) ultimately determine the dynamics of the populations. Population modeling in ecotoxicology is therefore served by models that are able to provide a description of the life-history traits of individuals over their entire life cycle, as a function of chemical exposure, food availability, temperature, etc. For this task, some form of energy budget is generally most suitable. The section below discusses the three most common population approaches in ecotoxicology. As with all models, the research question and the available data should drive selection of the most appropriate population model.

3.7.1 Individual-Based Models

Conceptually, the simplest approach to extrapolate from individuals to populations is using an individual-based model. In such models, all individuals in the population are modeled explicitly, and the population dynamics emerges as a consequence of the action of the individuals. This approach lends itself readily to an integration with energy-budget models for the individual (Martin et al., 2012) but can rapidly become calculation intensive (especially if we include multiple species interacting with each other). A thorough discussion on the use of individual-based approaches in ecotoxicology can be found elsewhere (Topping et al., 2009).

3.7.2 Matrix Models

Matrix population models have a long history in ecology and ecotoxicology. In these models, the population is divided over a limited number of discrete classes, based on age or stage (or body size). Energy-budget models can also be linked to matrix population models (eg, Klanjscek et al., 2006). However, corners have to be cut as simple matrix models only follow one state variable of the organism (age, size, or stage), which might rapidly become insufficient to capture chemical stress on the individual (eg, it is not straightforward to include toxicokinetics as a function of time and body size). A thorough discussion on the use of matrix models in ecotoxicology can be found elsewhere (Charles et al., 2009).

3.7.3 The Intrinsic Rate of Increase

As a last approach for population dynamics, there is the classical Euler–Lotka equation for calculating the intrinsic rate of population increase. In a constant environment, all populations will eventually grow exponentially, and it is this rate that can be calculated as a function of chemical exposure and food availability. Input is formed by the survival probability and reproduction rate over the life cycle of the organism, as continuous or discrete functions of time. Such output is provided by TKTD models as discussed in this chapter, making a link straightforward (see eg, Jager et al., 2006). In reality, the conditions in the environment will never remain constant for long, so the ecological realism of this calculation is limited. However, the intrinsic rate of increase integrates effects on growth, reproduction, and survival, and can be considered as an easy-to-interpret measure of population fitness (an inherent capability of the population to grow). As such, it can be a valuable statistic for environmental risk assessment (Forbes and Calow, 1999).

3.8 FUTURE POSSIBILITIES

This final section provides an outlook on the future of TKTD modeling. The supporting web page for this chapter (http://www.debtox.info/marecotox.html) provides links to papers that are pursuing some of these lines of thought and which could provide a starting point for future research.

3.8.1 Closer Collaboration Between Disciplines

Modeling is often viewed as a distinctly different activity from experimental work. However, both approaches strengthen each other; scientific progress in ecotoxicology requires a much closer integration of these two activities. This, for example, implies that modelers need to be included in the design stage of the experimental work. Most toxicity tests, for example, are performed using standardized test protocols. TKTD models, however, place very different demands on experimental design, which implies that additional observations are needed but, on the other hand, that some constraints can be removed. For example, there is no need to keep the exposure concentration constant, as long as the actual concentration–time pattern can be established. Furthermore, a closer integration requires the possibility for follow-up experiments after the first round of model analyses; inevitably, model analysis will yield new (and more fundamental) questions. Alternation of mechanistic modeling and experiments is the best way to progress our understanding of the effects of chemicals.

We also need a closer collaboration between modeling and statistics. Statisticians rely heavily on descriptive models, and modelers usually apply statistics that poorly match the problem at hand. As one particular example, the deviations between model and observation are routinely treated as random noise or measurement error, whereas, in fact, most of the deviation is caused by real differences between individuals. In TKTD models, differences between individuals translate to differences in model parameters. However, because these model parameters are linked to physiological processes, they are unlikely to vary independently in the population. More mechanistic models for the deviations between model and observation are therefore needed (see Chapter 2, Statistics).

Closer collaboration is also needed between the TKTD modelers and molecular biologists (eg, regarding biomarkers; see Chapter 5, Biomarkers). The last decade has seen an explosion of molecular research in ecotoxicology. However, the problem of linking the molecular level to life-history traits of individuals has not been solved. Linking the molecular level to parameters of TKTD models is a far more promising approach than linking molecules to traits directly. For example, linking changes at the molecular level to effects on reproduction is unlikely to yield efficient models because there are many indirect ways in which reproduction can be affected (eg, an increase in maintenance costs, or a decrease in feeding rate). However, linking the molecular level to a component of the energy budget such as assimilation or maintenance may be far more productive.

3.8.2 Topics for Future Research

The calibration of TKTD models requires more information on the organism and on the chemical effects than fitting a simple dose–response curve. Such information can be obtained from more extensive experimental effort, but there are other options to pursue. Because the parameters of TKTD models have a physiological meaning (unlike those of a dose–response curve), there are possibilities to predict parameters from statistical relationships or from first principles. Systematic work on how model parameters differ between species and between chemicals is needed.

Only in very few cases, body residues have been determined in conjunction with toxic effects, and in most of those cases, the elimination rate for the body residues does not match the one needed for the TD part of the model (ie, to capture the patterns of toxic effects over time). The whole-body concentration can be a poor substitute for the dynamics of the concentration at the target site. It is possible that the organism cannot be viewed as a well-mixed compartment, or that a metabolite should be followed, or that an additional stage of damage needs to be considered (or a combination of these factors). In any case, the k_e estimated from effects data (Tables 3.3 and 3.4) should not be interpreted as the whole-body elimination rate. Teasing out what causes the discrepancy between body residues and effect dynamics could substantially aid the development of predictive relationships as outlined in the previous paragraph.

Experimental work on TK generally attempts to avoid complicating factors such as growth, reproduction, and toxic effects. In TKTD modeling, however, we are particularly interested in the TK for growing and reproducing animals, and especially in the concentration range where toxicity occurs. With Eq. (3.5), a model extension for growing organisms was introduced, and it is certainly possible to include the effects of reproduction (ie, transfer to eggs). If toxicity affects growth and reproduction, it will thereby also affect TK. However, toxicity may have additional effects on TK, for example by affecting the activity and behavior of the animal. The performance of TK models under more realistic conditions needs to be tested.

In TK and TKTD modeling, eggs and embryos have received relatively little attention. Chemicals may be transferred to eggs by the mother, and eggs may exchange chemicals with their environment. Over the course of embryonic development, the total egg mass will decrease (as the embryo is burning calories to sustain itself and to develop), which affects TK. Furthermore, embryos can be affected by toxicant stress in the same way as juveniles and adults; they also have an energy budget and they can also die if the concentration is high enough. Deleterious effects on embryonic development may incur large effects on the population dynamics, and therefore, it is important to dedicate more attention to this life stage.

Organisms are never exposed to a single stressor in isolation; multiple stress is the norm. This does not only relate to mixtures of chemicals, but also to the combination of chemical stress and environmental factors such as food availability and temperature. Of particular interest for the marine environment are the effect of climate change and the associated acidification of the world's oceans (see Chapter 10, Global Change). This stressor affects the life histories of marine organisms and may well interact with chemical stress. Because TKTD models are based on mechanisms rather than on descriptions, they are well-suited to interpret and predict the effects of multiple stress. However, at this moment, there is a need for more systematic studies into the possibilities and limitations of TKTD models for this purpose.

TKTD models still need to find their way into risk assessment frameworks (see Chapter 9, Ecological Risk Assessment). Mechanistic models already feature prominently in the exposure assessment (as fate models) and for bioaccumulation (as TK models). The effects assessment, in contrast, still leans heavily on descriptive methods. This situation reflects the fact that TKTD modeling is not yet firmly established in ecotoxicology. However, even simple TKTD models have clear advantages over the current descriptive methods, for example by their ability to use all of the available data (all time points, all endpoints) into an integrated analysis, and by the possibility to make educated predictions for untested situations (eg, multiple stresses and fluctuating exposure concentrations). The most suitable models, and their implementation into a regulatory context, require further work.

Acknowledgments

I would like to thank my former colleagues at the Department of Theoretical Biology of the VU University in Amsterdam, and specifically Bob Kooi, Bas Kooijman, Jacques Bedaux, and Paul Doucet. Teaching courses with them, and discussing models and modeling in general, has shaped the thoughts expressed in this chapter.

References

Alda Álvarez, O., Jager, T., Marco Redondo, E., Kammenga, J.E., 2006. Physiological modes of action of toxic chemicals in the nematode *Acrobeloides nanus*. Environ. Toxicol. Chem. 25, 3230–3237.

Ashauer, R., Escher, B.I., 2010. Advantages of toxicokinetic and toxicodynamic modelling in aquatic ecotoxicology and risk assessment. J. Environ. Monit. 12, 2056–2061.

Ashauer, R., Hintermeister, A., Caravatti, I., Kretschmann, A., Escher, B.I., 2010. Toxicokinetic and toxicodynamic modeling explains carry-over toxicity from exposure to diazinon by slow organism recovery. Environ. Sci. Technol. 44, 3963–3971.

Barron, M.G., Stehly, G.R., Hayton, W.L., 1990. Pharmacokinetic modeling in aquatic animals. 1. Models and concepts. Aquat. Toxicol. 18, 61–86.

Bedaux, J.J.M., Kooijman, S.A.L.M., 1994. Statistical analysis of bioassays based on hazard modelling. Environ. Ecol. Stat. 1, 303–314.

Billoir, E., Delignette-Muller, M.L., Péry, A.R.R., Geffard, O., Charles, S., 2008. Statistical cautions when estimating DEBtox parameters. J. Theor. Biol. 254, 55–64.

Charles, S., Billoir, E., Lopes, C., Chaumot, A., 2009. Matrix population models as relevant modeling tools in ecotoxicology. In: Devillers, J. (Ed.), Ecotoxicology Modelling. Springer, New York, USA, pp. 261–298.

Doucet, P., Sloep, P.B., 1993. Mathematical Modelling in the Life Sciences. Ellis Horwood Limited, Chichester, UK.

Forbes, V.E., Calow, P., 1999. Is the per capita rate of increase a good measure of population-level effects in ecotoxicology? Environ. Toxicol. Chem. 18, 1544–1556.

Giese, A.C., 1959. Comparative physiology: annual reproductive cycles of marine invertebrates. Annu. Rev. Physiol. 21, 547–576.

Gobas, F.A.P.C., Opperhuizen, A., Hutzinger, O., 1986. Bioconcentration of hydrophobic chemicals in fish: relationship with membrane permeation. Environ. Toxicol. Chem. 5, 637–646.

Gomez, E., Bachelot, M., Boillot, C., Munaron, D., Chiron, S., Casellas, C., Fenet, H., 2012. Bioconcentration of two pharmaceuticals (benzodiazepines) and two personal care products (UV filters) in marine mussels (*Mytilus galloprovincialis*) under controlled laboratory conditions. Environ. Sci. Pollut. Res. 19, 2561–2569.

Hansen, B.H., Altin, D., Rørvik, S.F., Øverjordet, I.B., Olsen, A.J., Nordtug, T., 2011. Comparative study on acute effects of water accommodated fractions of an artificially weathered crude oil on *Calanus finmarchicus* and *Calanus glacialis* (Crustacea: Copepoda). Sci. Total Environ. 409, 704–709.

Hansen, F.T., Forbes, V.E., Forbes, T.L., 1999. Effects of 4-n-nonylphenol on life-history traits and population dynamics of a polychaete. Ecol. Appl. 9, 482–495.

Jager, T., 2011. Some good reasons to ban EC*x* and related concepts in ecotoxicology. Environ. Sci. Technol. 45, 8180–8181.

Jager, T., 2015a. DEBkiss. A Simple Framework for Animal Energy Budgets. Leanpub. https://leanpub.com/debkiss_book. Version 1.4, 12 August 2015.

Jager, T., 2015b. Making Sense of Chemical Stress. Applications of Dynamic Energy Budget Theory in Ecotoxicology and Stress Ecology. Leanpub. https://leanpub.com/debtox_book. Version 1.2, 11 August 2015.

Jager, T., Albert, C., Preuss, T.G., Ashauer, R., 2011. General unified threshold model of survival – a toxicokinetic-toxicodynamic framework for ecotoxicology. Environ. Sci. Technol. 45, 2529–2540.

Jager, T., Heugens, E.H.W., Kooijman, S.A.L.M., 2006. Making sense of ecotoxicological test results: towards application of process-based models. Ecotoxicology 15, 305–314.

Jager, T., Kooijman, S.A.L.M., 2005. Modeling receptor kinetics in the analysis of survival data for organophosphorus pesticides. Environ. Sci. Technol. 39, 8307–8314.

Jager, T., Ravagnan, E., 2015. Parameterising a generic model for the dynamic energy budget of Antarctic krill, *Euphausia superba*. Mar. Ecol. Prog. Ser. 519, 115–128.

Jager, T., Selck, H., 2011. Interpreting toxicity data in a DEB framework: a case study for nonylphenol in the marine polychaete *Capitella teleta*. J. Sea Res. 66, 456–462.

Jager, T., Zimmer, E.I., 2012. Simplified Dynamic Energy Budget model for analysing ecotoxicity data. Ecol. Modell. 225, 74–81.

Klanjscek, T., Caswell, H., Neubert, M.G., Nisbet, R.M., 2006. Integrating dynamic energy budgets into matrix population models. Ecol. Modell. 196, 407–420.

Kooijman, S.A.L.M., 1996. An alternative for NOEC exists, but the standard model has to be abandoned first. Oikos 75, 310–316.

Kooijman, S.A.L.M., 2001. Quantitative aspects of metabolic organization: a discussion of concepts. Phil. Trans. R. Soc. B 356, 331–349.

Kooijman, S.A.L.M., 2014. Metabolic acceleration in animal ontogeny: an evolutionary perspective. J. Sea Res. 94, 128–137.

Kooijman, S.A.L.M., Bedaux, J.J.M., 1996. Analysis of toxicity tests on *Daphnia* survival and reproduction. Water Res. 30, 1711–1723.

Lee, J.H., Landrum, P.F., Koh, C.H., 2002. Prediction of time-dependent PAH toxicity in *Hyalella azteca* using a damage assessment model. Environ. Sci. Technol. 36, 3131–3138.

Lee, R.F., Hagen, W., Kattner, G., 2006. Lipid storage in marine zooplankton. Mar. Ecol. Prog. Ser. 307, 273–306.

MacLeod, M., Scheringer, M., McKone, T.E., Hungerbuhler, K., 2010. The state of multimedia mass-balance modeling in environmental science and decision-making. Environ. Sci. Technol. 44, 8360–8364.

Martin, B.T., Zimmer, E.I., Grimm, V., Jager, T., 2012. Dynamic Energy Budget theory meets individual-based modelling: a generic and accessible implementation. Methods Ecol. Evol. 3, 445–449.

McCarty, L.S., Mackay, D., 1993. Enhancing ecotoxicological modeling and assessment. Body residues and modes of toxic action. Environ. Sci. Technol. 27, 1719–1728.

Meeker, W.Q., Escobar, L.A., 1995. Teaching about approximate confidence regions based on maximum likelihood estimation. Am. Stat. 49, 48–53.

Newman, M.C., McCloskey, J.T., 2000. The individual tolerance concept is not the sole explanation for the probit dose-effect model. Environ. Toxicol. Chem. 19, 520–526.

Nichols, J.W., McKim, J.M., Andersen, M.E., Gargas, M.L., Clewell, H.J., Erickson, R.J., 1990. A physiological based toxicokinetic model for the uptake and disposition of waterborne organic chemicals in fish. Toxicol. Appl. Pharmacol. 106, 433–447.

Nisbet, R.M., Muller, E.B., Lika, K., Kooijman, S.A.L.M., 2000. From molecules to ecosystems through dynamic energy budget models. J. Anim. Ecol. 69, 913–926.

Øverjordet, I.B., Altin, D., Berg, T., Jenssen, B.M., Gabrielsen, G.W., Hansen, B.H., 2014. Acute and sublethal response to mercury in Arctic and boreal calanoid copepods. Aquat. Toxicol. 155, 160–165.

Pechenik, J.A., 1999. On the advantages and disadvantages of larval stages in benthic marine invertebrate life cycles. Mar. Ecol. Prog. Ser. 177, 269–297.

Saraiva, S., van der Meer, J., Kooijman, S.A.L.M., Witbaard, R., Philippart, C.J.M., Hippler, D., Parker, R., 2012. Validation of a dynamic energy budget (DEB) model for the blue mussel *Mytilus edulis*. Mar. Ecol. Prog. Ser. 463, 141–158.

Topping, C.J., Dalkvist, T., Forbes, V.E., Grimm, V., Sibly, R.M., 2009. The potential for the use of agent-based models in ecotoxicology. In: Devillers, J. (Ed.), Ecotoxicology Modelling. Springer, New York, USA, pp. 205–235.

Van Haren, R.J.F., Schepers, H.E., Kooijman, S.A.L.M., 1994. Dynamic Energy Budgets affect kinetics of xenobiotics in the marine mussel *Mytilus edulis*. Chemosphere 29, 163–189.

Widdows, J., Johnson, D., 1988. Physiological energetics of *Mytilus edulis*: scope for growth. Mar. Ecol. Prog. Ser. 46, 113–121.

CHAPTER

4

Bioaccumulation and Biomonitoring

W.-X. Wang

Hong Kong University of Science and Technology, Kowloon, Hong Kong

Bioaccumulation is typically defined as the increase of concentrations of contaminants in aquatic organisms following uptake from the ambient environmental medium. Concentration is thus the central piece of any bioaccumulation study, and the significance of concentration must be understood. Numerous studies have therefore determined the concentrations of different contaminants in various species of aquatic organisms collected from different parts of the world, and the literature contains numerous examples of these data.

For aquatic organisms there are different sources of uptake, such as water (as waterborne uptake) and/or food particles (as foodborne uptake). Bioaccumulation focuses on the processes of contaminant uptake and elimination in organisms. Another important concept is bioavailability, which is defined as the fraction of contaminants potentially available for uptake or actually taken up from the environment. Although closely linked, these two concepts are inherently different.

Bioaccumulation examines the changes of concentrations of contaminants in the organisms, whereas bioavailability describes the portion of contaminants in the environment that is potentially available for bioaccumulation. Therefore, many bioavailability studies are concerned about the chemistry of contaminants in the environments. Also of concern is modification by biology of the chemistry of contaminants in the environments, and how this in turn affects bioavailability. Finally, bioaccumulation is also concerned with the controls of physiology and biochemistry on contaminant uptake and elimination.

In ecotoxicological studies, bioaccumulation and bioavailability are considered jointly. It would be impossible to study bioaccumulation without taking into account bioavailability, and vice versa. Thus, both are considered herein along with the use of bioaccumulation in biomonitoring.

Traditional ecotoxicology mainly encompasses three frameworks, namely, the environmental transport, bioaccumulation, and toxicity of contaminants interacting with organisms (Wang, 2011a). Environmental risk assessments build upon these frameworks and provide input to management of contaminants in the environment.

Bioaccumulation is the direct link between contaminants in environments and exposed organisms; toxicity can manifest after bioaccumulation occurs. Bioaccumulation directly links environmental chemistry/processes and organism physiology/biochemistry, and thus can be considered as an interface between environmental chemistry and biology.

Marine Ecotoxicology
http://dx.doi.org/10.1016/B978-0-12-803371-5.00004-7

There has been an enormous body of studies on the bioaccumulation of contaminants in aquatic organisms over the past few decades. It is impossible to summarize all these findings in one single book chapter and the readers are also referred to some earlier reviews of this topic (eg, Spacie et al., 1995). Rather, this chapter mainly discusses the basic principles of bioaccumulation and the methods used to quantify bioaccumulation, some of the important considerations involved in the bioaccumulation assessments, and their use in biomonitoring. Most of the focus of this chapter is on metals; however, organic contaminants are also briefly discussed.

4.1 GENERAL PRINCIPLES OF BIOACCUMULATION

Simply put, if an organism can be treated as one single compartment (or box) without further considering any internal transportation, the concentration (bioaccumulation) of a contaminant in the organism (box) is determined by the balance between influx and efflux, as shown in Fig. 4.1.

Fig. 4.1 provides a very basic illustration of the bioaccumulation of contaminants in organisms. Although simple, it helps to understand that bioaccumulation is a very dynamic process, affected by both the flows into the system (influx) and out of the system (efflux). This figure shows that it is critical to study the influx and efflux of contaminants in order to understand bioaccumulation and that only examining either influx or efflux is insufficient to fully appreciate the importance or process of bioaccumulation.

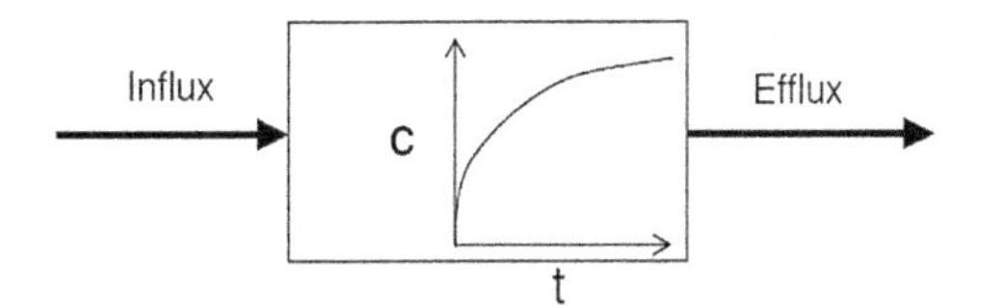

FIGURE 4.1 A simple illustration of bioaccumulation in an organism. *C*, net accumulated concentration in the organism; *t*, time of exposure.

Bioaccumulation is the net result of influx (uptake) and efflux. If influx is greater than efflux, contaminant concentration in the organism will increase with exposure time. When the influx is smaller than the efflux, there will be a net loss of contaminants and the concentration in the organism will decrease with exposure time. Under steady-state conditions, influx and efflux are equal, and contaminant concentration remains unchanged. In bioaccumulation modeling, steady-state condition is an important assumption (see below).

Influx is defined as the uptake of contaminants from different environmental matrices (water, food); efflux is defined as loss of contaminants from organisms due to metabolic processes (eg, excretion, molting, reproduction). Efflux is rather broadly defined; other terms (eg, depuration, elimination, or loss) are also frequently used to describe efflux. Subtle differences exist among these different terms, eg, depuration is generally defined as the loss of contaminants from organisms following transfer from contaminated to clean environments; elimination generally implies some metabolically controlled processes. Growth is another important term since it can act to dilute contaminant concentrations in organisms.

For most aquatic animals, both waterborne and dietborne contaminants contribute to overall bioaccumulation. Kinetic processes can be used to quantify uptake.

4.1.1 Absorption

Absorption is defined as uptake from the dissolved phase; in some cases internalization has been used to describe the absorption of contaminants. Contaminant uptake from the dissolved phase involves the initial sorption and the internalization process; however, any measurements of absorption should explicitly remove the initial sorption to the surface of the animals, as this is external, not internal. For some metals, chelates (eg, EDTA: ethylenediamine tetra-acetic acid, a crystalline acid with a strong tendency to form chelates with metal ions) can be used to wash off the surface sorption, which is defined as the weakly exchangeable fraction. For filter-feeding animals such as bivalves, absorption can be quantified by absorption efficiency, using the equation below:

$$\text{Influx} = k_u \times C_w = \alpha \times \text{FR} \times C_w \quad (4.1)$$

where k_u is the uptake rate constant from the dissolved phase (L/g/h), C_w is the contaminant concentration in the dissolved phase (μg/L), FR is the filtration (or clearance) rate of the animal (L/g/h), α is the absorption efficiency (%), considered as the first-order rate constant. Strictly speaking, absorption efficiency is independent of the C_w and FR if absorption is not diffusion limited.

4.1.2 Assimilation

In contrast to absorption from the dissolved phase, assimilation refers to uptake from a dietary source. When food is ingested by aquatic animals, digestion immediately occurs in the digestive system (eg, stomach). Undigested materials are subsequently egested in the form of feces, whereas the remaining components are absorbed across the gut linings. Following further metabolism, some of these components are lost from the body through elimination, respiration, or excretion; the remainder is finally incorporated into tissues, a process called assimilation (Fig. 4.2). The following equations summarize this energetically:

$$\text{IR} = \text{AR} + \text{Feces} \quad (4.2)$$

$$\text{AR} = \text{AE} + \text{Excretion} + \text{Respiration} \quad (4.3)$$

where IR is the total ingestion (feeding) rate, AR is the absorption rate, and AE is the assimilation rate. The assimilation efficiency (AE) is then calculated as the assimilation rate divided by the total ingestion.

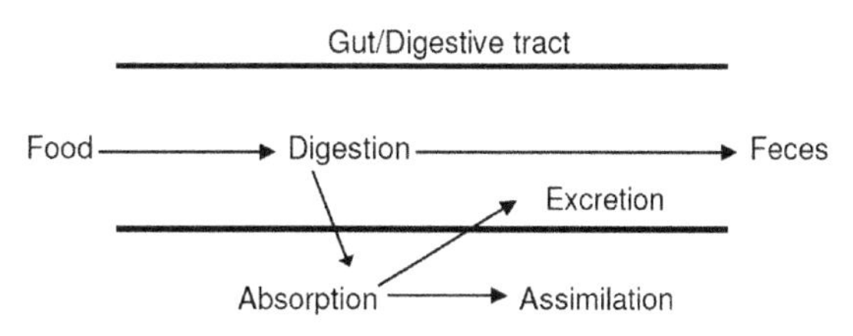

FIGURE 4.2 Processes involved when foods are digested.

4.1.3 Bioconcentration Factor (BCF)

BCF is defined as the uptake of contaminants from the dissolved phase; it can be calculated by the following equation:

$$\text{BCF} = C/C_w \quad (4.4)$$

where C is the contaminant concentrations in the organisms (μg/kg) under equilibrium condition, C_w is the contaminant concentration in the water (μg/L). A key assumption for the calculation or measurement of BCF is that of equilibrium between the contaminant and organism. Equilibrium can be easily achieved for small organisms with fast growth rates; however, equilibrium is very difficult to reach for large organisms such as fish, in which equilibrium may not be reached during their entire life history.

Numerous measurements of BCF for different contaminants (metals and organics) in different groups of aquatic organisms are now available, but many of these have potential problems. If the measurements are conducted using field populations, then bioaccumulation involves uptake from both water and food sources; therefore, the BCF should be supplemented with another important parameter, namely, the bioaccumulation factor (BAF), which is calculated as overall bioaccumulation in the organisms divided by the concentration in the environment (both dissolved and particulate phases). If the measurements are conducted in the laboratory, equilibrium between organism and water should be explicitly reached. In many laboratory studies, however, one can use concentration factor instead of BCF to quantify bioaccumulation if equilibrium is unknown.

BCF is an important concept in environmental risk assessment since it gives quantitative information regarding the ability of a contaminant to be taken up by organisms from the water. It is often used as one of the first screening parameters for persistent, bioaccumulative, and toxic substances. However, note that BCF is not a constant (or a factor as generally implied). Instead, BCF is a variable depending on different environmental and biological conditions. BCF is inversely dependent on contaminant concentrations in the water (for metals) or the octanol–water partitioning coefficient (K_{ow}) (for organic substances) (McGeer et al., 2003). Similar to BCF, the BAF is not a constant; it is also dependent on the ambient concentration (DeForest et al., 2007).

Toxicokinetics describes the kinetic process of contaminants in the organisms, including uptake, assimilation, storage, sequestration, transportation, and elimination. Thus, toxicokinetics is more broadly defined as compared to the bioaccumulation, since it also considers the internal processes of contaminants once they are taken up and accumulated. Toxicokinetics is an important research area in ecotoxicology, especially for organic contaminants due to their significant biotransformation within the body of organisms.

4.2 BIOACCUMULATION MODELING

Many models are used to simulate the bioaccumulation of contaminants. Overall, these can be categorized into two types: the equilibrium partitioning model (EqP) and the kinetic model.

4.2.1 Equilibrium Partitioning Model (EqP)

The EqP model is relatively simple, assuming that waterborne is the only source of accumulation (Fig. 4.3). Thus, the BCF can be used to quantify bioaccumulation of contaminants by organisms.

EqP modeling has been used extensively for organic contaminants mainly because the bioaccumulation of organic contaminants is related to the K_{ow} of the chemicals, thus BCF can be used to predict their bioaccumulation. For metals, this model can be used to predict bioaccumulation in small organisms such as bacteria and phytoplankton since equilibrium in these organisms occurs. However, such an approach is not reliable for larger organisms in which equilibrium does not occur. Table 4.1 summarizes some typical BCF values in marine organisms. Differences in BCF among metals or organisms are considerable and are related to the cell volume or the chemical property of metals (Fisher, 1986). BCF also varies greatly in different environments, and it is impossible to use one single BCF to represent different organisms for every chemical (metal or organic). BCF simply serves as the initial screening value for potential bioconcentration of different contaminants in organisms.

Another disadvantage of the EqP model is that it only considers exposure of organisms to contaminants in water and ignores the possible exposure of food source. Due to the importance of trophic transfer in contaminant accumulation in marine organisms (Wang, 2002; Wang and Rainbow, 2008), BCF cannot generally be used alone to model bioaccumulation of contaminants

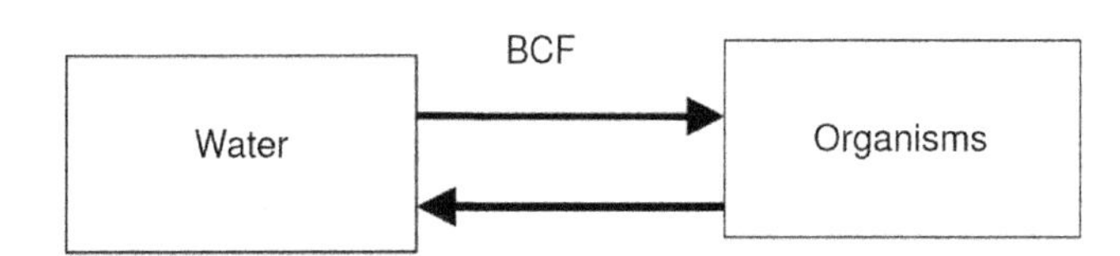

FIGURE 4.3 A simple equilibrium partitioning model. *BCF*, bioconcentration factor.

TABLE 4.1 Bioconcentration Factor (BCF) of Nine Metals in Different Groups of Marine Organisms (L/kg)

Group	Ag	Cd	Cs	Cr	Hg	Ni	Se	Pb	Zn
Phytoplankton	5×10^4	10^3	20	5×10^3	10^5	3×10^3	3×10^4	10^5	10^4
Macrophyte	5×10^3	2×10^4	50	6×10^3	2×10^4	2×10^3	10^3	10^3	2×10^3
Zooplankton	2×10^4	6×10^4	40	10^3	4×10^3	10^3	6×10^3	10^3	10^5
Mollusks	6×10^4	8×10^4	60	2×10^3	2×10^3	2×10^3	9×10^3	5×10^4	8×10^4
Crustaceans	2×10^5	8×10^4	50	10^2	10^4	10^3	10^4	9×10^4	3×10^5
Fish	10^4	5×10^3	100	2×10^2	3×10^4	10^3	10^4	2×10^2	10^3

Note that here and in this chapter the term "metals" includes metalloids such as arsenic (As) and nonmetals such as selenium (Se).
From IAEA (International Atomic Energy Agency), 2004. Sediment Distribution Coefficients and Concentration Factors for Biota in the Marine Environment. Technical Report Series, No. 422. Vienna, Austria.

in organisms. However, for organisms that accumulate contaminants only from the dissolved phase (eg, phytoplankton, macrophytes), BCF can still be a useful and practical approach to modeling bioaccumulation, especially for initial screening.

4.2.2 Kinetic Modeling

Kinetic modeling is not constrained by equilibrium considerations; it can be used to simulate the kinetic changes in contaminant bioaccumulation over time. There are several developed kinetic models, varying from the simplest one-compartmental model to multicompartmental models (Fig. 4.4).

The simple one-compartmental kinetic model can be simulated by the following equation:

$$dC/dt = k_u \times C_W - k_e \times C \quad (4.5)$$

where k_u is the uptake rate constant from water (L/g/h), C_W is the contaminant concentration in the water (μg/L), k_e is the efflux rate constant, and C is the accumulated concentration (μg/g).

After integration, the accumulated concentration of contaminant at time t can be described by the following equation:

$$C_t = k_u \times C_W / k_e \times \left[1 - e^{-k_e t}\right] \quad (4.6)$$

Under steady-state condition, $dC/dt = 0$, the steady-state concentration (C_{SS}) can be calculated as:

$$C_{SS} = k_u \times C_W / k_e \quad (4.7)$$

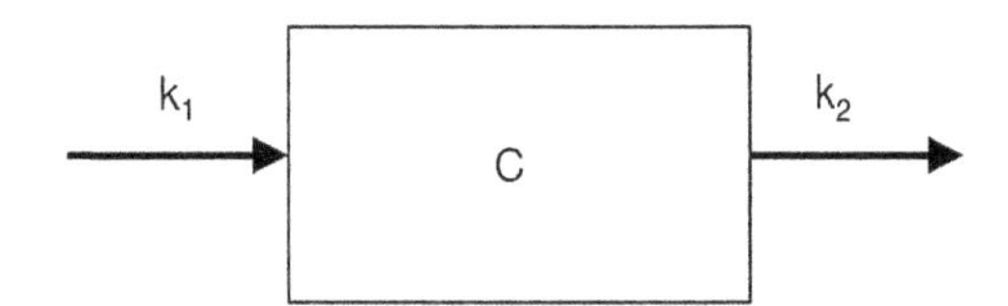

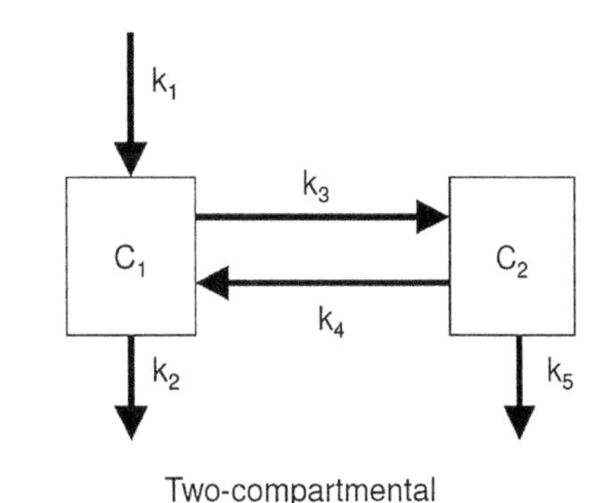

FIGURE 4.4 Schematic illustrations of one- and two-compartmental models with different rate constants involved (C as concentration, k as rate constant).

With known k_u, k_e, and C_w, it is then possible to predict the concentration of contaminants under steady-state conditions. Since BCF = C_{ss}/C_w, BCF can also be calculated using the two kinetic parameters:

$$BCF = k_u/k_e \tag{4.8}$$

The above equation directly quantifies the BCF, especially for large organisms. A simple determination of the k_u and k_e can then accurately predict the BCF. There are inherent links between the EqP and kinetic models, given the fact that both k_u and k_e are considered as the kinetic parameters.

For many aquatic animals, both dissolved and food sources contribute to contaminant accumulation. The simple kinetic model can incorporate the food exposure source (Fig. 4.5):

$$dC/dt = [k_u \times C_w + k_f \times C_f] - (k_e \times C) \tag{4.9}$$

where k_f is the uptake rate constant from the food source, C_f is the contaminant concentration in the food. Again, this model can be solved to predict the accumulated concentration at specific time t:

$$C_t = [k_u \times C_w + k_f \times C_f]/k_e \times \left[1 - e^{-k_e \times t}\right] \tag{4.10}$$

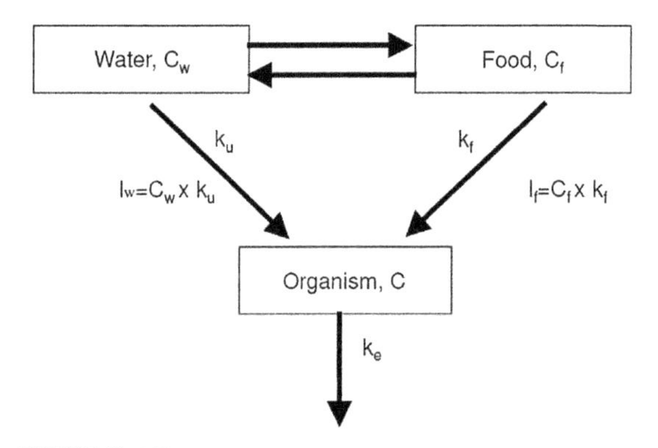

FIGURE 4.5 Kinetic model simultaneously considers uptake from the waterborne and dietborne phases, with different kinetic parameters used to quantify bioaccumulation.

Under steady-state conditions, $dC/dt = 0$, and the accumulated concentration can be calculated (C_{ss}) as:

$$C_{ss} = [k_u \times C_w + k_f \times C_f]/k_e \tag{4.11}$$

Kinetic modeling is a relatively simple concept that was used extensively in radioecology in the 1970s. Thomann (1981) further developed such modeling by introducing bioenergetic concepts into the equation:

$$k_u = \alpha \times FR \tag{4.12}$$

$$k_f = AE \times IR \tag{4.13}$$

where α is the absorption efficiency of contaminants from the water, FR is the filtration rate of the organisms (eg, amount of water filtered or cleared), AE is the dietary assimilation efficiency, IR is the ingestion rate of the organisms. The kinetic equation has therefore been sometimes termed as a bioenergetic-based model or biokinetic/biodynamic model. The C_{ss} can be calculated as:

$$C_{ss} = [\alpha \times FR \times C_w + AE \times IR \times C_f]/k_e \tag{4.14}$$

The above model ignores the growth of organisms, which can be an important parameter for organisms displaying high growth rates (eg, small organisms such as bacteria and phytoplankton). Thus a more complete kinetic model should also consider the growth rate constant (g) of the organisms:

$$C_{ss} = [\alpha \times FR \times C_w + AE \times IR \times C_f]/(k_e + g) \tag{4.15}$$

The above model treats an organism as one single compartment without further going into the kinetic processes within the organisms (eg, transportation, redistribution, storage, and sequestration). However, with this basic equation, it is then possible to further consider a few special cases. For example, water is the only source of uptake for marine phytoplankton in which g is much higher than k_e. The concentration of a contaminant in the phytoplankton can thus be calculated as:

$$C_{ss} = (k_u \times C_w)/g = I_w/g \quad (4.16)$$

where I_w is the influx from the water source. This equation has significant application in studying the uptake of contaminants in phytoplankton, which can be directly calculated with a known C_{ss} and g.

If trophic transfer is the only source of bioaccumulation, Eq. (4.15) can also be simplified as:

$$C_{ss} = (k_f \times C_f)/k_e \quad (4.17)$$

Trophic transfer factor (TTF) or biomagnification factor (BMF) is then calculated as:

$$TTF = BMF = C_{ss}/C_f = k_f/k_e \quad (4.18)$$

The concept of a TTF or BMF is very similar to that of BCF. A simple method to quantify the TTF or to examine whether a contaminant has the potential of being biomagnified in the food chain (ie, whether concentrations of a substance increase through three or more trophic levels via food uptake alone) is to compare the two kinetic parameters of k_f and k_e. A greater than 1 of TTF requires that k_f is greater than k_e.

The biokinetic model provides a complete picture of the overall bioaccumulation of contaminants in organisms. Biological or chemical parameters identified in the model are realistic, but their measurements require much more specialized methodology (eg, radiotracers or clean techniques). Currently, we need a more comprehensive understanding of these kinetic parameters, which will ultimately affect their measurement.

4.3 KINETIC PARAMETERS

Kinetic modeling plays an important role in studying the bioaccumulation of contaminants (especially metals) in marine organisms (as reviewed by Wang and Rainbow (2008)). There are quite a few parameters that need to be determined for the model simulation. Accurate measurements of these parameters are among the challenges faced by ecotoxicologists. For example, C_w and C_f require specialized analytical skills, especially in the measurements of low C_w in the water. FR, IR, and g are the subjects of physiological ecology studies (eg, feeding and growth processes of marine organisms under complex ecological conditions). AE, k_u, and α are the subjects of ecotoxicological study. Over the past decades, there has been significant progress in refining the methodology to quantify these parameters. Researchers now need to carefully consider the methods and caveats for different targeted organisms or contaminants. As methodology improves, the variation of these parameters under different environmental and biological conditions becomes more important.

4.3.1 Dissolved Uptake Rate Constant k_u

Table 4.2 summarizes some of the measured k_u values of metals in different groups of marine organisms, which generally suggests that (1) relatively small organisms tend to have a greater k_u value compared to larger organisms and (2) Class B metals have higher k_u values compared with Class A metals. SeO_3^{2-} and CrO_4^{2-} had the lowest k_u, most likely because they are transported through anionic channels.

In comparison with the k_u, α (absorption efficiency), the first-order kinetic parameter, is much less well known and investigated; α can be calculated as k_u/FR, whereas FR is often not simultaneously quantified in the majority of kinetic determinations of k_u. Strictly speaking, comparison of bioavailability among different organisms should be based on α instead of k_u since the FR can differ greatly among species of marine organisms.

TABLE 4.2 Waterborne Dissolved Uptake Rate Constants of Metals in Marine Animals (L/g/d) (Wang, 2011b)

		Ag	Cd	Zn	Se	Hg(II)	CH_3Hg
Copepods	*Temora* sp.	8.45–12.84	0.626–0.796	2.388–3.993	0.017–0.035		
Bivalves	*Mytilus edulis*	1.794	0.365	1.044	0.035		
	Perna viridis	0.638–8.212	0.206	0.637	0.019	10.51	99.6
	Ruditapes philippinarum	2.620	0.064	0.234			
	Crassostrea rivularis		0.719	2.050	0.060		
	Saccostrea glomerata		0.534	1.206	0.064	2.604	3.445
	Macoma balthica		0.032	0.091			
	Chlamys nobilis		0.455	0.677			
Polychaetes	*Nereis succinea*	1.853	0.028	0.359	0.006	1.27	2.58
Gastropods	*Thais clavigera*		0.030	0.069		0.079	0.108
	Haliotis diversicolor	1.78	0.056			0.32	
Sipunculans	*Sipuncula nudus*		0.0018	0.035			
Fish	*Acanthopagrus schlegeli*		0.002	0.0055			
	Plectorhinchus gibbosus					0.195	4.515
	Lutjanus argentimaculatus		0.005	0.0100	0.0008		
	Sparus auratus		0.005	0.004			

Numerous studies have determined the influences of different chemical and biological factors on the k_u. For example, Veltman et al. (2008) demonstrated the significant relationships between the covalent index representing the binding affinity of metals/biotic ligand and the k_u of 10 metals in 17 aquatic species. The covalent index can be calculated from $x^2_m\, r$, where x_m is the Pauling electronegativity value and r is the ionic radius. The α of metals was also significantly related to the covalent index. These relationships suggest that facilitated membrane transport is likely the main mechanism for the uptake of many of these metals.

The uptake rate, k_u, is also closely related to the biology of the organisms, eg, the body size of the animals. Relatively small animals display a higher k_u than larger animals (Zhang and Wang, 2007a; Wang and Dei, 1999). An interesting question in ecotoxicology is whether uptake is dependent on the growth rate of the organisms. Based on Eq. (4.16), if the growth rate is independent of the uptake rate (or k_u), a faster growth of organisms may lead to a reduced accumulated concentration (eg, the k_u remains constant). Conversely, if the uptake is related to g, bioaccumulation will be dependent on the relative magnitude of changes of these two parameters. Answering this question is fundamental in explaining the feedback mechanisms of metal accumulation in phytoplankton (Sunda and Huntsman, 1998). For example, a metal may inhibit the growth of phytoplankton, which then leads to a reduction in g and then a further increase in the metal concentration in the cells. This is considered as a positive feedback mechanism to phytoplankton.

Conversely, the limitation of an essential metal (eg, reduced availability of Cu or Zn) may reduce phytoplankton growth, leading to an increase in cellular metal concentration. Such feedback alleviates essential metal limitations in the cells. Previous studies specifically designed to test the dependence of metal uptake on phytoplankton showed that metal uptake by the cells increased with increasing phytoplankton growth (Miao and Wang, 2004; Wang et al., 2005). Therefore, with increasing cellular growth concentrations of accumulated metals in cells will be dependent on the relative degree of changes of these two parameters (uptake vs. growth).

Environmental factors can of course considerably affect the k_u of contaminants. Among these environmental factors, salinity, temperature, dissolved organic matter (DOM), other competing ions such as H^+, Ca^{2+}, Mg^{2+}, and dissolved oxygen have received the most attention. Most of such influences are due to changes in the speciation of contaminants as well as the physiological and biochemical processes of the animals. Salinity is probably the best studied environmental factor influencing the k_u (Wang et al., 1996; Wang and Dei, 1999).

In addition to its direct effect on speciation, salinity can also cause physiological changes. For example, Zhang and Wang (2007b) acclimated the marine seabream (*Acanthopagrus schlegeli*) to different salinities (including freshwater) and quantified uptake of Cd and Zn. With a decrease of salinity from 35 to 0 psu, the uptake of Cd increased by 31 times and of Zn increased by 16 times. A significant relationship was then between increased uptake and an increase in water-free ion concentration (for Cd uptake in the gills, the relationship was 1:1), strongly indicating that change of free [Cd^{2+}] could entirely explain the influence of salinity on Cd uptake in this marine fish. Transport of metals via the ion channel (Ca channel) was different at different salinities. At high salinity, the calcium channel was not involved in the uptake of Cd and Zn, which was primarily facilitated. At lower salinity (eg, freshwater), the calcium channel was actively involved in the uptake of Cd and Zn (transcellular uptake). This work illustrates the complexity of metal chemistry and fish physiology in affecting metal uptake in fish at different salinities.

Blackmore and Wang (2003a) compared metal uptake in marine green mussels *Perna viridis* from two contrasting salinity sites in Hong Kong waters after acclimation in the laboratory at different salinities. Mussels collected from a high-salinity site accumulated metals 1.2 to 2.2 times faster than mussels collected from a low-salinity site when they were acclimated at intermediate and high experimental salinities (>17 psu). This difference was not explained by the gill surface area (which was similar in both populations of mussels) and filtration rate. Instead, the apparent water permeability of the high-salinity population was on average about 1.6 times greater compared with the low-salinity population and may partially account for the difference in metal uptake between these two populations. This study also suggested that the effect of salinity on metal uptake is dependent on metal biogeochemistry as well as a range of physiological responses.

The role of DOM in metal uptake is contradictory for different organisms and presumably for different quantities and qualities of DOM. One school of thought is that DOM can complex metals in the water, thereby effectively reducing the bioavailable fraction of metals and reducing their uptake. However, there is also evidence that DOM–metal complexes may be directly available for organisms such as filter-feeding bivalves. Such cotransport of a DOM–metal complex has been demonstrated in bivalves (Roditi et al., 2000; Pan and Wang, 2004). Clearly the functional physiology of the animals should be considered in examining the influences of different factors on metal uptake in marine organisms. This is probably one of the most important considerations in future bioaccumulation studies.

4.3.2 Assimilation Efficiency

The concept of assimilation efficiency (AE) in bioaccumulation has been well developed. Numerous studies have quantified the dietary assimilation of contaminants (mostly metals) in marine animals, largely as a result of the availability of radiotracer techniques. Table 4.3 summarizes the available AEs of different metals in different marine animals. Similar to k_u, the AEs are influenced by many biological, chemical, and environmental factors (Wang and Fisher, 1999; Wang and Rainbow, 2008).

An active area of research on the bioaccumulation of dietary metals is the control of different cellular fractions of metals in the prey on the assimilation by their predators. In early studies, the distribution of metals in phytoplankton was divided into two fractions: the cell wall/membrane, and the cytoplasmic fraction. These studies found that metals bound with the cytosolic fraction displayed a much higher bioavailability to marine herbivores such as copepods and bivalves than to other marine organisms (Reinfelder and Fisher, 1991; Wang and Fisher, 1996). Later studies then focused on the controls of internal metal sequestration in prey on dietary bioavailability in predators. Wallace and Luoma (2003) proposed a concept of trophically available metal (TAM) to explain the trophic availability of Cd and Zn in the bivalve prey to grass shrimp *Palaemon macrodatylus*. This concept has been tested and verified in several species of marine predators such as gastropods and fish (Cheung and Wang, 2005; Zhang and Wang, 2006; Rainbow et al., 2007).

TABLE 4.3 Dietary Assimilation Efficiencies of Metals (%) in Different Marine Animals (Wang, 2011b)

Animals		Ag	Cd	Zn	Se	Hg(II)	CH_3Hg
Copepods	*Acartia* sp.		66	9	38		
	Temora sp.	8–19	33–53	52–64	50–59		
Bivalves	*Mytilus edulis*	4–34	28–34	32–45	56–72		
	Perna viridis	13–32	11–25	21–32	59		
	Ruditapes philippinarum	30–52	38–55	33–59		41–70	
	Crassostrea rivularis		58–75	68–80	56–74		
	Saccostrea glomerata		52–67	60–65	52–68		
	Macoma balthica		88	50	74		
	Chlamys nobilis		94	83			
Polychaetes	*Nereis succinea*	12–27	5–44	24–57	36–60	20	70
Gastropods	*Thais clavigera*		75	80		70	95
	Haliotis diversicolor	58–83	33–59			65–78	
Sipunculans	*Sipuncula nudus*		6–30	5–15			
Fish	*Terapon jarbua*		3–9	2–52	13–26	23–43	90
	Plectorhinchus gibbosus					20	80
	Lutjanus argentimaculatus		20	40	65		
	Sparus auratus		45	18			

Guo et al. (2013) investigated how the subcellular metal distribution and the metal burden in prey affected the transfer of metals to a marine fish, the grunt *Terapon jarbua*. Oysters, *Crassostrea hongkongensis*, which had different contaminated histories, were collected and separated into three subcellular fractions (metal-rich granules, cellular debris, and a combined fraction of organelles, heat-denatured proteins, and metallothionein-like proteins, defined as TAM). These purified fractions, representing a wide range of metal concentrations, were fed to the fish for a period of 7 days at a daily comparable feeding rate of 3% of fish body weight. Bioaccumulation of Cd, Cu, and Zn was quantified by the trophic transfer factor (TTF). All three subcellular fractions were bioavailable to the fish. With a certain degree of variation among metals, the TTFs showed a metal uptake sequence of cellular debris > TAM > metal-rich granules, indicating the impact of subcellular distribution in prey on metal bioavailability. However, significant inverse relationships between the TTFs and the metal concentrations in diets were also found in this study, especially for Cd and Zn, suggesting the high dependence of TTF on metal concentration in prey. In this case, the subcellular metal compartmentalization might be less important than the overall metal concentration in prey influencing the trophic transfer.

While the physicochemical form of the accumulated metal in prey is an important factor affecting the metal AE and trophic transfer (Luoma and Rainbow, 2008; Rainbow et al., 2011), physiological conditions certainly affect assimilation (Wang and Rainbow, 2005). Digestion processes such as the gut passage time, partitioning of extracellular and intracellular digestion, or induction of various binding ligands are all important considerations.

4.3.3 Efflux

Efflux is an important determinant of metal accumulation in marine organisms. The application of radiotracer technique offers an opportunity to quantify the efflux of contaminants in a variety of marine organisms (Table 4.4). However, efflux measurements must be done carefully with particular attention to the duration of radiolabeling (allowing sufficient equilibration of radioisotopes within the internal tissues). Fig. 4.6 illustrates the results of radiolabeling to measure metal efflux in marine animals. A predominant distribution in the digestive system (fast uptake compartment, eg, a short feeding time or radiolabeling) results in a relatively small partitioning of radiotracer in the physiological turnover (slow) compartment (involved mainly in metabolism) (k_2), which then leads to the predominance of fast compartment loss over the true efflux. Conversely, if the physiological turnover compartment becomes a dominant pool of metals, the potential influence of the fast metal uptake compartment on efflux would be small.

Compared to influx or uptake, the role of efflux in determining bioaccumulation has been less recognized, primarily due to the relatively conservative behavior of efflux in marine animals, as well as the long timeframe required to realistically determine efflux. In measuring efflux, organisms are exposed to radiotracers and then depurated for a longer duration (weeks to months for larger animals such as bivalves or fish). Therefore, k_e measurement is much more tedious as compared to measurements of dissolved uptake or dietary assimilation. A common consensus for efflux is that the difference among metals is smaller than the difference among organisms. For example, the efflux rate constant of different metals in marine bivalves ranges between 0.01 and 0.03 d^{-1}, whereas for small organisms such as copepods with a much higher weight-specific metabolic rate, the k_e can be much higher. Some of these animals even have a turnover rate of once per day (Table 4.4).

Efflux can lead to the hyperaccumulation of metals (Luoma and Rainbow, 2005). Oysters are the best-known marine organisms

TABLE 4.4 Typical Efflux Rate Constants of Metals in Different Marine Organisms (d^{-1}) (Wang, 2011b)

	Ag	Cd	Zn	Se	Hg(II)	CH_3Hg
COPEPODS						
Acartia sp.		0.590	0.620	0.890		
Temora sp.	0.173–0.294	0.108–0.297	0.079–0.108	0.155		
BIVALVES						
Mytilus edulis	0.034	0.011	0.020	0.026		
Perna viridis	0.032–0.087	0.020	0.029			
Ruditapes philippinarum		0.010	0.023			
Crassostrea rivularis		0.014	0.014	0.034		
Saccostrea glomerata		0.004	0.003	0.013		
Macoma balthica		0.018	0.012			
Chlamys nobilis		0.005–0.009	0.012–0.023			
POLYCHAETES						
Nereis succinea					0.027	0.014
Gastropods						
Haliotis diversicolor	0.003	0.011			0.011	
FISH						
Acanthopagrus schlegeli		0.089	0.016	0.043		
Plectorhinchus gibbosus					0.028–0.055	0.010–0.013
Lutjanus argentimaculatus		0.025–0.047	0.015	0.027–0.031		
Sparus auratus		0.016	0.006			

accumulating very high concentrations of Cu and Zn; barnacles are the typical hyperaccumulators of Zn. These high metal concentrations are primarily driven by the very low efflux of metals. For example, the k_e of Zn in the barnacle *Balanus amphitrite* can be as low as 0.001 d^{-1} (Table 4.4). Oysters also have a very low efflux of Cu and Zn. The physiological or biochemical mechanisms underlying the k_e are less well known, for instance, how metals are turned over in the organisms and whether their turnover is coupled with the turnover of their potential binding ligands (such as proteins).

Fig. 4.7 provides a conceptual framework for the possible subcellular systems involved in the efflux of metals although presently knowledge is lacking on the kinetic changes of metal distributions in the internal metal pools and how these will affect overall efflux. This is clearly an important future research area.

Ng and Wang (2005) examined the dynamics of subcellular distribution of Cd, Ag, and Zn in the green mussel *P. viridis* by partitioning the metals into the insoluble fraction (IF), heat-sensitive proteins, and metallothionein-like proteins (MTLP) during elimination. During efflux,

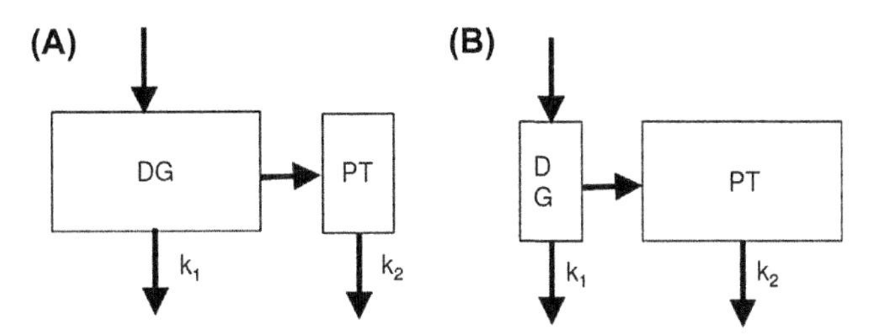

FIGURE 4.6 An illustration of the loss of metals from digestive (fast) compartment (DG) and physiological turnover (slow) compartment (PT). (A) dominated by digestion process; (B) dominated by physiological turnover process.

metals in the soluble fraction mediated depuration, whereas metals in the insoluble fraction acted as a final storage pool. A higher efflux rate of Ag and Cd was related to a higher partitioning in the MTLP and a lower partitioning in the IF.

Pan and Wang (2008a) investigated the dynamics of subcellular distribution in the scallop *Chlamys nobilis*, which accumulated very high concentrations of Cd and Zn. Storage in the nontoxic form both in organelles and MTLP accounted for the low efflux rate of Cd from scallops. During efflux, an increasing percentage of Cd was sequestered in organelles (eg, lysosomes), but that in the MTLP fraction remained stable, implying that MTLP may act simply as a sink in the metal detoxification process. Internal Cd may be eventually eliminated slowly from lysosomes or removed to and deposited in the MTLP fraction for storage, leading to the highest percentage of Cd in the MTLP fraction.

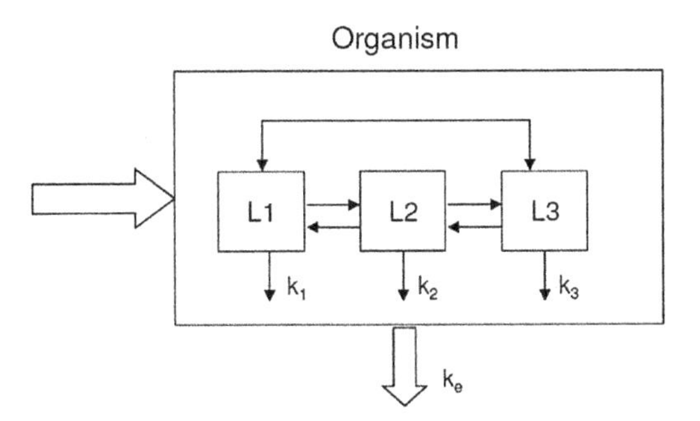

FIGURE 4.7 Hypothetical subcellular ligands binding with metals (L1, L2, L3) and possible efflux from different compartments. k_e is the overall efflux from the organism.

In contrast, Zn is eliminated more quickly than Cd from the scallops; its redistribution among each subcellular compartment was much faster than the redistribution of Cd, suggesting an effective regulation mechanism.

Efflux is influenced by environmental conditions such as temperature, tissue concentrations of the contaminants, route of exposure, as well as food conditions (Wang and Fisher, 1998) or internal sequestration. For example, once Ag was complexed by S in tissues of the mussel *P. viridis*, depuration basically did not occur (Shi et al., 2003). Buchwalter et al. (2008) also found that the efflux of Cd in 21 species of aquatic insects was directly related to its partitioning in the MTLP fraction. Poteat et al. (2013) explored species-specific traits in contributing to differences in environmental sensitivity among species. The efflux of Cd and Zn strongly covaried across species in aquatic insect families (Ephemerellidae and Hydropsychidae). Taxonomic groups (arthropods, mollusks, annelids, and chordates, 77 species total) exhibited marked variability in efflux, suggesting that some groups are more constrained than others in their abilities to eliminate metals.

4.4 APPLICATION OF BIOACCUMULATION MODELING

A bioaccumulation conceptual model, which provides a general framework for studying bioaccumulation, is reviewed by Luoma and Rainbow (2008) and Wang and Rainbow (2008).

The model cannot only be used to simulate and predict bioaccumulation, but also to examine the various exposure pathways, and generally to help understand the bioaccumulation processes. It is well known that Hg concentrations in marine fish increase with body size, in strong contrast with most of the other metals (ie, their concentrations decrease with body size). Such lack of growth dilution puzzled ecotoxicologists, whose speculations included shifts in dietary requirements (eg, a shift from a more herbivorous diet to a predatory diet) and slower elimination of Hg in larger fish. Dang and Wang (2012) addressed this interesting issue by first quantifying the size dependence of Hg concentrations in field-collected juvenile blackhead seabream *A. schlegeli*. Due to the size limit of kinetics measurements, only juvenile fish (3.8–12 cm) were examined. Hg measurements confirmed that total mercury (THg) and methylmercury (MeHg) concentrations were related to fish mass over a wide size range with a power coefficient of 0.19 and 0.33, respectively. Table 4.5 summarizes the body size dependence of different biokinetic parameters of Hg(II) and MeHg in the fish. Negative correlations between Hg biokinetics and body size were documented for k_u, k_e, and g, whereas the AE of Hg(II) increased with body size and MeHg had a comparable AE among different sizes of fish.

TABLE 4.5 Allometric Relationship Hg(II) and MeHg in Marine Fish *Acanthopagrus schlegeli*

Parameters	Hg(II)	MeHg
k_u (L/g/d)	$0.24W^{-0.68}$	$0.36W^{-0.54}$
AE (%)	$25.6W^{0.260}$	80–100
k_e (d^{-1})	$0.050W^{-0.36}$	$0.0062W^{-0.40}$
g (d^{-1})	$0.0077W^{-0.42}$	
Body concentration (ng/g)	$5.91W^{0.33}$	$21.5W^{0.19}$

Data from Dang, F., Wang, W.-X., 2012. Why mercury concentration increases with fish size? Biokinetic explanation. Environ. Pollut. 163, 192–198.

With these determined relationships, Dang and Wang (2012) then modeled the scaling exponents of Hg accumulation to be 0.21 for MeHg and 0.21 to 0.25 for THg, which were in fact very close to independent field measurements (0.33 for MeHg and 0.19 for THg). The allometric biokinetic parameters thus reasonably well explained the size-dependent mercury accumulation patterns observed in juveniles in natural systems. Sensitivity analysis further showed that a decrease of g and k_e with increasing body size effectively increased the Hg concentrations, and were the key drivers for such relationships.

Dang and Wang (2012) demonstrated that slower growth coupled with a lower mercury efflux rate increased both MeHg and THg concentrations and yielded positive size-dependent allometric correlations. To manage Hg contamination in fish, factors that enhance the g and k_e should be explored. Interestingly, farmed fish have a low Hg body concentration as a result of their rapid growth in the farming system, as well as the use of artificial diets with low Hg concentrations; Hg concentrations are also lower in eutrophic systems where the fish can grow rapidly than in oligotrophic waters. Chen and Folt (2005) analyzed fish Hg burdens and plankton densities from 38 lakes in the northeastern USA and found a negative correlation between zooplankton density and Hg concentrations in zooplankton and in both herbivorous and predatory fish. Zooplankton density alone explained more than 40% of the variation in predatory fish Hg concentrations across lakes. Rapid growth from high-quality food consumption can also significantly reduce the accumulation and trophic transfer of MeHg in freshwater food webs (Karimi et al., 2007).

Ward et al. (2010) also examined the effect of growth on trace element concentrations in fish by measuring the concentrations of seven metals (As, Cd, Cs, Hg, Pb, Se, Zn) in stream-dwelling

Atlantic salmon from 15 sites encompassing a 10-fold range in salmon growth. The fast-growing salmon had lower concentrations of all metals than slow-growing salmon, similarly suggesting that dilution of metals in larger biomass led to lower concentrations in fast-growing fish.

Another application of the biokinetic model is that of predicting the TTF of different contaminants in marine animals involved only in trophic transfer (dietary or food intake). Assuming that food is the only source for contaminant accumulation in marine animals, the TTF is calculated as:

$$\mathrm{TTF} = \mathrm{BMF} = C/C_f = \mathrm{AE} \times \mathrm{IR}/k_e \quad (4.19)$$

Clearly, potential biomagnification of contaminants is determined by three kinetic parameters, ie, AE, IR, and k_e. Wang (2002) predicted the TTF in different species of marine animals based on extensive measurements of these biokinetic parameters. Based on that prediction, contaminants and predators differ considerably in their potential biomagnification along food chains. Whereas it is well known that MeHg is biomagnified, other metals such as Cs, Se, or even Zn (for small fish) can potentially have a TTF > 1 in marine fish. Many benthic invertebrates such as bivalves and predatory gastropods show a high potential of TTF being greater than 1 for different reasons. For example, bivalves have a very high IR, as well as a high dietary metal AE, whereas predatory gastropods have a very high AE, although their IRs are typically low. Fish have a relatively low AE and IR for the majority of the metals, and their potential TTF is the lowest for the majority of metals. Ecotoxicologists need to appreciate the diversity of metal biology in studying the food chain transfer of contaminants. Such complexity has often been neglected in making the general conclusion regarding biomagnification of contaminants.

Strictly speaking, TTF is not a constant but rather is variable depending on the three kinetic parameters (AE, IR, and k_e) that are highly controlled by the environmental conditions as well animal physiology. In some cases, metal concentrations in marine animals are not related to food chain metal concentrations. Filter-feeding oysters hyperaccumulate Cu and Zn; scallops hyperaccumulate Cd; and barnacles hyperaccumulate Zn. The highest concentrations of Cu and Zn in oysters were around 2.5% of body tissue dry weights collected from a severely contaminated estuary in China (Wang et al., 2011; Wang and Wang, 2014). Wang et al. (2011) and Tan et al. (2015) have studied the mechanisms underlying such phenomenal metal concentrations.

The TTF for organic contaminants in marine animals has been relatively less well documented compared to metals; studies examining the biokinetic processes of organic contaminants in marine animals (eg, DDTs, PCBs, PAH) are limited. Some typical values for these contaminants are summarized in Table 4.6.

4.5 BIOMONITORING

The classical example of biomonitoring based on bioaccumulation is the Mussel Watch Program, first proposed in the 1970s (Phillips, 1976, 1977; Goldberg et al., 1978). Since the concentrations of many contaminants in marine waters are generally low and difficult to accurately measure, using biomonitors to reflect environmental contaminant concentrations is certainly an attractive approach. Bioconcentration of contaminants in the biomonitors can reach up to a few orders of magnitude higher than the ambient concentrations (Table 4.1) and can be measured with reasonable certainty. The most important reason for using the biomonitors is that these measured concentrations indicate the bioavailable fractions in the environments and are thus much more biologically relevant than simply measuring chemical concentrations in water (or sediments).

However, following significant improvements in analytical techniques, it is now possible to

TABLE 4.6 Biokinetics of Selected Organic Contaminants in Marine Animals

	k_u (L/g/h)	AE (%)	k_e (d^{-1})	TTF	References
BENZO(A)PYRENE					
Copepod, *Acartia erythraea*	1.20	2–25	0.82–1.66	<0.1	Wang and Wang (2006)
Fish, *Lutjanus argentimaculatus*	0.157	30–50			
DDT					
Copepod, *Acartia erythraea*	1.20	10–30	0.01–0.05	1–9	Wang and Wang (2005)
Fish, *Lutjanus argentimaculatus*	0.30	70–100	0.002	4–45	
DIOXINS					
Copepod, *Acartia spinicauda*	0.36	30–60	0.02–0.30	2–22	Zhang et al. (2011)
Snapper, *Acanthopagrus schlegeli*	0.08	30–70			

quantify very low environmental contaminant concentrations accurately and use these data in monitoring programs. Diffusion in gels is now a widely used technique to measure labile metal concentrations in water (Zhang and Davison, 1995, 2000); semipermeable membrane devices are used to measure organic contaminant concentrations (Preset et al., 1992; Huckins et al., 1993). However, the linkage between these chemical measurements and bioavailability to biota remains uncertain, in particular for pulse or long-term exposures. Biomonitors provide a time integrated measurement of bioavailable contaminant concentrations, allowing for "apples to apples" geographic and temporal comparisons.

4.5.1 Choice of Biomonitors

Practicality and scientific considerations govern the choice of biomonitors. The most practical consideration is distribution of species that allows for long-term monitoring including spatial and temporal comparisons. Common choices for biomonitors include sedentary invertebrates such as mussels, oysters, and barnacles. These species have been extensively used in different national and international monitoring programs, with numerous publications in the literature. At a national scale, the availability of a single species may be limited, and different species of biomonitors but with a wide range of availability may be needed. In the National Status and Trends Program (NS&T) in the USA, the mussels are employed on the east and west coasts, whereas the oysters are employed on the south coast (O'Connor and Lausenstein, 2006).

Most biomonitoring programs are implemented in coastal waters where environmental conditions such as salinity may be relatively stable. Monitoring programs conducted in transitional waters such as estuaries require euryhaline species that can tolerate a wide range of salinity. Estuarine oysters such as Hong Kong oysters, *C. hongkongensis*, have been used to monitor metal contamination due to their tolerance to a wide range of salinity (5–25 psu).

Sediment contamination in sediments has been monitored using deposit-feeding invertebrates such as polychaetes (Bryan and Langston, 1992). However, bioavailability of metals in sediment is also a complex issue and widely dependent on sediment geochemistry and environmental processes, such as different "speciation" of metals in the sediments, mobilization to interstitial waters, or biological activity (which can affect contaminant distribution by, eg, burrowing or which can affect contaminant exposure by behavior, eg, avoidance). The concept of biomonitoring using deposit-feeding animals has been much well developed, and most of the earlier studies attempted to understand the bioavailability and toxicity of contaminated sediments. In a most recent study, Fan et al. (2014) examined the relationship between metal accumulation in a deposit-feeding polychaete *Neanthes japonica* and metal concentration and geochemical fractionation (Cd, Cu, Pb, Zn, and Ni) in sediments of Jinzhou Bay from Northern China. Their study showed that metal accumulation in polychaetes was significantly influenced by Fe or Mn content and to a lesser degree by organic matter. Prediction of metal bioaccumulation in polychaetes was greatly improved by normalizing metal concentrations to Mn content in sediment. The geochemical fractionation of metals in sediments including the exchangeable, organic matter and Fe/Mn oxides were important in controlling the sediment metal bioavailability to polychaetes. In another study, Tan et al. (2013) measured the metal concentrations in a sipunculan worm, *Phascolosoma arcuatum*, and in sediments collected from intertidal zones of Xiamen, China. Significant correlations were found for the concentrations of chromium, nickel, copper, zinc, cadmium, and lead in sediments and their concentrations in both somatic tissue and coelomic fluid of the worms. This study proposed that it is possible to measure the coelomic-fluid metal concentrations as a rapid assessment of metal bioavailability in marine sediments.

4.6 PRINCIPLES OF BIOMONITORING AND SOME CONSIDERATIONS

Theoretically, bioaccumulation in biomonitors should be directly proportional to the bioavailable contaminant concentration in the environment being monitored. Measurements of contaminant concentration in the animals then indicate the bioavailable contaminant concentration in the environment. With the application of kinetic modeling, the relationship between C_{ss} and total concentration in the environment (C_t) can be described by the following equation (Wang et al., 1996):

$$C_{ss} = [k_u + AE \times IR \times K_d]/[k_e \times (1 + TSS \times K_d)] \times C_t \tag{4.20}$$

$$BAF = C_{ss}/C_t = [k_u + AE \times IR \times K_d]/[k_e \times (1 + TSS \times K_d)] \tag{4.21}$$

where K_d is the partitioning coefficient of metals in food and TSS is the total suspended solid load.

Eq. (4.21) can be further simplified to consider only C_w:

$$C_{ss}/C_w = [k_u + AE \times IR \times K_d]/k_e \tag{4.22}$$

Thus, comparison among stations or times assumes that BAF remains constant (Eq. (4.21)). If BAF is variable among stations or times, the data should be treated with caution. To ensure the constancy or homogeneity of BAF, the biokinetic parameters in Eq. (4.21) need to be comparable among stations and/or at different timeframes. This forms the basis for the principles underlying the biomonitoring program. Blackmore and Wang (2003b) compared the biokinetics of metals in mussels globally, verifying this premise. Mussels (*Mytilus* spp.) were collected from New York (*Mytilus edulis*, USA), Alaska (*Mytilus trossulus*, USA), Plymouth (*M. edulis*, UK), and Dalin (*Mytilus galloprovincialis*, China). Green mussels *P. viridis* were collected from different areas in Hong Kong. The results showed that the biokinetics of Cd and Zn (AE, k_u, and FR) in different populations of *Mytilus* spp. or *P. viridis* were comparable, suggesting that these kinetics were not significantly affected by the tissue concentrations of metals in the mussels. Therefore, BAF should be theoretically consistent among different stations, and it is possible to compare mussels at the global scale.

However, environmental conditions may significantly affect the biokinetics of contaminants and eventually lead to a change in BAF. For example, pre-exposure of organisms to metals may modify metal biokinetics (Wang and Rainbow, 2005). Another critical consideration is the growth rate of the biomonitors. Growth is controlled by environmental conditions such as temperature and food conditions, which may result in a change in bioaccumulation, as previously discussed. Growth is generally ignored in the bioaccumulation of many invertebrates because it is significantly lower than the efflux of contaminants. Pan and Wang (2008b) conducted a transplantation experiment using the scallop *C. nobilis* to investigate how scallops can accumulate metals differently from different marine environments with comparable ambient metal concentrations. Their study suggests that direct measurement of metal concentrations in aqueous and food phases cannot provide a full understanding of metal bioavailability in the marine environment. Instead, environmental conditions such as food availability and hydrology must be simultaneously considered when attempting to study metal bioaccumulation in marine bivalves. The growth rate does not solely act as a growth dilution term in the biokinetic model but is also related to other physiological parameters such as the ingestion rate, which subsequently affects metal concentrations in the animals. Overall, how the physiology of organisms can potentially modify interpretation of biomonitoring data remains an area of active research.

Given the potential influences of physiology and biochemistry on the biokinetics of contaminants and thus the BAF, Liu and Wang (2015) proposed a concept to link the concentrations of metals with the concentrations of macronutrients and macroelements such as K, Na, Ca, and Mg. Mussels (*M. edulis* and *P. viridis*) were collected from 21 sites in 8 countries/regions (South and North America, Europe, Australia, Asia), and the concentrations of macronutrients (C, N, P, S), major cations (Na, Mg, K, Ca), and trace elements (Al, V, Mn, Fe, Co, Ni, Cu, Zn, As, Se, Mo, Cd, Ba, Pb) in the whole soft tissues were analyzed. The mussels collected from different areas displayed significant variations in tissue Na and Mg, suggesting that there was a difference in the salinity among different stations (tissue concentrations of these and other major ions are a proxy for seawater salinity) (Liu and Wang, 2015). Salinity can affect the speciation of metals and thereby their bioavailability to mussels (and other marine animals), in addition to physiology. In this study, negative relationships between most trace cations (ie, Al, Mn, Fe, Co, Cu, Zn, Cd, and Ba) and major cations were observed in *P. viridis*, similar to many earlier findings of the negative relationships between tissue concentration of Cu, Zn, or Cd and salinity (Wright, 1995; Lee et al., 1998; Blackmore and Wang, 2003a). In contrast, positive relationships between major cations and trace oxyanions (ie, As, Se, and Mo) were documented in the mussels. Overall, about 12–84% of the variances in the trace elements were associated with major cations; therefore, salinity must be considered when mussels (and other marine animals) are employed as biomonitors.

A second major finding of Liu and Wang (2015) was the significant correlation between trace element concentrations in mussels and macronutrient concentrations; between 14% and 69% of the variances in the trace elements were associated with macronutrients. Macronutrient concentrations were strongly associated with animal growth and reproduction; correlations indicated that these biological processes strongly influenced the bioaccumulation of some trace elements. These correlations have important implications in the interpretation of mussel biomonitoring data. For instance, simultaneous measurement of tissue N and P is useful, because N and P are well-documented indicators of animal growth and reproduction, which can strongly affect the bioaccumulation of some trace elements. This is especially important in many coastal and estuarine waters where nutrient enrichment (eg, eutrophication) is common.

Macronutrients can significantly affect the metal uptake and trophic transfer in marine planktonic food chains (Wang and Dei, 2001; Wang et al., 2001). Coupling relationships among macronutrients, major cations, and trace elements should be a focus of future research, especially in regions where environmental variability is large. Mechanistic studies are also to experimentally test the influences of these environmental variabilities/stresses on actual metal bioaccumulation and bioavailability.

4.7 FUTURE POSSIBILITIES AND PROBABILITIES

Bioaccumulation has come a long way of study and modeling now provides an important approach in bioaccumulation study. Significant challenges remain for our understanding of the bioaccumulation under fluctuating environmental conditions, eg, metal releases under irregular patterns. It is important to appreciate the complexity and diversity of metal biology in aquatic organisms, without which it would be impossible to further move the field forward. In this regard, there are still ample questions in bioaccumulation study. Furthermore, the application of modeling in biomonitoring is still a great challenge for ecotoxicologists.

References

Blackmore, G., Wang, W.-X., 2003a. Inter-population differences in Cd, Cr, Se, and Zn accumulation by the green mussel *Perna viridis* acclimated at different salinities. Aquat. Toxicol. 62, 205–218.

Blackmore, G., Wang, W.-X., 2003b. Variations of metal accumulation in marine mussels at different local and global scales. Environ. Toxicol. Chem. 22, 388–395.

Bryan, G.W., Langston, W.J., 1992. Bioavailability, accumulation and effects of heavy metals in sediments with special reference to United Kingdom estuaries: a review. Environ. Pollut. 76, 89–131.

Buchwalter, D.B., Cain, D.J., Martin, C.A., Xie, L., Luoma, S.N., Garland Jr., T., 2008. Aquatic insect ecophysiological traits reveal phylogenetically based differences in dissolved cadmium susceptibility. Proc. Natl. Acad. Sci. USA 105, 8321–8326.

Chen, C.Y., Folt, C.L., 2005. High plankton densities reduce mercury biomagnification. Environ. Sci. Technol. 39, 115–121.

Cheung, M., Wang, W.-X., 2005. Influence of subcellular metal compartmentalization in different prey on the transfer of metals to a predatory gastropod. Mar. Ecol. Prog. Ser. 286, 155–166.

Dang, F., Wang, W.-X., 2012. Why mercury concentration increases with fish size? Biokinetic explanation. Environ. Pollut. 163, 192–198.

DeForest, D.K., Brix, K.V., Adams, W.J., 2007. Assessing metal bioaccumulation in aquatic environments: the inverse relationship between bioaccumulation factors, trophic transfer factors and exposure concentration. Aquat. Toxicol. 84, 236–246.

Fan, W., Xu, Z., Wang, W.-X., 2014. Metal pollution in a contaminated bay: relationship between metal geochemical fractionation in sediments and accumulation in a polychaete. Environ. Pollut. 191, 50–57.

Fisher, N.S., 1986. On the reactivity of meals for marine phytoplankton. Limnol. Oceanogr. 31, 443–449.

Goldberg, E.D., Bowen, V.T., Farrington, J.W., Harvey, G., Martin, J.H., Parker, P.L., Risebrough, R.W., Robertson, W., Schneider, E., Gamble, E., 1978. Mussel Watch. Environ. Conserv. 5, 101–125.

Guo, F., Yao, J., Wang, W.-X., 2013. Bioavailability of purified subcellular metals to a marine fish. Environ. Toxicol. Chem. 32, 2109–2116.

Huckins, J.N., Manuweera, G.K., Petty, J.D., Mackay, D., Lebo, J.A., 1993. Lipid containing semipermeable membrane devices for monitoring organic contaminants in water. Environ. Sci. Technol. 27, 2489–2496.

IAEA (International Atomic Energy Agency), 2004. Sediment Distribution Coefficients and Concentration Factors for Biota in the Marine Environment. Technical Report Series, No. 422. Vienna, Austria.

Karimi, R., Chen, C.Y., Pickhardt, P.C., Fisher, N.S., Folt, C.L., 2007. Stoichiometric controls of mercury dilution by growth. Proc. Natl. Acad. Sci. USA 104, 7477–7482.

Lee, B.G., Wallace, W.G., Luoma, S.N., 1998. Uptake and loss kinetics of Cd, Cr and Zn in the bivalves *Potamocorbula amurensis* and *Macoma balthica*: effects of size and salinity. Mar. Ecol. Prog. Ser. 175, 177–189.

Liu, F.J., Wang, W.-X., 2015. Linking trace element variations with macronutrients and major cations in marine mussels *Mytilus edulis* and *Perna viridis*. Environ. Toxicol. Chem. 34 (9), 2041–2050.

Luoma, S.N., Rainbow, P.S., 2005. Why is metal bioaccumulation so variable? Biodynamics as a unifying concept. Environ. Sci. Technol. 39, 1921–1931.

Luoma, S.N., Rainbow, P.S., 2008. Metal Contamination in Aquatic Environments: Science and Lateral Management. Cambridge University Press, London, UK.

McGeer, J.C., Brix, K.V., Skeaf, J.M., DeForest, D.K., Brigham, S.I., Adams, W.J., Green, A., 2003. Inverse relationship between bioconcentration factor and exposure concentration for metals: implications for hazard assessment of metals in the aquatic environment. Environ. Toxicol. Chem. 22, 1017–1037.

Miao, A.J., Wang, W.-X., 2004. Relationships between cell specific growth rate and uptake rate of cadmium and zinc by a coastal diatom. Mar. Ecol. Prog. Ser. 275, 103–113.

Ng, T.Y.T., Wang, W.-X., 2005. Dynamics of metal subcellular distribution and its relationship with metal uptake in marine mussels. Environ. Toxicol. Chem. 24, 2365–2372.

O'Connor, T.P., Lausenstein, G.G., 2006. Trends in chemical concentrations in mussels and oysters collected along the US coast: update to 2003. Mar. Environ. Res. 62, 261–285.

Pan, J.-F., Wang, W.-X., 2004. Differential uptake of particulate and dissolved organic carbon by the marine mussel *Perna viridis*. Limnol. Oceanogr. 49, 1980–1991.

Pan, K., Wang, W.-X., 2008a. The subcellular fate of cadmium and zinc in the scallop *Chlamys nobilis* during waterborne and dietary metal exposure. Aquat. Toxicol. 90, 253–260.

Pan, K., Wang, W.-X., 2008b. Validation of biokinetic model of metals in the scallop *Chlamys nobilis* in complex field environments. Environ. Sci. Technol. 42, 6285–6290.

Phillips, D.J.H., 1976. Common mussel *Mytilus edulis* as an indicator of pollution by zinc, cadmium, lead and copper. 1. Effects of environmental variables on uptake of metals. Mar. Biol. 38, 59–69.

Phillips, D.J.H., 1977. Use of biological indicator organisms to monitor trace metal pollution in marine and estuarine environments: review. Environ. Pollut. 13, 281–317.

Poteat, M.D., Garland, T., Fisher, N.S., Wang, W.-X., Buchwalter, D.B., 2013. Evolutionary patterns in trace metal (Cd and Zn) efflux capacity in aquatic organisms. Environ. Sci. Technol. 47, 7989–7995.

Preset, H.F., Jarman, W.M., Burns, S.A., Weismuller, T., Martin, M., Huckins, J.N., 1992. Passive water sampling via semipermeable-membrane devices (SMPDs) in concert with bivalves in the Sacramento San Joaquin River Delta. Chemosphere 25, 1811–1823.

Rainbow, P.S., Amiard, J.-C., Amiard-Triquet, C., Cheung, M.-S., Zhang, L., Zhong, H., Wang, W.-X., 2007. Trophic transfer of trace metals: subcellular compartmentalization in bivalve prey, assimilation by a gastropod predator and in vitro digestion simulations. Mar. Ecol. Prog. Ser. 348, 125–138.

Rainbow, P.S., Luoma, S.N., Wang, W.-X., 2011. Trophically available metal—a variable feast. Environ. Pollut. 159, 2347–2349.

Reinfelder, J.R., Fisher, N.S., 1991. The assimilation of elements ingested by marine copepods. Science 251, 794–796.

Roditi, H.A., Fisher, N.S., Sanudo-Wilhelmy, S.A., 2000. Uptake of dissolved organic carbon and trace elements by zebra mussels. Nature 407, 78–80.

Shi, D., Blackmore, G., Wang, W.-X., 2003. Effects of aqueous and dietary pre-exposure and resulting body burden on the biokinetics of silver in the green mussels, *Perna viridis*. Environ. Sci. Technol. 37, 936–943.

Spacie, A., McCarty, L.S., Rand, G.M., 1995. Bioaccumulation and bioavailability in multiphase systems. In: Rand, G.M. (Ed.), Fundamental of Aquatic Toxicology: Effects, Environmental Fates, and Risk Assessments. CRC Press, pp. 493–521.

Sunda, W.G., Huntsman, S.A., 1998. Processes regulating cellular metal accumulation and physiological effects: phytoplankton as model systems. Sci. Total Environ. 219, 165–181.

Tan, Q.G., Ke, C., Wang, W.-X., 2013. Rapid assessments of metal bioavailability in marine sediments using coelomic fluid of sipunculan worms. Environ. Sci. Technol. 47, 7499–7505.

Tan, Q.G., Wang, Y., Wang, W.-X., 2015. Synchrotron study of the speciation of Cu and Zn in two colored oyster species. Environ. Sci. Technol. 49, 6919–6925.

Thomann, R.V., 1981. Equilibrium-model of fate of microcontaminants in diverse aquatic food-chains. Can. J. Fish Aquat. Sci. 38, 280–296.

Veltman, K., Huijbregts, M., van Kolck, M., Wang, W.-X., Hendriks, J., 2008. Metal bioaccumulation in aquatic species: quantification of uptake and elimination kinetics using physico-chemical properties of metals. Environ. Sci. Technol. 42, 852–858.

Wallace, W.G., Luoma, S.N., 2003. Subcellular compartmentalization of Cd and Zn in two bivalves. II. Significance of trophically available metal (TAM). Mar. Ecol. Prog. Ser. 257, 125–137.

Wang, L., Wang, W.-X., 2014. Depuration of metals by the green-colored oysters (*Crassostrea sikamea*). Environ. Toxicol. Chem. 33, 2379–2385.

Wang, W.X., Dei, R.C.H., Xu, Y., 2001. Cadmium uptake and trophic transfer in coastal plankton under contrasting nitrogen regimes. Mar. Ecol. Prog. Ser. 211, 293–298.

Wang, W.-X., Dei, R.C.H., Hong, H., 2005. Seasonal study on the Cd, Se, and Zn uptake by natural coastal phytoplankton assemblage. Environ. Toxicol. Chem. 24, 161–169.

Wang, W.-X., Dei, R.C.H., 1999. Factors affecting trace element uptake in the black mussel *Septifer virgatus*. Mar. Ecol. Prog. Ser. 186, 161–172.

Wang, W.-X., Dei, R.C.H., 2001. Effects of major nutrient additions on metal uptake in phytoplankton. Environ. Pollut. 111, 233–240.

Wang, W.-X., Fisher, N.S., Luoma, S.N., 1996. Kinetic determinations of trace element bioaccumulation in the mussel, *Mytilus edulis*. Mar. Ecol. Prog. Ser. 140, 91–113.

Wang, W.-X., Fisher, N.S., 1996. Assimilation of trace elements and carbon by the mussel *Mytilus edulis*: effects of food composition. Limnol. Oceanogr. 41, 197–207.

Wang, W.-X., Fisher, N.S., 1998. Accumulation of trace elements in a marine copepod. Limnol. Oceanogr. 43, 273–283.

Wang, W.-X., Fisher, N.S., 1999. Assimilation efficiencies of chemical contaminants in aquatic invertebrates: a synthesis. Environ. Toxicol. Chem. 18, 2034–2045.

Wang, W.-X., Rainbow, P.S., 2005. Influence of metal exposure history on trace metal uptake and accumulation by marine invertebrates. Ecotoxicol. Environ. Saf. 61, 145–159.

Wang, W.-X., Rainbow, P.S., 2008. Comparative approach to understand metal accumulation in aquatic animals. Comp. Physiol. Biochem. 148C, 315–323.

Wang, W.-X., Yang, Y., Guo, X., He, M., Guo, F., Ke, C., 2011. Copper and zinc contamination in oysters: subcellular distribution and detoxification. Environ. Toxicol. Chem. 30, 1767–1774.

Wang, W.-X., 2002. Interactions of trace metals and different marine food chains. Mar. Ecol. Prog. Ser. 243, 295–309.

Wang, W.-X., 2011a. Incorporating exposure into aquatic toxicological studies: an imperative. Aquat. Toxicol. 105S, 9–15.

Wang, W.-X., 2011b. Trace Metal Ecotoxicology and Biogeochemistry. Science Press, Beijing, 322 p.

Wang, X.H., Wang, W.-X., 2005. Uptake, absorption efficiency, and elimination of DDT by marine phytoplankton, copepods, and fish. Environ. Pollut. 136, 453–464.

Wang, X.H., Wang, W.-X., 2006. Bioaccumulation and transfer of benzo(a)pyrene in a simplified marine food chain. Mar. Ecol. Prog. Ser. 312, 101–111.

Ward, D.M., Nislow, K.H., Chen, C.Y., Folt, C.L., 2010. Reduced trace element concentrations in fast-growing juvenile Atlantic salmon in natural streams. Environ. Sci. Technol. 44, 3245–3251.

Wright, D.A., 1995. Trace metal and major ion interactions in aquatic animals. Mar. Pollut. Bull. 31, 8–18.

Zhang, H., Davison, W., 1995. Performance characteristics of diffusion gradients in thin-films for the in situ measurement of trace metals in aqueous solution. Anal. Chem. 67, 3391–3400.

Zhang, H., Davison, W., 2000. Direct in situ measurements of labile inorganic and organically bound metal species in synthetic solutions and natural waters using diffusive gradients in thin films. Anal. Chem. 72, 4447–4457.

Zhang, L., Wang, W.-X., 2006. Significance of subcellular metal distribution in prey in influencing the trophic transfer of metals in a marine fish. Limnol. Oceanogr. 51, 2008–2017.

Zhang, L., Wang, W.-X., 2007a. Size dependence of the potential for metal biomagnification in early life stages of marine fish. Environ. Toxicol. Chem. 26, 787–794.

Zhang, L., Wang, W.-X., 2007b. Waterborne cadmium and zinc uptake in the euryhaline teleost acclimated at different salinities. Aquat. Toxicol. 84, 171–181.

Zhang, Q., Yang, L., Wang, W.-X., 2011. Bioaccumulation and trophic transfer of dioxins in marine copepods and fish. Environ. Pollut. 159, 3390–3397.

CHAPTER

5

Biomarkers and Effects

M. Hampel[1], J. Blasco[2], M.L. Martín Díaz[1]

[1]University of Cadiz, Puerto Real, Cádiz, Spain; [2]Institute for Marine Sciences of Andalusia (CSIC), Puerto Real, Cádiz, Spain

5.1 INTRODUCTION

The potential for adverse effects of environmental contaminants is a matter of growing concern, and the decline of ecosystems due to human pressure is widely recognized (United Nations, 2013). With over 30,000 substances produced above 1 ton per year on the European Union (EU) market and eventually discharged into the environment (European Commission, 2006), there is an urgent need to (1) evaluate the potential adverse effects of new and existing chemicals on ecosystem and human health and (2) monitor the environmental status of pollution-affected and sensitive ecosystems to guarantee the integrity of our planet.

This chapter describes how effects of exposure to contaminants manifest within organisms impairing physiological or cellular functioning and how these changes can be identified through the use of biomarkers. In ecotoxicology, a biomarker is generally a measurable indicator of some biological state or condition that links a specific environmental exposure to a health outcome. Biomarkers play an important role in understanding the relationships between exposure to environmental chemicals and the development of adverse effects at individual and population levels. Whereas during the last decades, scientists concentrated rather on molecular, cellular, or higher level effects in individuals, much progress has been made more recently in identifying and validating new biomarkers that can be used in population-based studies of contamination-induced effects by including behavioral and reproductive parameters in the environmental risk assessment (ERA) process.

Biomarkers have been used in environmental toxicology and monitoring since the early 1980s and have undergone a rapid development in terms of sensitivity and specificity ever since. Apart from effect evaluation after chemical challenge, biomarkers have also been extensively used in environmental monitoring (eg, MEDPOL and OSPAR programs) to identify the biological status of sentinel ecosystems. With the recent developments in molecular high-throughput techniques for the generation of massive gene and protein expression data and growing practical impracticability of traditional ERA, new promising perspectives are opening for the development of contaminant-specific biomarkers and an increased rationalization in the ERA process.

Here we give an overview of the most important biomarkers of contamination and their

Marine Ecotoxicology
http://dx.doi.org/10.1016/B978-0-12-803371-5.00005-9

application in the marine environment, followed by an introduction to the different molecular high-throughput or omic techniques which are elemental tools for the discovery of new specific biomarkers able to meet the future challenges of emerging contamination. Finally, future developments and trends within this field will be explored.

5.2 BIOMARKERS

The term biomarker has been defined as changes in biological responses—ranging from molecular through cellular, and from physiological to behavioral changes—that can be related to exposure to, or toxic effects of, environmental chemicals (van der Oost et al., 2005). Biomarker responses have been used in biomonitoring programs (Solé et al., 2009) and as tools for initial screening as early warning of toxic chemical effects on organisms (Cajaraville et al., 2000; Martín-Díaz et al., 2004). And they have been used for ERA of legacy and emergent pollutants (Blasco and DelValls, 2008). See Tables 5.1 and 5.2 for an overview of the biomarker groups and specific biomarkers described below.

5.2.1 Biotransformation Enzymes

Foreign chemicals, the so-called xenobiotics, that enter the body are generally lipophilic in nature and hence tend to be poorly excreted. However, most of these chemicals undergo metabolic conversion reactions in vivo in a process termed "biotransformation." These biochemical reactions are mediated by enzymes and result in the conversion of the parent chemical to more polar and readily excretable metabolites. Thus, one of the results of such biotransformation reactions is to facilitate the removal of toxic chemicals from an animal, which, unless excreted, could accumulate to harmful levels (Buhler and Williams, 1988). One of the most sensitive biomarkers is alterations in levels and activities of biotransformation enzymes. The activity of these enzymes may be induced or inhibited upon exposure to xenobiotics (Bucheli and Fent, 1995). Enzyme induction is an increase in the amount or activity of these enzymes, or both. The enzyme induction is mainly linked to the presence of bioavailable lipophilic xenobiotics in the environment. Biotic and abiotic factors interfere in the bioavailability of discharged xenobiotics. Once these xenobiotics are bioavailable for exposed organisms and enter them, a cascade of biotransformation reactions occurs aiming to transform the xenobiotic in a more water-soluble form easy to be excreted. These processes will avoid the bioaccumulation of the bioavailable xenobiotics as well as the adverse effects that these substances may cause to the organism. Two major types of enzymes involved in xenobiotic biotransformation are distinguished: Phase I enzymes and Phase II enzymes and cofactors.

5.2.1.1 Phase I Enzymes

Phase I of metabolism unmasking or adding reactive functional groups involves oxidation, reduction, or hydrolysis (Goeptar et al., 1995). These reactions introduce or expose a functional group ($-OH$, $-NH_2$, $-SH$) on the parent compound; metabolites formed are inactive but in some instances active metabolites are also formed.

The most studied of these latter systems is the cytochrome P450-dependent mixed-function oxidase (MFO) system. They are known as MFOs and monooxygenases as metabolism of a substrate consumes one molecule of molecular oxygen and produces an oxidized substrate and another molecule of oxygen appears in water as a by-product.

Cytochrome P450 monooxygenase enzymes (CYP) comprise an ancient and widely distributed protein superfamily. The latest published account gives more than 750 sequences belonging to more than 107 different families (Nelson, 1998). Many more CYP genes have

TABLE 5.1 Description of Different Biomarker Groups, Their Mechanisms of Action, Examples of Use, and Their Disadvantages in Marine Environmental Monitoring Application

Biomarker group	Effects	Examples of use	Disadvantages/inconveniences
Biotransformation enzymes	Biotransformation reactions occur aiming to transform the xenobiotic to a more water-soluble form easy to be excreted	Determination of the bioavailability of contaminants in the environment	• Absence of a standardized methodology • Induction of catalytic activity of enzymes not always is related to increasing gene expression • Inhibition of activity after exposure to contaminants • Not all isoform gens are identified in all marine species • Low ecological relevance
Inmunotoxicity	Immune dysfunction resulting from exposure of an organism to a xenobiotic	Determination of the impairment in the immunological resistance due to the presence of xenobiotics or eutrophication	• Most research on this system has been performed on mammalian species; it may be considered a promising field to search for new biomarkers • Not specific to a group of contaminants
Antioxidant enzymes	Changes in enzymatic activities as response to pollutants	Determination of changes in redox status and reduction of reactive oxygen species	• Not specific to pollutants • Difficult interpretation for moderate contamination • Interpretation should take in account the whole enzymatic system involved in antioxidant responses and other ROS scavengers
Biochemical indices of oxidative damage	Lipid peroxidation under which oxidants attack lipids especially polyunsaturated fatty acids	Determination of effects provoked by exposure to xenobiotics	• Not specific to a group of contaminants • This response needs long periods of exposure to be observed • It seems to be highly affected by the organisms' status and seasonality
Metallothionein	Induction by exposure to some metals (eg, Cd)	Determination of metal exposure and detoxification mechanisms	• Different methodologies involved, difficult for comparison, and different units • Changes of their levels can be related to other environmental variables • Seasonal variations related to nutritional or reproductive status • Lack of information about isoforms in biomonitoring studies

(Continued)

TABLE 5.1 Description of Different Biomarker Groups, Their Mechanisms of Action, Examples of Use, and Their Disadvantages in Marine Environmental Monitoring Application—cont'd

Biomarker group	Effects	Examples of use	Disadvantages/inconveniences
δ-aminolevulinic acid dehydratase (δ-ALA-D)	Decreasing of activity related to Pb exposure	Identification of lead exposure	• No clear relationship between inhibition and health status • Response of organisms can be species-specific • Can be responsive to other metals (−SH groups)
Reproduction alteration	Decreased reproductive capability due to the exposure to xenobiotics	Determination of endocrine disruption due to the exposure to xenoestrogenic contaminants	• Reproduction cycles of bioindicator species should be known before measuring biomarkers • The sex of the bioindicator species should be known before analysis
Neurotoxic effects	Inhibition of enzymes involved in neural functions	Detection of neurotoxic risk provoked by xenobiotics	• More research is needed to determine other possible physiological alterations in the enzyme inhibition
Genotoxicity	Changes in the genetic material of biota	The detection and quantification of various events in this sequence may be used as biomarkers of exposure and effects in organisms environmentally exposed to genotoxic substances	• Absence of a standardized methodology • Not specific to a group of contaminants

been described in the last several years and a running total can be found in regular updates (see Dr. Nelson's webpage, http://drnelson.utmem.edu:homepage.html). P450 proteins are found in a diverse array of organisms including bacteria, plants, fungi, and animals. Most oxidative Phase I biotransformation reactions are catalyzed by cytochrome P450 isoforms which belong to the microsomal monooxygenase (MO) enzymes.

In the marine environment, the best-studied member of the cytochrome P450 superfamily is CYP1A1, the major form induced by dioxins, polycyclic aromatic hydrocarbons (PAHs), and polychlorinated biphenyls (PCBs). Antisera have been produced against the CYP1A1 of a number of fish species and successfully used to detect changes in enzyme protein levels by immunoblotting and enzyme-linked immunosorbent assay (ELISA) (Goksøyr and Larsen, 1991; Goksøyr and Förlin, 1992). However, this promising strategy has not yet been expanded significantly into other xenobiotic-affected CYP forms in marine organisms, especially invertebrates (Snyder, 2000).

Variations in CYP profiles are commonly used as indicators of exposure to environmental contaminants (Andersson and Lars, 1992; Stegeman and Lech, 1991). These CYP profiles can be determined by different methodologies which

TABLE 5.2 Description of the Analytical Methodologies of the Most Used Biomarkers in Marine Environmental Monitoring Programs

Biomarker	Type	Mechanisms and methodology	References
Cytochrome 450	Exposure	Induction of CYP450 isoforms catalytic activities Spectrophotometric methods	Gagnè and Blaise (1999) and Gagnè et al. (2007)
Glutathione S transferase activity (GST)	Exposure	Increase of enzyme activity Spectrophotometric methods	McFarland et al. (1999)
Antioxidant enzymes	Exposure/effect	Changes in enzymatic activities Spectrophotometric methods	Regoli et al. (2012)
Metallothionein	Exposure/detoxification	Changes in protein levels Spectrophotometric method Polarography-DPP Silver Saturation assay Chromatography MT gene expression	Viarengo et al. (1997), Olafson and Olsson (1991), Scheuhammer and Cherian (1991), Romero et al. (2008), and Russo et al. (2003)
α-aminolevulinic acid dehydratase (α-ALA-D)	Exposure	Enzymatic activity	Berlin and Schaler (1974) and Company et al. (2011)
Phagocytosis Lysosomal membrane stability (LMS) NO Cyclooxygenase (COX) activity	Exposure/effect	Decrease of phagocytic activity Decrease of neutral red retention time Increase of nitrite levels Increase of enzyme activity	Blaise et al. (2002), Lowe and Pipe (1994), Martínez-Gómez et al. (2008), Verdon et al. (1995), and Fulimoto et al. (2005)
VTG levels	Exposure/effect	Increase or decrease of levels of VTG: **a.** Enzyme Linked Inmunosorbent Assay **b.** Alkali-Labile Phosphate Levels	Pateraki and Stratakis (1997) and Gagnè et al. (2003)
DNA strands break	Exposure/effect	Increase of DNA strands break	Gagnè et al. (1995) and Olive (1988)
Lipid peroxidation	Exposure/effect	Increase of lipid peroxidation products	Janero (1990) and Wills (1987)
Acetyl cholinesterase activity	Exposure	Increase of enzyme activity	Ellman et al. (1961) and Guilhermino et al. (1996)

Biomarker of effect: Biomarker that indicates an effect of a compound on the measured parameter. Biomarker of exposure: A measured parameter that indicates exposure to a compound. It need not be associated with any toxic effect. Biomarker of detoxification. A measured parameter involved in mechanisms of reduction of toxic effects.

Adapted from Nikimann, M., 2014. An Introduction to Aquatic Toxicology. Academia Press, Oxford, 240 p.

reach different responses to address the same objective, which is the measurement of the biotransformation levels and bioavailability of xenobiotics in polluted sites: (1) mRNA expression of CYP-related genes; (2) CYP protein levels determined immunologically, using mono or polyclonal antibodies with ELISA, Western blotting or histochemical techniques (Bucheli and Fent, 1995); and (3) CYP isoenzymes catalytic activities. CYP1A determinations may be used in various steps of the ERA process, such as quantification of impact and exposure of various organic trace pollutants, environmental monitoring of organism and ecosystem health, identifying subtle early toxic effects, triggering of regulatory action, identification of exposure to specific compounds, toxicological screening, and the research on toxic mechanisms of xenobiotics (Stegeman et al., 1992). In this sense, ethoxyresorufin-O-deethylase (EROD) appeared to be the most sensitive catalytic probe for determining the inductive response of the cyt P450 system in fish (Goksøyr and Förlin, 1992). The EROD activity is measured by following the increase in fluorescence of the reaction product, resorufin (Burke and Mayer, 1974). An extensive review has been published, compiling and evaluating existing scientific information on the use, limitations, and procedural considerations for EROD activity in fish as a biomarker of chemical exposure (Whyte et al., 2000). Generally, a good correlation is observed between CYP1A protein levels and EROD activity (eg, van der Oost et al., 1996). Numerous field studies demonstrated a strong and significant increase of hepatic CYP1A protein levels and activity in many species of fish from polluted environments. EROD activities in fish liver are very sensitive biomarkers and may thus be of great value in ERA processes. Although certain chemicals may inhibit EROD induction or activity, this interference is generally not a drawback to the use of EROD as a biomarker (Whyte et al., 2000). Together with levels of CYP1A protein and mRNA, the induction of CYP1A catalytic activities may be used both for the assessment of exposure and as early-warning sign for potentially harmful effects of many organic trace pollutants. Research on mechanisms of CYP1A-induced toxicity suggests that EROD activity may not only indicate chemical exposure but may also predict effects at various levels of biological organization (Whyte et al., 2000). Certain confounding variables, which may affect the enzyme activities, however, will have to be considered when interpreting the responses in these parameters. In the marine environment, the best-studied member of the cytochrome P450 superfamily is CYP1A1, the major form induced by dioxins, PAHs, and PCBs, followed by CYP3A, always in fish species. However, this promising strategy has not yet been expanded significantly into other xenobiotic-affected CYP forms in marine organisms, especially for invertebrates. CYP1A and CYP3A-like proteins in fish accounted for the main part of the measured activities and results indicate a lack of CYP2B-like enzymes. In contrast, the crustacean *Aristeus antennatus* appeared to lack CYP1A- and CYP3A-like biotransformation capacities, while exhibiting activities for enzymes related to the mammalian CYP2 family. These differences in metabolism between fish and crustacea were consistent with PCB bioaccumulation profiles. *Aristeus antennatus* clearly accumulated CYP1A inducing congeners, but metabolized congeners that are known as mammalian CYP2B inducers. However, additional studies are necessary to further identify the isoenzymes present in these animal groups and to corroborate that these results can be extrapolated to other crustacean species. The fact that most of the existing 209 PCB congeners are mainly metabolized by CYP2B-like enzymes supports the idea that crustacea generally accumulate lower levels of PCBs than fish. Furthermore, their greater capacity to metabolize PCBs also indicates a greater potential responsiveness of their CYPs to PCBs (Koenig et al., 2012). EROD catalytic activity, which represents catalytic activity of CYP1A1, 1A2, and 1B1, was significantly induced in the crustacean *Carcinus maenas* and the clam *Ruditapes philippinarum* after exposure to dredged material (Martín-Díaz et al., 2007, 2008), oil spills (Morales-Caselles et al., 2008a,b), different pharmaceutical products (Aguirre-Martínez et al., 2016; Maranho et al., 2015a–c), and wastewater treatment plant effluents (Maranho et al., 2012). The measurement of the catalytic isoenzyme activity of the isoform CYP3A-like activities, known as dibenzylfluorescein dealkylase (DBF) activity, has been measured in crustaceans and bivalves as *C. maenas* and *R. philippinarum*, being found activation after the exposure to different pharmaceutical products and exposure to wastewater treatment plant effluents in marine environments (Aguirre-Martínez et al., 2016; Maranho et al., 2012; 2015a–c). Studies have observed increased transcription rates of different CYP genes in bivalves treated with different chemicals. Zanette et al. (2010) observed that mussels *Mytilus edulis* treated with AhR agonists showed different patterns of CYP transcription. Mussels treated with beta-naphthoflavone (25 μg/g) showed an upregulation of CYP3-like-1 and CYP3-like-2 genes in gills, and treated with 3,3′,4,4′,5-polychlorinated biphenyl 126 (2 μg/g) showed increased rates of CYP3-like-2 gene transcription in the digestive diverticula, suggesting that distinct mechanisms of CYP gene activation could be present in *M. edulis*. Different responses on CYP gene transcription were observed in clam *R. philippinarum* exposed to ibuprofen (Milan et al., 2013). While an upregulatory pattern was observed in CYP3A65 and CYP2U1 after exposure of the clams for 3 and 5 days of exposure to 1 mg/L ibuprofen, other gene encoding CYP subfamily 4 (CYP4) proteins displayed opposite responses. In the Pacific oyster, *Crassostrea gigas*, a gene belonging to the CYP subfamily, CYP356A1, was identified in sanitary sewage-exposed specimens (Rodrigues-Silva et al., 2015).

5.2.1.2 *Phase II Enzymes*

Phase II enzymes and cofactors involve a conjugation of the xenobiotic parent compound or its metabolites with an endogenous ligand. Conjugations are addition reactions in which large and often polar chemical groups or compounds such as sugars and amino acids are covalently added to xenobiotic chemical compounds and drugs (Lech and Vodicnik, 1985). The majority of the Phase II-type enzymes catalyze these synthetic conjugation reactions, thus facilitating the excretion of chemicals by the addition of more polar groups. Glutathione S-transferase (GST) catalyzes the conjugation of glutathione with xenobiotic compounds containing electrophilic centers. Oxides produced by the action of cytochrome P450 systems on aromatic compounds can be conjugated to glutathione by GST. It is important for organisms to deal with active electrophiles since they can react with macromolecules controlling cell growth such as DNA, RNA, and proteins. Many, if not all, chemical carcinogens are electrophiles (Miller and Miller, 1979). Thus GST plays an important role in detoxifying strong electrophiles having toxic, mutagenic, and carcinogenic properties (Miller and Miller, 1979; Lee, 1988).

GSTs, which belong to the superfamily of Phase II detoxification enzymes (Boutet et al., 2004), are essential enzymes that have been found in all kingdoms of life. In fact, GSTs are multifunctional isoenzymes for cellular defense against xenobiotics and provide protection for organisms (Blanchette et al., 2007). The subfamily of GSTs is further distinguished into at least 14 classes, namely alpha (α), beta (β), delta (δ), epsilon (ε), zeta (ζ), theta (θ), kappa (κ), lambda (λ), mu (μ), pi (π), sigma (σ), tau (τ), phi (φ), and omega (Ω) (Navaneethaiyer et al., 2012) based on N-terminal amino acid sequence, substrate specificity, antibody cross-reactivity, and sensitivity to inhibitors (Kim et al., 2009). Each GST contains a G-site (glutathione substrate binding site) and an H-site (hydrophobic substrate binding site) (Mannervik and Danielson,

1988). The G-site is conserved in the N-terminal region among different classes, while the H-site is highly diverse, characterized with significant variation in sequence and topology accounting for the variability of enzyme activity in GSTs (Navaneethaiyer et al., 2012; Liu et al., 2015).

These enzymes are mainly located in the cytosolic fraction of the liver/hepatopancreas/digestive gland tissues (Sijm and Opperhuizen, 1989). Most studies determine the total GST activity using the artificial substrate 1-chloro-2,4-dinitrobenzene (CDNB), which is conjugated by all GST isoforms with the exception of the q-class enzymes (George, 1994; Van der Aar et al., 1996). The toxicity of many exogenous compounds can be modulated by induction of GSTs. Effects of inducing agents on total hepatic GST activity, measured by CDNB conjugation, have been observed in several fish species (George, 1994). Levels of GST activity were significantly increased after exposure to primary-treated municipal effluent plumes containing pharmaceuticals and personal care products in crustaceans and bivalves.

Although metals are not natural substrates for these enzymes, studies have shown GST induction in aquatic invertebrates exposed to several metals (Moreira et al., 2006). Verlecar et al. (2007) showed that Hg exposure led to an increase of GST activity in mussels; Martín-Díaz et al. (2007) found its activity significantly induced by the presence of As, Cd, Cr, Cu, Fe, Hg, Mn, and Pb in crabs and by Cd, Cu, Mn, and Ni in clams; Moreira et al. (2006) got the same results in polychaeta after sediments contaminated with Cd, Pb, Zn, Cu, Cr, Hg, and As exposure. An increase in hepatic GST activity has been reported in several studies after exposure of fish to PAHs, PCBs, OCPs, and PCDDs (van der Oost et al., 2003); nevertheless, total GST activity in fish does not seem to be feasible as a biomarker for ERA, since increased and decreased activities are observed for the same xenobiotic exposure in the same species. However, more research on this parameter, which is of paramount importance for major detoxification processes, may elucidate specific isoenzymes that have a more sensitive and selective response to pollutants.

5.2.2 Oxidative Stress Parameters

5.2.2.1 *Antioxidant Enzyme Activity*

Eukaryotic life is dependent upon molecular oxygen (O_2) for the provision of energy through the coupling of oxidation energy transfer via the phosphorylation of ADP. This process is handled by the mitochondrial electron transport chain in which O_2 is four electrons reduced to water. Partial reduction by various endogenous processes results in the constant production of small amounts of highly reactive oxygen species (ROS, so-called oxyradicals) including the superoxide anion radical and hydrogen peroxide (Fig. 5.1). These processes can produce structural damage in DNA, proteins, carbohydrate, and lipids. The oxidative damage caused by ROS has been called "oxidative stress." The production of oxyradical species can be increased by xenobiotics: which include redox reactions with

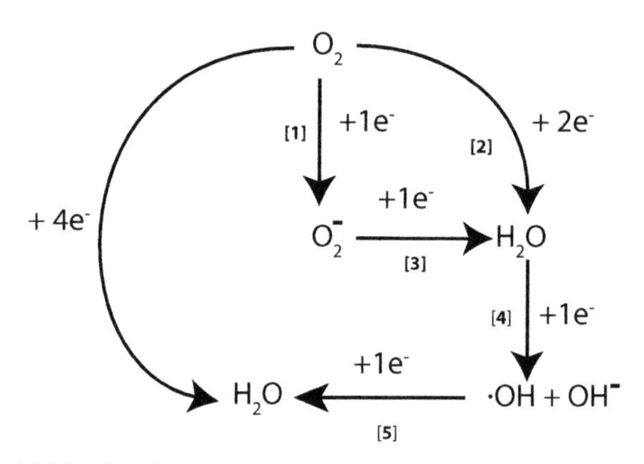

FIGURE 5.1 Reactions involved in the oxygen reduction metabolism: (1) production of superoxide anion radical, (2) formation of hydrogen peroxide by reduction (2e) of molecular oxygen, (3) formation of hydrogen peroxide by reduction (1e) of superoxide anion, (4) formation of hydroxyl radical by reduction (1e) of hydrogen peroxide, (5) reduction (1e) of hydroxyl radical to water.

transition metals (eg, Fe, Cu, Ni, and Co) and organic free radicals and redox cycling of xenobiotics (Livingstone et al., 2000). In fact, ROS can be considered as a mechanism where the toxicity of many pollutants is evidenced.

To prevent undesirable effects of oxyradicals (eg, changes in redox balance and intracellular free calcium, enzyme inactivation, lipid peroxidation, etc.), several actions are possible, among them low molecular weight compounds (glutathione; vitamins A, C, and E; and uric acid) and antioxidant enzyme activities: superoxide dismutase (SOD), catalase, glutathione peroxidase, and glutathione reductase, all of them represent mechanisms to remove ROS and to protect the organisms against oxidative stress.

SOD (superoxide: superoxide oxidoreductase, EC 1.15.1.1) dismutes two molecules of superoxide radicals (O^{2-}) generated during the monovalent reduction of oxygen to H_2O_2. Superoxide radical does not cross biological membranes and it should be detoxified in the same compartment where it is produced, as a consequence more than SOD can be found. Cu, Zn-SOD is found in the cytosol and it has been found in peroxisomes, nucleus, and peroxisomes. On the other hand, Mn-SOD is a mitochondrial enzyme, although this is found in peroxisomal membrane (Dhanunsi et al., 1993; Singh, 1996). Another extracellular SOD form has been found in connective tissue and prokaryotic (Fridovich, 1995).

GPX (glutathione: hydrogen peroxide oxidoreductase, EC 1.11.1.9) is the most important peroxidase for detoxification hydroperoxides. The enzyme catalyzes the glutathione dependent reduction of H_2O_2 and organic hydroperoxides. Several forms of glutathione peroxidase exist (eg, Se and non-Se dependent). At the ultrastructural level, cellular GPX has been located in the nucleus and mitochondria, although it has been found in cytoplasmic matrix, lysosomes, and peroxisomal matrix (cited by Orbea et al., 2000).

Catalase (hydrogen peroxide: hydrogen peroxide oxidoreductase, EC 1.11.1.6): The main function of the catalase is to split hydrogen peroxide into water and oxygen. This enzyme is located in the peroxisomes and is used as a marker of this cellular compartment (Fahimi and Cajaraville, 1995). However, other locations for this enzyme have been found in mammals and invertebrates.

Box 5.1 shows the reactions for these outlined enzyme activities and methodologies used for their quantification.

BOX 5.1

MAIN ANTIOXIDANT ENZYME ACTIVITIES AND SPECTROPHOTOMETRIC MEASUREMENT METHODS

Superoxide dismutase (SOD) EC.1.15.1.1 (McCord and Fridovich, 1976).

$$2O_2^- + 2H \rightarrow H_2O_2 + O$$

Catalase (CAT) EC 1.11.1.6 (Aebi, 1984).

$$2H_2O_2 \rightarrow 2H_2O + O$$

Glutathione peroxidase Se dependent (Se-GPX) EC 1.11.1.9 (Lawrence and Burk, 1976).

$$H_2O_2 + 2GSH \rightarrow 2H_2O + GSSG$$

Orbea et al. (2000) immunolocalized SODs and GPX in the digestive gland of hepatopancreas or mollusks and crustaceans, with similar location for Cu, Zn-SOD and Mn-SOD; in invertebrates they are present in the ducts, although in mussels and crabs they were found indigestive tube of cells. In crabs, connective tissue was stained for Cu, Zn-SOD and GPX. The same authors found catalase in peroxisomes in the digestive gland of mussels and hepatopancreas of crabs and liver of fish.

Besides, these ones, other enzyme activities depending on glutathione can be involved in the antioxidant system. Regoli et al. (1997) characterized in the scallop, *Adamussium colbecki*, scallop *Pecten japonicus* and mussel *Mytilus galloprovincialis* antioxidant enzyme activities including glyoxylase I (EC 4.4.1.5 transforms toxic α-ketoaldehydes formed in cellular oxidative process in an intermediate thiol ester using GSH as cofactor) and glyoxylase II (EC 3.1.2.6 hydrolyzes the glyoxylase I-catalyzed reaction product to the corresponding D-hydroxyacid with regeneration of GSH), GSTs (EC 2.5.1.18 involved in the conjugation of GSH with electrophilic centers of xenobiotic compounds), and glutathione reductase (EC 1.6.4.2 converts oxidized glutathione GSSG to the reduced form GSH). These species showed a similar antioxidant system, although in general terms, the scallop, *A. colbecki* was more efficient to lower temperature.

Trace metals can be involved in the generation of ROS by means of its participation in Haber Weiss and Fenton reactions.

Haber Weiss reaction	$H_2O_2 + O^{2-}$	$\bullet OH + OH^- + O_2$
$Fe(III) + O^{2-}$	$Fe(II) + O_2$	
Fenton reaction	$Fe(II) + H_2O_2$	$Fe(III) + \bullet OH + OH$

Although Fe(III)/Fe(II) has been selected as an example, other metals can catalyze the reaction with different efficiencies (eg, Cu(I), Cr(III)(IV), and (VI)) (Regoli, 2012).

Organic pollutants, like PAHs, PCBs, dioxins, and dioxin-like compounds can increase ROS production by cytochrome P450 (see in more detail in Section 5.2.1 of this chapter) that catalyzes several oxidative reactions to produce more hydrophilic compounds. These metabolites can suffer additional transformation by conjugation reactions in order and to make excretion easier and detoxify the compounds. In some cases, produced metabolites can exert a prooxidant effect activated by quinones, diols, and transition metal chelates (Livingstone, 2001).

The use of antioxidant enzyme activities as stress biomarkers for marine organisms in exposure experiments to legacy and emergent pollutants is frequent in the scientific literature (Chandurvelan et al., 2013; Hariharan et al., 2014; Franzelliti et al., 2015; Katsumiti et al., 2015; Macías-Mayorga et al., 2015; Volland et al., 2015) although their use in field studies should be considered with precaution because biological variability (eg, nutritional status) (González-Fernández et al., 2015) can affect biomarker responses. Campillo et al. (2013) carried out caged experiments in Mar Menor lagoon and reported a decrease in catalase (CAT) activity associated to increase of pollution levels. In the mussel, *M. galloprovincialis* and fish, *Mullus barbatus* collected from Italian coast (Lionetto et al., 2003), the use of antioxidant enzymes (CAT and GPX) was useful for detecting the exposure effect induced by pollutants, although the responses were highly specific of the considered species. This specific species sensitive to changes in antioxidant enzymes was found by Liu and Wang (2016) for two oyster species transplanted into a metal-contaminated estuary. The response of the enzyme activities to toxic chemicals showed a bell-shaped trend, initially increasing due to the activation of enzyme synthesis and decreasing due to the catabolic rate of toxic direct effect. Viarengo et al. (2007) pointed out that enzyme assays should be carried out in parallel with other biomarkers to find the meaning of physiological changes.

In summary, antioxidant enzymes are interesting from the point of view of their potential use as biomarkers of pollution and they have been used as biomarkers in monitoring program in estuaries and coastal areas. The response was not consistent in all cases and ranged from increasing levels of these enzyme activities in polluted sites with respect to reference sites to little or no consistent variation between sites subjected to different pollutant inputs. In fact, the use of certain enzyme activities as stress biomarkers involves the necessity of a better knowledge of their seasonal and natural variations (Viarengo and Canesi, 1991; Blasco et al., 1993; Solé et al., 1995). It is fundamental to define the range of normal values for enzymatic activities in sentinel species before implementing the use of these biomarkers in field studies.

5.2.2.2 Lipid Peroxidation

Cytotoxic ROS, also referred to as reactive oxygen intermediates (ROIs), oxygen free radicals, or oxyradicals (Di Giulio et al., 1989) are known to produce oxidative toxicity. These oxygen molecular species are the superoxide anion radical (O^{2-}), hydrogen peroxide (H_2O_2), and the hydroxyl radical (OH^+), an extremely potent oxidant capable of reacting with critical cellular macromolecules, possibly leading to enzyme inactivation, lipid peroxidation (LPO), DNA damage, and ultimately cell death (Winston and Di Giulio, 1991).

ROS formation occurs in many biological processes; it may be naturally produced or induced by the exposure to different environmental stress agents as the exposure to xenobiotics. They are naturally produced during several cellular pathways of aerobic metabolism including oxidative phosphorylation, electron transport chains in mitochondria and microsomes, the activity of oxidoreductase enzymes producing ROS as intermediates or final products, or even immunological reactions such as active phagocytosis (Halliwell and Gutteridge, 1999). In environmental ecotoxicology, the immediate interest is mainly focused on the ability of a number of xenobiotics compounds to increase intracellular production of ROS through the process of redox cycling.

ROS is of particular interest since it may react with critical cellular macromolecules, possibly leading to enzyme inactivation, lipid peroxidation (LPO), DNA damage, and ultimately cell death (Winston and Di Giulio, 1991). Currently, lipid peroxidation is considered as the main molecular mechanisms involved in the oxidative damage to cell structures and in the toxicity process that lead to cell death (Repetto et al., 2012).

LPO can be described generally as a complex process known to occur in both plants and animals, under which oxidants such as free radicals or nonradical species attack lipids containing carbon–carbon double bond(s), especially polyunsaturated fatty acids (PUFAs) that involve hydrogen abstraction from a carbon, with oxygen insertion resulting in lipid peroxyl radicals and hydroperoxides (Yin et al., 2011). Lipid peroxidation has been included in monitoring studies using vertebrate and invertebrate marine organisms as bioindicators.

Lipid peroxidation is a free radical-mediated chain of reactions that, once initiated, results in an oxidative deterioration of polyunsaturated lipids. The most common targets are components of biological membranes. When oxidant compounds target lipids, they can initiate the lipid peroxidation process, a chain reaction that produces multiple breakdown molecules and degradation products such as aldehydes, acetone, and malondialdehyde (MDA). New trends in the demonstration of LPO by measurement of degradation products such as aldehydes, acetone, and MDA have been described by De Zwart et al. (1997).

Lipid peroxidation can also be assessed by measurement of conjugated dienes, ethane and pentane gases, isoprostanes, and 4-hydroxynonenal (4-HNE). Another way to measure the oxidative damage is by protein and DNA modifications, but these markers many times can be formed by pathways other that from free radicals. Thus, MDA is by far the most popular indicator of oxidative damage to cells and tissues. MDA, formed by the rupture of unsaturated fats, is widely used as an index for lipid peroxidation. Since the 1960s, several methods have been developed to assess MDA, including quantitative methods using spectrophotometry or fluorimetric detection, high performance liquid chromatography (HPLC), gas chromatography, and immunological techniques. MDA is one of several low molecular weight end products formed via the decomposition of certain primary and secondary lipid peroxidation products. At low pH and elevated temperature, MDA readily participates in nucleophilic addition reaction with 2-thiobarbituric acid (TBA), generating a red, fluorescent 1:2 MDA:TBA adduct. These facts, along with the availability of facile and sensitive methods to quantify MDA (as the free aldehyde or its TBA derivative), have led to the routine use of MDA determination and, particularly, the "TBA test" to detect and quantify lipid peroxidation in a wide array of sample types (Janero, 1990).

Anthropogenic contaminants such as metals, PAHs, PCBs, and residues from pesticides and the cellulose industry induce antioxidant and lipid peroxidation defenses (van der Oost et al., 2003; McDonagh and Sheehan, 2008). Trace metals and organic xenobiotics are typical classes of environmental pollutants with prooxidant effects (Regoli et al., 2014). Numerous studies have demonstrated enhancements of LPO in various tissues from fish species exposed to metals as sea bass (Romeo et al., 2000) exposed to Cd and Cu and bivalve species as *R. philippinarum* (Ramos Gómez et al., 2011) exposed to metal-contaminated marine sediments.

Other classes of typically prooxidant chemicals include aromatic xenobiotics, like PAHs, PCBs, halogenated hydrocarbons, dioxin (TCDD), and dioxin-like chemicals, which increase the intracellular generation of ROS through the induction of the cytochrome P450 pathway. Increase in lipid peroxidation has been detected in marine mussels, *M. galloprovincialis* after exposure to benzo(a) pyrene (Maria and Bebianno, 2011).

Nevertheless, not many studies in marine environments describe the enhancement of LPO after the exposure to xenobiotics. This response needs long periods of exposure to be observed and it seems to highly affect the organisms' status and seasonality.

5.2.3 Metallothioneins

Although metallothioneins were discovered at the same time as the first zinc metalloenzyme, the lack of evidence on scientific activity proves that MTs were forgotten for a long time. They can be considered as specific group of stress proteins (van der Oost et al., 2005). Later, two special characteristics: (1) high content of cysteine residues and (2) high metal content (7 g-atom of metal per mole) (Kojima et al., 1976) attracted the attention of the scientific community. Other special characteristics of these molecules are small size and inductibility to some metals (essential and nonessentials) in vivo, although not specifically because other chemical compounds can induce them (eg, glucocorticoids, progesterone, estrogen, ethanol, interferon (Vallee, 1991)). A relevant property of these proteins is the regulation of essential metals (Cu and Zn) and their role in the detoxification process of Cd or Hg (Roesijadi and Robinson, 1994). These molecules were isolated in the equine kidney initially, but their occurrence is widely distributed in the animal kingdom (plants, invertebrates, and vertebrates).

According to the nomenclature of metallothioneins, MT proteins have been divided into three classes (Fowler et al., 1987): Class I comprises all proteinaceous MTs with locations of cysteine closely related to those in mammals. Some molluscan and crustacean MTs belong to this class, such as those characterized in mussels (Mackay et al., 1990), oysters (Roesijadi et al., 1989), crabs (Lerch et al., 1982), and lobsters (Brouwer et al., 1989). Class II includes proteinaceous MTs which lack this close similarity to mammalian MTs and Class III MTs consist of nonproteinaceous MTs also known as phytochelatins. A new classification proposed by Binz and Kagi (1999) taking in account phylogenetic features distributed Class I and II MTs in 15 families among the animal, plant, prokaryote, and fungi kingdoms.

MTs are not single proteins because several isoforms have been isolated and different biological properties can be associated to different isoforms. MTs probably play different physiological roles, and the dependence on MT in detoxification processes varies environmentally and between zoological groups. Multiple isoforms have been identified and polymorphism appears to be particularly important in invertebrates compared to mammals (Amiard et al., 2006).

Although MTs have got two main roles in the cell process related to Zn, Cu homeostasis and binding to nonessential metals to reduce toxic effects, they can act as protective agents against oxidative stress. Viarengo et al. (1999) showed the protective effect of MTs in mussels which this protein had induced; MT induction can be assessed at transcriptional and translational level and information of increasing of metal at cellular level is conveyed to MT gene via metal-activated transcriptional factor that initiates expression following binding specific metals (Roesijadi, 1994). In *Crassostrea virginica*, the hypothesis that displaced Zn is the basis for MT induction has been tested (Roesijadi, 1996).

MTs have been identified in approximately 50 different species of aquatic invertebrates, the majority of which are mollusks or crustaceans. Therefore they have become of great interest for assessing pollution in the marine environment and have been seen as potential biomarkers of metal exposure in mollusks and fish (Langston et al., 1998). MTs occur mainly in the cytosol and have also been detected in the nucleus and lysosomes following laboratory exposure or exposure in the field to essential and nonessential metal ions. MT induction has been demonstrated in organisms from polluted populations or following laboratory exposure to class B metals such as Ag, Cd, Cu, Hg, and Zn. The involvement of MT in metal sequestration is more evident in the gills, digestive gland, and kidney, reflecting the significance of these tissues in uptake, storage, and excretion of metals (Bebianno et al., 1993, 1994). The degree of MT induction can vary between species and between tissues. On the other hand, aquatic species are subjected to mixture of pollutants (eg, metals) as a result of industrial activities, agricultural runoffs, and urban wastewater discharges; in a study carried out with the clam, *Ruditapes decussatus* higher induction of MT from the exposure to metal mixture tan single exposure, was denoted to metal interactions, depending on their affinity for this protein (Serafim and Bebianno, 2010).

MT levels are influenced by several factors of an environmental (salinity, season, location in the intertidal zone, etc.) and biological (sexual maturity, weight, etc.) nature (Hamza-Chaffai et al., 1999; Mouneyrac et al., 1998). This is especially relevant in field studies where changes in the concentration of MTs due to metal contamination need to be distinguished from those related to natural variations. Trombini et al. (2010) found for the oyster, *Crassostrea angulata,* that metallothionein pattern was not associated exclusively to metal contamination and other environmental variables should be taken into account for explaining seasonal variations in MT concentrations. Although MT induction is affected by other environmental stress agents and not only by metal occurrence, it is predominantly used as a general marker for metal exposure (eg, MEDPOL and OSPAR programs). Information about metals bound to MTs can give additional information about the agent that is responsible for the induction. Amiard et al. (2006) pointed out that MTs do have a role to play as biomarkers, if used wisely in well-designed sampling programs. And the selection of the organism, choice of tissue, and method of analysis should be carefully selected. In fact, Oaten et al. (2015) have reviewed the influence of sampling and sample treatment protocols for MT results, indicating that storage, tissue-type selection and pretreatment of the samples can affect the observed response and questioned their widespread use in biomonitoring approaches.

The detection and quantification of MTs are not simple due to the unique primary structure and their relatively low molecular mass. The proposed methods do not provide an absolute value, and several methods of MT quantification have been proposed (Cosson and Amiard, 2000; Dabrio et al., 2002), with additional later improvements (El Hourch et al., 2003, 2004). Comparative assays of different techniques have been carried out, showing a good correlation between results obtained for instance using differential pulse polarography (DPP) and a metal saturation assay (Onosaka and Cherian, 1982), and between DPP and spectrophotometric determination (Romeo et al., 1997) and RP-HPLC-FL and DPP (Romero et al., 2008). However, the various techniques produce concentration values that do differ and the results are often expressed in different units. Adam et al. (2010) reviewed MT analysis methodology and pointed out that hyphenated spectroscopic instruments will reveal new transport mechanisms and interactions of MTs with other biologically active compounds.

5.2.4 δ-Aminolevulinic Acid Dehydratase

The δ-aminolevulinic acid dehydratase (ALA-D) enzyme, also called porphobilinogen (PBG) synthase (PBGS, EC 4.2.1.24) or aminolevulinate dehydratase (δ-ALA-D), catalyzes the formation of one molecule of PBG, a precursor of hemoglobin, from two molecules of aminolevulinic acid (ALA). This enzyme is widely distributed in nature and requires Zn^{2+} as cofactor to be active; although the high occurrence of thiol groups in its structure can permit binding other metals to it and inhibiting its activity and PBG levels (Finelli, 1977). The higher affinity of Pb^{2+} by ALA-D than Zn^{2+} can modify the enzymatic properties and be a critical step in hemoglobin (Hb) synthesis. The possible modulation of δ-ALA-D by other metals, such as Hg, Cu, and Ag can be rather unspecific from intoxication by other metals (Rocha et al., 2012). The relationship between ALA-D inhibition and anemia has not been found in fish. A decrease in this enzyme activity has been related to Pb levels in *M. barbatus* collected in the Mediterranean Spanish coast (Fernández et al., 2015) while for other fish species (toadfish) significant changes for lead exposure in kidney, liver, and blood cells were not reported (Campana et al., 2003). In *M. barbatus*, no significant relationship was found between sex, gonad status, or length. Nevertheless, other authors found a relationship between size and ALA-D, negative (Tudor, 1984) and positive in the case of *Gadus morhua* (Hylland et al., 2009) associated with Zn hepatic levels and its increasing with size. In the fish, *Prochilodus lineatus*, no significant differences for sex were found and negative correlation between blood samples and lead levels was observed. The single effect of Zn and dithiothreitol and the combined effect on reactivation of enzyme activity were checked; they found that enzyme activity was reactivated by Zn at 15 μM and the relationship between reactivation index in blood samples versus blood level concentrations was statically significant ($p < .001$) (Lombardi et al., 2010). In fact, these authors proposed that the use of reactivation index can be a more reliable and sensitive biomarker of lead exposure.

For mollusks, in different bivalve species, *Chamelea gallina* (Kalman et al., 2008), *Corbicula fluminea* (Company et al., 2008), *M. galloprovincialis* (Company et al., 2011), an inhibition for this enzyme activity has been recorded for Pb levels. On the other hand, some studies have shown the role of 5-aminolevulinic acid or δ-ALA as prooxidant which can affect the redox status (Rocha et al., 2012). In the frame of multibiomarker approach for environmental monitoring, δ-ALA-D has been used; thus Cravo et al. (2009) found in *M. galloprovincialis* collected in different sampling sites of south coast of Portugal, seasonal variation, lower in summer than in winter. This pattern was related to increasing of lead levels associated with boat traffic increasing, because leaded petrol could be used in some boat engines. On the contrary, Cravo et al. (2013) in *R. decussatus* did not find seasonal variation and inhibition of δ-ALA-D with respect to Pb levels. The authors suggested that Pb levels were so low to provoke the inhibition.

The methodology used for measuring this enzyme activity is based on the European standardized method for the determination of delta-ALA-D activity in blood (Berlin and Schaller, 1974), although Hylland (2004) has published an adaptation to fish blood and for implementation in the frame of OSPAR monitoring.

5.2.5 Immunological Parameters

A large number of environmental chemicals have the potential to impair components of the immune system. Both antibody- and cell-mediated immunity may be depressed by certain pollutants, as reviewed by Vos et al. (1989). Although most research on this system has been performed on mammalian species, it may be considered a promising field to search for new biomarkers (Wester et al., 1994). Immunotoxicology is the study of immune dysfunction resulting from exposure of an organism to a xenobiotic. The immune dysfunction may take the form of immunosuppression or alternatively, allergy, autoimmunity, or any number of inflammatory-based diseases or pathologies. Because the immune system plays a critical role in host resistance to disease as well as in normal homeostasis of an organism, identification of immunotoxic risk is significant in the protection of human, animal, and wildlife health (van der Oost et al., 2003). Although major changes in the immune system are rapidly expressed in significant morbidity and even mortality of the organisms involved, they are often preceded by subtle changes in some of the components of the immune system, which could be used as early indicators of immunotoxicity or as biomarkers (Fournier et al., 2000; Brousseau et al., 1997; Dean and Murray, 1990). This aspect has stimulated great interest because these effects generally occur at levels that are lower than those associated with acute toxicity (Brousseau et al., 1997; Koller and Exon, 1985). The immune system would be characterized at both the cellular and humoral levels: phagocytosis, NK-like cytotoxic activity, cyclooxygenase (COX) activity, NO, and lysozyme secretion. Phagocytosis before intracellular degradation of invasive antigens comprises an integral part of cellular defense in organisms. Environmental stressors are postulated to be capable of reducing immunocompetence by altering the phagocytic activity of hemocytes. Moreover, phagocytic activity is a well-conserved function maintained throughout evolution and therefore is present in all living species. In more evolved species, this function is also central to the development of more complex immune responses such as humoral- and cellular-mediated immunity (Fournier et al., 2000). For the majority of the species of concern, phagocytic cells can be collected from peripheral blood or circulatory fluid using noninvasive techniques. The main activities of these cells are phagocytosis of foreign particles or production of biocide molecules and these functions can be assessed using standardized methodologies in a wide variety of species (Brousseau et al., 1999). Depressed phagocytic activity by hemocytes of various species of fish has been found associated with high levels of PAHs (Weeks et al., 1989, 1990a,b; Seeley and Weeks-Perkins, 1991). Pipe (1992) demonstrated that in situ exposures of *M. galloprovincialis* to chlorinated hydrocarbons and trace metals resulted in impairment in the total and differential hemocyte counts, phagocytosis, and the generation and release of ROIs. These responses were all correlated with the contaminant burdens. On the other hand, a hormetic stimulation of phagocytosis was noted after exposure to phenolic compounds and low doses of some heavy metals (Pipe et al., 1995).

The uptake of neutral red by mussel hemocytes may occur by pinocytosis or passive diffusion across the cell membrane. Neutral red is a cationic dye which accumulates in the lysosomal compartment of cells. Alteration in its uptake may reflect damage to the plasma membrane of the hemolymph cells, changes in the volume of the lysosomal compartment and damage of the lysosomal membrane. Immunotoxic effects of environmental contaminants can be evaluated by monitoring neutral red retention time (NRTT), as described earlier. Hemolymph can be easily collected from different organisms and lysosomal membrane stability (LMS) analyzed by NRRT method (Lowe and Pipe, 1994). Monitoring studies have validated the applicability of the NRTT assay in wild mussel populations from different localities of the Iberian Peninsula (Martínez-Gómez et al., 2008) and in the Mediterranean Sea (Gorbi et al., 2008; Dagnino et al., 2007; Viarengo et al., 2007). LMS is a sensitive methodology to analyze inmunotoxicity of new emergent contaminants as pharmaceuticals, as it has been shown after exposure to the clam *R. philippinarum* and crab *C. maenas* (Aguirre-Martínez et al., 2013a,b) and *M. edulis* (Martín-Díaz et al., 2009).

The production of lysozymes is induced primarily by bacteria (Chu and Peyre, 1989; Hong et al., 2006; Li et al., 2008). Notwithstanding this, lysozyme activity has been described to be stimulated by estrogenic compounds, triclosan, and blood-lowering pharmaceuticals, probably through interaction with cell membranes (Canesi et al., 2007a,b,c) in *Mytilus* sp. For example, some PCB congeners induced the spontaneous release of lysozyme and prevented the bactericidal activity toward *E. coli* (Canesi et al., 2003). Oliver et al. (2001) reported that lysozyme activity was positively correlated with copper and PCB contents in wild oysters *C. virginica*, but negatively correlated with cadmium and mercury (Dondero et al., 2006) contents, showing the complexity of effects at play in contaminated environments by urban wastewaters. Lysozyme activity is measured in plasma according to the method of Lee and Yang (2002).

The production of nitric oxide is estimated by measuring the levels of nitrite concentration in the plasma. Because NO readily reacts with oxygen to give nitrite (NO^{2-}) and nitrates (NO^{3-}), nitrate reductase is added to measure the total nitrite concentration in the plasma (Verdon et al., 1995). NO is produced to assist the destruction of ingested bacteria in the so-called phagosomes during phagocytosis (Gourdon et al., 2001). Induction of phagocytosis and NO secretion by estradiol-17β has been shown in *Mytilus* sp. (Canesi et al., 2007c).

NO synthesis is closely linked with inflammatory processes and induces the expression of the COX activity (Grisham et al., 1999). Municipal effluents have shown to induce COX activity in gonad tissues of caged *R. philippinarum* (Maranho et al., 2015b) and in the marine polychaete *Hediste diversicolor* (Maranho et al., 2015c). The activity of COX can be measured in hemocytes and other biological tissue such as gonads by the oxidation of the 2,7-dichlorofluorescein in the presence of arachidonic acid method (Fujimoto et al., 2002).

Due to the fundamental physiological role of the immune system, any kind of impairment in the immunological resistance of marine bivalves may be interpreted as an important signal in ERA. More research should be performed in including immunological resistance of marine bivalves in ERA.

5.2.6 Genotoxic Parameters

More than 1000 individual carcinogenic compounds have been detected in marine sediments from polluted areas around the world and many of these chemicals accumulate in a variety of organisms living in impacted aquatic areas. There is today growing evidence on the increased risk of disease in marine organisms, especially fish that inhabit contaminated waters. Different types of tumors have been evidenced in fish and shellfish populations (Bolognesi et al., 1996). The exposure of an organism to genotoxic chemicals may induce a cascade of events (Shugart et al., 1992): formation of structural alterations in DNA, procession of DNA damage, and subsequent expression in mutant gene. Genetic ecotoxicology can be defined as the study of pollutant-induced changes in the genetic material of biota. The detection and quantification of various events in this sequence may be used as biomarkers of exposure and effects in organisms environmentally exposed to genotoxic substances (van der Oost et al., 2003). The endpoints are different effects at the molecular and cellular level such as gene mutation, chromosome alteration, and induction of DNA damage and repair.

Different genotoxicity parameters are being used nowadays as biomarkers of genotoxicity in marine environments. Studies are mainly focused on fish and shellfish species. In this sense, the study of DNA adducts is an important approach for determining mechanisms of how chemical and physical agents cause genetic disorders and frequency of damage. A more general approach involves the detection of DNA strand breaks that are produced, either directly by the toxic chemical (or its metabolite) or by the processing of structural damage (Shugart et al., 1992). Regarding irreversible effects, chromosomal mutation has also been used as a biomarker of marine pollution.

In mollusks, a cause–effect relationship has been clearly demonstrated in field and laboratory studies between exposure to pollutants (PAHs, chlorinated hydrocarbons, PCBs, heavy metals) and the development of histopathological lesions and tumors (Yevich et al., 1987; Auffret, 1988; Gardner et al., 1991, 1992). More extensive studies in fish have extended the theory of mammalian chemical carcinogenesis (Barrett, 1992) to marine organisms. In benthic fish, positive correlation has been found in situ between the development of hepatic neoplasms and PAH concentration in sediment (Landahl et al., 1990; Myers et al., 1994). Cancerous lesions have been induced in the laboratory following exposure to some PAHs, including benzo[a]pyrene (B[a]P) (Hendricks et al., 1998; Fong et al., 1993), and K-ras oncogenes have been activated in induced neoplasms (Wirgin et al., 1989; MacMahon et al., 1990; Fong et al., 1993). As PAHs are predominant pollutants in the environment, PAH-induced DNA damage has been extensively studied in vertebrates for the model compound B[a]P (Colapietro et al., 1993; Chen et al., 1996; De Vries et al., 1997).

5.2.6.1 DNA Adducts

In molecular genetics, a DNA adduct is a piece of DNA covalently bonded to a (cancer-causing) chemical. This process could be the start of a cancerous cell or carcinogenesis. DNA adducts in scientific experiments are used as biomarkers of exposure (La et al., 1996). In addition to their use as biomarkers for exposure and effects of genotoxins, DNA adducts may provide information about the biological effect and potential risk of a chemical, since it has been suggested that any chemical that forms DNA adducts, even at very low levels, should be considered to have carcinogenic and mutagenic potential (Maccubbin, 1994). DNA adduct formation has been mainly studied in ecotoxicological approaches regarding exposure to PAHs. They have been observed not only by the binding of a chemical like PAH to DNA but also by the binding to reactive forms. The same metabolic processes (Phase I and Phase II) that are responsible for the efficient elimination of biodegradable substances such as PAHs from the organism are also able to activate environmental carcinogens to DNA-reactive forms (Dunn, 1991). Greater Phase I or Phase II induction might be associated with a greater risk of carcinogenesis. It is, nevertheless, most likely that liver PAH–DNA levels in marine organisms reflect the extent of exposure to carcinogenic or mutagenic PAHs. DNA adducts are mainly measured in liver/hepatopancreas/digestive gland, although studies have also been focused on gill tissues. After DNA extraction (Helbock et al., 1998), DNA adducts are determined applying the ^{32}P postlabeling technique as described in Genevois et al. (1998). Significant DNA adducts formation has been observed after exposure to PAHs in mussels (Solé et al., 1996; Akcha et al., 2000), in fish (Ericson et al., 1996), and in crustaceans (James et al., 1992).

5.2.6.2 Secondary DNA Modifications

Damage to DNA can be one of the most significant consequences of chemical insults to organisms since this has the potential to affect not only individual organisms but populations of organisms. DNA damage is an alteration in the chemical structure of DNA, such as a break in a strand of DNA, a base missing from the backbone of DNA, or a chemically changed base such as 8-OHdG. Damage to DNA that occurs naturally can result from metabolic or hydrolytic processes. Metabolism releases compounds that damage DNA including ROS, reactive nitrogen species, reactive carbonyl species, lipid peroxidation products, and alkylating agents, among others, while hydrolysis cleaves chemical bonds in DNA. DNA damage could reflect a recent pollution status of the marine environment. DNA breaks can be repaired by different mechanisms (Turner and Parry, 1989). The activity of DNA repair can vary greatly among different species and individuals. These differences are correlated with the maximum life span of the species. In addition the time taken for DNA repair differs also with the damaging agents. DNA damage is distinctly different from mutation, although both are types of error in DNA. DNA damage is an abnormal chemical structure in DNA, while a mutation is a change in the sequence of standard base pairs. Alkaline unwiding and DNA precipitation allow the determination of damaged DNA by fluorometric measurements (Brunk, 1979): a more simple and rapid method due to the fact that it is based on precipitation (Olive et al., 1986, 1988). The first two assays, alkaline unwinding and precipitation approaches allow the determination of damaged DNA by fluorometry (Brunk, 1979): a more simple and rapid method as it is based on precipitation (Olive et al., 1986, 1988). The Comet assay provides a rapid screening for potential DNA damage (Frenzilli et al., 2009; Picado et al., 2007) compared to more laborious processes. The Comet assay (single-cell gel electrophoresis) is a simple method for measuring deoxyribonucleic acid (DNA) strand breaks in eukaryotic cells. Cells embedded in agarose on a microscope slide are lysed with detergent and high salt to form nucleoids containing supercoiled loops of DNA linked to the nuclear matrix. Electrophoresis at high pH results in structures resembling comets, observed by fluorescence microscopy; the intensity of the comet tail relative to the head reflects the number of DNA breaks (Nandhakumar et al., 2011). DNA damage has been observed in fish such as *Serranus scriba*, mussels such as *M. galloprovincialis*, crab such as *Maja crispata*, sea cucumber such as *Holothuria tubulosa* (Bihari and Fafan, 2004), and in bottlenose dolphins (Lee et al., 2013), when applied to assess environmental contamination in ERA process.

5.2.6.3 Irreversible Genotoxic Events

DNA damage and mutation have different biological consequences. While most DNA damages can undergo DNA repair, such repair is not 100% efficient. Unrepaired DNA damages accumulate in nonreplicating cells, such as cells in the brains or muscles of adult mammals and can cause aging (Bolognesi et al., 1990). In replicating cells, errors occur upon replication of past damages in the template strand of DNA or during repair of DNA damages. These errors can give rise to mutations or epigenetic alterations (Brunnetti et al., 1988). Both of these types of alteration can be replicated and passed on to subsequent cell generations. These alterations can change gene function or regulation of gene expression and possibly contribute to progression to cancer. Micronucleus frequencies provide an index of accumulated genetic damage during the life span of the cells. Animals exposed to mutagenic compounds may exhibit increased micronuclei frequency after the exposure has ceased (Mayone et al., 1990). These considerations suggest the suitability of this test to monitor genotoxic damage induced long before the sampling time.

Micronuclei are characterized in the cells that have some sort of DNA damage. This includes damage caused by radiation, harmful chemicals, and random mutations that occur throughout the genome. Micronuclei are small bodies that can be seen budding off of a newly divided daughter cell and can contain a whole chromosome or part of a chromatid. The increased formation of micronuclei is usually an indication of increased DNA damage or mutation. It is characteristically found in cancer cells, or cells that have been exposed to increased risk factors. Another mechanism to micronuclei formation is by a double-strand break in the DNA, creating a separate linear fragment. Micronuclei are often overlooked in cancer. If observed under a microscope, they are viewable and often next to other larger nuclei. This approach is under investigation and research regarding whether or not they can be used to predict future cancer risks. They are easy to analyze compared to chromosome aberrations.

Of these test systems, micronuclei (MN) test is one of the most reliable techniques used to determine genetic changes in the organisms in contaminated waters and complex mixtures. In recent years, this test has been improved using many aquatic organisms. MN experiments are a fast method in detecting the chromosomal damage because it makes it possible to determine the remaining chromosomes and broken chromosomes due to its several advantages such as (1) giving more objective results than other tests in detecting chromosomal impairments; (2) being easy to learn; (3) not requiring to count the chromosomes to investigate the chromatids and chromosomal damage hard to detect and see in the metaphase stage; (4) its fast preparation stage; and (5) making it possible to count thousands of cells, not hundreds of cells in each experiment (OECD, 2004).

In environmental mutagenesis, MN tests yield quite practical results in monitoring clastogenic and genotoxic effects of the pollutants. Fish and mussels are main indicators of health of the aquatic environment. To obtain these results, aquatic organisms are usually used such as bivalve *M. galloprovincialis*, *C. gigas*, and *Chamelea galina*, and fish European sea bass *Dicentrarchus labrax* (Hooftman and Raat, 1982; Manna et al., 1985; Metcalfe, 1988; Rodriguez-Ariza et al., 1992; Al-Sabti et al., 1994; Arslan et al., 2010; Tsarpalias and Dailianis, 2012).

5.2.7 Reproductive and Endocrine Parameters

The seriousness of the problem related to the alteration of reproduction on organisms exposed to xenobiotics has led international bodies such as the Organization for Economic Cooperation and Development (OECD) and the European Union (EU) to initiate large research programs and developments toward new guidelines and regulations. Endocrine Disruption Chemicals (EDCs) have both synthetic and natural sources. Synthetic sources include plastics, detergents, drugs (such as oral contraceptives), cosmetics, flame retardants, herbicides, and pesticides, and others, entering the environment through industrial and sewage discharges, active application, and runoff. Natural sources include human and animal hormones and phyto and mycoestrogens found in sewage, animal husbandry runoff, and intentionally or accidentally as food and feed ingredients (Goksøyr, 2006).

Hormone regulation may be impaired as a consequence of exposure to environmental pollutants (Spies et al., 1990). Decreased reproductive capability in feral organisms is considered as one of the most damaging effects of persistent pollutants released by man. A number of xenobiotics with widespread distribution in the environment are reported to have endocrine activity which might affect reproduction and thus might threaten the existence of susceptible species (Colborn et al., 1993; Peterson et al., 1993; White et al., 1994). Animals at high trophic levels, generally having limited reproduction rates, are likely to be the most vulnerable in this regard.

The mechanisms of action of these xenobiotics can be divided into (1) agonistic/antagonistic effect ("hormone mimics"), (2) disruption of production, transport, metabolism, or secretion of natural hormones, and (3) disruption of production and/or function of hormone receptors (Rotchell and Ostrander, 2003; Goksøyr et al., 2003). A wide spectrum of potential biomarkers could be applied to the study of endocrine disruption in the aquatic environment. In marine organisms, these include gamete morphology alteration, oocyte maturation, ovulation and spawning, egg numbers and viability, vitellogenesis, hypothalamic and pituitary hormones, gamete steroids, and hepatic catabolism.

The stage at which a marine organism is exposed to and the duration of such exposure will determine to a large extent the effect on the ovary. Long-term exposure, which generally begins early in the reproductive cycle and finishes toward oocyte maturation, has often used gonadosomatic index (GSI), oocyte stage, histological examination, or number and viability of ovulated eggs as measures of pollutant effect. Shorter-term exposures, or in vitro studies, are better for the examination of the specific mechanisms involved, such as vitellogenesis, steroidogenesis, and pituitary activity. Long-term exposure to pollutants almost invariably leads to a decrease in GSI, smaller, less-developed oocytes and fewer large, mature oocytes, and an increase in the numbers of atretic follicles, as occurred. Oocytes frequently contained less yolky granules (Sukumar and Karpagaganapathy, 1992), ruptured oocyte walls, damaged yolk vesicles (Kulshrestha and Arora, 1984) and nucleoli, and cytoplasm which had undergone major changes (Kirubagaran and Joy, 1988; Murugesan and Haniffa, 1992). Cadmium and mercury exposure resulted in extensive vacuolation in the oocortex, necrosis of oolemma, and hypertrophy of the follicular cells and, while mercury had little effect on the incorporation of vitellogenin, cadmium appeared to inhibit the transfer of nutrients (Victor et al., 1986).

Since the incorporation of vitellogenin into the developing oocyte accounts for the major part of the growth during ovarian recrudescence, the lower GSI in marine organisms exposed to pollutants is probably due to inhibition of vitellogenesis. A more direct measure is obtained from plasma, hepatic, and ovarian vitellogenin concentrations.

EDCs have been defined as chemicals that can alter function(s) of the endocrine system and consequently cause adverse health effects in an intact organism or its progeny or (sub) populations (World Health Organization (WHO/IPCS, 2002)/International Programme on Chemical Safety (IPCS), 2002). EDCs described as xenoestrogens can bind to ERs and provoke induction of vitellogenin (Vtg) expression, which has been proposed as a specific biomarker of estrogenicity in fish (Goksøyr et al., 2003; WHO/IPCS, 2002).

Induction of vitellogenin, the precursor molecule of yolk proteins, in oviparous males or juveniles is a well-known effect of xenoestrogenic contaminants in fish and has been extensively used as a biomarker both in laboratory and field studies (Matthiessen and Sumpter, 1998; WHO/IPCS, 2002; Arukwe and Goksøyr, 2003; Goksøyr et al., 2003; Ortiz-Zarragoitia and Cajaraville, 2005). The levels of Vtg have also been determined as a biomarker of pollution in marine invertebrates. No induction of Vtg has been detected in male invertebrates after exposure to xenoestrogens; therefore studies have been focused on the determination of the disruption of normal Vtg levels in female marine invertebrates.

The liver plays an important role in ovarian recrudescence in its production of the yolk protein, vitellogenin, under the stimulation of ovarian estradiol. Vitellogenin is a glycolipophosphoprotein derived from hepatic lipids. Pollutant effects on liver lipid synthesis may, therefore, affect ovarian development and cholesterol is, of course, also a precursor of all of the steroid hormones (Kime et al., 1995). Changes in hepatic, ovarian, and plasma content of lipids may be a reflection of pollutant inhibition of the mobilization of hepatic lipids into the plasma and its incorporation into developing oocytes of the ovary (Lal and Singh, 1987; Singh, 1992). Since such mobilization is maximal during the vitellogenic phase, it is at this period that pollutants might have maximal effect on lipid distribution. Changes in nutritional input for developing oocytes via vitellogenin incorporation in ovaries may be correlated with a decrease in the gonadosomatic index, in number of ovulated eggs as well as a decrease in spawning and hatching success.

Vtg concentration can be measured in blood/hemolymph in vertebrates/invertebrates species, respectively; consequently, the determination of the disruption of Vtg levels as biomarker of xenoestrogens pollution may constitute a noninvasive methodology. Nevertheless, this could be possible in marine fish and crustaceans, since the sex of the organisms can be determined observing the external body, but not for bivalves, for which a smear of gonads has to be performed to find out the individual sex under the microscope. Vtg concentrations can also be measured in hepatopancreas, digestive gland, liver, and gonad tissues of the bioindicator species. The concentration of this lipoprotein could vary depending on the reproduction stage of the organisms. It is important to verify if the organisms is in the reproduction period or not, and when being in the reproduction period, to determine the ovarian maturation stage. Confounding results may be obtained if levels of Vtg are compared between individuals belonging to different ovarian maturation stages or reproduction periods. Moreover, a study over time is advised, since concentrations of these lipoproteins may vary over time in the biological tissue which is being focused. Note that Vtg is transported from the hepatopancreas/liver/digestive gland through the circulatory fluids (blood and hemolymph) to the ovaries.

Several methodologies have been developed for determination of Vtg: immunotechniques based on the use of specific antibodies such as radioimmunoassays, enzyme-linked immunosorbent assays (ELISAs), Western blot, and immunohistochemistry, molecular tools such as RNA protection assays and transcript analysis by Northern blotting or various variants of polymerase chain reaction (PCR), and protein expression studies by proteomic approaches (Denslow et al., 1999; Arukwe and Goksøyr, 2003; Marin and Matozzo, 2004). Among these methods, the measurement of phosphoproteins by the alkali-labile phosphate (ALP) method has been widely used in different aquatic organisms such as fish and bivalve mollusks (Kramer et al., 1998; Blaise et al., 1999; Verslycke et al., 2002; Marin and Matozzo, 2004). In fish, ALP levels have been shown to associate with Vtg levels measured using specific immune techniques and gene expression tools (Versonnen et al., 2004; Robinson et al., 2004). There are no described specific antibodies for Vtg-like proteins in marine mussels like *M. edulis* and *M. galloprovincialis*. Thus, the ALP method could have potential as a simple cost-effective biomarker of endocrine disruption in mussels and other widely used molluscan sentinel species. Seasonality studies in different bivalve mollusks have shown that ALP levels follow the same trend of the gametogenic cycle in females (Blaise et al., 1999, 2002; Ortiz-Zarragoitia, 2005). ALP levels in female *M. galloprovincialis* increased during active gametogenesis reaching maximum levels at gonad maturation (Porte et al., 2006). ALP levels are measured for each animal in gonad homogenates according to the protocol of Gagné et al. (2003).

5.2.8 Neurotoxic Parameters

Cholinesterases (ChE) are considered enzymes of interest when studying neural functions (Payne et al., 1996). Acetyl cholinesterase (AChE) is involved in the termination of impulse transmission by rapid hydrolysis of the neurotransmitter acetylcholine in numerous cholinergic pathways in the central and peripheral nervous systems.

AChE belongs to the family of ChEs, which are specialized carboxylic ester hydrolases, that break down esters of choline. ChE class includes AChE which hydrolyzes the neurotransmitter acetylcholine and pseudocholinesterase or butyrylcholinesterase (BChE) which uses butyrylcholine as substrate (Lionetto et al., 2013). It is essential for the normal functioning of the central and peripheral nervous system. Therefore, there are two types of ChE recognized; firstly, those with a high affinity for acetylcholine (AChE), and secondly, those with affinity for butyrylcholine (BChE), also known as nonspecific esterases or pseudocholinesterases (Walker and Thompson, 1991; Sturm et al., 2000). Nowadays, more information is available about AChE, about its mode of action, inhibition, and activation, than about BChE. AChE hydrolyzes acetylcholine into choline and acetate after activation of acetylcholine receptors at the postsynaptic membrane. AChE activity serves to terminate synaptic transmission, preventing continuous nerve firings at nerve endings. Therefore, it is essential for the normal functioning of the central and peripheral nervous system. BChE physiological function is still unknown (Daniels, 2007).

AChE inhibitors inhibit the ChE enzyme from breaking down acetylcholine, increasing both the level and duration of the neurotransmitter action. Taking into consideration the mode of action, AChE inhibitors can be divided into two groups: irreversible and reversible. Reversible inhibitors, competitive or noncompetitive, mostly have therapeutic applications, while toxic effects are associated with irreversible AChE activity modulators (Lionetto et al., 2013). The enzyme inactivation, induced by various inhibitors, leads to acetylcholine accumulation, hyperstimulation of nicotinic and muscarinic receptors, and disrupted neurotransmission (Colovic et al., 2013). Lately, the inhibition of AChE from several chemical species has been increasingly reported in humans and other animals (Goldstein, 1992; Lionetto et al., 2004; Jebali et al., 2006; Vioque-Fernández et al., 2007). The interest is focused in the application of AChE inhibition as a biomarker in ERA.

Organophosphorus and carbamate pesticides are known to be specific inhibitors of acetyl cholinesterase catalytic activity and this physiological inactivation has been widely studied. They have become the most widely used pesticides today since the removal of organochlorine pesticides from use. They inactivate the enzyme by binding the esteratic site by phosphorylation or decarbamylation. Organophosphorus compounds are considered to be functionally irreversible inhibitors of AChE, since the time necessary to liberate the enzyme from inhibition may be in excess of the time required for synthesis of new AChE. Carbamates, on the other hand, have a fairly rapid decarbamylation step so that substantial recovery of the enzyme can occur in a finite period of time (Lionetto et al., 2013). AChE inhibition has been commonly used as a biomarker of pollution by pesticides including neurotoxic organophosphates and carbamates as effective compounds.

Regarding the species, although a number studies have successfully used AChE inhibition in fish as a biomarker, there is considerably less knowledge about effects of multitoxicants on the AChE activity of marine invertebrates. Nevertheless, in the study performed by Bocquené et al. (1990), AChE activity was determined in the fish *Solea solea* and in the invertebrates *Palaemon serratus*, *Crangon crangon*, *M. edulis*, *C. gigas*, and the polychaete *Nereis* sp.

Most of the studies have been performed using AChE activity as a biomarker in ERA to assess pesticides as carbamates; nevertheless, metals, PAHs, detergents, and components of complex mixtures of contaminants have been increasingly reported in marine environments and AChE activity inhibition used as the biomarker (Pérez et al., 2004; Lionetto, 2004; Jebali et al., 2006).

Due to the growing interest in nanomaterials in various applications, different classes of nanoparticles, including metals, oxides, and carbon nanotubes (SiO_2, TiO_2, Al_2O_3, Al, Cu, carbon-coated copper, multiwalled carbon nanotubes, gold nanoparticles, single-walled carbon nanotubes), have shown high affinity for AChE (Wang et al., 2009a,b), in ragworm and bivalve mollusk (Buffet et al., 2014) and in fish (Ferreira et al., 2016).

The measurement of AChE activity is performed spectrophotometrically using the methodology developed by Ellman et al. (1961). The principle of the method is the measurement of the rate of production of thiocholine as acetylthiocholine is hydrolyzed. Acetylthiocholine is used as the substrate. This photometric method is used for determining acetyl cholinesterase activity of tissue extracts, homogenates, cell suspensions, etc. The enzyme activity is measured by following the increase of yellow color produced from thiocholine when it reacts with dithiobisnitrobenzoate ion. Enzyme inhibition is expressed as acetylcholine hydrolyzed min^{-1} mg^{-1} protein.

A number studies have successfully used AChE inhibition a biomarker of pesticide exposure; more research is required before this parameter can be used in ERA programs. Additional research is needed to better explain the species-specific differences in the relationship between AChE inhibition and mortality and to investigate other physiological perturbations associated with AChE inhibition (Fulton and Key, 2001). The usefulness of this biomarker could be that of providing an integrative measurement of the overall neurotoxic risk posed by the whole burden of bioavailable contaminants present in the environment.

5.3 HIGH-THROUGHPUT SCREENING TECHNIQUES OR "OMICS"

Since the discovery and description of DNA and its functioning in the early 1950s (Watson and Crick, 1953), unthinkable advances have been achieved in the understanding of molecular mechanisms underlying cellular processes. Molecular biology is particularly concerned with the flow of biological information and its consequences at the level of genes and proteins. Although not completely free of controversy about its true author, the discovery of the double helix structure of DNA made clear that genes are functionally defined parts of DNA molecules of a certain sequence of nucleotides and that there must be a way for cells to make use of their genes to make proteins. Gene transcription and translation were later discovered as the drivers of gene expression and subsequent protein synthesis and higher level processes. With the increasing knowledge on genetic regulation and functioning of cells mainly driven by medical research, toxicologists also began to focus on the evaluation of contaminant-triggered molecular effects of exposure. This trend has significantly increased the scientific effort in the evaluation of molecular effects of contamination and the application of molecular and cellular approaches in ecotoxicology. For a comprehensive list of terminologies arising from this relatively new field, readers are referred to the Collection of Working Definitions of the OECD (2012).

There are several reasons for the increasing importance that is given to the analysis and understanding of molecular effects caused by a contaminant in toxicology.

- One is the fact that traditional ERA of contaminants is based on apical effects such as death and histopathologies, among others, which were the effects measured during the last century when thoughtlessly released huge levels of contaminants lead to massive mortalities and other instantaneously alarming responses. However, nowadays the environmental exposure to contaminants occurs rather at relatively low concentrations but during an increased time or chronically. Chronic low-dose effects, however, may not cause overt toxicities but may cause adverse ecological outcomes in terms of population levels and biodiversity through rather subtle changes in the health and physiology (eg, behavior) of the organisms (Relyea and Hoverman, 2006). These can be detected and monitored using molecular biology tools.
- Molecular and cellular tools are also useful in field monitoring for diagnosing actual pollution impacts on wildlife and for the detection of sublethal, chronic effects and their relation to alterations at ecosystem level (Relyea and Hoverman, 2006). As effects at DNA and protein level are considered the first response to the challenge, molecular and cellular tools are particularly relevant in the (early) detection of such subtle toxicant effects, as highlighted by the utility of biomarkers in environmental assessment.
- Finally, the knowledge of molecular and cellular processes underlying a toxic event triggered by a certain contaminant also helps for the prioritization of chemicals for hazard testing as well as for interspecies extrapolation. With over 30,000 substances produced above 1 ton per year on the EU market and eventually discharged, in one form or another, into the environment, the traditional descriptive, compound-by-compound approach is coming to its limits, from both an economic and time-consuming point of view. Thus, effect evaluation requires an approach that provides mechanistic understanding of the elicited toxicity processes and allows the elucidation of molecular pathways and networks involved in the toxic response and thus the Mode of Action (MoA) of the substance. A MoA describes functional or anatomical changes, at the cellular level, resulting from the exposure of a living organism to a substance. Therefore, molecular and cellular approaches help to overcome the above-mentioned limitations by replacing the current testing approach that relies on phenotypic responses in animals. By exploring principal molecular processes or pathways that are somehow altered by the interaction with the contaminant, the specific origin of the pathology can be identified, and chemicals can be grouped according to their MoA (Diamond et al., 2011; Segner, 2011). With the Adverse Outcome Pathway (AOP) concept suggested by Ankley et al. (2010), effects from lower levels of organization (ie, reduced or increased expression of certain genes or proteins) should be linked to higher levels of organization through the use of, eg, high-throughput screening, toxicogenomic assays (that reveal molecular effects) in combination with histopathological and/or behavioral assessments, known as

physiological anchoring, to identify the organismal or populational effectiveness of the chemical insult as well as the molecular origins of its manifestation. As such, an AOP conceptually links a direct molecular initiating event (eg, a molecular interaction between a contaminant and a specific biomolecule) and an adverse outcome (histopathology, death, etc.) at a biological level of organization relevant to risk assessment (Fig. 5.2). Chemicals with specific MoAs may also target specific ecological receptors or specific life stages of a given species and thereby cause toxic effects that are not predicted from standard ecotoxicological risk assessment. Here, molecular and cellular approaches can inform which ecological receptor groups, which life stages, or which functions and traits are likely to be at risk by a given group of chemicals (Hutchinson et al., 2006; Yadetie et al., 2012; Segner et al., 2013; Brown et al., 2014). Overall, mechanism-based assessment approaches support evidence-based risk assessment where uncertainty is explicit and improve our ability to diagnose and predict adverse effects of chemical pollution.

The increasing application of molecular and cellular approaches in ecotoxicology is also related to the huge technical progress of the relevant methodologies over the last decades. With the development of the PCR in 1983 (Mullis et al., 1986), it was possible to achieve a great number of copies of particular DNA fragments coding for a specific feature (Fig. 5.3). This technique, based on denaturation and hybridization of DNA double strands by playing with the temperature, introduced a revolution in biological and medical research, which led Mullis to

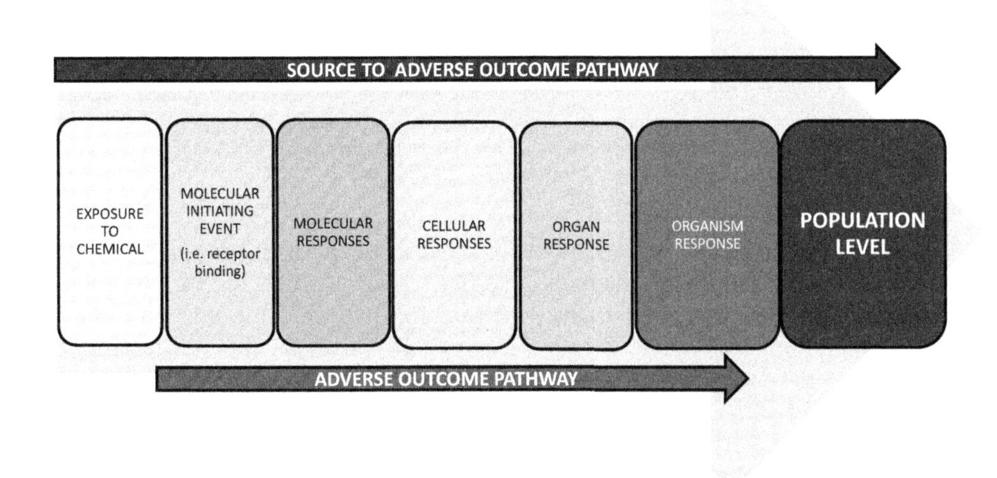

FIGURE 5.2 Adverse outcome pathway (AOP). An AOP is the sequential progression of events from the molecular initiating event to the in vivo outcome of interest (community or population level). Generally, it refers to a broader set of pathways that would: (1) proceed from the MIEs, in which a chemical interacts with a biological target (eg, DNA binding, protein oxidation), (2) continue on through a sequential series of biological activities (eg, gene activation, altered tissue development), and (3) ultimately culminate in the final adverse effect relevance to human or ecological risk assessors (eg, mortality, disrupted reproduction, cancer, extinction) (OECD, 2011).

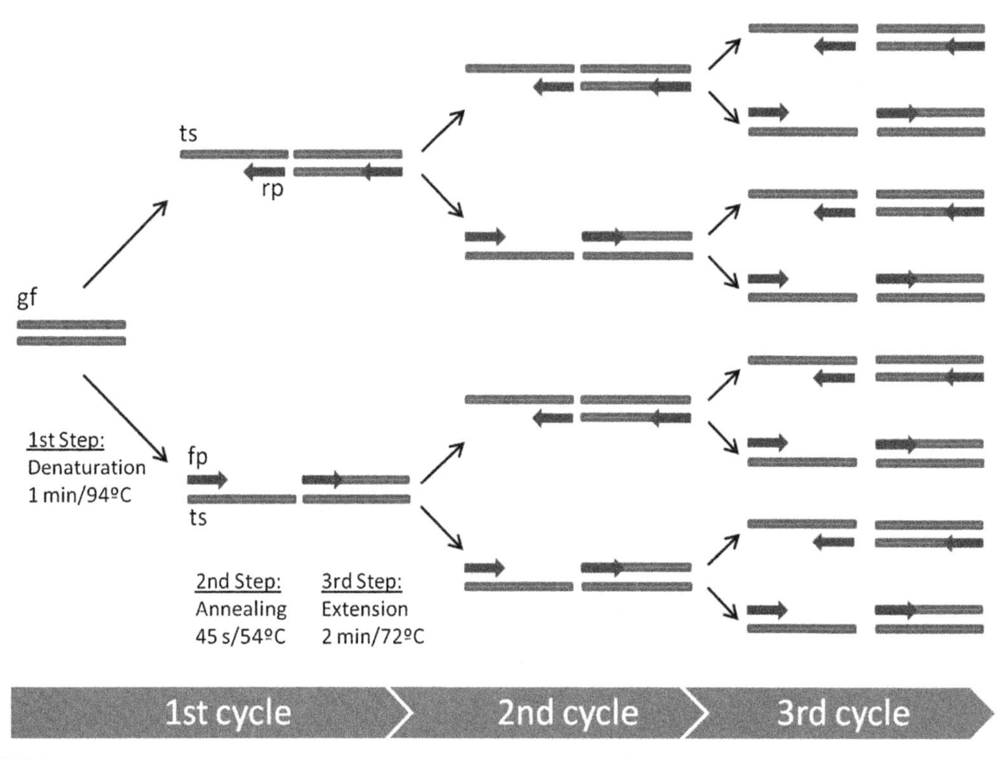

FIGURE 5.3 Schematic representation of the polymerase chain reaction (PCR). In the first step, a double strand gene fragment of interest (gf) is physically separated at a high temperature into two single template strands (ts) (denaturalization). In the second step, the temperature is lowered allowing the union of the specific forward (fp) and reverse primers (rp) (annealing). During the third step, the polymerase introduces consecutively nucleotides into the growing DNA strand according to the complementarity of the template strand (ts). Successive PCR cycles amplify exponentially the region of interest or gene fragment. The selectivity of PCR results from the use of primers that are complementary to the DNA region targeted for amplification under specific thermal cycling conditions. Given temperatures are orientative.

receive the Nobel Prize in Chemistry in 1993 and was further developed in the 1990s with the real-time polymerase chain reaction (RT-PCR). The difference between conventional PCR and real-time PCR is that RT-PCR monitors the amplification of a targeted DNA molecule during the PCR, ie, in real-time, and not at its end, as in conventional PCR. Real-time PCR can be quantitative (quantitative real-time PCR), semi-quantitative, ie, with regard to a certain amount of DNA molecules (semi-quantitative real-time PCR), or qualitative (qualitative real-time PCR). For further information on RT-PCR procedures and data normalization steps, readers are referred to Gause and Adamovicz (1994), Bustin (2002), and Bustin et al. (2005). Aspects regarding quantification and amplification efficiency are extensively addressed in Pfaffl (2001), Pfaffl et al. (2002), and Pfaffl (2004), among others.

RT-PCR allows toxicologists to detect differential expression of single genes or DNA fragments in individuals or populations submitted to different (environmental or experimental) conditions, eg, between contaminant-exposed and nonexposed organisms. In toxicology, advances in transcript analysis have led to the recognition that altered gene expression is potentially an early, rapid, and sensitive means of stress response detection. This technique, however, is based on informed guesswork of expected effects targeting a specific transcript and does not provide information on the interaction of a huge number of genes at the time nor it is useful for the detection of unexpected and unknown processes and pathways. It is therefore not useful for the discovery of new biomarkers but has rather been used in marine ecotoxicology to confirm genetical induction or silencing of certain biomarker enzymes or other features due to its extremely high sensitivity, besides forming the conceptual base of the more recent high-throughput techniques. Some examples on the use of RT-PCR techniques applied to marine (eco-)toxicology are Seo et al. (2004) who studied the expression of glutathione reductase in the intertidal copepod *Tigriopus japonicas* submitted to different sources of oxidative stress (salinity, heavy metals, H_2O_2). Won et al. (2013) used the marine polychaete *Perinereis nuntia* exposed to crude oil to measure the expression of three novel cytochromes P450 (CYP) and antioxidative genes and reported that crude oil-exposed *P. nuntia* was properly responding to this kind of chemical stress. Another example of the application of the RT-PCR technique in marine environments is the work provided by Rhee et al. (2013). These authors studied the co-expression of antioxidant enzymes together with the expression of p53, DNA repair, and heat shock protein genes in the marine fish *Kryptolebias marmoratus* after gamma ray irradiation. Du et al. (2013) addressed the problem of housekeeping genes as internal controls for studying gene expression in Pacific oyster (*C. gigas*). These are only some examples for the wide application of this useful technique in marine ecotoxicology.

During this last decade, several different techniques have been developed to analyze simultaneously more than only one gene or feature of interest, including serial analysis of gene expression (SAGE). SAGE is a technique to produce a snapshot of the messenger RNA population in a sample in the form of small tags that correspond to fragments of those transcripts (Velculescu et al., 1995), being a tag or an expressed sequence tag a short fragment of a cDNA sequence that can be used to identify gene transcripts through complementarity (Adams et al., 1991). Different variants have been developed since, ie, LongSAGE (Saha et al., 2002), RL-SAGE (Gowda et al., 2004), and SuperSAGE (Matsumura et al., 2005) improving the original technique by capturing longer tags which enables a more confident identification of the source gene. However, it was not until the development of high-throughput or omic techniques such as DNA microarray and RNAseq that it was possible to analyze simultaneously the expression of thousands of genes under a certain condition.

The term "omics" indicates "a totality of some sort." In biology, it is used for very large-scale data collection and analysis and can be divided into three main categories: genomics/transcriptomics, proteomics, and metabolomics/metabonomics (Gehlenborg et al., 2010). Omic profiling is the measurement of the activity or expression of thousands of genes (transcriptomics), proteins (proteomics), or metabolites (metabolomics/metabonomics) at the same time to create a global picture of cellular function. By measuring these expression levels, omics allows to study the detailed mode of action of compounds (Dulin et al., 2012). Besides, advances in DNA sequencing technologies have facilitated the knowledge of the sequences of many genomes, including human (Venter et al., 2001; Lander et al., 2001), and coupled with our ability to quantify changes in mRNA expression using DNA microarrays and other, more sophisticated technologies allow us to determine the onset or silencing of specific genes (transcriptomics) or the increased or reduced abundance of proteins (proteomics) or metabolites (metabolomics) involved in the toxic event (Table 5.3).

TABLE 5.3 Summary of the Most Used Omic Techniques in Ecotoxicology, the Analyzed Targets, and Expected Information

Technique	Target	Information about
Transcriptomics	All transcripts (mRNA) in sample	Which genes are turned on/off
Proteomics	All proteins in sample	Which genetic information is translated into proteins
Metabolomics	Metabolites of cell	Which proteins are functioning in the cell

5.3.1 Transcriptomics

Generally, transcriptomics is the most frequently used omics technique in environmental toxicology, being the predominant platform cDNA microarrays or oligo arrays (Hartung and McBride, 2011). A DNA microarray is a collection of microscopic DNA spots attached to a solid surface similar to a microscope slide. Each DNA spot contains a specific DNA sequence, known as probes (or reporters or oligos). Basic principles and applications of cDNA microarrays have been extensively described in Pollack et al. (1999) and Carter (2007). Microarray techniques require that the targets, mRNA molecules of the sample, have to be transcribed to copy DNA (cDNA) that binds to the probes spotted on the chip. An important drawback of microarray techniques is that obtained results are limited by the features of interest (number and nature) spotted onto the array, and the possibility of "loosing" interesting responses in unexpected features. This is now being overcome by new promising techniques to measure gene expression such as next generation and RNA sequencing tools (next-generation sequencing (NGS) and RNA-seq). RNAseq uses the capabilities of NGS to reveal a snapshot of RNA presence and quantity from a genome at a given moment in time (Chu and Corey, 2012). RNAseq produces thousands or millions of sequences at once. RNAseq is probably to substitute microarray-based transcriptomics in the future when this technique will become more and more affordable. Although RNAseq is still a technology under active development, it offers several key advantages over existing technologies. Its advantage over hybridization-based approaches such as cDNA microarrays lies in the fact that the information that can be obtained in cDNA microarray studies is limited to the probes spotted onto the surface corresponding to existing genome sequences, whereas RNAseq identifies all RNAs present in a sample by sequencing and is thus completely independent from any predesigned platform. This makes RNAseq particularly attractive for nonmodel organisms with genomic sequences that are yet to be determined (Wang et al., 2009a,b). However, arrays still have a place for targeted identification of gene expression, making them ideal for regulatory monitoring purposes.

5.3.2 Proteomics

Similar to transcriptomics, progress in the field of mass spectrometry (MS) has allowed the quantification and identification of proteins (proteomics) and metabolites (metabolomics) under different conditions and gives an insight into the organisms' mechanisms to fight against the challenge. Proteomics is a bundle of techniques that attempt to separate, quantify, and identify many hundreds of proteins in a sample simultaneously to analyze cell- and tissue-wide protein expression. Thus, while genomics is mainly concerned with gene expression, proteomics analyses the protein products of the genes. While determination of expression levels of individual proteins was traditionally made through union to a specific antibody (=identification) with a technique called Western blot (Towbin et al., 1979), in current proteomic techniques, proteins are identified and quantified by MS, with (2D-PAGE, 2D-DIGE) or without (iTRAQ) previous separation by gel electrophoresis. Two-dimensional polyacrylamide gel electrophoresis (2D-PAGE) and MS has been used since the late 1960s as a powerful tool for protein separation and identification, respectively, and to compare protein expression between controls and diseased or chemically exposed biological samples (Martyniuk et al., 2012). The most significant innovation for proteomics was coupling MS with 2D gels as a way to identify differentially expressed proteins (Lilley et al., 2001; Beranova-Giorgianni, 2003; Herbert et al., 2001). With this technique, proteins of a sample are first separated according to their charge by isoelectric focusing (IEF, first dimension) and subsequently according to their molecular weight. The separation in the first dimension allows resolution of proteins that differ by a single positive or negative charge making this method ideal for distinguishing protein isoforms and proteins that are posttranslationally modified, in addition to quantifying protein expression changes. The introduction of Differential in Gel Electrophoresis (DIGE) methods has later increased the standard for this proteomics approach, allowing co-separation of a control set of proteins with proteins isolated from a treatment or disease within the same gel through differential fluorescent staining of the samples (Tonge et al., 2001). More recently, gel-independent techniques are being used based on MS with differently labeled protein fragments. In this context, protein quantification through incorporation of stable isotopes has become a central technology in modern proteomics research. Quantification by iTRAQ (isobaric tags for relative and absolute quantification) is one of the several techniques in toxicology to monitor relative changes in protein abundance across perturbed biological systems (Ross et al., 2004). Application of iTRAQ method enables comparing of up to eight different samples in one MS-based experiment and avoids the much less sensitive in-gel separation of proteins. A rapidly developing discipline within proteomics is redox proteomics. Redox proteomics is especially concerned with modifications of the proteome under redox-active conditions that are frequently found under contamination scenarios. There are many chemicals that trigger the formation of ROS within a cell and that can further evoke inflammatory processes. The redox proteome consists of reversible and irreversible covalent modifications that link redox metabolism to biologic structure and function. These modifications function at the molecular level in protein folding and maturation, catalytic activity, signaling, and macromolecular interactions and at the macroscopic level in control of secretion and cell shape. Interaction of the redox proteome with redox-active chemicals is central to macromolecular structure, regulation, and signaling during the life cycle and has a central role in the tolerance and adaptability to diet and environmental challenges (McDonagh and Sheehan, 2006; Sheehan, 2006).

5.3.3 Metabolomics

The metabolites, or small molecules, within a cell, tissue, organ, biological fluid, or entire organism constitute the metabolome (Miller, 2007). Metabolomics aims to measure the global, dynamic metabolic response of living systems to biological stimuli by identifying and quantifying all the small molecules in a biological system (cell, tissue, organ, and organism). Most commonly used metabolomics techniques are liquid chromatography–mass spectrometry (LC-MS), gas chromatography–mass spectrometry (GC-MS), and nuclear magnetic resonance (NMR) (Nicholson and Lindon, 2008). Though metabolomics is viewed as complementary to other omic techniques, it may actually provide a solution to the many shortcomings that are encountered with the other omic methods (Griffin and Bollard, 2004; Bilello, 2005; van Ravenzwaay et al., 2007). Measurements of changes in gene expression and protein production are subjected to a variety of homeostatic controls and feedback mechanisms. This may result in metabolomics being a more sensitive indicator of the external stress than other omic technologies (Ankley et al., 2006; van Ravenzwaay et al., 2007). Although methods have been developed to detect changes in genomic, transcriptomic, and proteomic profiles, the basic information required to make meaningful interpretations based on these data are sometimes not readily available (Ankley et al., 2006). Alternatively, the structure and function of most metabolites is fairly well characterized and may be lower in number than genes and proteins (van Ravenzwaay et al., 2007).

Analogically, there are other, existing or emerging omic techniques such as lipidomics (to describe the complete lipid profile within a cell, tissue, or organism and is a subset of the "metabolome" which also includes the three other major classes of biological molecules: proteins/amino-acids, sugars, and nucleic acids (Wenk, 2005)), metallomics (to describe the distribution of free metal ions in every one of the cellular compartments (Banci and Bertini, 2013)), and metallometabolomics (to describe the distribution of metabolites of a metallodrug in a sample (Ge et al., 2016)), among others.

5.3.4 Marine Genomic Resources and Examples

Most of the currently available (eco-)toxicogenomic data regarding the molecular effects of exposure to pollutants are limited to species used in regulatory testing or freshwater species (Handy et al., 2008a,b; Blaise et al., 2008; Federici et al., 2007; Warheit et al., 2007; Lovern and Klaper, 2006). The lack of previous genetic information on most marine fish and invertebrate species frequently used in toxicity testing and risk assessment has been a major drawback for a more general application of the different omic technologies currently available. Due to the difficulty of identifying transcripts and proteins in nonmodel organisms, many environmental omic studies rely on identifying differently expressed features by homology search (Hampel et al., 2015; Silvestre et al., 2006; Romero-Ruiz et al., 2006; Vioque-Fernández et al., 2009). This approach was successfully applied in mussels exposed to Cu or salinity stress (Shepard et al., 2000) and *C. gallina* clams exposed to graded concentration of Cu, As, Aroclor, and tributyltin (Rodríguez-Ortega et al., 2003). At present, advances in genome sequencing of marine nonmodel species are being made, and part or complete genomes are publicly available for the following marine organisms frequently used in classical toxicity testing: the Japanese puffer, *Fugu rubripes* (http://www.fugu-sg.org/); the Killifish or Mummichog, *Fundulus heteroclitus* (http://www.ccs.miami.edu/cgi-bin/Fundulus/Fundulus_home.cgi); the Pacific oyster, *C. gigas* (http://www.oysterdb.com/FrontHome Action.do?method=home), the Mediterranean mussel, *M. galloprovincialis* (http://mussel.cribi.unipd.it); the Japanese clam, *R. philippinarum* (http://compgen.bio.unipd.it/ruphibase); the European flounder, *Platichthys flesus*; the Sea bream, *Sparus aurata* (http://www.nutrigroup-iats.org/seabreamdb); the sea bass, *Dicentrarchus labrax*, among others, including some anadromous species such as Atlantic salmon, *Salmo salar* and the stickleback, *Gasterosteus aculeatus*.

Despite the reduced number of genomic resources for marine organisms, efforts are being carried out to advance in their availability and application. Transcriptomic approaches have also been used in environmental monitoring exercises. For example, Milan et al. (2011) sequenced the transcriptome of the Manila clam, *R. philippinarum* and developed a microarray which they later applied for environmental monitoring of the Venice lagoon (Milan et al., 2015). Similarly, the transcriptome of the mussel, *M. galloprovincialis* was obtained by pyrosequencing and has provided extensive genomic information for this organism and generated novel observations on expression of different tissues, mitochondria, and associated microorganisms and will also facilitate the much needed production of an oligonucleotide microarray for the organism (Craft et al., 2010).

In Atlantic salmon (Salmo salar, exposed to environmentally relevant concentrations of the anti-epileptic drug Carbamazepine, Hampel et al. (2015) observed changes in the transcriptome expression of the brain, altering the expression of different metabolic pathways, including the negative regulation of apoptosis, heme biosynthesis, iron – and inorganic catión homeostasis, among others.

Analogously, Knigge et al. (2004) analyzed the proteomic profile of blue mussels (*M. edulis*) from polluted marine habitats surrounding the island of Karmøy, Norway. Differentially expressed proteins/peptides were found in two differently contaminated sites (heavy metals versus PAHs) showing a specific induction or a general suppression associated with the field site of origin. By combining sets of protein markers in a tree-building algorithm, the authors were able to correctly classify samples from these sites with an accuracy of 90%.

5.4 FUTURE POSSIBILITIES AND PROBABILITIES

Omic techniques are experiencing an incredible boom in all sorts of science disciplines. Such are the advances in these relatively new but rapidly expanding techniques, that the massive amount of data that are being generated represents another new challenge for scientists. Apart from disposing of the physical storage site of the amount of information, it is also the actual use and summarized output of this wealth of data that are giving scientists a hard time. Omic data obtained by high-throughput techniques offer a great potential for shedding more light onto the mechanisms triggered by the exposure to environmental pollutants (among many others). However, often the huge amount of data generated is not fully exploited. Recent advances in "omic" technologies have created unprecedented opportunities for biological research, but current software and database resources are extremely fragmented (Henry et al., 2014) and there is an urgent need of organizing the bioinformatics resources (Cannata et al., 2005). Moreover, the emerging technologies in transcriptomics, proteomics, metabolomics, and other life science areas are generating an increasing amount of complex data and information that cannot be interpreted in a traditional manner. Recent changes related to these emerging technologies have made the role of computer science much more critical, and the discipline of bioinformatics is a rapidly developing field needed for the interpretation of the vast amount of data. To tackle the growing complexity associated with emerging and future life science challenges, bioinformatics and computational biology researchers and developers need to explore, develop, and apply novel computational concepts, methods, tools, and systems. In addition, tool details and access often change following the original publication (Dellavalle et al., 2003), rendering it more and more challenging for research groups to stay current (for example, van der Ven et al., 2005, 2006). Moreover, design and performance of these studies vary significantly in exposure times and regimes, doses applied, and development stages used of a relatively high number of different model organisms. Thus, these data are difficult to interpret and unify for making joint conclusions. Therefore, initiatives should be encouraged that allow the wide variety of data to be eventually integrated into a single model of biological function that would act as a "simulation space." An "omic" simulation would contain for each gene an estimate of each critical parameter, as well as rules for interactions at each level to provide networks of interactions and allow estimations of biological functions like an in silico biological laboratory (Evans, 2000).

References

Adam, V., Fabrik, I., Kizek, R., Adam, V., Eckschlager, T., Stiborova, M., Trnkova, L., 2010. Vertebrate metallothioneins as target molecules for analytical techniques. TrAC – Trends Anal. Chem. 29 (5), 409–418.

Adams, M.D., Kelley, J.M., Gocayne, J.D., Dubnick, M., Polymeropoulos, M.H., Xiao, H., Merril, C.R., Wu, A., Olde, B., Moreno, R.F., et al., 1991. Complementary DNA sequencing: expressed sequence tags and human genome project. Science 252 (5013), 1651–1656.

Aebi, H., 1984. Catalase in vitro. In: Sies, H., Kaplan, N., Colowick, N. (Eds.), Methods in Enzymology, vol. 105, pp. 121–126.

Aguirre-Martinez, G.V., Buratti, S., Fabbri, E., DelValls, T.A., Martin-Diaz, M.L., 2013a. Using lysosomal membrane stability of haemocytes in *Ruditapes philippinarum* as a biomarker of cellular stress to assess contamination by caffeine, ibuprofen, carbamazepine and novobiocin. J. Environ. Sci. China 25, 1408–1418.

Aguirre-Martinez, G.V., Buratti, S., Fabbri, E., DelValls, T.A., Martin-Diaz, M.L., 2013b. Stability of lysosomal membrane in *Carcinus maenas* acts as a biomarker of exposure to pharmaceuticals. Environ. Monit. Assess. 185, 3783–3793.

Aguirre-Martinez, G.V., DelValls, T.A., Martin-Diaz, M.L., 2016. General stress, detoxification pathways, neurotoxicity and genotoxicity evaluated in *Ruditapes philippinarum* exposed to human pharmaceuticals. Ecotoxicol. Environ. Saf. 124, 18–31.

Akcha, F., Izuel, C., Venier, P., Budzinski, H., Burgeot, T., Narbonne, J.F., 2000. Enzymatic biomarker measurement and study of DNA adduct formation in B[a] P contaminated mussels, *Mytilus galloprovincialis*. Aquat. Toxicol. 49, 269–287.

Al-Sabti, K., Franko, M., Andrijanic, B., Knez, S., Stegnar, P., 1994. Chromium induced micronuclei in fish. J. Appl. Toxicol. 14, 333–336.

Amiard, J.C., Amiard-Triquet, C., Barka, S., Pellerin, J., Rainbow, P.S., 2006. Metallothioneins in aquatic invertebrates: their role in metal detoxification and their use as biomarkers. Aquat. Toxicol. 76 (2), 160–202.

Andersson, T., Lars, F., 1992. Regulation of the cytochrome P450 enzyme system in fish. Aquat. Toxicol. 24, 1–20.

Ankley, G.T., Daston, G.P., Degitz, S.J., Denslow, N.D., Hoke, R.A., Kennedy, S.W., Miracle, A.L., Perkins, E.J., Snape, J., Tillitt, D.E., Tyler, C.R., Versteeg, D., 2006. Toxicogenomics in regulatory ecotoxicology. Environ. Sci. Technol. 40 (13), 4055–4065.

Ankley, G.T., Bennett, R.S., Erickson, R.J., Hoff, D.J., Hornung, M.W., Johnson, R.D., Mount, D.R., Nichols, J.W., Russom, C.L., Schmieder, P.K., Serrrano, J.A., Tietge, J.E., Villeneuve, D.L., 2010. Adverse outcome pathways: a conceptual framework to support ecotoxicology research and risk assessment. Environ. Toxicol. Chem. 29 (3), 730–741.

Arslan, O.C., Parlak, H., Katalay, S., Boyacioglu, M., Karaaslan, M.A., Guner, H., 2010. Detecting micronuclei frequency monitoring pollution of Izmir Bay (Western Turkey). Environ. Monit. Assess. 165, 55–66.

Arukwe, A., Goksøyr, A., 2003. Eggshell and egg yolk proteins in fish: hepatic proteins for the next generation: Oogenetic population, and evolutionary implications of endocrine disruption. Comp. Hepatol. 2, 4.

Auffret, M., 1988. Histopathological changes related to chemical contamination in *Mytilus edulis* from field and experimental conditions. Mar. Ecol. Prog. Ser. 46, 101–107.

Banci, L., Bertini, I., 2013. Metallomics and the cell: some definitions and general comments. In: Banci, L. (Ed.), Metallomics and the Cell. Metal Ions in Life Sciences, vol. 12, Chapter 1. Springer.

Barrett, J.C., 1992. Mechanisms of action of known human carcinogens. In: Vainio, H., Magee, P.N., McGregor, D.B., McMichael, A.J. (Eds.), Mechanisms of Carcinogenesis in Risk Assessment. International Agency for Cancer Research, Lyon, France, pp. 115–134.

Bebianno, M.J., Nott, J.A., Langston, W.J., 1993. Cadmium metabolism in the clam *Ruditapes decussata*: the role of metallothioneins. Aquat. Toxicol. 27, 315–334.

Bebianno, M.J., Serafim, M.A., Rita, M.F., 1994. Involvement of metallothionein in cadmium accumulation and elimination in the clam *Ruditapes decussata*. Bull. Environ. Contam. Toxicol. 53, 726–732.

Beranova-Giorgianni, S., 2003. Proteome analysis by two dimensional gel electrophoresis and mass spectrometry: strengths and limitations. TrAC – Trends Anal. Chem. 5, 273–281.

Berlin, A., Schaller, K.N., 1974. European standardized method for determination of ∂-ALAD activity in blood. Z. Klin. Chem. Klin. Biochem. 12, 389–390.

Bihari, N., Fafandel, M., 2004. Interspecies differences in DNA single strand breaks caused by benzo(a)pyrene and marine environment. Mutat. Res. 552, 209–217.

Bilello, J.A., 2005. The agony and ecstasy of "OMIC" technologies in drug development. Curr. Mol. Med. 5 (1), 39–52.

Binz, P.A., Kagi, J.H.R., 1999. Metallothionein: molecular evolution classification. In: Klaassen, C. (Ed.), Metallothionein IV. Birkhäuser Verlag, Basel, pp. 7–13.

Blaise, C., Gagné, F., Pellerin, J., Hansen, P.D., 1999. Measurement of a vitellogenin-like protein in the hemolymph of *Mya arenaria* (Saguenay Fjord, Canada): a potential biomarker for endocrine disruption. Environ. Toxicol. 14 (4), 455–465.

Blaise, C., Trottier, S., Gagné, F., Lallement, C., Hansen, P.-D., 2002. Immunocompetence of bivalve hemocytes as evaluated by a miniaturized phagocytosis assay. Environ. Toxicol. 17, 160–169.

Blaise, C., Gagné, F., Férard, J.F., Eullaffroy, P., 2008. Ecotoxicity of selected nano-materials to aquatic organisms. Environ. Toxicol 23, 591–598.

Blanchette, B., Feng, X., Singh, B.R., 2007. Marine glutathione S-transferases. Mar. Biotechnol. 9, 513–542.

Blasco, J., DelValls, T.A., 2008. Impact of emergent contaminants in the environment: environmental risk assessment. In: Barceló, D., Petrovic, M. (Eds.), Water Pollution, Handbook of Environmental Chemistry, vol. 5 S1. Springer-Verlag, Heiderberg, pp. 169–188.

Blasco, J., Puppo, J., Sarasquete, M.C., 1993. Acid and alkaline phosphatase activities in the clam *Ruditapes philippinarum*. Mar. Biol. 115 (1), 113–118.

Bocquené, G., Galgani, F., Truquet, P., 1990. Characterization and assay conditions for use of AChE activity from several marine species in pollution monitoring. Mar. Environ. Res. 30, 75–89.

Bolognesi, C., Rabboni, R., Roggieri, I., 1996. Genotoxicity biomarkers in *M. galloprovincialis* as indicators of marine pollutants. Comp. Biochem. Phys. C 2, 319–323.

Bolognesi, C., 1990. Carcinogenic and mutagenic effects of pollutants in marine organisms: a review. In: Grandjean, E. (Ed.), Carcinogenic, Mutagenic, and Teratogenic Marine Pollutants: Impact on Human Health and the Environment. Portfolio Publishing Company, The Woodland, TX, USA, pp. 67–83.

Boutet, I., Tanguy, A., Moraga, D., 2004. Response of the Pacific oyster *Crassostrea gigas* to hydrocarbon contamination under experimental conditions. Gene 329, 147–157.

Brousseau, P., Dunier, M., De Guise, S., Fournier, M., 1997. Marqueurs immunologiques. In: Lagadic, L., et al. (Eds.), Biomarqueurs en ecotoxicologie: aspects fondamentaux. Masson, Paris, pp. 287–315.

Brousseau, P., Payette, Y., Blakley, B.R., Boermans, H., Tryphonas, H., Fournier, M., 1999. Manual of Immunological Methods. CRC Press, Boston, USA.

Brouwer, M., Winge, D.R., Gray, W.R., 1989. Structural and functional diversity of copper-metallothioneins from the American lobster *Homarus americanus*. J. Inorg. Biochem. 35, 289–303.

Brown, A.R., Gunnarsson, L., Kristianson, E., Tyler, C.R., 2014. Assessing variation in the potential susceptibility of fish to pharmaceuticals, considering evolutionary differences in their physiology and ecology. Philos. Trans. R. Soc. B369, 20130576.

Brunetti, R., Majone, F., Gola, I., Beltrame, C., 1988. The micronucleus test: examples of application to marine ecology. Mar. Ecol. Prog. Ser. 44, 65–68.

Brunk, C.F., 1979. Assay for nanogram quantities of DNA in cellular homogenates. Anal. Biochem. 92, 497–500.

Bucheli, T.D., Fent, K., 1995. Induction of cytochrome P450 as a biomarker for environmental contamination in aquatic ecosystems. Crit. Rev. Environ. Sci. Technol. 25, 201–268.

Buffet, P.E., Zalouk-Vergnoux, A., Châtel, A., Berthet, B., Métais, I., Perrein-Ettajani, H., Poirier, L., Luna-Acosta, A., Thomas-Guyon, H., Risso-de Faverney, C., Guibbolini, M., Gilliland, D., Valsami-Jones, E., Mouneyrac, C., 2014. A marine mesocosm study on the environmental fate of silver nanoparticles and toxicity effects on two endobenthic species: the ragworm *Hediste diversicolor* and the bivalve mollusc *Scrobicularia plana*. Sci. Tot. Environ. 470–471, 1151–1159.

Buhler, D.R., Williams, D.E., 1988. The role of biotransformation in the toxicity of chemicals. Aquat. Toxicol. 11, 19–28.

Burke, M.D., Mayer, R.T., 1974. Ethoxyresorufin: direct fluorometric assay of a microsomal O-dealkylation which is preferentially inducible by 3 methylcholanthrene. Drug Metab. Dispos. 2, 583–588.

Bustin, S.A., Benes, V., Nolan, T., Pfaffl, M.W., 2005. Quantitative real-time RT-PCR–a perspective. J. Mol. Endocrinol. 34 (3), 597–601.

Bustin, S.A., 2002. Quantification of mRNA using real-time reverse transcription PCR (RT-PCR): trends and problems. J. Mol. Endocrinol. 29, 23–39.

Cajaraville, M.P., Bebianno, M.J., Blasco, J., Porte, C., Sarasquete, C., Viarengo, A., 2000. The use of biomarkers to assess the impact of pollution in coastal environments of the Iberian Peninsula: a practical approach. Sci. Total Environ. 247 (2–3), 295–311.

Campana, O., Sarasquete, C., Blasco, J., 2003. Effect of lead on ALA-D activity, metallothionein levels, and lipid peroxidation in blood, kidney, and liver of the toadfish *Halobatrachus didactylus*. Ecotoxicol. Environ. Saf. 55 (1), 116–125.

Campillo, J.A., Albentosa, M., Valdés, N.J., Moreno-González, R., León, V.M., 2013. Impact assessment of agricultural inputs into a Mediterranean coastal lagoon (Mar Menor, SE Spain) on transplanted clams (*Ruditapes decussatus*) by biochemical and physiological response. Aquat. Toxicol. 142–143, 365–379.

Canesi, L., Ciacci, C., Betti, M., Scarpato, A., Citterio, B., Pruzzo, C., Gallo, G., 2003. Effects of PCB congeners on the immune function of Mytilus hemocytes: alterations of tyrosine kinase-mediated cell signaling. Aquat. Toxicol. 63, 293–306.

Canesi, L., Ciacci, C., Lorusso, L.C., Betti, M., Gallo, G., Pojana, G., Marcomini, A., 2007a. Effects of triclosan on *Mytilus galloprovincialis* hemocyte function and digestive gland enzyme activities: possible modes of action on non target organisms. Comp. Biochem. Physiol. C Toxicol. Pharmacol. 145, 464–472.

Canesi, L., Lorusso, L.C., Ciacci, C., Betti, M., Regoli, F., Poiana, G., Gallo, G., Marcomini, A., 2007b. Effects of blood lipid lowering pharmaceuticals (bezafibrate and gemfibrozil) on immune and digestive gland functions of the bivalve mollusc, *Mytilus galloprovincialis*. Chemosphere 69, 994–1002.

Canesi, L., Lorusso, L.C., Ciacci, C., Betti, M., Rocchi, M., Pojana, G., Marcomini, A., 2007c. Immunomodulation of Mytilus hemocytes by individual estrogenic chemicals and environmentally relevant mixtures of estrogens: in vitro and in vivo studies. Aquat. Toxicol. 81, 36–44.

Cannata, N., Merelli, E., Altman, R.B., 2005. Time to organize the bioinformatics resourceome. PLoS Comput. Biol 1, 76–81.

Carter, N.P., 2007. Methods and strategies for analyzing copy number variation using DNA microarrays. Nat. Genet. 39, S16–S21.

Chandurvelan, R., Marsden, I.D., Gaw, S., Glover, C.N., 2013. Biochemical biomarker responses of green-lipped mussel, *Perna canaliculus,* to acute and subchronic waterborne cadmium toxicity. Aquat. Toxicol. 140–141, 303–313.

Chen, L., Devanesan, P.D., Higginbotham, S., Ariese, F., Jan-kowiak, R., Small, G.J., Rogan, E.G., Cavalieri, E.L., 1996. Expanded analysis of benzo[a]pyrene-DNA adducts formed in vitro and in mouse skin: their significance in tumor initiation. Chem. Res. Toxicol. 9, 897–903.

Chu, Y., Corey, D.R., 2012. RNA sequencing: platform selection, experimental design, and data interpretation. Nucl. Acid Ther. 22 (4), 271–274.

Chu, F.E., Peyre, J.F.L., 1989. Effect of environmental factors and parasitism on hemolymph lysozyme and protein in American oysters (*Crassostrea virginica*). J. Invertebr. Pathol. 54, 224–232.

Colapietro, A.M., Goodell, A.L., Smart, R.C., 1993. Characterization of benzo[a]pyrene-initiated mouse skin papillomas for Ha-ras mutations and protein kinase C levels. Carcinogenesis 14, 2289–2295.

Colborn, T., Saal, F.V.S., Soto, A.M., 1993. Development effects of endocrine-disrupting chemicals in wildlife and humans. Environ. Health Perspect. 101, 378–384.

Colovic, M.B., Krstic, D.Z., Lazarevic-Pasti, T.D., Bondzic, A.M., Vasic, V.M., 2013. Acetylcholinesterase inhibitors: pharmacology and toxicology. Curr. Neuropharmacol. 11, 315–335.

Company, R., Serafim, A., Lopes, B., Cravo, A., Shepherd, T.J., Pearson, G., Bebianno, M.J., 2008. Using biochemical and isotope geochemistry to understand the environmental and public health implications of lead pollution in the lower Guadiana River, Iberia: a freshwater bivalve study. Sci. Total Environ. 405 (1–3), 109–119.

Company, R., Serafim, A., Lopes, B., Cravo, A., Kalman, J., Riba, I., DelValls, T.A., Blasco, J., Delgado, J., Sarmiento, A.M., Nieto, J.M., Shepherd, T.J., Nowell, G., Bebianno, M.J., 2011. Source and impact of lead contamination on δ-aminolevulinic acid dehydratase activity in several marine bivalve species along the Gulf of Cadiz. Aquat. Toxicol. 101 (1), 146–154.

Cosson, R.P., Amiard, J.C., 2000. Use of metallothionein as biomarkers of exposure to metals. In: Lagadic, L., Caquet, T., Amiard, J.-C., Ramade, F. (Eds.), Use of Biomarkers for Environmental Quality Assessment. Science Publishers, Inc., Enfield, NH, pp. 79–111.

Craft, J.A., Gilbert, J.A., Temperton, B., Dempsey, K.E., Ashelford, K., et al., 2010. Pyrosequencing of *Mytilus galloprovincialis* cDNAs: tissue-specific expression patterns. PLoS One 5 (1), 8875.

Cravo, A., Lopes, B., Serafim, A., Company, R., Barreira, L., Gomes, T., Bebianno, M.J., 2009. A multibiomarker approach in *Mytilus galloprovincialis* to assess environmental quality. J. Environ. Monit. 11 (9), 1673–1686.

Cravo, A., Lopes, B., Serafim, A., Company, R., Barreira, L., Gomes, T., Bebianno, M.J., 2013. Spatial and seasonal biomarker responses in the clam *Ruditapes decussatus*. Biomarkers 18 (1), 30–43.

Dabrio, M., Rodriguez, A.R., Bordin, G., Bebianno, M.J., De Ley, M., Sestakova, I., Vasak, M., Nordberg, M., 2002. Recent developments in quantification methods for metallothionein. J. Inorg. Biochem. 88, 123–134.

Dagnino, A., Allen, J.I., Moore, M.N., Broeg, K., Canesi, L., Viarengo, A., 2007. Development of an expert system for the integration of biomarker responses in mussels into an animal health index. Biomarkers 12, 155–172.

Daniels, D., 2007. Functions of red cell surface proteins. Vox Sang. 93, 331–340.

De Vries, A., Dollé, M.E.T., Broekhof, J.L.M., Muller, J.J.A., Dinant Kroese, E., Van Kreijl, C.F., Capel, P.J.A., Vijg, J., Van Steeg, H., 1997. Induction of DNA adducts and mutations in spleen, liver and lung of XPA-deficient/lac Z transgenic mice after oral treatment with benzo[a]pyrene: correlation with tumor development. Carcinogenesis 18, 2327–2332.

De Zwart, L.L., Venhorst, J., Groot, M., Commandeur, J.N.M., Hermanns, R.C.A., Meerman, J.H.M., van Baar, B.L.M., Vermeulen, N.P.E., 1997. Simultaneous determination of eight lipid peroxidation degradation products in urine of rats treated with carbon tetrachloride using gas chromatography with electron capture detection. J. Chromatogr. B 694, 227–287.

Dean, J.H., Murray, M.J., 1990. Toxic responses of the immune system. In: Klaassen, C.D., Amdur, M.O., Doull, J. (Eds.), Toxicology: The Basic Science of Poisons, vol. 4. McMillan, New York, pp. 282–333.

Dellavalle, R.P., Hester, E.J., Heilig, L.F., et al., 2003. Information science. Going, going, gone: lost Internet references. Science 302, 787–788.

Denslow, N.D., Chow, M.C., Kroll, K.J., Green, L., 1999. Vitellogenin as a biomarker of exposure for estrogen or estrogen mimics. Ecotoxicology 8, 385–398.

Dhaunsi, G.S., Singh, I., Hanevld, C.D., 1993. Peroxisomal participation in cellular responses to the oxidative stress of endotoxin. Mol. Cell. Biochem. 126, 25–35.

Di Giulio, R.T., Washburn, P.C., Wenning, R.J., Winston, G.W., Jewell, C.S., 1989. Biochemical responses in aquatic animals: a review of determinants of oxidative stress. Environ. Toxicol. Chem. 8, 1103–1123.

Diamond, J.M., Latimer, H.A., Munkittrick, K.R., Thorntn, K.W., Bartell, S.M., Kidd, K.A., 2011. Prioritizing contaminants of emerging concern for ecological screening assessments. Environ. Toxicol. Chem. 30, 2385–2394.

Dondero, F., Piacentini, L., Marsano, F., Rebelo, M., Vergani, L., Venier, P., Viarengo, A., 2006. Gene transcription profiling in pollutant exposed mussels (*Mytilus* spp.) using a new low-density oligonucleotide microarray. Gene 376, 24–36.

Du, Y., Zhang, L., Xu, F., Huang, B., Zhang, G., Li, L., 2013. Validation of housekeeping genes as internal controls for studying gene expression during Pacific oyster (*Crassostrea gigas*) development by quantitative real-time PCR. Fish Shellfish Immunol. 34, 939–945.

Dulin, D., Lipfert, J., Moolman, M.C., Dekker, N.H., 2012. Studying genomic processes at the single-molecule level: introducing the tools and applications. Nat. Rev. Genet. 14 (1), 9–22.

Dunn, B.P., 1991. Carcinogen adducts as an indicator for the public health risks of consuming carcinogen-exposed fish and shellfish. Environ. Health Perspect. 90, 111–116.

El Hourch, M., Dudoit, A., Amiard, J.C., 2003. Optimization of new voltammetric method for the determination of metallothionein. Electrochim. Acta 48, 4083–4088.

El Hourch, M., Dudoit, A., Amiard, J.C., 2004. An optimization procedure for the determination of metallothionein by square wave cathodic stripping voltammetry. Application to marine worms. Anal. Bioanal. Chem. 378 (3), 776–781.

Ellman, G.L., Courtney, K.D., Andres Jr., V., Featherstone, R.M., 1961. A new and rapid colorimetric determination of acetylcholinesterase activity. Biochem. Pharmacol. 7, 88–95.

Ericson, G., Akerman, G., Liewenborg, B., Balk, L., 1996. Comparison of DNA damage in the early life stages of cod, *Gadus morhua*, originating from the Barents Sea and Baltic Sea. Mar. Environ. Res. 42, 119–123.

European Commission, 2006. Regulation (EC) No 1907/2006 of the European Parliament and of the Council of 18 December 2006 Concerning the Registration, Evaluation, Authorisation and Restriction of Chemicals (REACH), Establishing a European Chemicals Agency.

Fahimi, H.D., Cajaraville, M.P., 1995. Induction of peroxisome proliferation by some environmental pollutants and chemicals. In: Cajaraville, M.P. (Ed.), Cell Biology in Environmental Toxicology. Servicio Editorial de UPV/EHU, Leioa, pp. 221–255.

Federici, G., Shaw, B.J., Handy, R.D., 2007. Toxicity of titanium dioxide nanoparticles to rainbow trout, (Oncorhynchus mykiss): gill injury, oxidative stress, and other physiological effects. Aquat. Toxicol 84, 415–430.

Fernández, B., Martínez-Gómez, C., Benedicto, J., 2015. Delta-aminolevulinic acid dehydratase activity (ALA-D) in red mullet (Mullus barbatus) from Mediterranean waters as biomarker of lead exposure. Ecotoxicol. Environ. Saf 115, 209–216.

Ferreira, P., Fonte, E., Elisa Soares, M., Carvalho, F., Guilhermino, L., 2016. Effects of multi-stressors on juveniles of the marine fish *Pomatoschistus microps*: gold nanoparticles, microplastics and temperature. Aquat. Toxicol. 170, 89–103.

Finelli, V., 1977. Lead, zinc and ∂-aminolevulinate dehydratase. In: Lee, S., Peirano, B. (Eds.), Biochemical Effects of Environmental Pollutants. Ann Arbor Science Publishers, Ann Arbor, Michigan, US, pp. 351–364.

Fong, A.T., Dashwood, R.H., Cheng, R., Mathews, C., Ford, B., Hen-dricks, J.D., Bailey, G.S., 1993. Carcinogenicity, metabolism and Ki-ras proto-oncogene activation by 7-12-dimethyl-benz[a]anthracene in rainbow trout. Carcinogenesis 14, 629–635.

Fournier, M., Cyr, D., Blakley, B., Boermans, H., Brousseau, P., 2000. Phagocytosis as a biomarker of immunotoxicity in wildlife species exposed to environmental xenobiotics. Am. Zool. 40, 412–420.

Fowler, B.A., Hildebrand, C.E., Kojima, Y., Webb, M., 1987. Nomenclature of metallothioneins. In: Kagi, J.H.R., Kojima, Y. (Eds.), Metallothionein II. Birhäuser Verlag, Basel, pp. 19–22.

Franzellitti, S., Buratti, S., Du, B., Haddad, S.P., Chambliss, C.K., Brooks, B.W., Fabbri, E., 2015. A multibiomarker approach to explore interactive effects of propranolol and fluoxetine in marine mussels. Environ. Pollut. 205, 60–69.

Frenzilli, G., Nigro, M., Lyons, B.P., 2009. The comet assay for the evaluation of genotoxic impact in aquatic environments. Mutat. Res. 681, 80–92.

Fridovich, I., 1995. Superoxide radical and superoxide dismutases. Ann. Rev. Biochem. 64, 97–112.

Fujimoto, Y., Sakuma, S., Inoue, T., Uno, E., Fujita, T., 2002. The endocrine disruptor nonylphenol preferentially blocks cyclooxygenase-1. Life Sci. 70, 2209–2214.

Fujimoto, Y., Usa, K., Sakuma, S., 2005. Effects of endocrine disruptors on the formation of prostaglandin and arachidonoyl-CoA formed from arachidonic acid in rabbit kidney medulla microsomes. Prost. Leuko. Essent. Fatty Acids 73, 447–452.

Fulton, M.H., Key, P.B., 2001. Acetylcholinesterase inhibition in estuarine fish and invertebrates as an indicator of organophosphorus insecticide exposure and effects. Environ. Toxicol. Chem. 20, 37–45.

Gagné, F., Blaise, C., 1999. Toxicological effects of municipal wastewaters to rainbow trout hepatocytes. Bull. Environ. Contam. Toxicol. 63, 503–510.

Gagné, F., Blaise, C., Pellerin, J., Pelletier, E., Douville, M., Gauthier-Clerc, S., Viglino, L., 2003. Sex alteration in soft-shell clams (*Mya arenaria*) in an intertidal zone of the St. Lawrence River (Québec, Canada). Comp. Biochem. Phys. C 134, 189–198.

Gagné, F., Blaise, C., André, C., Gagnon, C., Salazar, M., 2007. Neuroendocrine disruption and health effects in *Elliptio complanata* mussels exposed to aeration lagoons for wastewater treatment. Chemosphere 68, 731–743.

Gardner, G.R., Yevich, P.P., Harshbarger, J.C., Malcolm, A.R., 1991. Carcinogenicity of Black Rock Harbor sediment to the eastern oyster and trophic transfer of Black Rock Harbor carcinogens from the blue mussel to the winter flounder. Environ. Health Perspect 90, 53–66.

Gardner, G.R., Pruell, R.J., Malcolm, A.R., 1992. Chemical induction of tumors in oysters by a mixture of aromatic and chlorinated hydrocarbons, amines and metals. Mar. Environ. Res. 34, 59–63.

Gause, W.C., Adamovicz, J., 1994. The use of the PCR to quantitate gene expression. PCR Methods Appl. 3 (6), S123–S135.

Ge, R., Sun, X., He, Q.Y., 2016. Overview of the metallometabolomic methodology for metal-based drug metabolism. Curr. Drug Metab. 12 (3), 287–299.

Gehlenborg, N., O'Donoghue, S., Baliga, N.S., Goesmann, A., Hibbs, M.A., Kitano, H., Kohlbacher, O., Neuweger, H., Schneider, R., Tenenbaum, D., Gavin, A.C., 2010. Visualization of omics data for systems biology. Nat. Methods 7 (3), 56–68.

Genevois, C., Pfohl-Leszkowicz, A., Boillot, K., Brandt, H., Castegnaro, M., 1998. Implication of cytochrome P450 1A isoforms and the Ah receptor in the genotoxicity of coal tar fume condensate and bitumen fumes condensates. Environ. Toxicol. Pharmacol. 5, 283–294.

George, S.G., 1994. Enzymology and molecular biology of phase II xenobiotic-conjugating enzymes in fish. In: Malins, D.C., Ostrander, G.K. (Eds.), Aquatic Toxicology Molecular, Biochemical and Cellular Perspectives. Lewis Publishers, CRC Press, p. 37.

Goeptar, A.R., Scheerens, H., Vermeulen, N.P.E., 1995. Oxygen reductase and substrate reductase activity of cytochrome P450. Crit. Rev. Toxicol. 25, 25–65.

Goksøyr, A., Förlin, L., 1992. The cytochrome P450 system in fish, aquatic toxicology and environmental monitoring. Aquat. Toxicol. 22, 287–312.

Goksøyr, A., Larsen, E., 1991. The cytochrome P450 system of Atlantic salmon (*Salmo salar*): I. Basal properties and induction of P450 1A1 in liver of immature and mature fish. Fish Physiol. Biochem. 9, 339–349.

Goksøyr, A., Arukwe, A., Lasrsson, J., Cajaraville, M.P., Hauser, L., Nilsen, B.M., Lowe, D., Matthiessen, P., 2003. Molecular/cellular processes and the impact on reproduction. In: Lawrence, J.A., Hemingway, K.L. (Eds.), Effects of Pollution on Fish: Molecular Effects and Population Responses. Oxford: Blackwell Science, pp. 179–220.

Goksøyr, A., 2006. Endocrine disruptors in the marine environment: mechanisms of toxicity and their influence on reproductive processes in fish. J. Toxicol. Environ. Health A 69, 175–184.

Goldstein, W., 1992. Neurologic concepts of lead poisoning in children. Pediatr. Ann. 21, 384–388.

González-Fernández, C., Albentosa, M., Campillo, J.A., Viñas, L., Fumega, J., Franco, A., Besada, V., González-Quijano, A., Bellas, J., 2015. Influence of mussel biological variability on pollution biomarkers. Environ. Res. 137, 14–31.

Gorbi, S., Virno Lamberti, C., Notti, A., Benedetti, M., Fattorini, D., Moltedo, G., Regoli, F., 2008. An ecotoxicological protocol with caged mussels, *Mytilus galloprovincialis*, for monitoring the impact of an offshore platform in the Adriatic Sea. Mar. Environ. Res. 65, 34–49.

Gourdon, I., Guerin, M.C., Torreilles, J., Roch, P., 2001. Nitric oxide generation by hemocytes of the mussel *Mytilus galloprovincialis*. Nitric Oxide Biol. Chem. 5, 1–6.

Gowda, M., Jantasuriyarat, C., Dean, R.A., Wang, G.L., 2004. Robust-LongSAGE (RL-SAGE): a substantially improved LongSAGE method for gene discovery and transcriptome analysis. Plant Physiol. 134 (3), 890–897.

Griffin, J.L., Bollard, M.E., 2004. Metabonomics: its potential as a tool in toxicology for safety assessment and data integration. Curr. Drug Metab. 5 (5), 389–398.

Grisham, M.B., Jourd'heuil, D., Wink, D.A., 1999. Nitric oxide. I. Physiological chemistry of nitric oxide and its metabolites: implications in inflammation. Am. J. Physiol. G 276, 315–321.

Guilhermino, L., Lopes, M.C., Carvalho, A.P., Soares, A.M.V.M., 1996. Inhibition of acetylcholinesterase activity as effect criterion in acute tests with juvenile Daphnia magna. Chemosphere 32 (4), 721–738.

Halliwell, B., Gutteridge, 1999. Free Radicals in Biology and Medicine, third ed. Oxford University Press, Oxford, UK.

Hampel, M., Alonso, E., Aparicio, I., Santos, J.L., Leaver, M.J., 2015. Hepatic proteome analysis of Atlantic salmon (Salmo salar) after exposure to environmental concentrations of human pharmaceuticals. Mol. Cell. Prot 14 (2), 371–381.

Hamza-Chaffai, A., Amiard, J.C., Cosson, R.P., 1999. Relationship between metallothioneins and metals in a natural population of the clam *Ruditapes decussatus* from Sfax coast: a non-linear model using Box-Cox transformation. Comp. Biochem. Physiol. C 123, 153–163.

Handy, R.D., Henry, T.B., Scown, T.M., Johnstone, B.D., Tyler, C.R., 2008a. Manufactured nanoparticles: their uptake and effects on fish—a mechanistic analysis. Ecotoxicology 17, 396–409.

Handy, R.D., Von der Kammer, F., Lead, J.R., Hassellöv, M., Owen, R., Crane, M., 2008b. The ecotoxicology and chemistry of manufactured nanoparticles. Ecotoxicology 17, 287–314.

Hariharan, G., Purvaja, R., Ramesh, R., 2014. Toxic effects of lead on biochemical and histological alterations in green mussel (*Perna viridis*) induced by environmentally relevant concentrations. J. Toxicol. Environ. Health A 77 (5), 246–260.

Hartung, T., McBride, M., 2011. Food for thought. On mapping the human toxome. ALTEX 28, 83–93.

Helbock, H.J., Beckman, K.B., Shigenaga, M.K., Walter, P.B., Woodall, A.A., Yeo, H.C., Ames, B.N., 1998. DNA oxidation matters: the HPLC-electrochemical detection assay of 8-oxo-deoxy-guanosine and 8-oxo-guanine. Proc. Natl. Acad. Sci. U.S.A. 95, 288–293.

Hendriks, A.J., Pieters, H., de Boer, J., 1998. Accumulation of metals, polycyclic (halogenated) hydrocarbons, and biocides in zebra mussel and eel from the Rhine and Meuse rivers. Environ. Toxicol. Chem. 17, 1885–1898.

Henry, V.,J., Bandrowski, A.E., Pepin, A.S., Gonzalez, B.J., Desfeux, A., 2014. OMICtools: an informative directory for multi-omic data analysis. Database. http://dx.doi.org/10.1093/database/bau069. Article ID: bau069.

Herbert, B.R., Harry, J.L., Packer, N.H., Gooley, A.A., Pederson, S.K., Williams, K.L., 2001. What place for polyacrylamide in proteomics? Trends Biotechnol. 19 (10), S3–S9.

Hong, X.T., Xiang, L.X., Shao, J.Z., 2006. The immunostimulating effect of bacterial genomic DNA on the innate immune responses of bivalve mussel,Hyriopsis cumingii. Fish Shellfish Immunol 21, 357–364.

Hooftman, R.N., Raat, W.K., 1982. Induction of nuclear anomalies (micronuclei) in the peripheral blood erythrocytes of the eastern mudminnow *Umbra pygmaea* by ethyl methanesulphonate. Mutat. Res. 104, 147–152.

Hutchinson, T.H., Ankley, G.T., Segner, H., Tyler, C.R., 2006. Screening and testing for endocrine disruption in fish—biomarkers as signposts, not traffic lights, in risk assessment. Environ. Health Perspect. 114, 106–114.

Hylland, K., Ruus, A., Grung, M., Green, N., 2009. Relationships between physiology, tissue contaminants, and biomarker responses in Atlantic cod (*Gadus morhua* L.). J. Toxicol. Environ. Health 72A, 226–233.

Hylland, K., 2004. Biological Effects of Contaminants: Quantification of ∂-aminolevulinic Acid Dehydratase (ALA-D) Activity in Fish Blood, vol. 34. ICES Tech. Mar. Environ. Sci., 9 pp.

James, M.O., Altman, A.H., Li, C.L.J., Boyle, S.M., 1992. Dose- and time-dependent formation of benzo[a]pyrene metabolite DNA adducts in the spiny lobster, Panulirus argus. Mar. Environ. Res 34, 299–302.

Janero, D.R., 1990. Malondialdehyde and thiobarbituric acid-reactivity as diagnostic indices of lipid peroxidation and peroxidative tissue injury. Free Radic. Biol. Med. 9, 515–540.

Jebali, J., Banni, M., Guerbej, H., Almeida, E.A., Bannaoui, A., Boussetta, H., 2006. Effects of malathion and cadmium on acetylcholinesterase activity and metallothionein levels in the fish *Seriola dumerili*. Fish Physiol. Biochem. 32, 93–98.

Kalman, J., Riba, I., Blasco, J., DelValls, T.A., 2008. Is δ-aminolevulinic acid dehydratase activity in bivalves from southwest Iberian Peninsula a good biomarker of lead exposure? Mar. Environ. Res. 68 (1), 38–40.

Katsumiti, A., Gilliland, D., Arostegui, I., Cajaraville, M.P., 2015. Mechanisms of toxicity of Ag nanoparticles in comparison to bulk and ionic Ag on mussel hemocytes and gill cells. PLoS One 10 (6).

Kim, M., Ahn, I.Y., Cheon, J., Park, H., 2009. Molecular cloning and thermal stress-induced expression of a pi-class glutathione S-transferase (GST) in the Antarctic bivalve *Laternula elliptica*. Comp. Biochem. Physiol. A 152, 207–213.

Kime, D.E., 1995. The effects of pollution on reproduction in fish. Rev. Fish Biol. Fish. 5, 52–96.

Kirubagaran, R., Joy, K.P., 1988. Inhibition of testicular 313-hydroxy-As-steroid dehydrogenase (313-HSD) activity in catfish *Clarias batrachus* (L.) by mercurials. Ind. J. Exp. Biol. 26, 907–908.

Knigge, T., Monsinjon, T., Andersen, O.K., 2004. Surface-enhanced laser desorption/ionization-time of flight-mass spectrometry approach to biomarker discovery in blue mussels (Mytilus edulis) exposed to polyaromatic hydrocarbons and heavy metals under field conditions. Proteomics 4, 2722–2727.

Koenig, S., Fernández, P., Solé, M., 2012. Differences in cytochrome P450 enzyme activities between fish and crustacea: relationship with the bioaccumulation patterns of polychlorobiphenyls (PCBs). Aquat. Toxicol. 108, 11–17.

Kojima, Y., Berger, C., Vallee, B.L., Kagi, J.H.R., 1976. Amino-acid sequence of equine renal metallothionein-IB. Proc. Natl. Acad. Sci. U.S.A. 73 (10), 3413–3417.

Koller, L.D., Exon, J.H., 1985. The rat as a model for immunotoxicity assessment. In: Dean, J.H., Luster, M.I., Munson, A.E., Amos, H. (Eds.), Immunotoxicology and Immunopharmacology. Raven Press, New York, pp. 99–111.

Kramer, V.J., Miles-Richardson, S., Pierens, S.L., Giesy, J.P., 1998. Reproductive impairment and induction of alkaline-labile phosphate, a biomarker of estrogen exposure, in fathead minnows (*Pimephales promelas*) exposed to waterborne 17β-estradiol. Aquat. Toxicol. 40, 335–360.

Kulshrestha, S.K., Arora, N., 1984. Impairments induced by sublethal doses of two pesticides in the ovaries of a freshwater teleost *Channa striatus* Bloch. Toxicol. Lett. 20, 93–98.

La Rocca, C., Conti, L., Crebelli, R., Crochi, B., Iacovella, N., Rodriguez, F., Turrio-Baldassarri, L., Di Domenico, A.,

1996. PAH content and mutagenicity of marine sediments from the Venice lagoon. Ecotoxicol. Environ. Saf. 33, 236–245.

Lal, B., Singh, T.P., 1987. The effect of malathion and BHC on the lipid metabolism in relation to reproduction in the tropical teleost, *Clarias batrachus*. Environ. Pollut. 48, 37–47.

Landahl, J.T., McCain, B.B., Myers, M.S., Rhodes, L.T., Brown, D.W., 1990. Consistent associations between hepatic lesions in English sole (*Parophrys vetulus*) and polycyclic aromatic hydrocarbons in sediment. Environ. Health Perspect. 89, 195–203.

Lander, E.S., Linton, L.M., Birren, B., Nusbaum, C., Zody, M.C., Baldwin, J., Devon, K., International Human Genome Sequencing Consortium, et al., February 15, 2001. Initial sequencing and analysis of the human genome. Nature 409 (6822), 860–921.

Langston, W.J., Bebianno, M.J., Burt, G.R., 1998. Metal handling strategies in molluscs. In: Batley, G.E., Langston, W.J., Bebianno, M.J. (Eds.), Metal Metabolism in Aquatic Environments. Chapman and Hall, London, pp. 219–283.

Lawrence, R.A., Burk, R.F., 1976. Glutathione peroxidase activity in selenium-deficient rat liver. Biochem. Biophys. Res. Comm. 71, 952–958.

Lech, J.J., Vodicnik, M.J., 1985. Biotransformation. In: Rand, G.M., Petrocelli, S.R. (Eds.), Fundamentals of Aquatic Toxicology; Methods and Applications. Hemisphere Publishing Corporation, New York, USA, pp. 526–557.

Lee, R.F., 1988. Gluthatione-S-transferase in marine invertebrates from Langsundford. Mar. Ecol. Progr. Ser 46, 33–36.

Lee, Y.C., Yang, D., 2002. Determination of lysozyme activities in a microplate format. Anal. Biochem. 310, 223–224.

Lee, R.F., Bulski, K., Adams, J.D., Peden-Adams, M., Bossart, G.D., King, L., Fair, P.A., 2013. DNA strand breaks (comet assay) in blood lymphocytes from wild bottlenose dolphins. Mar. Pollut. Bull. 77, 355–360.

Lerch, K., Ammer, D., Olafson, R.W., 1982. Crab metallothionein: primary structures of metallothionein 1 and 2. J. Biol. Chem. 257, 2420–2426.

Li, H., Parisi, M.G., Toubiana, M., Cammarata, M., Roch, P., 2008. Lysozyme gene expression and hemocyte behaviour in the Mediterranean mussel, *Mytilus galloprovincialis*, after injection of various bacteria or temperature stresses. Fish Shellfish Immunol. 25, 143–152.

Lilley, K.S., Razzaq, A., Dupree, P., 2001. Two-dimensional gel electrophoresis: recent advances in sample preparation, detection and quantitation. Curr. Opin. Chem. Biol. 6, 46–50.

Lionetto, M.G., Caricato, R., Giordano, M.E., Pascariello, M.F., Marinosci, L., Schettino, T., 2003. Integrated use of biomarkers (acetylcholinesterase and antioxidant enzymes activities) in *Mytilus galloprovincialis* and *Mullus barbatus* in an Italian coastal marine area. Mar. Pollut. Bull. 46 (3), 324–330.

Lionetto, M.G., Caricato, R., Giordano, M.E., Schettino, T., 2004. Biomarker application for the study of chemical contamination risk on marine organisms in the Taranto marine coastal area. Chem. Ecol. 20, 333–343.

Lionetto, M.G., Caricato, R., Calisi, A., Giordano, M.E., Schettino, T., 2013. Acetylcholinesterase as a biomarker in environmental and occupational medicine: new insights and future perspectives. Biomed. Res. Int 2013, 321213.

Liu, X., Wang, W.X., 2016. Time changes in biomarker responses in two species of oyster transplanted into a metal contaminated estuary. Sci. Total Environ. 544, 281–290.

Liu, H., Hea, J., Zhao, R., Chi, C., Bao, Y., 2015. A novel biomarker for marine environmental pollution of pi-class glutathione S-transferase from *Mytilus coruscus*. Ecotoxicol. Environ. Saf. 118, 47–54.

Livingstone, D.R., Chipman, J.K., Lowe, D.M., Minier, C., Mitchelmore, C.L., Moore, M.N., Peters, L.D., Pipe, R.K., 2000. Development of biomarkers to detect the effects of organic pollution on aquatic invertebrates: recent molecular, genotoxic, cellular and immunological studies on the common mussel (*Mytilus edulis* L.) and other mytilids. Int. J. Environ. Pollut. 13 (1), 56–91.

Livingstone, D.R., 2001. Contaminated-stimulated reactive oxygen species production and oxidative damage in aquatic organisms. Mar. Pollut. Bull. 42, 656–666.

Lombardi, P.E., Peri, S.I., Verrengia Guerrero, N.R., 2010. ALA-D and ALA-D re-activated as biomarkers of lead contamination in the fish *Prochilodus lineatus*. Ecotoxicol. Environ. Saf. 73, 1704–1711.

Lovern, S.B., Klaper, R., 2006. Daphnia magna mortality when exposed to titanium dioxide and fullerene nanoparticles. Environ. Toxicol. Chem 25 (4), 1132–1137.

Lowe, D.M., Pipe, R.K., 1994. Contaminant induced lysosomal membrane damage in marine mussel digestive cells: an in vitro study. Aquat. Toxicol. 30, 357–365.

Maccubbin, A.E., 1994. DNA adduct analysis in fish: laboratory and field studies. In: Malins, D.C., Ostrander, G.K. (Eds.), Aquatic Toxicology; Molecular, Biochemical and Cellular Perspectives. Lewis Publishers, CRC Press, pp. 267–294.

Macías-Mayorga, D., Laiz, I., Moreno-Garrido, I., Blasco, J., 2015. Is oxidative stress related to cadmium accumulation in the mollusc Crassostrea angulata? Aquat. Toxicol 161, 231–241.

Mackay, E.A., Dunbar, B., Davidson, I., Fothergill, J.E., 1990. Polymorphism of cadmium-induced mussel metallothionein. Experentia 46, A36.

MacMahon, G., Huber, L.J., Moore, M.J., Stegeman, J.J., Wogan, G.N., 1990. Mutations in c-Ki-ras oncogenes in diseased livers of winter flounder from Boston harbor. Proc. Natl. Acad. Sci. U.S.A. 87, 841–845.

Manna, G.K., Banerjee, G., Gupta, S., 1985. Micronucleus test in the peripheral erythrocytes of the exotic fish. Nucleus 23, 176–179.

Mannervik, B., Danielson, U.H., 1988. Glutathione transferases structure and catalytic activity. CRC Crit. Rev. Biochem. 23, 283–337.

Maranho, L.A., Seabra Pereira, C.D., Brasil Choueri, R., Cesar, A., Kachel Gusso-Choueri, P., José Torres, R., de Souza Abessa, D.M., Davino Morais, R., Mozeto, A.A., DelValls, T.A., Martín-Díaz, M.L., 2012. The application of biochemical responses to assess environmental quality of tropical estuaries: field surveys. J. Environ. Monit 14, 2608–2615.

Maranho, L.A., Moreira, L.B., Baena-Nogueras, R.M., Lara, P., DelValls, T.A., Martin-Diaz, M.L., 2015a. A candidate short-term toxicity test using *Ampelisca brevicornis* to assess sublethal responses to pharmaceuticals bound to marine sediments. Arch. Environ. Contam. Toxicol. 68, 237–258.

Maranho, L.A., André, C., DelValls, T.A., Gagné, F., Martin-Diaz, M.L., 2015b. In situ evaluation of wastewater discharges and the bioavailability of contaminants to marine biota. Sci. Total Environ. 15, 876–887.

Maranho, L.A., André, C., DelValls, T.A., Gagné, F., Martin-Diaz, M.L., 2015c. Toxicological evaluation of sediment samples spiked with human pharmaceuticals products: energy status and neuroendocrine effects in marine polychaetes *Hediste diversicolor*. Ecotoxicol. Environ. Saf. 118, 27–36.

Maria, V.L., Bebianno, M.J., 2011. Antioxidant and lipid peroxidation responses in Mytilus galloprovincialis exposed to mixtures of benzo(a)pyrene and copper. Comp. Biochem. Physiol. C 154 (1), 56–63.

Marin, M.G., Matozzo, V., 2004. Vitellogenin induction as a biomarker of exposure to estrogenic compounds in aquatic environments. Mar. Pollut. Bull. 48, 835–839.

Martín-Díaz, M.L., Blasco, J., Sales, D., DelValls, T.A., 2004. Biomarkers as tools to assess sediment quality. Laboratory and field surveys. TrAC – Trends Anal. Chem. 23 (10–11), 807–818.

Martín-Díaz, M.L., Blasco, J., Sales, D., DelValls, T.A., 2007. Biomarkers study for sediment quality assessment in Spanish ports using the crab *Carcinus maenas* and the clam *Ruditapes philippinarum*. Arch. Environ. Contam. Toxicol. 53, 66–76.

Martín-Díaz, M.L., Blasco, J., Sales, D., DelValls, T.A., 2008. Field validation of a battery of biomarkers to assess sediment quality in Spanish ports. Environ. Pollut. 151, 631–640.

Martín-Díaz, M.L., Franzellitti, S., Buratti, S., Valbonesi, P., Capuzzo, A., Fabbri, E., 2009. Effects of environmental concentrations of the antiepileptic drug carbamazepine on biomarkers and cAMP-mediated cell signaling in the mussel *Mytilus galloprovincialis*. Aquat. Toxicol. 94, 177–185.

Martínez-Gómez, C., Benedicto, J., Campillo, J.A., Moore, M., 2008. Application and evaluation of the neutral red retention (NRR) assay for lysosomal stability in mussel populations along the Iberian Mediterranean coast. J. Environ. Monit. 10, 490–499.

Martyniuk, C.J., Alvarez, S., Denslow, N.D., 2012. DIGE and iTRAQ as biomarker discovery tools in aquatic toxicology. Ecotoxicol. Environ. Saf. 76, 3–10.

Matsumura, H., Ito, A., Saitoh, H., Winter, P., Kahl, G., Reuter, M., Krüger, D.H., Terauchi, R., 2005. SuperSAGE. Cell. Microbiol. 7 (1), 11–18.

Matthiessen, P., Gibbs, P.E., 1998. Critical appraisal of the evidence for tributyltin-mediated endocrine disruption in mollusks. Environ. Toxicol. Chem. 17, 37–43.

Mayone, F., Brunetti, R., Fumagalli, 0, Gabriele, M., Levis, A.G., 1990. Induction of micronuclei by mitomycin C and colchicine in the marine mussel *Mytilus galloprovincialis*. Mutat. Res. 244, 147–151.

McCord, J.M., Fridovich, I., 1976. Superoxide dismutase: an enzymatic function for erythrocuprein (hemiocuprein). J. Biol. Chem. 244, 6049–6055.

McDonagh, B., Sheehan, D., 2006. Redox proteomics in the blue mussel *Mytilus edulis*: carbonylation is not a prerequisite for ubiquitination in acute free radical-mediated oxidative stress. Aquat. Toxicol. 79 (4), 325–333.

McDonagh, B., Sheehan, D., 2008. Effects of oxidative stress on protein thiols and disulphides in *Mytilus edulis* revealed by proteomics: actin and protein disulphide isomerase are redox targets. Mar. Environ. Res. 66, 193–195.

McFarland, V.A., Inouye, S.L., Lutz, C.H., Jarvis, A.S., Clarke, J.U., McCant, D.D., 1999. Biomarkers of oxidative stress and genotoxicity in livers of field collected brown bullhead, *Ameiurus nebulosus*. Arch. Environ. Contam. Toxicol. 37, 236–241.

Metcalfe, C.D., 1988. Induction of micronuclei and nuclear abnormalities in the erythrocytes of mudminnows (*Umbra limi*) and brown bullheads (Ictalurusnebulosus). Bull. Environ. Contam. Toxicol. 40, 489–495.

Milan, M., Coppe, A., Reinhardt, R., Cancela, L.M., Leite, R.B., Saavedra, C., Ciofi, C., Chelazzi, G., Patarnello, T., Bortoluzzi, S., Bargelloni, L., 2011. Transcriptome sequencing and microarray development for the Manila clam, *Ruditapes philippinarum*: genomic tools for environmental monitoring. BMC Genomics 12, 234.

Milan, M., Ferraresso, S., Ciofi, C., Chelazzi, G., Carrer, C., Ferrari, G., Pavan, L., Patarnello, T., Bargelloni, L., 2013. Exploring the effects of seasonality and chemical pollution on the hepatopancreas transcriptome of the Manila clam. Mol. Ecol 22, 2157–2172.

Milan, M., Pauletto, M., Boffo, L., Carrer, C., Sorrentino, F., Ferrari, G., Pavan, L., Patarnello, T., Bargelloni, L., 2015. Transcriptomic resources for environmental risk assessment: a case study in the Venice lagoon. Environ. Pollut. 197, 90–98.

Miller, J.A., Miller, E.C., 1979. Perspectives on the metabolism of chemical carcinogens. In: Emmelot, P., Kriek, E. (Eds.), Environmental Carcinogenes. Elsevier, Amsterdam, pp. 25–100.

Miller, M.G., 2007. Environmental metabolomics: a SWOT analysis (strengths, weaknesses, opportunities, and threats). J. Proteome Res. 6 (2), 540–545.

Morales-Caselles, C., Martín-Díaz, M.L., Riba, I., Sarasquete, C., DelValls, T.A., 2008a. The role of biomarkers to assess oil-contaminated sediment quality using toxicity tests with clams and crabs. Environ. Toxicol. Chem. 27, 1309–1316.

Morales-Caselles, C., Martín-Díaz, M.L., Riba, I., Sarasquete, C., DelValls, T.A., 2008b. Sublethal responses in caged organisms exposed to sediments affected by oil spills. Chemosphere 72, 819–825.

Moreira, S.M., Moreira-Santos, M., Guilhermino, L., Ribeiro, R., 2006. An in situ post exposure feeding assay with *Carcinus maenas* for estuarine sediment-overlying water toxicity evaluations. Environ. Pollut. 139, 318–329.

Mouneyrac, C., Amiard, J.C., Amiard-Triquet, C., 1998. Effects of natural factors (salinity and body weight) on cadmium, copper, zinc and metallothionein-like protein levels in resident populations of oysters *Crassostrea gigas* from a polluted estuary. Mar. Ecol. Progr. Ser. 162, 125–135.

Mullis, K.F., Faloona, F., Scharf, S., Saiki, R., Horn, G., Erlich, H., 1986. Specific enzymatic amplification of DNA in vitro: the polymerase chain reaction. Cold Spring Harb. Symp. Quant. Biol. 51, 263–273.

Murugesan, A.G., Haniffa, M.A., 1992. Histopathological and histochemical changes in the oocytes of the air-breathing fish *Heteropneus tesfossilis* (Bloch) exposed to textile-mill effluent. Bull. Environ. Contam. Toxicol. 48, 929–936.

Myers, M.S., Stehr, C.M., Olsen, O.P., Johnson, L.L., McCain, B.B., Chan, S.L., Varanasi, U., 1994. Relationships between toxicopathic hepatic lesions and exposure to chemical contaminants in English sole (Pleuronectus vetulus), starry flounder (Platichthys stellatus), and white croaker (Genyonemus lineatus) from selected marine sites on the Pacific coast, USA. Environ. Health Perspect 102, 200–215.

Nandhakumar, S., Parasuraman, S., Shanmugam, M.M., Rao, K.R., Chand, P., Bhat, B.V., 2011. Evaluation of DNA damage using single-cell gel electrophoresis (Comet Assay). J. Pharmacol. Pharmacother. 2, 107–111.

Navaneethaiyer, U., Kasthuri, S.R., Youngdeuk, L., Ilson, W., Cheol, Y.C., Jehee, L., 2012. A novel molluscan sigma-like glutathione S-transferase from Manila clam, *Ruditapes philippinarum*: cloning, characterization and transcriptional profiling. Comp. Biochem. Physiol. C 155, 539–550.

Nelson, D.R., 1998. Metazoan cytochrome P450 evolution. Biol. Comp. Biochem. Physiol. C 121, 15–22, 27, 605–615.

Nicholson, J.K., Lindon, J.C., 2008. Systems biology: metabonomics. Nature 455 (7216), 1054–1056.

Nikimann, M., 2014. An Introduction to Aquatic Toxicology. Academia Press, Oxford, 240 p.

Oaten, J.F.P., Hudson, M.D., Jensen, A.C., Williams, L.D., 2015. Effect of organism preparation in metallothionein and metal analysis in marine invertebrates for biomonitoring marine pollution. Sci. Total Environ. 518-519, 238–247.

OECD, 2004. Oecd Guideline For The Testing Of Chemicals. Draft Proposal For A New Guideline 487: In Vitro Micronucleus Test, 487.

OECD, 2011. Report of the Workshop on Using Mechanistic Information in Forming Chemical Categories. OECD Environment, Health and Safety Publications Series on Testing and Assessment. No. 138. ENV/JM/MONO(2011)8.

OECD, 2012. Workshop on Using Mechanistic Information in Forming Chemical Categories. Appendix I: Collection of Working Definitions.

Olafson, E.W., Olsson, P.E., 1991. Electrochemical detection of metallothionein. In: Riordan, J.F., Valle, B.L. (Eds.), Methods in Enzymology, vol. 205. Academic Press Inc., San Diego, pp. 205–213.

Olive, P.L., Hilton, J., Durand, R.E., 1986. DNA conformation of Chinese hamster V79 cells and sensitivity to ionizing radiation. Radiat. Res. 107, 115–124.

Olive, P.L., Chan, A.P.S., British, C.S.C., 1988. Comparison between the DNA precipitation and alkali unwinding assays for detecting DNA strand breaks and cross-links. Can. Res. 48, 6444–6449.

Oliver, L.M., Fisher, W.S., Winstead, J.T., Hemmer, B.L., Long, E.R., 2001. Relationships between tissue contaminants and defense-related characteristics of oysters (*Crassostrea virginica*) from five Florida bays. Aquat. Toxicol. 55, 203–222.

Onosaka, S., Cherian, G., 1982. Comparison of metallothionein determination by polarographic and cadmium-saturation methods. Toxicol. Appl. Pharmacol. 63, 270–274.

Orbea, A., Fahimi, H.D., Cajaraville, M.P., 2000. Immunolocalization of four antioxidant enzymes in digestive glands of mollusks and crustaceans in fish liver. Histochem. Cell Biol. 114, 393–404.

Ortiz-Zarragoitia, M., 2005. Effects of Endocrine Disruptors on Peroxisome Proliferation, Reproduction and Development of Model Organisms, Zebrafish and Mussel (Ph.D. thesis). University of the Basque Country, p. 164.

Pateraki, L.E., Stratakis, E., 1997. Characterization of vitellogenin and vitellin from land crab *Potamon potamios*: identification of a precursor polypeptide in the molecule. J. Exp. Zool. 279, 597–608.

Payne, J.F., Mathieu, A., Melvin, W., Fancey, L.L., 1996. Acetylcholinesterase, an old biomarker with a new future? Field trials in association with two urban rivers and a paper mill in Newfoundland. Mar. Pollut. Bull. 32, 225–231.

Pérez, E., Blasco, J., Montserrat, S., 2004. Biomarker responses to pollution in two invertebrate species: *Scrobicularia plana* and *Nereis diversicolor* from the Cádiz bay (SW Spain). Mar. Environ. Res. 58, 275–279.

Peterson, R.E., Theobald, H.M., Kimmel, G.L., 1993. Developmental and reproductive toxicity of dioxin and related compounds: cross-species comparisons. Crit. Rev. Toxicol. 23, 283–335.

Pfaffl, M.W., Horgan, G.W., Dempfle, L., 2002. Relative expression software tool (REST) for group-wise comparison and statistical analysis of relative expression results in real-time PCR. Nucl. Acids Res. 30–36.

Pfaffl, M.W., 2001. A new mathematical model for relative quantification in real-time RT-PCR. Nucl. Acids Res. 29 (9), 45–53.

Pfaffl, M.W., 2004. Quantification strategies in real-time PCR. In: Bustin, S.A. (Ed.), A–Z of Quantitative PCR, IUL Biotechnology Series. International University Line, La Jolla, CA, pp. 87–120.

Picado, A., Bebianno, M.J., Costa, M.H., Ferreira, A., Vale, C., 2007. Biomarkers: a strategic tool in the assessment of environmental quality of coastal waters. Hydrobiologia 587, 79–87.

Pipe, R.K., Coles, J.A., 1995. Environmental contaminants influencing immune function in marine bivalve molluscs. Fish Shellfish Immunol. 5, 581–595.

Pipe, R.K., 1992. Generation of reactive oxygen metabolites by the haemocytes of the mussel *Mytilus edulis*. Dev. Comp. Immunol. 16, 111–122.

Pollack, J.R., Perou, C.M., Alizadeh, A.A., Eisen, M.B., Pergamenschikov, A., Williams, C.F., Jeffrey, S.S., Botstein, D., Brown, P.O., 1999. Genome-wide analysis of DNA copy-number changes using cDNA microarrays. Nat. Gen. 23, 41–46.

Porte, C., Janer, G., Lorusso, L.C., Ortiz-Zarragoitia, M., Cajaraville, M.P., Fossi, M.C., Canesi, L., 2006. Endocrine disruptors in marine organisms: approaches and perspectives. Comp. Biochem. Phys. C 143, 303–315.

Ramos-Gómez, J., Coz, A., Viguri, J.R., Luque, A., Martín-Díaz, M.L., DelValls A, T., 2011. Biomarker responsiveness in different tissues of caged *Ruditapes philippinarum* and its use within an integrated sediment quality assessment. Environ. Pollut. 159, 1914–1922.

van Ravenzwaay, B., Cunha, G.C.P., Leibold, E., Looser, R., Mellert, W., Prokoudine, A., Walk, T., Wiemer, J., 2007. The use of metabolomics for the discovery of new biomarkers of effect. Toxicol. Lett. 172 (1–2), 21–28.

Regoli, F., Giuliani, M.E., 2014. Oxidative pathways of chemical toxicity and oxidative stress biomarkers in marine organisms. Mar. Environ. Res. 93, 106–117.

Regoli, F., Principato, G.B., Bertoli, E., Nigro, M., Orlando, E., 1997. Biochemical characterization of the antioxidant system in the scallop *Adamussium colbecki*, a sentinel organism for monitoring the Antarctic environment. Polar Biol. 17, 251–258.

Regoli, F., Bocchetti, R., Filho, D.W., 2012. Spectrophotometric assays of antioxidants. In: Abele, D., Vázquez-Medina, J.P., Zenteno-Savín, T. (Eds.), Oxidative Stress in Aquatic Ecosystems. Wiley-Blackwell, Chichester, pp. 367–380.

Regoli, F., 2012. Chemical pollutants and the mechanisms of reactive oxygen species generation in aquatic organisms. In: Abele, D., Vázquez-Medina, J.P., Zenteno-Savín, T. (Eds.), Oxidative Stress in Aquatic Ecosystems. Wiley-Blackwell, Chichester, pp. 308–316.

Relyea, R., Hoverman, J., 2006. Assessing the ecology in ecotoxicology: a review and synthesis in freshwater systems. Ecol. Lett. 9, 1157–1171.

Repetto, M., Semprine, J., Boveris, A., 2012. Lipid peroxidation: chemical mechanism, biological implications and analytical determination. In: Catala, A. (Ed.), Biochemistry, Genetics and Molecular Biology. CC BY 3.0 license, 1–30 pp.

Rhee, J.S., Kim, B.M., Kim, R.O., Seo, S.J., Kim, I.C., Lee, Y.M., Lee, J.S., 2013. Co-expression of antioxidant enzymes with expression of p53, DNA repair, and heat shock protein genes in the gamma ray-irradiated hermaphroditic fish Kryptolebias marmoratus larvae. Aquat. Toxicol 140–141, 58–67.

Robinson, C.D., Brown, E., Craft, J.A., Davies, I.M., Moffat, C.F., 2004. Effects of prolonged exposure to 4-tert-octylphenol on toxicity and indices of oestrogenic exposure in the sand goby (*Pomatoschistus minutus*, Pallas). Mar. Environ. Res. 58, 19–38.

Rocha, J.B.T., Saraiva, R.A., Garcia, S.C., Gravina, F.S., Nogueira, C.W., 2012. Aminolevulinate dehydratase (δ-ALA-D) as marker protein of intoxication with metals and other pro-oxidant situations. Toxicol. Res. 1 (2), 85–102.

Rodrigues-Silva, C., Flores-Nunes, F., Vernal, J.I., Cargnin-Ferreira, E., Bainy, A.C.D., 2015. Expression and immunohistochemical localization of the cytochrome P450 isoform 356A1 (CYP356A1) in oyster Crassostrea gigas. Aquat. Toxicol 159, 267–275.

Rodriguez-Ariza, A., Abril, N., Navas, J.I., Dorado, G., Lopez-Barea, J., Pueyo, C., 1992. Metal mutagenicity and biochemical studies on bivalve mollusks from Spanish Coasts. Environ. Mol. Mutagen 19, 112–124.

Rodríguez-Ortega, M., Grosvik, B.E., Rodríguez-Ariza, A., Goksoyr, A., López-Barea, J., 2003. Change in protein expression profiles in bivalve molluscs (*Chamaelea gallina*) exposed to four model environmental pollutants. Proteomics 3, 1535–1543.

Roesijadi, G., Robinson, W.E., 1994. Metal regulation in aquatic animals: mechanisms of uptake, accumulation and release. In: Malins, D.C., Ostrander, K.G. (Eds.), Aquatic Toxicology: Molecular, Biochemical and Cellular Perspectives. CRC/Lewis Publisher, Boca Raton (FL), pp. 387–420.

Roesijadi, G., Kielland, S., Klerks, P., 1989. Purification and properties of novel molluscan metallothioneins. Arch. Biochem. Biophys. 273 (2), 403–413.

Roesijadi, G., 1994. Behavior of metallothionein-bound metals in a natural population of an estuarine mollusc. Mar. Environ. Res. 38 (3), 147–168.

Roesijadi, G., 1996. Metallothionein and its role in toxic metal regulation. Comp. Biochem. Physiol. C 113 (2), 117–123.

Roméo, M., Cosson, R.P., Gnassia-Barelli, M., Risso, C., Stien, X., Lafaurie, M., 1997. Metallothionein determination in the liver of the sea bass *Dicentrarchus labrax* treated with copper and B(a)P. Mar. Environ. Res. 44 (3), 275–284.

Romeo, M., Bennani, N., Gnassia-Barelli, M., Lafaurie, M., Girard, J.P., 2000. Cadmium and copper display different responses towards oxidative stress in the kidney of the sea bass. Aquat. Toxicol. 48, 185–194.

Romero-Ruiz, A., Carrascal, M., Alhama, J., Gómez-Ariza, J.L., Abian, J., López-Barea, J., 2006. Utility of proteomics to assess pollutant response of clams from the Doñana bank of Guadalquivir Estuary (SW Spain). Proteomics 6, S245–S255.

Romero-Ruiz, A., Alhama, J., Blasco, J., Gómez-Ariza, J.L., López-Barea, J., 2008. New metallothionein assay in *Scrobicularia plana*: heating effect and correlation with other biomarkers. Environ. Pollut. 156 (3), 1340–1347.

Ross, P.L., Huang, Y.N., Marchese, J.N., Williamson, B., Parker, K., Hattan, S., Khainovski, N., Pillai, S., Dey, S., Daniels, S., Purkayastha, S., Juhasz, P., Martin, S., Bartlet-Jones, M., He, F., Jacobson, A., Pappin, D.J., 2004. Multiplexed protein quantitation in *Saccharomyces cerevisiae* using amine-reactive isobaric tagging reagents. Mol. Cell. Proteomics 3, 1154–1169.

Rotchel, J.M., Ostrander, G.K., 2003. Molecular markers of endocrine disruption in aquatic organisms. J. Toxicol. Environ. Health B 6, 453–495.

Russo, R., Bonaventura, R., Zito, F., Schröder, H.C., Müller, I., Müller, W.E.G., Matranga, V., 2003. Stress to cadmium monitored by metallothionein gene induction in *Paracentrotus lividus* embryos. Cell Stress Chaperones 8 (3), 232–241.

Saha, S., Sparks, A.B., Rago, C., Akmaev, V., Wang, C.J., Vogelstein, B., Kinzler, K.W., Velculescu, V.E., 2002. Using the transcriptome to annotate the genome. Nat. Biotechnol. 20 (5), 508–512.

Scheuhammer, A.M., Charian, M.G., 1991. Quantification of metallothionein by silver saturation. In: Riordan, J.F., Valle, B.L. (Eds.), Methods in Enzymology, vol. 205. Academic Press Inc., San Diego, pp. 78–83.

Seeley, K.R., Weeks-Perkins, B.A., 1991. Altered phagocytic activity of macrophages in oyster toadfish from a highly polluted subestuary. J. Aquat. Anim. Health 3, 224–227.

Segner, H., Casanova-Nakayama, A., Kase, R., Tyler, C.R., 2013. Impact of environmental estrogens on fish considering the diversity of estrogen signaling. Gen. Comp. Endocrinol. 191, 190–201.

Segner, H., 2011. Moving beyond a descriptive aquatic toxicology: the value of biological process and trait information. Aquat. Toxicol. 105, 50–55.

Seo, K.H., Valentin-Bon, I.E., Brackett, R.E., Holt, P.S., 2004. Rapid, specific detection of Salmonella Enteritidis in pooled eggs by real-time PCR. J. Food Prot 67, 864–869.

Serafim, A., Bebianno, M.J., 2010. Effect of a polymetallic mixture on metal accumulation and metallothionein response in the clam *Ruditapes decussatus*. Aquat. Toxicol. 99 (3), 370–378.

Sheehan, D., 2006. Detection of redox-based modification in two-dimensional electrophoresis proteomic separations. Biochem. Biophys. Res. Commun. 349 (2), 455–462.

Shepard, J.L., Olsson, B., Tedengren, M., Bradley, B.P., 2000. Protein expression signature identified in *Mytilus edulis* exposed to PCBs, copper and salinity stress. Mar. Environ. Res. 50, 337–340.

Shugart, L.R., Bickham, J., Jackim, G., McMahon, G., Ridley, W., Stein, J., 1992. DNA alterations. In: Huggett, R.J., Kimerly, R.A., Mehrle Jr., P.M., Bergman, H.L. (Eds.), Biomarkers: Bio-Chemical, Physiological and Histological Markers of Anthropogenic Stress. Lewis Publishers, Chelsea, MI, USA, pp. 155–210.

Shugart, L.R., 1990. Biological monitoring: testing for genotoxicity. In: McCarthy, J.F., Shugart, L.R. (Eds.), Biomarkers of Environmental Contamination. Lewis Publishers, Boca Raton, FL, USA, pp. 205–216.

Sijm, D.T.H.M., Wever, H., Opperhuizen, A., 1989. Influence of biotransformation on the accumulation of PCDDs from flyash in fish. Chemosphere 19, 475–480.

Silvestre, F., Dierick, J.F., Dumont, V., Dieu, M., Raes, M., Devos, P., 2006. Differential protein expression profiles in anterior gills of *Eriocheir sinensis* during acclimation to cadmium. Aquat. Toxicol. 76, 46–58.

Singh, R.B., Singh, T.P., 1992. Modulatory actions of ovine luteinizing hormone-releasing hormone and Mystus gonadotropin on y-BHC-induced changes in lipid levels in the freshwater catfish, *Heteropneustes fossilis*. Ecotoxicol. Environ. Saf. 24, 192–202.

Singh, I., 1996. Mammalian peroxisomes: metabolism of oxygen and reactive oxygen species. Ann. N.Y. Acad. Sci. 804, 612–627.

Snyder, M., 2000. Cytochrome P450 enzymes in aquatic invertebrates: recent advances and future directions. Aquat. Toxicol. 48, 529–547.

Solé, M., Porte, C., Albaigés, J., 1995. Seasonal variation in the mixed function oxygenase system and antioxidant enzymes of the mussel *Mytilus galloprovincialis*. Environ. Toxicol. Chem. 14, 157–164.

Solé, M., Porte, C., Biosca, X., Mitchelmore, C.L., Chipman, J.K., Livingstone, D.R., Albaigès, J., 1996. Effects of the Aegean Sea oil spill on biotransformation enzymes, oxidative stress and DNA-adducts in digestive gland of the mussel (*Mytilus edulis* L.). Comp. Biochem. Physiol. C 113, 257–265.

Solé, M., Kopecka-Pilarczyk, J., Blasco, J., 2009. Pollution biomarkers in two estuarine invertebrates, *Nereis diversicolor* and *Scrobicularia plana*, from a Marsh ecosystem in SW Spain. Environ. Int. 35 (3), 523–531.

Spies, R.B., Stegeman, J.J., Rice, D.W., Woodlin Jr., B., Thomas, P., Hose, J.E., Cross, J.N., Prieto, M., 1990. Sublethal responses of Platichtus stellatus to organic contamination in San Francisco Bay with emphasis on reproduction. In: McCarthy, J.F., Shugart, L.R. (Eds.), Biomarkers of Environmental Contamination. Lewis Publishers, Boca Raton, FL, USA, pp. 87–122.

Stegeman, J.J., Lech, J.J., 1991. Cytochrome P450 monooxygenase systems in aquatic species: carcinogen metabolism and biomarkers for carcinogen and pollutant exposure. Environ. Health Perspect. 90, 101–109.

Stegeman, J.J., Brouwer, M., Richard, T.D.G., Förlin, L., Fowler, B.A., Sanders, B.M., van Veld, P.A., 1992. Molecular responses to environmental contamination: enzyme and protein systems as indicators of chemical exposure and effect. In: Huggett, R.J., Kimerly, R.A., Mehrle Jr., P.M., Bergman, H.L. (Eds.), Biomarkers: Biochemical, Physiological and Histological Markers of Anthropogenic Stress. Lewis Publishers, Chelsea, MI, USA, pp. 235–335.

Sturm, A., Wogram, J., Segner, H., Liess, M., 2000. Different sensitivity to organophosphates of acetylcholinesterase and butyrylcholinesterase from three-spined stickleback (*Gasterosteus aculeatus*): application in biomonitoring. Environ. Toxicol. Chem. 19, 1607–1615.

Sukumar, A., Karpagaganapathy, P.R., 1992. Pesticide-induced atresia in ovary of a freshwater fish, Colisa ialia (Hamilton-Buchanan). Bull. Environ. Contam. Toxicol. 48, 457–462.

Tonge, R., Shaw, J., Middleton, B., Rowlinson, R., Rayner, S., Young, J., Pognan, F., Hawkins, E., Currie, I., Davison, M., 2001. Validation and development of fluorescence two-dimensional differential gel electrophoresis proteomics technology. Proteomics 1, 377–396.

Towbin, H., Staehelin, T., Gordon, J., 1979. Electrophoretic transfer of proteins from polyacrylamide gels to nitrocellulose sheets: procedure and some applications. Proc. Natl. Acad. Sci. U.S.A. 76 (9), 4350–4354.

Trombini, C., Fabbri, E., Blasco, J., 2010. Temporal variations in metallothionein concentration and subcellular distribution of metals in gills and digestive glands of the oyster *Crassostrea angulata*. Sci. Mar. 74 (Suppl. 1), 143–152.

Tsarpalias, V., Dailianis, S., 2012. Investigation of landfill leachate toxic potency: an integrated approach with the use of stress indices in tissues of mussels. Aquat. Toxicol. 125, 58–65.

Tudor, M., 1984. Preliminary Evaluation of 6-aminolevulinic Acid Dehydratase in Blood of Lesser Spotted Dogfish (*Sciliorhinus canicula* L.) from the Middle Adriatic. Institut ZA Oceanografijuiribarstvo-Splitsfr Jugosla Vij AN° 55v. Bil-jeske-Notes. 55.

Turner, J.E., Parry, J.M., 1989. The induction and repair of DNA damage in the mussel Mytilus edulis. Mar. Environ. Res 28, 346–347.

United Nations, 2013. The Millennium Development Goals Report.

Vallee, B.L., Coleman, J.E., Auld, D.S., 1991. Zinc fingers, zinc clusters and zinc twists in DNA-binding protein domains. Proc. Natl. Acad. Sci. U.S.A. 88, 999–1003.

van der Aar, E.M., Buikema, D., Commandeur, J.N.M., te Koppele, J.M., van Ommen, B., van Bladeren, P.J., Vermeulen, N.P.E., 1996. Enzyme kinetics and substrate selectivities of rat glutathione S-transferase isoenzymes towards a series of new 2-substituted 1-chloro-4-nitrobenzenes. Xenobiotica 26, 143–155.

van der Oost, R., Goksøyr, A., Celander, M., Heida, H., Vermeulen, N.P.E., 1996. Biomonitoring aquatic pollution with feral eel (*Anguilla anguilla*): II. Biomarkers: pollution-induced biochemical responses. Aquat. Toxicol. 36, 189–222.

van der Oost, R., Beyer, J., Vermeulen, N.P.E., 2003. Fish bioaccumulation and biomarkers in environmental risk assessment: a review. Environ. Toxicol. Pharmacol. 13, 57–149.

van der Oost, R., Porte-Visa, C., Van den Brink, N.W., 2005. Biomarkers in environmental assessment. In: den Besten, P.J., Munawar, M. (Eds.), Ecotoxicological Testing of Marine and Freshwater Ecosystems: Emerging Techniques, Trends and Strategies. Taylor & Francis, Boca Raton, FL, pp. 87–152.

van der Ven, K., DeWit, M., Keil, D., Moens, L., Van Leemput, K., Naudts, B., De Coen, W., 2005. Development and application of a brain-specific cDNA microarray for effect evaluation of neuro-active pharmaceuticals in zebrafish (Danio rerio). Comp. Biochem. Physiol. B 141, 408–417.

van der Ven, K., Keil, D., Moens, L.N., VanHummelen, P., Van Remortel, P., Maras, M., De Coen, W.M., 2006. Effects of the antidepressant mianserin in zebrafish: molecular markers of endocrine disruption. Chemosphere 65, 1836–1845.

Velculescu, V.E., Zhang, L., Vogelstein, B., Kinzler, K.W., 1995. Serial analysis of gene expression. Science 270 (5235), 484–487.

Venter, J.C., Adams, M.D., Myers, E.W., Li, P.W., Mural, R.J., Sutton, G.G., Smith, H.O., et al., 2001. The sequence of the human genome. Science 291 (5507), 1304–1351.

Verdon, C.P., Burton, B.A., Prior, R.L., 1995. Sample pretreatment with nitrate reductase and glucose-6-phosphate dehydrogenase quantitatively reduces nitrate while avoiding interference by NADP+ when the Griess reaction is used to assay for nitrite. Anal. Biochem. 224, 502–508.

Verlecara, X., Jena, K., Chainy, G., 2007. Biochemical markers of oxidative stress in Perna viridis exposed to mercury and temperature. Chem. Biol. Interact. 167, 219–226.

Verslycke, T., Vandenbergh, G.F., Versonnen, B., Arijs, K., Janssen, C.R., 2002. Induction of vitellogenesis in 17a-ethinylestradiol-exposed rainbow trout (*Oncorhynchus mykiss*): a method comparison. Comp. Biochem. Physiol. Part C 132, 483–492.

Versonnen, B.J., Goemans, G., Belpaire, C., Janssen, C.R., 2004. Vitellogenin content in European eel (*Anguilla anguilla*) in Flanders, Belgium. Environ. Pollut. 128, 363–371.

Viarengo, A., Canesi, L., 1991. Mussels as biological indicators of pollution. Aquaculture 94, 225–243.

Viarengo, A., Ponzano, E., Dondero, F., Fabbri, R., 1997. A simple spectrophotometric method for metallothionein evaluation in marine organisms: an application to Mediterranean and Antarctic molluscs. Mar. Environ. Res. 44 (1), 69–84.

Viarengo, A., Burlando, B., Cavaletto, M., Marchi, B., Ponzano, E., Blasco, J., 1999. Role of metallothionein against oxidative stress in the mussel *Mytilus galloprovincialis*. Am. J. Physiol. 277, R1612–R1619.

Viarengo, A., Lowe, D., Bolognesi, C., Fabbri, E., Koehler, A., 2007. The use of biomarkers in biomonitoring: a 2-tier approach assessing the level of pollutant-induced stress syndrome in sentinel organisms. Comp. Biochem. Physiol. C 146, 281–300.

Victor, B., Mahalingam, S., Sarojini, R., 1986. Toxicity of mercury and cadmium on oocyte differentiation and vitellogenesis of the teleost, *Lepidocephalichthys thermalis* (Bleeker). J. Environ. Biol. 7, 209–214.

Vioque-Fernández, A., de Almeida, E.A., Ballesteros, J., García-Barrera, T., Gómez-Ariza, J.L., López-Barea, J., 2007. Doñana National Park survey using crayfish (*Procambarus clarkii*) as bioindicator: esterase inhibition and pollutant levels. Toxicol. Lett. 168, 260–268.

Vioque-Fernández, A., Alves de Almeida, E., López-Barea, J., 2009. Assessment of Doñana National Park contamination in *Procambarus clarkii*: integration of conventional biomarkers and proteomic approaches. Sci. Total Environ. 407, 1784–1797.

Volland, M., Hampel, M., Martos-Sitcha, J.A., Trombini, C., Martínez-Rodríguez, G., Blasco, J., 2015. Citrate gold nanoparticle exposure in the marine bivalve *Ruditapes philippinarum*: uptake, elimination and oxidative stress response. Environ. Sci. Pollut. Res. 23, 17414–17424.

Vos, J., van Loveren, H., Wester, P., Vethaak, D., 1989. Toxic effects of environmental chemicals on the immune system. Trends Pharmacol. Sci. 10, 289–292.

Walker, C.H., Thompson, H.M., 1991. Phylogenetic distribution of cholinesterases and related esterases. In: Mineau, P. (Ed.), Cholinesterase-Inhibiting Insecticides, Chemicals in Agriculture. Elsevier, Amsterdam, pp. 1–17.

Wang, Z., Zhao, J., Li, F., Gao, D., Xing, B., 2009a. Adsorption and inhibition of acetylcholinesterase by different nanoparticles. Chemosphere 77, 67–73.

Wang, Z., Gerstein, M., Snyder, M., 2009b. RNA-Seq: a revolutionary tool for transcriptomics. Nat. Rev. Genet. 10 (1), 57–63.

Warheit, D.B., Borm, P.J.A., Hennes, C., Lademann, J., 2007. Testing strategies to establish the safety of nanomaterials: conclusions of an ECETOC workshop. Inhal. Toxicol 19, 631–643.

Watson, J.D., Crick, F.H., 1953. Molecular structure of nucleic acids; a structure for deoxyribose nucleic acid. Nature 171 (4356), 737–738.

Weeks, B.A., Warinner, J.E., Rice, C.D., 1989. Recent advances in the assessment of environmentally-induced immunomodulation. In: Oceans '89 Proceedings, vol. 2. Institute of Electrical and Electronics Engineers, New York, pp. 408–441.

Weeks, B.A., Warinner, J.E., Mathews, E.S., Wishkovsky, A., 1990a. Effects of toxicants on certain functions of the lymphoreticular system of fish. In: Perkins, F.O., Cheng, T.C. (Eds.), Pathology in Marine Science. Academic Press, San Diego, pp. 369–374.

Weeks, B.A., Huggett, R.J., Warinner, J.E., Mathews, E.S., 1990b. Macrophage responses of estuarine fish as bioindicators of toxic contamination. In: McCarthy, J.F., Shugart, L.R. (Eds.), Biomarkers of Environmental Contamination. Lewis Publishers, CRC Press, Boca Raton, pp. 193–220.

Wenk, M.R., 2005. The emerging field of lipidomics. Nat. Rev. Drug Discov. 4 (7), 594–610.

Wester, P.W., Vethaak, D., van Muiswinkel, W.B., 1994. Fish as biomarkers in immunotoxicology. Toxicology 86, 213–232.

White, R., Jobling, S., Hoare, S.A., Sumpter, J.P., Parker, M.G., 1994. Environmentally persistent alkylphenolic compounds are estrogenic. Endo 36, 175–182.

Whyte, J.J., Jung, R.E., Schmitt, C.J., Tillitt, D.E., 2000. Ethoxyresorufin-O deethylase (EROD) activity in fish as a biomarker of chemical exposure. Crit. Rev. Toxicol. 30, 347–570.

Wills, E.D., 1987. Evaluation of lipid peroxidation in lipids and biological membranes. In: Snell, K., Mulloc, B. (Eds.), Biochemical Toxicology: A Practical Approach. IRL Press, USA, pp. 127–150.

Winston, G.W., Di Giulio, R.T., 1991. Prooxidant and antioxidant mechanisms in aquatic organisms. Aquat. Toxicol. 19, 137–161.

Wirgin, I., Currie, D., Garte, S.J., 1989. Activation of the K-ras oncogene in liver tumors of Hudson River tomcod. Carcinogenesis 10, 2311–2315.

Won, F.J., Rhee, J.S., Shin, K.H., Jung, J.H., Shim, W.J., Lee, Y.M., Lee, J.S., 2013. Expression of three novel cytochrome P450 (CYP) and antioxidative genes from the polychaete, Perinereis nuntia exposed to water accommodated fraction (WAF) of Iranian crude oil and Benzo[α]. Mar. Environ. Res 90, 75–84.

World Health Organization/International Programme on Chemical Safety, 2002. Global assessment of the state-of-the-science of endocrine disruptors. In: Damstra, T., Barlow, S., Bergman, A., Kavlock, R., Van der Kraak, G. (Eds.), World Health Organization (WHO/PCS/EDC/02.2). World Health Organization, Geneva, Switzerland.

Yadetie, F., Butcher, S., Forde, H.E., Campstejn, C., Bouquet, J.M., Karlsen, O.A., Denoued, F., Metpally, R., Thompson, E.M., Manak, J.R., Goksoyr, A., Chourrot, D., 2012. Conservation and divergence of chemical defense system in the tunicate *Oikopleura dioica* revealed by genome wide responses to two xenobiotics. BMC Genomics 13, 55–63.

Yevich, P.P., Yevich, C., Pesch, G., 1987. Effects of Black Rock Harbor Dredged Material on the Histopathology of the Blue Mussel *Mytilus edulis* and Polychaete Worm *Nephtys incisa* After Laboratory and Field Exposures. US EPA Technical Report D-87-8, Narragansett, RI.

Yin, H., Xu, L., Porter, N.A., 2011. Free radical lipid peroxidation: mechanisms and analysis. Chem. Rev. 111, 5944–5972.

Zanette, J., Goldstone, J.V., Bainy, A.C.D., Stegeman, J.J., 2010. Identification of CYP genes in *Mytilus* (mussel) and *Crassostrea* (oyster) species: first approach to the full complement of cytochrome P450 genes in bivalves. Mar. Environ. Res. 69, 1–3.

CHAPTER

6

Saltwater Toxicity Tests

B. Anderson, B. Phillips

University of California, Davis, CA, United States

6.1 INTRODUCTION

Scientists responsible for assessing the ecological integrity of aquatic resources rely on a number of tools including chemical analysis of water, sediment, and tissue; biological assessments; and toxicity tests. Toxicity tests are an important component for assessing the impact of chemicals in aquatic ecosystems because they indicate toxic effects of complex chemical mixtures. In aquatic toxicity tests, groups of selected organisms are exposed to test materials (in this case seawater samples) under defined conditions to determine potential adverse effects. Tests are conducted under controlled laboratory conditions or in situ. A number of standardized toxicity test protocols have been developed for determining toxicity of chemicals to aquatic species. Detailed guidance manuals for marine toxicity tests are available from the United States Environmental Protection Agency (U.S. EPA) and other entities such as the American Society for Testing and Materials (ASTM) and the American Public Health Association (APHA). These protocols provide guidance on application of toxicity tests for assessing toxicity of single chemicals, complex effluents, and ambient water samples.

This chapter is intended to provide an overview of the various standardized aquatic toxicity test protocols available for marine and estuarine water quality assessments. This document is intended to familiarize readers with one of the tools used by ecotoxicologists for environmental assessments but is not intended to be a comprehensive review of all marine aquatic toxicity testing methods. Although it is recognized that a variety of other nonstandardized toxicity test methods have been reported, emphasis is placed on standardized protocols provided by the U.S. EPA, ASTM, and other entities because these are the tests most commonly used in regulatory applications. The majority of these tests have been developed in response to several US and international environmental regulations. In the United States, these include the Federal Insecticide, Fungicide, and Rodenticide Act (FIFRA), the Clean Water Act (CWA), the Marine Protection, Research, and Sanctuaries Act (MPRSA), and the Toxic Substances and Control Act (TSCA) (Ward, 1995). Species and protocols relevant to the Atlantic, Gulf, and Pacific Coasts of the United States are emphasized in this chapter, as are methods used in Canadian waters, and application of specific methods for toxicity assessments in transitional coastal environments. Some examples of the application of selected saltwater toxicity tests in field assessments are presented. Standardized protocols

Marine Ecotoxicology
http://dx.doi.org/10.1016/B978-0-12-803371-5.00006-0

for South America and Europe are also discussed. In addition, methods developed for a number of tropical species are briefly described. Finally, future research needs for development and application of toxicity tests in marine ecotoxicology are summarized.

6.2 TERMINOLOGY

Toxicity tests are categorized by test duration, life stage, and endpoints. Acute, short-term tests are usually 48- or 96-h exposures and measure mortality to determine the median lethal concentration (LC50), ie, the concentration at which 50% of the exposed test population dies. Chronic toxicity tests are conducted for longer periods and incorporate sublethal endpoints. Life stages used vary depending on regulatory application. U.S. EPA acute toxicity test protocols are initiated with the same life stages as those used in short-term chronic tests (U.S. EPA, 2002a). These include early life-stage tests that have been developed for fish and invertebrates (Goodman et al., 1985; ASTM, 2008b). Long-term chronic tests can be conducted for 28 days or longer and are designed to measure embryo development through larval growth and juvenile development. Full life-cycle tests have also been developed with invertebrates such as mysid shrimp and these incorporate measures of early development, growth, and reproduction (ASTM, 2008b). In response to legislative requirements to screen effluents for chronic toxicity, abbreviated chronic tests have been developed with fish, invertebrates, and macroalgae. These tests have been developed to be short-term indicators of chronic toxicity and incorporate sublethal endpoints such as measures of germination and reproduction using macroalgae, egg fertilization using echinoids, embryo development using echinoids and mollusks, and larval survival and growth using fish and mysid neonates. These tests are conducted for periods from 1 h to 7 days (U.S. EPA, 1995, 2002b). Species, procedures, and endpoints used in the various standardized acute and chronic toxicity test protocols are described below.

6.3 GENERAL SALTWATER TOXICITY TEST METHODOLOGY AND PROCEDURES

The standardized methods used to conduct saltwater toxicity tests are summarized in Tables 6.1–6.3 (after Ward, 1995). These include standardized methods by ASTM, APHA/AWWA/WPFC (Standard Methods), the U.S. EPA, and Environment Canada.

6.3.1 Fish

The fish species most commonly used for effluent and ambient monitoring in Atlantic and Gulf Coast marine waters are the sheepshead minnow (*Cyprinodon variegatus*) and silversides species (*Menidia* sp.). Most testing with *C. variegatus* and *Menidia beryllina* is conducted using larval fish provided by commercial suppliers or using fish from in-house cultures. Standardized acute and chronic protocols have been developed for both species (U.S. EPA, 2002a,b). The acute protocols for *C. variegatus*, *Menidia menidia*, *M. beryllina*, and *Menidia peninsulae* are described in U.S. EPA (2002a). Acute tests with these species may be conducted for 24–96 h and use 20 larval fish per replicate container (two replicates minimum). These species are euryhaline and may be tested at salinities between 5‰ and 32‰ (*C. variegatus*) or between 1 and 32‰ (*M. beryllina*). The tests may be conducted under static, static-renewal, or flow-through conditions using 1 or 2-L test volumes. Short-term chronic toxicity test procedures for these species are described in U.S. EPA (2002b). These tests use larval sheepshead minnows (*C. variegatus*) or inland silversides (*M. beryllina*) and both

TABLE 6.1 Saltwater Toxicity Test Guidelines

STANDARD METHODS, PART 8000 (APHA, 2012)	
8111	Biostimulation (Algal Productivity)
8112	Phytoplankton
8113	Marine Macroalgae
8310	Ciliated Protozoa
8410	Scleractinian Coral
8610	Mollusks
8710	Arthropods
8810	Echinoderms
8910	Fish
ASTM (ASTM, 2007, 2008a,b, 2012a–e, 2013)	
E–1191–90	Guide for Conducting Life-Cycle Toxicity Tests with Mysids
E–729–88a	Guide for Conducting Acute Toxicity Tests with Fishes, Macroinvertebrates, and Amphibians
E–1241–92	Guide for Conducting Early Lifestage Toxicity Tests with Fishes
E–1440–91	Guide for Acute Toxicity Test with the Rotifer, *Brachionus* (and Estuarine and Marine Rotifers)
E–724–89	Guide for Conducting Static Acute Toxicity Tests Starting with Embryos of Four Species of Saltwater Bivalve Mollusks
E–1463–92	Guide for Conducting Static and Flow-Through Acute Toxicity Tests with Mysids from the West Coast of the United States
E–1192–88	Guide for Conducting Acute Toxicity Tests on Aqueous Effluents with Fishes, Macroinvertebrates, and Amphibians
E–1498–92	Guide for Conducting Sexual Reproduction Tests with Seaweeds
E–1218–90	Guide for Conducting Static 96-Hour Toxicity Tests with Microalgae
EPA ACUTE EFFLUENT TESTS (U.S. EPA, 2002a)	
	Acute toxicity test with *Americamysis (Mysidopsis) bahia* (*Holmesimysis costata* = west coast alternative)
	Acute toxicity test with *Menidia* sp. (*Atherinops affinis* = west coast alternative)
	Acute toxicity test with *Cyprinodon variegatus*
EPA CHRONIC EFFLUENT TESTS (U.S. EPA, 1995, 2002b)	
	Sheepshead minnow (*Cyprinodon variegatus*) Larval Survival and Growth Test Method
	Sheepshead minnow (*Cyprinodon variegatus*) Embryo-Larval Survival and Teratogenicity Test Method
	Topsmelt (*Atherinops affinis*) Larval Survival and Growth Test Method[b]
	Inland silverside (*Menidia beryllina*) Larval Survival and Growth Test Method
	Mysid (*Americamysis (Mysidopsis) bahia*) Survival, Growth, and Fecundity Test Method

Continued

TABLE 6.1 Saltwater Toxicity Test Guidelines—cont'd

	Mysid (*Holmesimysis costata*) Survival and Growth Test Method[b]
	Sea urchin (*Arabacia punctulata*) Fertilization Test Method
	Sea urchin (*Strongylocentrotus purpuratus*) Fertilization Test Method[b]. Also may be used with the Sand dollar (*Dendraster excentricus*)
	Sea urchin (*Strongylocentrotus purpuratus*) Embryo-Larval Development Test Method[b]
	Mussel (*Mytilus galloprovincialis*) Embryo-Larval Development Test Method[b] Also may be used with oyster (*Crassostrea gigas*)
	Red abalone (*Haliotis rufescens*) Embryo-Larval Development Test Method[b]
	Algal (*Champia parvula*) Sexual Reproduction Test Method
	Algal (*Macrocystis pyrifera*) Gametophyte Germination and Growth Test Method[b]
EPA TSCA TEST GUIDELINES (CODE OF FEDERAL REGULATIONS (1990))	
797.1050	Alga Acute Toxicity Test
797.1075	Freshwater and Marine Algae Acute Toxicity Test
797.1400	Fish Acute Toxicity Test
797.1440	Fish Acute Toxicity Test
797.1600	Fish Early Life-Stage Toxicity Test
797.1800	Oyster Acute Toxicity Test
797.1930	Mysid Shrimp Acute Toxicity Test
797.1950	Mysid Shrimp Chronic Toxicity Test
797.1970	Penaeid Shrimp Acute Toxicity Test
EPA-FIFRA TEST GUIDELINES (U.S. EPA, 1985a–d; 1986a–c)	
	SEP[a]: Acute Toxicity Test for Estuarine and Marine Organisms (Estuarine Fish 96h Acute Toxicity) (EPA-540/9-85-009, 1985)
	SEP: Acute Toxicity Test for Estuarine and Marine Organisms (Shrimp 96h Acute Toxicity) (EPA-540/9-85-010, 1985)
	SEP: Acute Toxicity Test for Estuarine and Marine Organisms (Mollusk 96h Flow-Through Shell Deposition Study) (EPA-540/9-85-011, 1985)
	SEP: Acute Toxicity Test for Estuarine and Marine Organisms (Mollusk 48h Embryo Larvae Study) (EPA-540/9-85-012, 1985)
	SEP: Fish Early Life-Stage (EPA-540/9-86-138, 1986)
	SEP: Fish Life-Cycle Toxicity Tests (EPA-540/9-86-137, 1986)
	SEP: Nontarget Plants: Growth and Reproduction of Aquatic Plants – Tiers 1 and 2 (EPA-540/9-86-134, 1986)
ENVIRONMENT CANADA (ENVIRONMENT CANADA, 2011)	
	Fertilization Assay with Echinoids (Sea Urchins and Sand Dollars) (2011)

[a] *SEP = Standard Evaluation Procedure.*
[b] *Eastern Pacific test species.*

TABLE 6.2 Test Species Commonly Used for Saltwater Toxicity Tests

VERTEBRATES
Sheepshead minnow, *Cyprinodon variegatus* (**Atlantic and Gulf Coast Regions**)
Mummichog, *Fundulus heteroclitus*
Longnose killifish, *Fundulus similis*
Silverside, *Menidia* sp. (**Atlantic and Gulf Coast Regions**)
Topsmelt, *Atherinops affinis* (**Eastern Pacific and Canadian Regions**)
Three-spine stickleback, *Gasterosteus aculeatus*
Pinfish, *Lagodon rhomboides*
Spot, *Leiostomus xanthurus*
Shiner surfperch, *Cymatogaster aggregata*
Tidepool sculpin, *Oligocottus maculosus*
Sanddab, *Citharichthys stigmaeus*
Flounder, *Paralichthys dentatus, P. lethostigma*
Starry flounder, *Platichthys stellatus*
English sole, *Parophrys vetulus*
Herring, *Clupea harengus* (**Eastern Pacific Region**)
Tasmanian blenny, *Parablennius tasmanians*[b]
INVERTEBRATES
Copepods, *Acartia clausi, Acartia tonsa*
Shrimp, *Penaeus setiferus, P. duorarum, P. aztecus*
Grass shrimp, *Palaemonetes pugio, P. intermedius, P. vulgaris*
Sand shrimp, *Crangon* sp.
Shrimp, *Pandalus jordani, P. danae*
Bay shrimp, *Crangon nigricauda*
Mysid, *Americamysis (Mysidopsis) bahia, M. bigelowi, M. intii, M. almyra, Holmesimysis costata, M. juniae, Mysidium gracile*[a]
Blue crab, *Callinectes sapidus*
Shore crab, *Hemigrapsus* sp., *Pachygrapsus* sp.
Green crab, *Carcinus maenas*
Fiddler crab, *Uca* sp.
Oyster, *Crassostrea virginica, C. gigas* (**Eastern Pacific, Atlantic, and Gulf Coast Regions**)
Clam, *Mercenaria mercenaria, Mulinia lateralis* (**Atlantic and Gulf Coast Regions**)
Abalone, *Haliotis rufescens* (**Eastern Pacific Region**)
Mussel, *M. edulis, M. galloprovincialis, M. californianus* (**Eastern Pacific and Canadian Regions**)
Sea urchin, *Strongylocentrotus purpuratus* (**Eastern Pacific and Canadian Regions**), *S. droebachiensis, Arabacia punctulata* (**Atlantic and Gulf Coast Regions**), *Lytechinus* sp., *Echinometra lucunter*[a], *Heliocidaris tuberculata*[b]
Sand dollar, *Dendraster excentricus* (Eastern Pacific and Canadian Regions)

[a] *Species used in Brazil.*
[b] *Species used in Australian temperate waters.*
After Ward, G.S., 1995. Saltwater tests. In: Rand, G.M. (Ed.), Fundamentals of Aquatic Toxicology: Effects, Environmental Fate, and Risk Assessment, second ed. Taylor and Francis, Washington, DC.

TABLE 6.3 Test Species Used for Saltwater Toxicity Tests in Tropical Waters

INVERTEBRATES
Sea urchin, *Tripneustes gratilla*[a]
Copepod, *Acartia sinjiensis*[b], *Glandioferans imparipes*[b]
Coral, *Acropora millepora*[b], *A. formosa*[b], *Montipora digitata*[b], *Porites cylindrica*[b], *Seriatopora hystrix*[b]
Prawn, *Penaeus monodon*[b]

[a] *Species used in Hawaiian waters.*
[b] *Species used in Australian tropical waters.*

procedures assess larval growth and survival after a 7-day exposure. The short-term chronic test with *C. variegatus* is a 7-day static-renewal protocol that is initiated with newly hatched larvae (<24-h-old) that are hatched from embryos shipped by commercial suppliers or through in-house cultures. The embryos are incubated at 25°C (30‰) for 5–6 days before test initiation. As hatching occurs, early hatching larvae are separated. As the majority of larvae hatch over the next 24 h, these are used in the test, unless additional early-hatching fish are required to provide sufficient numbers for the experimental design. This test is conducted in 600-mL or 1-L beakers, and solutions are renewed daily. Ten larvae are tested in each beaker, and each concentration is replicated four times. Larvae are fed newly hatched *Artemia* daily. After 7 days, larval survival and growth (as dry wt) are determined. Procedures for the short-term chronic test with *M. beryllina* are similar to those for *C. variegatus*, except that the test is initiated with 7- to 11-day-old fish, the age when larvae are able to eat *Artemia* nauplii.

In addition to tests with larvae, an embryo-larval development test and an early life-stage test has been developed for *C. variegatus* and *Menidia* sp. Like the 7-day larval test, the 9-day embryo-larval test with *C. variegatus* is initiated with <24-h-old embryos, hatched from in-house cultures or provided by commercial suppliers. In this test, 15 embryos per replicate (four replicates) are exposed in beakers, crystallization dishes, or tissue culture ware, and development is recorded daily by microscopic observation of the transparent embryos. Embryos begin hatching on days 5–6 at 25°C, and the test is terminated after 9 days. Endpoints measured include various morphological deformities, swimming behavior, and percent hatching success. Test solutions are renewed daily, and no feeding is required (U.S. EPA, 2002b).

Early life-stage (28 days) and full life-cycle (~120 days posthatch) chronic tests with *C. variegatus* were described in Rand and Petrocelli (1985) and summarized by Ward (1995). These tests are initiated with 40–50 embryos per replicate (minimum of two replicates), and both tests are conducted under flow-through conditions. Embryo development is monitored daily, and once larvae hatch, these are fed *Artemia* nauplii in excess three times daily. In the early life-stage test, fry survival and development are monitored, and the test is terminated after 28 days, when survival and growth are recorded (length and dry wt). In the life-cycle test, fry growth is measured photographically at 28–30 days, then fry density is thinned to 25 fish per replicate, and the exposure continues until the fish become sexually mature (3–4 months posthatch). Once the fish are sexually mature, egg productivity is monitored in three, two-week spawning periods. Each spawning period uses a different spawning group consisting of three females and two males. These are held in screened spawning chambers designed to allow eggs extruded by the female fish and fertilized by the males to drop through a large mesh screen onto a finer mesh screen. This prevents egg cannibalism by the adults. The eggs are collected and counted daily throughout the two-week spawning period. In addition, the eggs are collected during one of the three spawning periods and these are placed in incubation chambers to evaluate embryo larval development and hatching success. Ward (1995) discussed difficulties associated with the life-cycle test for *C. variegatus*. These include problems with bacterial growth associated with carrier solvents, and pathogen infections transferred from adult fish to the fry.

The topsmelt, *Atherinops affinis*, is a west coast atherinid species commonly found in bays and estuaries. The topsmelt is one of the most ecologically important fish species in California estuaries, often representing the greatest fish biomass in these systems. Adult topsmelt are reared by a single commercial lab (Aquatic Biosystems, Fort Collins, CO) that provides embryos and larvae for toxicity testing. This species is used for effluent and ambient monitoring in California, Washington State, and in Canada. A 7-day larval growth and survival test with this species is described in U.S. EPA (1995), and research with *A. affinis* is summarized in Middaugh and Anderson (1993), McNulty et al. (1994), and Anderson et al. (1995).

The 7-day larval growth and survival protocol with this species is analogous to the test protocol for *M. beryllina* and other atherinid species and was designed to be used in place of the *M. beryllina* protocol in West Coast testing. In addition to the larval growth and survival protocol with *A. affinis*, Anderson et al. (1991) developed a 12-day embryo-larval development test with this species. This test was not considered practical for routine effluent testing but may be appropriate in situations where teratogens are of particular concern. In a study of nickel water quality criteria, Hunt et al. (2002) also described a 40-day early life-stage toxicity test with *A. affinis*. This test was initiated with 30 early-gastrula-stage embryos exposed to nickel solutions in 10 L aquaria under flow-through conditions at 20°C and 34‰. The experiment proceeded through embryo development, and larval hatching success was determined at 12 days. The exposure continued for an additional 28 days, and surviving larvae were counted on day 40, and length and dry weight were measured at this time.

One of the strengths of the protocol with *A. affinis* is that topsmelt are euryhaline and therefore tolerant of a wide range of salinities. Toxicity Identification Evaluation (TIE) procedures have been developed for topsmelt and other atherinid larvae (U.S. EPA, 1996). Although topsmelt demonstrate comparable or greater sensitivity relative to other atherinid species (Middaugh and Anderson, 1993), their use in water quality assessments may be limited by lack of sensitivity relative to other fish and invertebrate species. Rose et al. (2005) recently showed that otolith growth is a more sensitive indicator of topsmelt growth than the larval weight endpoint recommended in the standard protocol. Topsmelt larvae may be useful as an indicator of unionized ammonia toxicity in estuarine situations (eg, Phillips et al., 2005), and ancillary data suggest topsmelt larvae are particularly sensitive to low dissolved oxygen conditions (Middaugh D, personal communication). Although studies have not been conducted with topsmelt, investigations using other atherinid species, and *C. variegatus*, suggest that larval fish are sensitive to ionic imbalances, and this may confound results of tests with these species (Pillard et al., 2000). Anderson et al.

(1995) found that although topsmelt embryos and larvae are tolerant of salinities ranging from 5‰ to 35‰, experimental evidence suggests that larvae tested at lower salinities (≤17‰) may be more sensitive to contaminants due to osmotic stress.

Two other marine species commonly used in marine toxicological research are mummichog (*Fundulus heteroclitus*; eg, Weis and Weis, 1982) and herring (*Clupea* sp.). Both have been studied extensively although neither has been used for routine National Pollution Discharge Elimination System (NPDES) monitoring. Dinnel et al. (2005) developed three separate protocols with *Clupea pallasi* intended for use in whole effluent toxicity testing in Washington State. One protocol is an 18-day embryo development test which incorporates sublethal endpoints such as heart rate, embryo movement, time to 50% hatch, embryo survival, and larval length at hatch. Larval tests developed with *C. pallasi* incorporate survival and growth endpoints after a 10-day exposure. All procedures use either field-collected herring embryos, or herring gametes harvested from wild-caught fish, or larvae developed from these sources.

Pacific herring embryos and larvae have also been used to assess effects of oil and oil dispersants and these experiments have included standard and biomarker endpoints in laboratory (Carls et al., 1999) and in situ exposures (Kocan, 1996; Kocan et al., 1996). Additional biomarker endpoints have also been reported with other fish embryos (Nacci et al., 1998). There is growing recognition of the sensitivity of fish embryo development to polynuclear aromatic hydrocarbons, for example from oil spills. A number of studies have shown that blue sac disease is a common developmental response when fish embryos are exposed to hydrocarbons (Barron et al., 2004; Incardona et al., 2004, 2005; Hicken et al., 2011). Incardona et al. (2005) found that the tricyclic polycyclic aromatic hydrocarbon (PAH) components of weathered Alaskan North Slope (ANS) crude oil comprised the major toxic fraction to developing zebra fish embryos. Their data suggested phenanthrene and dibenzothiophene were particularly important toxic constituents in weathered oil. These authors reported that the earliest and most pronounced effect of weathered ANS crude oil was impaired cardiac function during cardiac morphogenesis. A recent perspective article summarized the effects of tricyclic PAHs on a diverse array of species from subtropical to sub-Arctic habitats (Scholz and Incardona, 2015) and showed that the characteristics of the cardiac syndrome are consistent across species. In addition to weathered crude oil from oil spills, tricyclic PAHs from multiple sources including vessel fuel spills, creosote, and urban runoff have been shown to cause the same effect. Marine fish embryo development is a particularly important endpoint to consider in environmental settings where PAHs are of concern (Fig. 6.1).

6.3.2 Mollusks

Embryo-larval development tests using gastropod and bivalve mollusks have been used in water and sediment quality assessments for decades. These tests are particularly useful for toxicity monitoring purposes because they require relatively short-term exposures (≤48 h), yet incorporate sensitive, sublethal endpoints that represent critical life stages of ecologically and economically important marine and estuarine species. Bivalve species used in Atlantic and Gulf Coast testing include the eastern oyster *Crassostrea virginica*, the hard shell clam *Mercenaria mercenaria* (ASTM, 2012c) and the coot clam *Mulinia lateralis* (Morrison and Petrocelli, 1990). Protocols have also been developed for several Pacific Coast mollusk species. These include the red abalone (*Haliotis rufescens*; Fig. 6.2), a gastropod mollusk found in marine rocky bottom environments along the Pacific Coast, the estuarine mussel species *Mytilus galloprovincialis* and *Mytilus edulis* (U.S. EPA, 1995), and the coastal mussel *Mytilus*

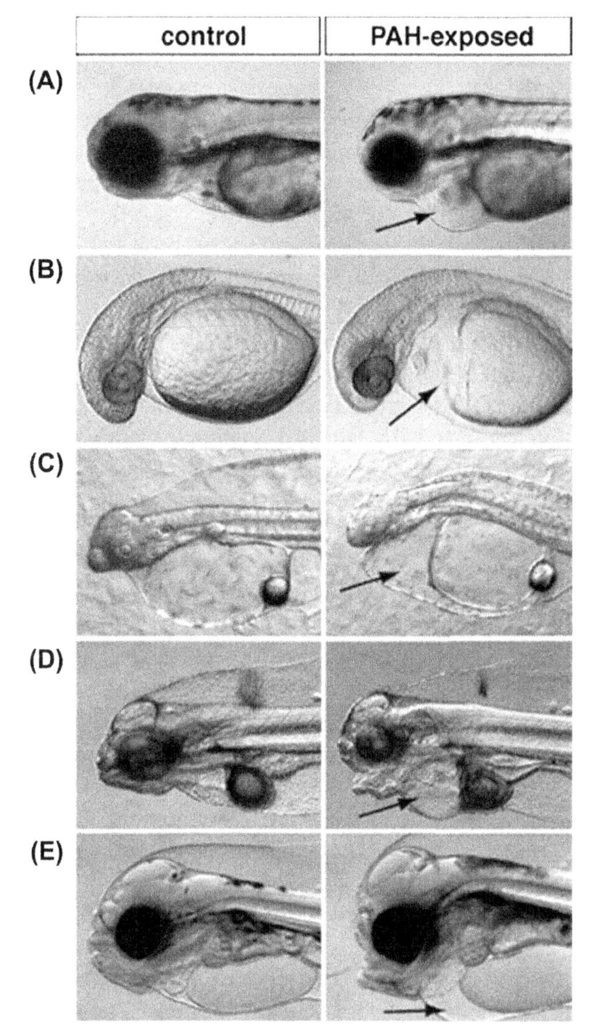

FIGURE 6.1 Polycyclic aromatic hydrocarbons (PAHs) derived from crude oils from different geological sources worldwide cause a consistent heart failure injury phenotype across a diversity of fish species. All images are hatching-stage larvae (A, C–E) or late embryos (B) exposed to crude oil starting shortly after fertilization, with representative controls in the left column and oil-exposed fish in the right column. *Arrows* indicate pericardial fluid accumulation or edema as an indication of circulatory failure: (A) zebrafish (*Danio rerio*) and Louisiana (USA) sweet crude oil; (B) Pacific herring (*Clupea pallasi*) and Alaska North Slope (USA) crude oil, (C) olive flounder (*Paralichthys olivaceus*) and Iranian heavy crude oil, (D) yellowfin tuna (*Thunnus albacares*) and Louisiana (USA) sweet crude oil, and (E) Atlantic haddock (*Melanogrammus aeglefinus*) and Norwegian Sea crude oil.

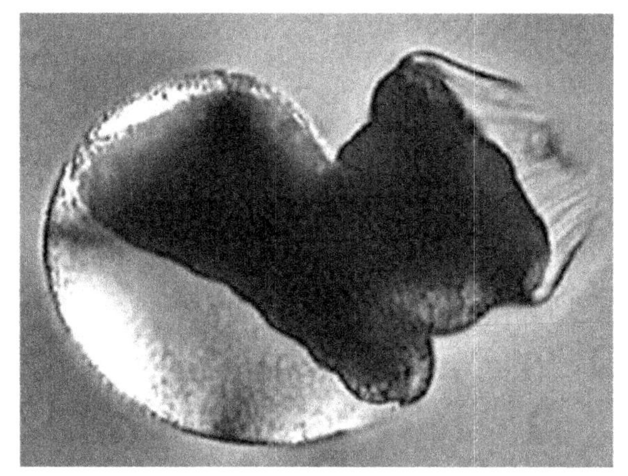

FIGURE 6.2 Photomicrograph of normally developed red abalone (*Haliotis rufescens*) veliger larva at 48 h (400× magnification).

californianus (Cherr et al., 1990). These species are important prey items in their respective habitats and can form large aggregations that provide habitat for other organisms. *M. californianus* and *H. rufescens* are also valued by humans as food items and are harvested commercially and for sport. A review of toxicity testing with marine mollusks is provided in Hunt and Anderson (1993).

Adults of these species are used as brood stock for embryo-larval toxicity tests. Brood stock can be collected from wild populations but are typically provided by commercial suppliers who can culture reliably reproductive individuals. Embryo-larval tests are initiated by inducing gamete release (spawning) in male and female brood stock, and combining the eggs and sperm to produce the embryos used in testing. Spawning is induced differently in these species: *H. rufescens* is spawned in cool,

PAH, Polycyclic aromatic hydrocarbon. *From Scholz, N.L., Incardona J.P., 2015. In response: scaling polyaromatic hydrocarbon toxicity to fish early life stages: a governmental perspective. Environ. Toxicol. Chem. 34, 459–461; Copyright 2015 SETAC; with permission.*

aerated seawater to which hydrogen peroxide and Tris reagent are added, and *Crassostrea* sp. and *Mytilus* sp. are subjected to a warmwater treatment that has been exposed to ultraviolet light. Once a sufficient number of gametes have been produced, eggs and sperm are combined, fertilization occurs, and embryo densities are determined. Tests are initiated within 4 h of fertilization when embryos are in the 2–8 cell stage. Tests with *C. virginica* are conducted at 20°C, tests with *M. mercenaria* and *M. lateralis* are conducted at 25°C, and tests with *Mytilus* sp., *Crassostrea gigas*, and *H. rufescens* are conducted at 15°C.

Tests are typically performed in small, covered, glass containers to which test solution (10–200 mL) is added. Test solutions can consist of marine samples, salted fresh or estuarine samples, seawater/saltwater controls, and reference toxicant controls. Four to five replicate containers of each test solution are inoculated with a known density of embryos that develop into motile larvae over the duration of the test. The final density of these embryos ranges from 10 to 25 per mL, depending on the species being tested. These static, nonrenewal tests are terminated after 48 h by the addition of buffered formalin. The endpoints, percent normal development, and percent survival for *Mytilus* are determined by counting normally and abnormally developed larvae using an inverted compound microscope (Fig. 6.3).

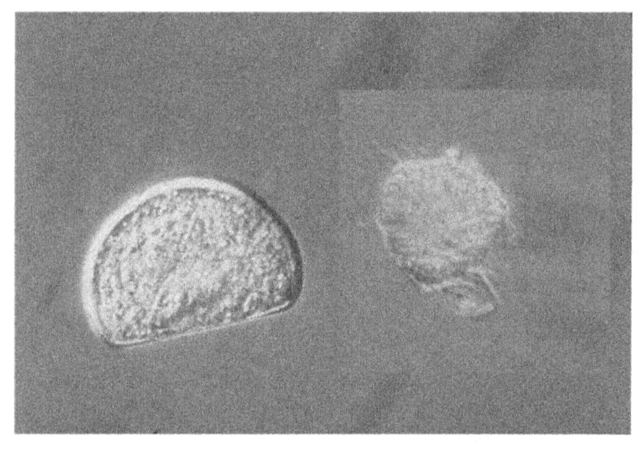

FIGURE 6.3 Photomicrograph of normal (left) and abnormal (right) mussel (*Mytilus galloprovincialis*) veliger larvae at 48 h (400× magnification).

A limited number of studies have investigated the environmental consequences of abnormal embryo-larval development in mollusks. Using the red abalone embryo-larval development test as an example, Hunt and Anderson (1989) demonstrated the ecological consequences of the shell development endpoint. In these experiments, zinc-exposed embryos that did not develop normally shaped veliger larval shells were shown to be incompetent to proceed to the next developmental stage, settlement, and metamorphosis (Fig. 6.4). In similar experiments Conroy et al. (1996) used continuous and pulse-recovery exposures with zinc and bleached kraft mill effluent to show that larval shell abnormalities preclude survival beyond the planktonic stage. In an evaluation of water quality criteria for nickel, Hunt et al. (2002) showed that the abalone embryo-larval development and larval metamorphosis tests were among the most sensitive saltwater tests reported for this metal. An additional strength of the embryo-larval development tests is that they are amenable to TIE procedures (U.S. EPA, 1996). One potential consideration associated with the use of embryo-larval tests in water column toxicity assessments is their sensitivity to unionized ammonia toxicity (Phillips et al., 2005).

6.3.3 Echinoderms

Echinoids (eg, sea urchins and sand dollars) are found in the majority of habitats in the world's oceans and are ecologically important as grazers of marine algae and as food for mammals, fish, and predatory invertebrates. A number of species are also valued by humans as food items and are harvested commercially

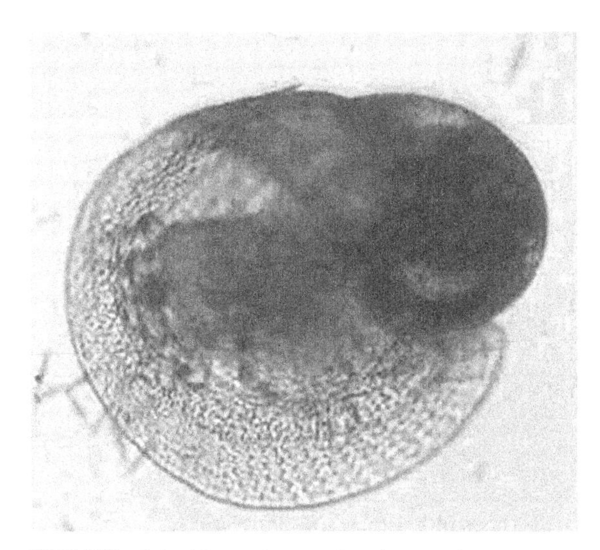

FIGURE 6.4 Photomicrograph of recently metamorphosed juvenile red abalone (*Haliotis rufescens*) after ~12–14 days in culture (400× magnification).

and for sport. Since the introduction of procedures for marine pollution assessment by Kobayashi et al. (1972), toxicity test protocols with sea urchins are now among the most commonly used tests in aquatic toxicology. Two tests are used, one evaluates embryo-larval development, and one evaluates egg fertilization success after the sperm have been exposed to test solutions. These have been summarized (Dinnel et al., 1989; Nacci et al., 1991; Bay et al., 1993), and more recently, the fertilization and development test methods have been described in detail by U.S. EPA (1995).

Embryo-larval tests with sea urchins are conducted following procedures similar to those described above for bivalve mollusks, where adults are induced to spawn, and developmental abnormalities are observed in developing echinoplutei larvae after a 72- to 96-h exposure. Spawning is induced differently in these species: *Arabacia punctulata* is induced to spawn with mild electrical stimulus, whereas *Strongylocentrotus purpuratus* and the sand dollar *Dendraster excentricus* are injected with potassium chloride and the resulting osmotic shock results in gamete release. Once gametes are obtained, eggs are fertilized by sperm using specific sperm-to-egg ratios, defined for each species. Fertilized eggs are then inoculated into test containers, as described above for molluskan embryo development tests. Test containers are usually small-volume glass vials, such as scintillation vials, allowing observation of developed embryos using an inverted microscope.

Echinoid embryo-larval development tests are conducted under static conditions and are terminated after 48 h for *A. punctulata* (20°C) or 72–96 h for *S. purpuratus* (15°C) and *Heliocidaris tuberculata* (Woodworth et al., 1999) by the addition of buffered formalin. The endpoint, percent normal development, is determined by counting normally and abnormally developed larvae (Fig. 6.5).

The sea urchin fertilization test was first developed for eastern Pacific species by Dinnel et al. (1983). This procedure has proven to be a sensitive indicator of effluent and ambient water toxicity. The fertilization test is among the most sensitive to certain chemicals, particularly metals, and is particularly useful as a screening test for large numbers of samples because it can be conducted quickly (eg, Bay et al., 1999). This attribute also makes the fertilization test useful for investigating toxicity of highly volatile or transient chemicals (eg, chlorine; Bay et al., 1993). A number of different echinoid species have been used for this test. On the Atlantic and Gulf Coasts the red urchin *A. punctulata* is used. Although the purple sea urchin (*S. purpuratus*) is the most commonly used species, a number of alternative echinoid species have been employed on the West Coast (*Strongylocentrotus franciscanus*, *Strongylocentrotus droebachiensis*, *D. excentricus*, *Lytechinus pictus*). In addition to these species, a fertilization test has been developed for use

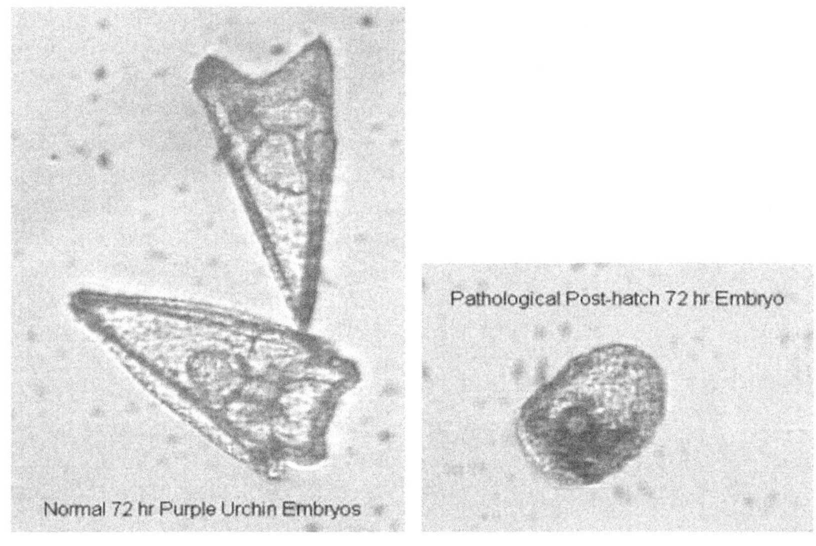

FIGURE 6.5 Photomicrographs of normal (left) and abnormal (right) echinoplutei larva of purple sea urchin (*Strongylocentrotus purpuratus*) after 72 h (400× magnification).

in Hawaiian waters using the tropical sea urchin *Tripneustes gratilla*. A test with this species was first adapted by Nacci and Morrison (1993), updated by Vazquez (2003), and detailed by U.S. EPA (2012). In Australia, a fertilization test has been developed using the temperate species *H. tuberculata* (Woodworth et al., 1999).

This protocol is amenable to TIE procedures and several studies have used the sea urchin fertilization test to identify causes of ambient toxicity (eg, sediment toxicity due to PCBs in New Bedford Harbor (Ho et al., 1997)) and stormwater toxicity due to cationic metals in Southern California coastal waters (Bay et al., 1999). Another positive attribute of this protocol is the high tolerance of echinoid sperm to elevated unionized ammonia concentrations. This characteristic makes this test particularly useful in situations where ammonia may mask toxicity of other contaminants.

Bay et al. (1993) listed a number of limitations with this method. These authors noted toxicity artifacts associated with commercial sea salts and hypersaline brines with tests conducted with *S. purpuratus* and *D. excentricus* and also described the sensitivity of this test to pH extremes. These authors also discussed the occurrence of an unusually high rate of false-positive toxicity results with this method when it has been used to assess ambient toxicity. False-positive results occur when apparently nontoxic samples are identified as toxic.

One additional attribute of tests with echinoid embryos is that these tests have been demonstrated to be amenable to measuring genotoxic effects in addition to the developmental endpoints described above. For example, Anderson et al. (1994) used exposures of the purple sea urchin (*S. purpuratus*) to show that the embryo-development protocol with this species is amenable to inclusion of endpoints that directly measure chromosome damage. These include analysis of anaphase aberrations resulting from chromosome breakage which can be observed in stained embryos using light microscopy. Others have adapted DNA-unwinding protocols (eg, the Comet Assay) for use with echinoid embryos (Shugart et al., 1992).

6.3.4 Mysids

Mysids are small shrimp-like crustaceans that are primary forage prey for marine and estuarine fish. In addition to their ecological importance, they have many attributes that make them ideal test organisms for water quality monitoring. They are sensitive to a wide variety of contaminants; they have relatively short generation times and are amenable to laboratory culture. Since the initial development of protocols (Nimmo et al., 1977), tests with the Atlantic/Gulf coast species *Americamysis* (formerly *Mysidopsis*) *bahia* have been used for regulatory applications for decades. Culture methods for this species are described by Lussier et al. (1988), and life history and ecological descriptions are provided in Rodgers et al. (1986) and U.S. EPA (2002a). Procedures developed for this species include acute, short-term chronic, and chronic life-cycle test protocols. The 7-day short-term chronic test is used for NPDES regulatory monitoring (Lussier et al., 1999; U.S. EPA, 2002b). This test is initiated with 7-day-old preadult mysid neonates (adults usually provided by commercial suppliers or through in-house cultures) and incorporates survival, growth, and fecundity endpoints. In this test, five mysids per replicate test container are exposed to test solutions, and survival is recorded daily. The test is conducted at 26 ± 1°C, mysid neonates are fed twice daily newly hatched *Artemia* nauplii, and test solutions are changed once per day after containers are cleaned. The number of surviving mysids is recorded daily, and the test is terminated after 7 days (survival at 4 days in this test may be used to provide 96 h acute mortality data). Upon test termination, the live animals are examined using a stereomicroscope to determine the number of immature animals, the sex of mature animals, and the presence or absence of eggs in the oviducts or brood sacs of mature females (Fig. 6.6). Of the three endpoints measured, evidence suggests that laboratories have the most difficulty with the fecundity endpoint. Although quantification of fecundity is optional, U.S. EPA notes that this is often the most sensitive endpoint in this test (U.S. EPA, 2002b). Lussier et al. (1999) found that when the mysid holding temperature is maintained at 26–27°C during the pretest period, and holding densities are less than 10 organisms per liter, the percentage of tests meeting the acceptability criteria for fecundity (egg production by ≥50% of control females) increased from 60% to 97%.

A 28-day life-cycle test with *A. bahia* was first described by Nimmo et al. (1977). This test is initiated with newly hatched mysid neonates (<24-h-old), which under flow-through conditions, develop into adults and reproduce during

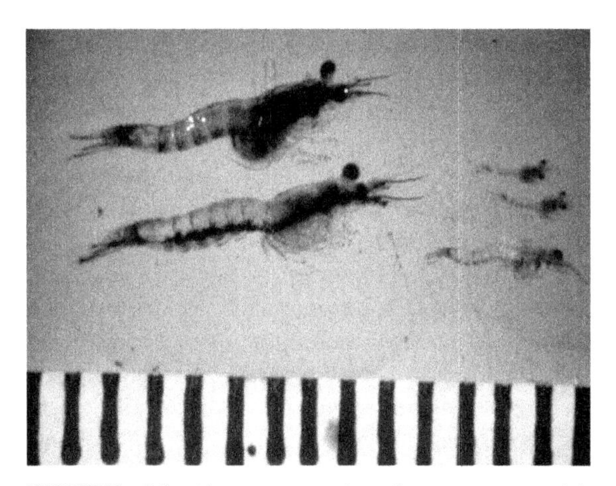

FIGURE 6.6 Photomicrographs of representative life stages and sexes of the mysid *(A. bahia)*. Top left is a mature gravid female mysid with distended marsupium, bottom left is a mature male; top right are two mysid neonates; bottom right is a juvenile. *Photo courtesy of Kay Ho; U.S. EPA.*

the test. The test is terminated following evaluation of reproductive success and survival of the F_1 generation. In this test, mysid neonates develop into sexually mature adults between 10 and 12 days, at which time, females develop brood sacs (marsupia). Eggs are deposited in the brood sacs and as these develop into larvae, the brood sacs grow and darken. Reproductive success is determined by isolating brooding females and counting the number of neonates released. Survival of released neonates may also be determined by isolating them for an additional 96 h. The test is terminated after 28-day exposure of the F_0 mysids, unless reproduction data are desired on the F_1 population. Growth of first generation mysids can also be evaluated by measuring total length and/or dry weight at test termination. Because of size differences between male and female mysids, sex-specific growth is determined. Detailed test guidance and test acceptability criteria for a standardized 28-day life-cycle test with *A. bahia* are provided in ASTM (2008b).

Standardized test protocols have been developed with several other mysid species. *Holmesimysis costata* is a mysid crustacean found in the surface canopy of giant kelp beds off the Pacific Coast, where it serves as an important food source for fish. A 7-day short-term chronic test has been developed for NPDES monitoring in California (U.S. EPA, 1995). This test is initiated with mysid neonates aged 3–4 days old, which are released from adult mysids collected from the field. Five mysids per replicate are exposed to test solutions, each replicated five times, and test solutions are renewed after 48 and 96 h. Daily observations of survival are made so that 96 h and 7 days mortality endpoints can be determined. At the end of 7 days, the test organisms are dried and weighed on a microbalance for determination of a growth endpoint. Other species are often used in mysid short-term tests. *Neomysis mercedis*, another west coast mysid, can be substituted in tests conducted at lower salinities (ASTM, 2008b). *N. mercedis* has been demonstrated to be useful in fresh and brackish water ambient monitoring studies where receiving water conductivities are beyond the range tolerated by other crustacea such as cladocerans (Finlayson et al., 1991; Hunt et al., 1999). Its optimum test salinity range is 1–3‰, but it can survive in the wild in salinities up to 18‰. In addition, Hunt et al. (2002) reported using the Pacific Coast mysid (*Mysidopsis intii*) in acute (96 h) and chronic 28-day whole life-cycle exposures in an evaluation of nickel water quality criteria for marine organisms. Low reproductive success was observed in this experiment, so chronic lowest observed effect concentrations and no observed effect concentrations were based on growth and mortality endpoints. A 7-day short-term chronic test protocol with *M. intii* has been developed, but this test has not been implemented in NPDES monitoring (Harmon and Langdon, 1996; Langdon et al., 1996). Short-term acute tests have been developed for use in Brazilian waters using the species *Mysidopsis juniae* and *Mysidium gracile* (Prosperi et al., 1998).

TIE methods have been developed for *A. bahia* (U.S. EPA, 1996) and *N. mercedis* (Hunt et al., 1999), and tolerances of *H. costata* to various TIE manipulations have been assessed as part of the California State Water Resources Control Board's Marine Bioassay Project (Phillips et al., 2003a). The 7-day growth and survival test with *H. costata* may be limited by test organism availability because commercial suppliers have reported limited availability during winter (Hunt J, personal communication). In addition, some researchers have reported difficulty meeting control performance using this species in 7-day tests (Phillips et al., 2003a). Mysid tests may be confounded by ionic concentrations above or below specific effect thresholds, particularly in certain effluents (eg, produced water and agricultural drain water; Ho and Caudle, 1997). Pillard et al. (2000) developed models to predict the toxicity of elevated major ion concentrations and effects related to their deficiencies

(K^+, Ca^{2+}, Mg^{2+}, $Br^-SO_4^{2-}$, HCO_3^-, $B_4O_7^{2-}$). Use of mysids in ambient toxicity testing is discussed below.

6.3.5 Copepods

Planktonic estuarine and marine copepods are used widely in toxicity testing because they are amenable to laboratory culture conditions, sensitive to toxicants, and ecologically important. Standardized test protocols have been developed for calanoid and harpacticoid species. General guidelines for conducting static acute tests with the calanoid species *Acartia tonsa* were described by Gentile and Sosnowski (1978). In this test, adult copepods are exposed to toxicants in 250 mL crystallizing dishes containing 100 mL of test solution and mortality is recorded after 96 h. Chronic tests with *A. tonsa* have been described by Ward et al. (1979) and more recently by Kusk and Petersen (1997). Ward et al. (1979) reported results of 30-day multiphase toxicity tests with this species that was initiated with adult copepods. After the adult copepods reproduced in the first phase, the resulting nauplii were exposed under flow-through conditions for 20 days in the second phase. The nauplii were allowed to develop to sexual maturity and reproduce, and reproductive success was quantified by counting the number of hatched nauplii under static conditions. Kusk and Petersen (1997) measured mortality of adult *A. tonsa* in acute (48 h) exposures to tributyltin and linear alkylbenzene sulfonate. Results of these tests were compared to 8-day exposures where embryo-larval development was measured. In this test, larval development rate was expressed as the ratio of nauplii hatched to the total number of larvae. Standardized acute and chronic test protocols have also been developed for marine water testing with harpacticoid copepods such as the estuarine species *Tigriopus brevicornis* (Lassus et al., 1984). A protocol to assess copepod immobilization after short-term exposure have been developed using the tropical Australian species *Acartia sinjiensis* and *Gladioferens imparipes* (Evans et al., 1996; Tsvetnenko et al., 1996).

6.3.6 Decapods

Numerous decapod crustaceans have been used in toxicity testing and these are listed in Table 6.4. Previous studies have shown decapods to be sensitive to a wide variety of toxicants, in particular pesticides, and various life stages of this group have been used in ecotoxicity studies. Life stages tested include larvae, megalops, juveniles, and adults. The latter two stages are more commonly used in acute exposures (ASTM, 2007; APHA, 2012). Tests with the earlier life stages are used less commonly because they are less amenable to laboratory testing

TABLE 6.4 Summary of Marine Decapod Crustacea Most Frequently Used in Toxicity Testing

Family	Species
Palaemonidae	*Palaemon adspersus*
	P. macrodactylus
	Palaemonetes pugio
	P. vulgaris
Penaeidae	*Penaeus aztecus*
	P. duorarum
	P. setiferus
	P. stylirostris
Caridea	*Crangon* sp.
Canceridae	*Cancer irroratus*
	C. magister
	C. productus
Portunidae	*Callinectes sapidus*
	Carcinus maenas
Nephropidae	*Homarus americanus*

conditions (Gentile et al., 1984). Grass shrimp (*Palaemonetes* sp.) are among the decapod crustacea most commonly used in toxicology studies and both acute and chronic protocols have been developed using species of this genus. Larvae are obtained by isolating ovigerous (gravid) female shrimp in glass culture bowls and collecting larvae as they are released (Tyler-Schroeder, 1978a). In acute tests, this author exposed shrimp larvae to test solutions in 1-L glass culture dishes under static conditions at 25°C. Larvae were fed newly hatched *Artemia* nauplii daily after test solutions were renewed. Larval mortality was recorded at 96 h. A 145-day life-cycle test with *Palaemonetes* has also been described (Tyler-Schroeder, 1978b, 1979). This test was conducted under flow-through conditions at 20‰ salinity and 25°C and is initiated with 100 juvenile shrimp per treatment (<15 mm rostrum-telson length). In this procedure, juvenile shrimp growth is monitored every 4 weeks until test termination. As shrimp reach the age of sexual maturity at approximately 3 weeks under an 8 h light:16 h dark photoperiod, photoperiod is altered incrementally to reach a final ratio of 14 h light:10 h dark and intensity is increased from 15 to 100 W. As sexual maturation occurs, ovigerous females are isolated from the culture. Reproductive endpoints measured in this test include the number of ovigerous females produced, egg production, and embryo hatching success. Because chronic tests with grass shrimp are labor intensive, Ward (1995) suggested that short-term chronic tests with mysids have largely replaced testing with this group. In Australia, the tropical prawn *Penaeus monodon* has been used in short-term acute toxicity tests (Evans et al., 1996; Tsvetnenko et al., 1996).

Because of their ecological importance in estuarine food webs, grass shrimp are relevant sentinel species in studies of effects of emerging contaminants of concern. A number of these studies have incorporated biomarker or physiological measures using grass shrimp. McKenney (1998) evaluated effects of the growth regulator methoprene on *Palaemonetes pugio* using a combination of developmental, growth, and metabolic responses. In this study, larval grass shrimp were exposed to methoprene and developmental rate was observed by measuring molting success (indicated as presence of exuvia). In addition, selected larvae were sealed in glass syringes during the course of the exposure and oxygen consumption rates, ammonia excretion rates, and growth (as dry wt) were monitored. Results with grass shrimp were compared to similar studies using mud crab (*Rhithropanopeus harrisii*). Others have incorporated biomarker endpoints in studies using grass shrimp. These include measures of DNA damage (Hook and Lee, 2004) and measures of cytochrome P450 (CYP1A) vitellin induction (Oberdorster et al., 2000).

6.3.7 Rotifers

Rotifers are small microscopic invertebrates characterized by a relatively short life cycle which may incorporate a dormant (cyst) stage under stress conditions. These attributes make them amenable to laboratory toxicity testing. An acute toxicity test with the estuarine rotifer (*Brachionus plicatilis*) was described by Snell and Janssen (1998). Cysts of this species are induced to hatch in warm dilute seawater (15‰) 24 h before test initiation, and tests are initiated with newly hatched (neonate) rotifers <2 h old. Acute tests are conducted in multiwell polystyrene tissue culture containers with 10 neonates per well, and mortality is recorded after a 24-h exposure. No chronic test protocols have been developed with this species but a number of sublethal responses have been investigated including behavioral endpoints such as swimming and feeding rate. In addition, biochemical and molecular responses such as enzyme activity and induction of stress proteins have been reviewed by Snell and Janssen (1998).

6.3.8 Cnidarians

The Phylum Cnidaria includes hydras, jellyfish, sea anemones, and corals, and tests with this group were summarized by Ward (1995) (see review by Stebbing and Brown, 1984). The most commonly used tests are those with hydroids (colonial hydrozoans) and corals. One protocol described for colonial marine hydroids (eg, *Laomedea flexuosa, Eirene viridula*) assesses colony growth rate and morphological changes in developing polyps (Karbe et al., 1984; Ward, 1995). In this procedure, colonies are established on glass plates and these are exposed for a minimum of 2 weeks. Colonies are fed *Artemia* twice a week and colony size is determined at the start of the test and on feeding days. Endpoints measured include the number of feeding polyps (hydranths) produced and colony budding rates. Ward (1995) suggested that tests with hydroids have not been emphasized for regulatory applications because their testing and culturing are labor intensive, and most species are not considered to be as ecologically or economically important as other species used in routine testing.

Additional tests with Cnidarians include those developed for scleractinian corals (anthozoans). Laboratory tests with corals are described in Standard Methods (APHA, 2012), and these procedures have been used with a number of species. Microscale tests with corals are discussed in Branton (1998). The primary test species is *Acropora cervicornis* for tropical Atlantic waters because of its wide distribution, demonstrated sensitivity, and similarity to the Indo-Pacific species *Acropora formosa* (Ward, 1995). In the standard test procedure, 20 colonies of uniform initial size (10 g wet wt) are exposed under flow-through conditions, and endpoints measured include survival and several sublethal responses (eg, zooxanthellae extrusion, polyp contraction). Colony growth can be measured as buoyant weighing (Branton, 1998). In addition to tests with the species described above, protocols with a number of coral species have been developed for use in Australian tropical waters. These include a procedure to assess fertilization and larval metamorphosis using *Acropora millepora* (Negri and Heyward, 2000), and a test of chlorophyll fluorescence in the symbiotic dinoflagellates in the coral species *A. formosa, Montipora digitata, Porites cylindrica,* and *Seriatopora hystrix* (Jones et al., 2003).

6.3.9 Marine Algal Toxicity Tests

Thursby et al. (1993) reviewed use of marine and estuarine micro- and macroalgae in toxicity testing and noted that one of the main reasons for including algal test protocols in water quality assessments is that, as primary producers, algae represent the foundation of aquatic food webs. Because of the prevalence of herbicides, fungicides, and other chemicals specifically designed to affect algae in many effluent and ambient samples, algal tests are a necessary component of water quality assessment programs. Lewis (1995) provides a summary of marine and freshwater toxicity tests with microalgae, macroalgae, and vascular plants. The following discussion briefly summarizes species commonly used in standardized protocols.

As part of U.S. EPA efforts to develop marine toxicity test protocols for whole effluent toxicity testing, macroalga species were evaluated, including the red alga *Champia parvula* and the brown alga *Laminaria saccharina*. This research led to development of a standardized sexual reproduction test using *C. parvula* (U.S. EPA, 2002b). This test exposes sexually mature male and female branches of the alga to test solutions for two days, followed by a 5- to 7-day recovery period in control seawater to allow reproduction to occur. This test is included in the U.S. EPA marine toxicity testing manual (U.S. EPA, 2002b) but was not promulgated by U.S. EPA for regulatory monitoring and is not routinely used.

While the U.S. EPA was developing East and Gulf Coast test methods, a 48-h test protocol using spores of the giant kelp, *Macrocystis pyrifera*, was developed to provide an algal toxicity test protocol for marine effluent monitoring on the West Coast of the United States. This protocol is among the most commonly used test protocols in California regulatory monitoring. The test procedure is similar to those developed for *Ulva* species and rockweeds (briefly discussed below) so the methods described for giant kelp can be considered representative. *M. pyrifera* is a large marine alga that forms extensive forests in nearshore areas on the eastern Pacific Coast. These forests are structurally complex and provide habitat and food for numerous species. This kelp has a two-phase life cycle that alternates between the large, spore-forming stage (sporophyte) and the microscopic, gamete-producing stage (gametophyte). Spore-producing fronds are collected from the base of wild plants. These sporophylls are then subjected to cool, dry conditions in the laboratory for 24 h, followed by immersion in seawater, resulting in spore release. Spores are collected, diluted to a known concentration, and inoculated into 200 mL of test solution in 600-mL containers. Test solutions can consist of marine samples, salted fresh or estuarine samples, seawater/saltwater controls, and reference toxicant controls. The static, nonrenewal test proceeds for 2 days, during which time the spores settle and germinate developing into gametophytes. Two endpoints are measured: spore germination success and length of gametophyte germ tubes (Anderson and Hunt, 1988; Anderson et al., 1990). As part of efforts to develop marine toxicity test methods for California, both short-term chronic (48-h) and longer-term reproductive tests (21-day) were developed. Because of the practicalities involved, the 48-h test was adopted for routine effluent testing, while the reproductive test was used to calibrate the relative sensitivity and ecological significance of the 48-h test. A number of studies demonstrated the ecological significance of the 48-h test endpoints with *M. pyrifera* spores. These experiments showed that toxicants that inhibited spore germination and growth also inhibited kelp reproduction (sporophyte production (Anderson et al., 1997), summarized in Thursby et al. (1993) and U.S. EPA (1995)). TIE methods have also been developed for this test (U.S. EPA, 1996).

One potential confounding factor associated with the 48-h test with *M. pyrifera* has been reported. In this test, kelp sporophylls are collected from the field the day before the test is initiated. Sporophylls are shipped to the testing laboratory, where they are then immersed in seawater to induce spore release. The time between sporophyll collection and spore release is typically <24 h. Gully et al. (1999) found that sporophyll storage affected response of the spore germination endpoint in reference toxicant tests with copper. While these authors found no effect on the germ-tube growth endpoint, they suggested possible effects on the germination endpoint may confound interpretation of effluent tests with this protocol by increasing the relative sensitivity of the germ-tube growth endpoint.

One other limitation of this protocol is that it may be less appropriate for testing estuarine samples. *Macrocystis* is a coastal species restricted to rocky subtidal areas. An alternative test for studies concerning algal toxicity in estuarine habitats has been reported by Hooten and Carr (1998). This test is analogous to the protocol for *M. pyrifera* but uses zoospores of the estuarine alga (*Ulva fasciata*). These authors evaluated this test for sediment porewater testing and suggest that because *U. fasciata* spores are relatively sensitive to a number of toxicants and are tolerant of unionized ammonia, this test may be useful in situations where elevated NH_3 is a potential confounding factor. In addition, similar protocols have been developed using rockweed germlings *Fucus vesiculosus* (Brooks et al., 2008), *Fucus spiralis* (Girling et al., 2015), and the green alga (*Ulva intestinalis*) (Girling et al., 2015). These have been shown to be sensitive to metals, and the *U. intestinalis* test is sensitive to antifouling biocides containing mixtures of metals and herbicides (Girling et al., 2015).

6.3.10 Application of Marine Water Column Toxicity Tests in Ambient Monitoring

The protocols listed above have been used extensively for effluent toxicity monitoring and to a lesser extent in ambient water monitoring. In 1995, the Society of Environmental Toxicology and Chemistry (SETAC) convened a workshop in Pellston, Michigan, to evaluate current methods for using whole effluent toxicity tests in effluent and ambient water quality assessments. This workshop consisted of experts from government, industry, and academia who were experienced in issues concerning the use of toxicity tests for these applications. The consensus of the workshop participants was that these test protocols are technically sound when conducted according to U.S. EPA methods. Although the workshop participants concluded that these tests provide useful information on the potential for effluents to impact receiving waters, the application of these tests for marine and estuarine ambient water toxicity monitoring has not been as thoroughly evaluated as in freshwater systems (Grothe et al., 1996). The workshop proceedings identified several areas where more research is needed. Schimmel and Thursby (1996) noted that for a variety of reasons, no studies have been conducted to link ambient toxicity in marine or estuarine receiving waters with impacts on water column or benthic communities in those systems. The relationship between ambient toxicity and receiving system ecological impacts are more difficult to ascertain in these systems because of the complex biotic and abiotic factors that may interact with chemical stressors in estuaries, but some studies have demonstrated links between toxicity and benthic impacts in estuarine systems (eg, Anderson et al., 2014). The Pellston Workshop participants concluded that water column toxicity tests such as the standardized tests should be used in concert with biological assessments and chemical analyses for integrated decision-making.

For effluent testing purposes, the U.S. EPA (1991) recommends a minimum of three of the standardized test protocols listed in their guidance documents be used to screen effluent samples for toxicity. If possible, the test species shall include a fish, an invertebrate, and an aquatic plant, because these groups may respond differently to different classes of toxicants. The practice of including a suite of test species representing different phyla and groups also applies to ambient toxicity studies (U.S. EPA, 1991).

As discussed above, state and federal guidance on application of water column toxicity tests for ambient water quality monitoring suggests that a toxicity screening phase be conducted with a minimum of three species representing a variety of groups including invertebrates, fish, and plants. Subsequent testing can then be done with the most sensitive species. Because protocol sensitivities vary both between and within these groups, selection of appropriate protocols for use in effect characterizations in hazard assessments depends on the chemicals of concern identified in the problem formulation stage of the assessment. For example, relative to mysids, embryo-larval development tests with echinoids and mollusks, and fertilization tests with echinoids are not particularly sensitive to cadmium; therefore, screening with an invertebrate test other than mysids might underestimate ecological risk posed by this metal. Conversely, echinoid and molluskan embryo-larval development tests and fertilization tests are the most sensitive of the standardized protocols to copper and zinc. In situations where these metals are the primary chemicals of concern, these protocols would be more appropriate.

Note that although the embryo-larval development (*M. galloprovincialis*, *S. purpuratus*, *H. rufescens*) tests are often grouped together because they incorporate similar endpoints, these protocols may not respond similarly to all toxicants. For example, Phillips et al. (2003b) found considerable difference between mussel and sea urchin embryos in response to cadmium, copper, zinc, and nickel. Similar differences between sea urchin and bivalve embryos have been reported by others (Gries, 1998).

Like other crustacea (eg, amphipods), mysids are also sensitive to many general biocides (eg, sodium azide, pentachlorophenol) and pesticides, particularly organochlorine, organophosphorus, and pyrethroid pesticides (Clark et al., 1989; Cripe, 1994; Hunt et al., 1997). Mysids are also relatively sensitive to other organochlorine compounds, such as PCBs. The test using *A. bahia* or an alternative species (*N. mercedis* or *H. costata*) would be appropriate when these are the primary contaminants of concern. The 7-day growth and survival test using *H. costata* does not include a reproductive endpoint. If reproductive effects on mysids or other crustacea are of concern in a particular risk assessment, testing with the Gulf Coast species (*A. bahia*) is an appropriate surrogate. Given their sensitivity, mysids, sea urchin fertilization tests, and embryo-larval development tests with some species are also appropriate for assessments associated with some organochlorine pesticides (eg, DDT) and metalloid compounds (eg, TBT). As discussed previously, the sea urchin fertilization test (*S. purpuratus* or *D. excentricus*) is sensitive to a wide variety of toxicants and is particularly useful for screening highly volatile or transient chemicals (eg, chlorine, stormwater; Bay et al., 1993).

In some cases there is insufficient data to determine relative sensitivity of marine water column toxicity test protocols to certain contaminant classes. For example, few comparative studies have been conducted to assess the relative sensitivity of these protocols to PAHs, though recent research has demonstrated that fish embryo cardiovascular development is particularly sensitive to tricyclic PAHs (discussed above). Ancillary research indicates that, because of their apparent sensitivity, protocols using larval marine fish are appropriate for hazard assessments where petroleum hydrocarbons are of concern. For example, Schiff et al. (1992) found that silverside larvae (*M. beryllina*) were among the most sensitive of five protocols tested with produced water (*S. purpuratus* fertilization > *M. beryllina* larval survival > *A. bahia* neonate survival > Microtox > *Neanthes arenaceodentata* survival). In tests assessing the interactive effects of chemical dispersants and oil, Singer et al. (1998) found that topsmelt larvae (*A. affinis*) were sometimes the most sensitive species to the water-accommodated fraction (WAF) of Prudhoe Bay crude oil, compared to abalone embryos (*H. rufescens*) and mysid neonates (*H. costata*). When dispersants were used to chemically enhance the preparation of the Prudhoe Bay WAF, mysids were more sensitive than both other species. It should be noted that fish embryo development was not included in these comparisons and given the recent evidence of the cardiovascular toxicity due to PAHs, these endpoints should be included in evaluations of dispersant effects on oil spills (Carls et al., 1999; Couillard et al., 2005; McIntosh et al., 2010).

Many of these protocols are sensitive to noncontaminant factors and naturally occurring compounds that may confound interpretation of toxicity test results. For example, mysids, fish larvae, and in some cases sea urchin sperm may be affected by ion concentrations above or below effect thresholds (Pillard et al., 2000; Bay et al., 2003). In situations where ambient waters may be influenced by produced water, agricultural drain water, or other sources that may increase ion concentrations (eg, K^+, Ca^{2+}, Mg^{2+}, $Br^-SO_4^{2-}$, HCO_3^-, $B_4O_7^{2-}$), these constituents should be measured and compared to established effect models. In addition, many of these test protocols are sensitive to elevated unionized ammonia. Because all of these protocols are amenable to TIE procedures, they are useful to confirm causes of toxicity, particularly when noncontaminant factors affect the results.

As discussed previously, because of the variable sensitivities of these protocols to contaminants, U.S. EPA recommends testing with multiple protocols representing a variety of phyla and groups. This is especially important where ambient waters may be impacted by complex chemical mixtures. Schimmel et al. (1989) assessed the toxicity of seven different effluents and their receiving waters using five different Atlantic Coast toxicity test protocols (*C. parvula, A. bahia, A. punctulata, M. beryllina, C. variegatus*). Sensitivity to effluents and receiving waters varied between protocols and no one protocol was the most sensitive to every effluent or receiving water sample. In addition to using multiple species in standardized protocols, additional endpoints may be assessed with many of these protocols to provide ancillary information regarding ecological risk. For example, cytogenetic endpoints have been assessed with sea urchin sperm and embryos, and with fish embryos and larvae (Anderson et al., 1994; Kocan et al., 1996).

The integration of toxicity testing, Toxicity Identification Evaluation and Toxicity Reduction Evaluation (TIE/TRE) procedures in NPDES and other monitoring has resulted in significant reduction of toxic inputs from point source pollution (Norberg-King et al., 2005). As point source pollution is reduced, a logical evolution is the application of toxicity tests in programs designed to monitor ambient toxicity resulting from nonpoint pollution. The following provides some examples of regional and statewide ambient monitoring programs that incorporate saltwater toxicity testing and discusses results of more recent studies of stormwater toxicity in California.

6.3.10.1 Ambient Monitoring

Saltwater toxicity tests have been incorporated into regional ambient monitoring programs in the Chesapeake Bay and the San Francisco Estuary, and in a state-wide surface water monitoring program in California. In a review of regional monitoring in Chesapeake Bay conducted between 1990 and 1994, Hall and Alden (1997) discussed toxicity test results using a number of Atlantic Coast species. These included 8-day survival and growth tests with sheepshead minnows (*C. variegatus*), grass shrimp (*Palaemonetes* sp.), copepods (*Eurytemora affinis*), and mysids (*M. bahia*), and 48-h embryo/larval development tests with coot clams (*M. lateralis*). Toxicity test results were synthesized into a "TOX-INDEX" to allow for screening of a large number of sites for ranking of their relative degradation. Results of the water column toxicity tests showed some degree of toxicity in the nine rivers and harbors monitored over the 5-year testing period. Results of water column toxicity tests were combined with sediment toxicity tests and water and sediment chemical analyses and illustrated the utility of using multispecies water and sediment toxicity tests to identify sites most at risk. This approach was used to recommend sites for more comprehensive follow-up studies. In addition to its use in ambient monitoring in the Chesapeake Bay, *A. bahia* has been used (as *M. bahia*) in a number of other studies of ambient toxicity using both laboratory and in situ exposures (Clark, 1989; Schimmel et al., 1989; Kahn et al., 1993).

In an early example of ambient testing with mollusks, Woelke (1967) used in situ exposures with oyster (*C. gigas*) embryos to demonstrate receiving water toxicity in the vicinity of pulp and paper mills in Puget Sound, Washington, and showed a rapid elimination of ambient water toxicity when effluent discharges were stopped. In California, saltwater toxicity testing has been incorporating into the San Francisco Estuary Regional Monitoring Program since

1993. Water column toxicity is monitored using mysid survival and growth (*A. bahia*) and mussel embryo larval development (*M. galloprovincialis*). As with other monitoring programs, results of this testing is integrated with chemical analyses in water and sediments and toxicity testing in sediments to provide long-term trends in water quality in the estuary. Few instances of water toxicity to mysids or bivalve embryos have been observed in the central part of the estuary since testing was initiated. Toxicity testing with mysids has documented decreasing toxicity in stormwater inputs at the margins of the estuary, and this has coincided with decreased use of organophosphate pesticides in watersheds entering the estuary (Anderson et al., 2007).

To meet Clean Water Act requirements and provide comprehensive information on the status of beneficial uses of California's surface waters, The State Water Resources Control Board and the Regional Water Quality Control Boards (collectively referred to as the California Water Boards) introduced the Surface Water Ambient Monitoring Program (SWAMP) in 2001 (http://www.waterboards.ca.gov/swamp/docs/cw102swampcmas.pdf). The program was designed to coordinate a statewide framework of high-quality, consistent and scientifically defensible methods and strategies to improve the monitoring, assessment, and reporting of California's water quality. This monitoring program emphasizes characterizations of watersheds, and most toxicity testing use protocols for water column and sediment species. Saltwater toxicity testing is incorporated into the SWAMP program where freshwater inputs enter estuaries and bays. The saltwater toxicity tests most commonly used in this program include survival and growth tests with larval topsmelt (*A. affinis*), and embryo/larval development tests with bivalves (*M. galloprovincialis*).

6.3.10.2 Stormwater Monitoring

Regulatory concerns in California have recently emphasized nonpoint source pollution, and ambient testing in the marine environment has included monitoring of stormwater entering the nearshore coastal habitats, such as bays and estuaries. A number of recent studies have documented stormwater toxicity using Pacific Coast species. Schiff et al. (2003) monitored toxicity of stormwater entering San Diego Bay, California, using the purple sea urchin (*S. purpuratus*) fertilization test. Laboratory testing of field-collected samples in conjunction with field mapping of the stormwater plume was used to indicate the extent of toxicity in the bay. Chemical analyses and TIEs showed that toxicity was most likely due to zinc. In a similar study using sea urchin fertilization tests, Bay et al. (1999) mapped the extent of sea surface toxicity where Ballona Creek entered the ocean in southern California. These authors found that toxic plumes extending over 4 km^2 were measured in Santa Monica Bay following large storm events. Chemical analyses and TIEs indicated toxicity to sea urchin sperm was again likely due to zinc. These authors conducted synoptic studies of benthic macroinvertebrate communities but did not detect impacts in the offshore environment.

Two studies have used tests with fish embryos and larvae to document stormwater toxicity in California. As part of NPDES stormwater monitoring requirements, Skinner et al. (1998) conducted tests with embryos of inland silversides (*M. beryllina*) and medaka (*O. latipes*) to investigate effects of stormwater from a number of creeks in coastal San Diego County. In addition to egg mortality and larval hatching success, a number of teratogenic endpoints were evaluated. Developmental impacts and/or larval mortality were observed in 74% of the samples tested. Impacts on fish embryos were correlated with metals in these samples, especially copper, lead, and zinc.

Phillips et al. (2004) tested samples from 15 stations entering Monterey Bay in central California as part of the Monterey Bay National Marine Sanctuary First Flush stormwater sampling program. Tests were conducted with larval topsmelt

(*A. affinis*) and bivalves (*M. galloprovincialis*). First flush samples entering this system caused widespread toxicity to topsmelt, and TIEs and chemical analyses of stormwater suggested toxicity was due to high concentrations of copper and zinc. Bivalve (*M. galloprovincialis*) embryo development was also inhibited by the majority of these stormwater samples and in those samples where TIEs were conducted, metals (copper and zinc) were implicated as the cause of toxicity. Toxicity testing of stormwater is becoming a required component of NPDES monitoring in California. When combined with chemical analyses and TIEs, marine toxicity tests provide important data on potential hazards of stormwater discharge, and this information may be used to help identify contaminants of concern, a key step before source control.

6.3.10.3 Transitional Environments

Coastal estuaries are among the most ecologically important and critically threatened habitats in the world. As transitional environments are contaminated by freshwater runoff from coastal watersheds that enter marine systems, they pose a particular challenge for ecotoxicologists. A primary concern for laboratory testing of estuarine water samples is ensuring that the sample salinity is within the salinity tolerance range of the test species. Several approaches are used and these depend on study objectives.

Standard marine test procedures such as those developed by the U.S. EPA include methods for adjusting the salinity of samples using either hypersaline brine from frozen seawater or artificial sea salts (U.S. EPA, 1995). An alternate approach is to use euryhaline marine species for testing waters with low or variable salinity. Examples of euryhaline fish include silversides (eg, *Menidia* sp., *A. affinis*), herring (*Clupea* sp.), and killifish (*Fundulus* sp.). Some silversides may be used for testing samples with salinities between 0‰ and 34‰. Examples of euryhaline invertebrates appropriate for variable salinity water testing include mysids (eg, *A. bahia*), bivalve mussels (eg, *M. galloprovincialis*), and some oyster species (eg, *C. virginica*). Stransky et al. (2014) used laboratory and in situ exposures of mysids and mussels to monitor the transient nature of toxicity caused by urban stormwater in a marine receiving system. Similar studies were conducted by Tait et al. (2014) in a rocky intertidal system impacted by stormwater. In both cases grab samples tested in the laboratory were less variable than in situ exposures, and the in situ exposures better characterized the interaction of variable salinity and contaminant concentrations in the receiving systems. These studies illustrate how the application of standard euryhaline laboratory test organisms for in situ tests can be used to more realistically evaluate the relative toxicity of episodic storm events on nearshore marine systems.

It is also conceivable that some species will be useful for evaluating the interaction of contaminants and elevated salinity conditions, such as might occur when hypersaline brine from ocean desalination plants are discharged into coastal waters. A recent study using eastern Pacific species showed that the embryo-larval stages of bivalves (*M. galloprovincialis*) and red abalone (*H. rufescens*) were particularly sensitive to elevated salinity from brine discharges (Voorhees et al., 2013). This study also showed that two of the estuarine species evaluated (*A. bahia* and *A. affinis*) were more tolerant to elevated salinity than the strictly marine species.

Another approach for assessing toxicity in estuarine habitats is to use freshwater species with higher tolerance to elevated salinity. Studies by Werner et al. (2010) and Deanovic et al. (2013) used laboratory exposures with the freshwater amphipod *Hyalella azteca* to assess impacts of runoff contaminated with pyrethroid pesticides high-conductivity laboratory waters and in ambient samples from the northern San Francisco Estuary. Similar studies with *H. azteca* were conducted in a coastal central

California Estuary impacted by pyrethroid and organophosphate pesticides from agricultural runoff (Anderson et al., 2014). This species is tolerant of salinities up to 15‰ and because of its sensitivity to pesticides is particularly appropriate for monitoring toxicity in transitional environments impacted by nonpoint source pollution.

6.3.10.4 Future Possibilities and Probabilities

Oceanic uptake of atmospheric carbon dioxide causes ocean acidification by altering seawater carbonate chemistry. This process lowers seawater pH and carbonate ion concentration. Ocean temperatures and carbonate chemistry are predicted to change in the coming decades, and ocean acidification has been described as one of the most pervasive environmental changes in the ocean (Feely et al., 2009; Hofmann et al., 2014). Natural ocean circulation brings deep low-pH water to the surface at the coast, but the cause of the reduced pH can be both natural and anthropogenic (Boehm et al., 2015). There is evidence that acidification of coastal waters extends into the nearshore environments and is reaching the rocky intertidal zone (Evans et al., 2013).

Ocean acidification has the potential to impact coastal waters either directly or indirectly by affecting species that rely on the uptake of calcium carbonate for development and growth. Corals, bivalve and gastropod mollusks, and echinoderms are examples of marine organisms which may be impacted by ocean acidification through impacts on exoskeleton and shell development (Hofmann et al., 2014). Toxicity tests with species from these groups are adaptable for assessing the effects of these phenomena.

Recent studies with echinoids and mollusks have been used to establish tolerance thresholds for pH effects on embryo larval development. A number of studies have been conducted with adult organisms (Miller et al., 2014; Sanford et al., 2014; Moulin et al., 2015), but there has also been a focus on the early life stages that are used in toxicity tests. Larval purple urchins (*S. purpuratus*) developed normally under current and predicted ranges of pCO_2 conditions (Kelly et al., 2013; Padilla-Gamino et al., 2013), but the mechanical integrity of larval mussel (*M. californianus*) shells was significantly impacted at low pH (Gaylord et al., 2011). Olympia oysters (*Ostrea lurida*) reared in pH 7.8 water had slower shell growth rates and smaller shells (Hettinger et al., 2012). A study conducted on the endangered northern abalone (*Haliotis kamtschatkana*) demonstrated that a doubling of the current atmospheric concentration of CO_2 caused significant larval mortality, and surviving larvae had approximately 40% abnormal shell development (Crim et al., 2011). Research conducted on marine algal species indicates that the sporophyte stage of giant kelp (*M. pyrifera*) will likely be unaffected by changes in oceanic carbon chemistry but will likely be more affected by changes in temperature and nutrients (Hepburn et al., 2011; Fernández et al., 2015).

Test protocols that incorporate the initial stages of molluskan embryo-larval development with the process of metamorphosis are particularly useful for establishing ecologically relevant effects (Crim et al., 2011). For example, the standard 48-h embryo-larval development test with the gastropod red abalone *(H. rufescens)* can be extended an additional 10–12 days to incorporate metamorphosis (Hunt and Anderson, 1993; Conroy et al., 1996). This has been shown to be a sensitive endpoint to contaminants, particularly metals (Hunt et al., 2002). Because this test incorporates the complex physiological processes of planktonic embryo development, shell formation, larval metamorphosis, and benthic recruitment, this species will also be likely very sensitive to shifts in pH. Because the effects of low pH may also exacerbate effects of anthropogenic contaminants such as cationic metals, future research should address the interactive effects of ocean acidification using marine species sensitive to both metals (and other contaminants) and low pH. This will likely disproportionately impact corals, mollusks, and echinoderms.

There is increasing interest in adapting toxicity tests to account for chemicals of emerging concern (CEC) in freshwater and marine systems. Recent advances in the development of fish and invertebrate cell lines for use in bioanalytical tools have encouraged the application of this approach for screening the potential for chemicals to impact receiving systems. For example, in the United States this approach is being used to screen for endocrine disrupting chemicals in wastewater as well as the potential for these chemicals to impact freshwater ecosystems. This approach is being expanded to nearshore marine systems in the San Francisco Bay (Jayasinghe et al., 2014) and is being recommended elsewhere in California (Dodder et al., 2015). As these methods are adopted for screening of CECs, concurrent studies have been recommended to evaluate how results of cell-based screening assays correlate with acute and chronic whole organism toxicity tests. Correlation studies have been recommended for acute and chronic toxicity tests with marine fish and invertebrate species (Dodder et al., 2015).

References

Anderson, B.S., Hunt, J.W., 1988. Bioassay methods for evaluating the toxicity of heavy metals, biocides, and sewage effluent using microscopic stages of giant kelp *Macrocystis pyrifera* (Agardh): a preliminary report. Mar. Environ. Res. 26, 113–134.

Anderson, B.S., Hunt, J.W., Piekarski, W.J., Phillips, B.M., Englund, M.A., Tjeerdema, R.S., Goetzl, J.D., 1995. Influence of salinity on copper and azide toxicity to larval topsmelt *Atherinops affinis* (Ayres). Arch. Environ. Contam. Toxicol. 29, 366–372.

Anderson, B.S., Hunt, J.W., Turpen, S.L., Coulon, A.R., Martin, M., 1990. Copper toxicity to microscopic stages of *Macrocystis pyrifera*: interpopulation comparisons and temporal variability. Mar. Ecol. Prog. Ser. 68 (1–2), 147–156.

Anderson, B.S., Middaugh, D.P., Hunt, J.W., Turpen, S.L., 1991. Copper toxicity to sperm, embryos, and larvae of topsmelt *Atherinops affinis*, with notes on induced spawning. Mar. Environ. Res. 31, 17–35.

Anderson, B.S., Hunt, J.W., Piekarski, W.J., 1997. Recent advances in toxicity testing methods using kelp gametophytes. In: Wells, P.G., Lee, K., Blaise, C. (Eds.), Microscale Aquatic Toxicology – Advances, Techniques and Practice. Lewis Publishers, Boca Raton, FL, pp. 255–268.

Anderson, B.S., Hunt, J.W., Phillips, B.M., Thompson, B., Lowe, S., Taberski, K., Carr, R.S., 2007. Patterns and trends in sediment toxicity in the San Francisco estuary. Environ. Res. 105 (1), 145–155.

Anderson, B.S., Phillips, B.M., Hunt, J.W., Siegler, K., Voorhees, J.P., Smalling, K., Kuivila, K., Hamilton, M., Ranasinghe, J.A., Tjeerdema, R.S., 2014. Impacts of pesticides in a central California estuary. Environ. Monitor. Assess. 186, 1801–1814.

Anderson, S.L., Hose, J.E., Knezovich, J.P., 1994. Genotoxic and developmental endpoints effects in sea urchins are sensitive indicators of effects of genotoxic chemicals. Environ. Toxicol. Chem. 13, 1033–1041.

APHA, 2012. Standard Methods for the Examination of Water and Wastewater, twenty-second ed. Washington, DC.

ASTM, 2007. Standard guide for conducting acute toxicity tests on test materials with fishes, macroinvertebrates, and amphibians. E729–96. In: Annual Book of American Society of Testing and Materials Standards, vol. 11.06. West Conshohocken, PA, USA.

ASTM, 2008a. Standard guide for conducting acute toxicity tests on aqueous ambient samples and effluents with fishes, macroinvertebrates, and Amphibians. E1192–97. In: Annual Book of American Society of Testing and Materials Standards, vol. 11.06. West Conshohocken, PA, USA.

ASTM, 2008b. Standard guide for conducting life-cycle toxicity tests with saltwater mysids. E 1191–03a. In: Annual Book of American Society of Testing and Materials Standards, vol. 11.06. West Conshohocken, PA, USA.

ASTM, 2012a. Standard guide for acute toxicity test with the rotifer Branchionus. E 1440–91. In: Annual Book of American Society of Testing and Materials Standards, vol. 11.06. West Conshohocken, PA, USA.

ASTM, 2012b. Standard guide for conducting sexual reproduction tests with seaweeds. E1498–92. In: Annual Book of American Society of Testing and Materials Standards, vol. 11.06. West Conshohocken, PA, USA.

ASTM, 2012c. Standard guide for conducting static acute toxicity tests starting with embryos of four species of saltwater bivalve molluscs. E724–98. In: Annual Book of American Society of Testing and Materials Standards, vol. 11.06. West Conshohocken, PA, USA.

ASTM, 2012d. In: Standard Guide for Conducting Static and Flow-Through Acute Toxicity Tests With Mysids From the West Coast of the United States. E1463–92, vol. 11.06, pp. 828–854. West Conshohocken, PA, USA.

ASTM, 2012e. Standard guide for conducting static toxicity tests with microalgae. E1218–1304. In: Annual Book of American Society of Testing and Materials Standards, vol. 11.06. West Conshohocken, PA, USA.

ASTM, 2013. Standard guide for conducting early life-stage toxicity tests with fishes. E12411–1305. In: Annual Book of American Society of Testing and Materials Standards, vol. 11.06. West Conshohocken, PA, USA.

Barron, M.G., Carls, M.G., Heintz, R., Rice, S.D., 2004. Evaluation of fish early lifestage toxicity models of chronic embryonic exposures to complex polycyclic aromatic hydrocarbon mixtures. J. Toxicol. Sci. 78, 60–67.

Bay, S.M., Burgess, R.M., Nacci, D., 1993. Status and applications of echinoid (Phylum Echinodermata) toxicity test methods. In: Landis, W.G., Hughes, J.S., Lewis, M.A. (Eds.), Environmental Toxicology and Risk Assessment, ASTM STP 1179. ASTM, Philadelphia, PA, pp. 281–302.

Bay, S., Jones, B.H., Schiff, K., 1999. Study of the Impact of Stormwater Discharge on Santa Monica Bay. Technical Publication USCSG-TR-02-99. Sea Grant Program, Wrigley Institute of Environmental Studies, University of Southern California, Los Angeles, CA, p. 16.

Bay, S.M., Anderson, B.S., Carr, R.S., 2003. Relative performance of porewater and solid-phase toxicity tests: characteristics, causes, and consequences. In: Carr, R.S., Nipper, M. (Eds.), Porewater Toxicity Testing: Biological Chemical and Ecological Considerations – Methods, Applications, and Recommendations for Future Areas of Research. SETAC Press, Pensacola, FL, USA, pp. 11–36.

Boehm, A.B., Jacobson, M.Z., O'Donnell, M.J., Sutula, M., Wakefield, W.W., Weisberg, S.B., Whiteman, E., 2015. Ocean acidification science needs for natural resource managers of the North American west coast. Oceanography 28, 170–181.

Branton, M., 1998. Microscale bioassays for corals. In: Wells, P.G., Lee, K., Blaise, C. (Eds.), Microscale Testing in Aquatic Toxicology: Advances, Techniques, and Practice. CRC Press, Boca Raton, FL, pp. 371–382.

Brooks, S.J., Bolam, T., Tolhurst, L., Bassett, J., La Roche, J., Waldock, M., Barry, J., Thomas, K.V., 2008. Dissolved organic carbon reduces the toxicity of copper to germlings of the macroalgae, *Fucus vesiculosus*. Ecotoxicol. Environ. Saf. 70, 88–98.

Carls, M.G., Rice, S.D., Hose, J.E., 1999. Sensitivity of fish embryos to weathered crude oil: Part I. Low level exposure during incubation causes malformation, genetic damage, and mortality in larval Pacific herring (*Clupea pallasi*). Environ. Toxicol. Chem. 18, 481–493.

Cherr, G.N., Shoffner-Mcgee, J., Shenker, J.M., 1990. Methods for assessing fertilization and embryonic-larval development in toxicity tests using the California (USA) mussel *Mytilus californianus*. Environ. Toxicol. Chem. 9, 1137–1146.

Clark, J.R., 1989. Field studies in estuarine ecosystems: a review of approaches for assessing contaminant effects. In: Cowgill, U.M., Williams, L.R. (Eds.), Aquatic Toxicology and Hazard Asssessment, vol. 12. American Society for Testing and Materials, Philadelphia, PA, pp. 120–133. STP 1027.

Clark, J.R., Goodman, L.R., Borthwick, P.W., Patrick, J.M., Cripe, G.M., Moody, P.M., Moore, J.C., Lores, E.M., 1989. Toxicity of pyrethroids to marine invertebrates and fish: a literature review and test results with sediment-sorbed chemicals. Environ. Toxicol. Chem. 8, 393–401.

Code of Federal Regulations, 1990. Environmental Effects Testing Guidelines (TSCA). Part 797-Environmental Effects Testing Guidelines, Subpart B – Aquatic Guideline. 40 CFR, pp. 298–392.

Conroy, P.T., Hunt, J.W., Anderson, B.S., 1996. Validation of a short-term toxicity test endpoint by comparison with longer-term effects on larval red abalone *Haliotis rufescens*. Environ. Toxicol. Chem. 15, 1245–1250.

Couillard, C.M., Lee, K., Legare, B., King, T.L., 2005. Effect of dispersant on the composition of the water-accommodated fraction of crude oil and its toxicity to larval marine fish. Environ. Toxicol. Chem. 24, 1496–1504.

Crim, R.N., Sunday, J.M., Harley, C.D.G., 2011. Elevated seawater CO_2 concentrations impair larval development and reduce larval survival in endangered northern abalone (*Haliotis kamtschatkana*). J. Exp. Mar. Biol. Ecol. 400, 272–277.

Cripe, G.M., 1994. Comparative acute toxicities of several pesticides and metals to *Mysidopsis bahia* and postlarval *Penaeus duorarum*. Environ. Toxicol. Chem. 13, 1867–1872.

Deanovic, L., Markewicz, D., Stillway, M., Fong, S., Werner, I., 2013. Comparing the effectiveness of chronic water column tests with the crustaceans *Hyalella azteca* (Order: Amphipoda) and *Ceriodaphnia dubia* (Order: Cladocera) in detecting toxicity of current-use pesticides. Environ. Toxicol. Chem. 32, 707–712.

Dinnel, P.A., Stober, Q.J., Link, J.M., Letourneau, M.W., Roberts, W.E., Felton, S.P., Nakatani, R.E., 1983. Methodology and Validation of a Sperm Cell Toxicity Test for Testing Toxic Substances in Marine Waters. Final Report FRI-UW-8306. Fish. Res. Inst., Schl. of Fish., Univ. of Washington, Seattle, WA, p. 208.

Dinnel, P.A., Link, J.M., Stober, Q.J., Letourneau, M.W., Roberts, W.E., 1989. Comparative sensitivity of sea urchin sperm bioassays to metals and pesticides. Arch. Environ. Contam. Toxicol. 18, 748–755.

Dinnel, P.A., Farren, H.M., Marko, L., Morales, L., 2005. *Clupea pallasi*, Bioassay Protocols: Phase IV. Final technical report. Washington Department of Ecology, Olympia, WA, 26 pp with appendices.

Dodder, N.G., Mehinto, A.C., Maruya, K.A., 2015. Monitoring of Constituents of Emerging Concern (CECs) in

California's Aquatic Ecosystems – Pilot Study Design and QA/QC Guidance. Southern California Coastal Water Research Project Technical Report No. 854, pp. 1–93.

Environment Canada, 2011. Biological Test Method: Fertilization Assay Using Echinoids (Sea Urchins and Sand Dollars). Method Development and Applications Unit Science and Technology Branch Environment Canada Ottawa. Ontario Report EPS 1/RM/27 Second Edition February 2011.

Evans, L.H., Bidwell, J.R., Spickett, J., Rippingdale, R.J., Tsvetnenko, Y., Tsvetnenko, E., 1996. Ecotoxicological Studies in North West Australian Marine Organisms. Final technical report. Curtin University of Technology.

Evans, T.G., Chan, F., Menge, B.A., Hofmann, E., 2013. Transcriptomic responses to ocean acidification in larval sea urchins from a naturally variable pH environment. Mol. Ecol. 22, 1609–1625.

Feely, R.A., Doney, S.C., Cooley, S.R., 2009. Ocean acidification: present conditions and future changes in a high-CO_2 world. Oceanography 22, 37–47.

Fernández, P.A., Roleda, M.Y., Hurd, C.L., 2015. Effects of ocean acidification on the photosynthetic performance, carbonic anhydrase activity and growth of the giant kelp *Macrocystis pyrifera*. Photosynth. Res. 124, 293–304.

Finlayson, B.J., Harrington, J.M., Fojimura, R., Isaacs, G., 1991. Identification of methyl parathion toxicity in Colusa Basin drain water. Environ. Toxicol. Chem. 12, 291–303.

Gaylord, B., Hill, T.M., Sanford, E., Lenz, E.A., Jacobs, L.A., Sato, K.N., Russell, A.D., Hettinger, A., 2011. Functional impacts of ocean acidification in an ecologically critical foundation species. J. Exp. Biol. 214, 2586–2594.

Gentile, J.H., Sosnowski, S.L., 1978. Methods for the culture and short term bioassay of the calanoid copepod (*Acartia tonsa*). In: EPA, U.S. (Ed.), Bioassay Procedures for the Ocean Disposal Permit Program. EPA 600/9–78/010. United States Environmental Protection Agency, Gulf Breeze, FL.

Gentile, J.H., Johns, D.M., Cardin, J.A., Heltshe, J.F., 1984. Marine ecotoxicological testing with crustaceans. In: Persoone, G., Jaspers, E., Claus, C. (Eds.), Ecotoxicological Testing for the Marine Environment. State University of Ghent and Institute for Marine Scientific Research, Bredene, Belgium.

Girling, J.A., Thomas, K.V., Brooks, S.J., Smith, D.J., Shahsavari, E., Ball, A.S., 2015. A macroalgal germling bioassay to assess biocide concentrations in marine waters. Mar. Pollut. Bull. 91, 82–86.

Goodman, L.R., Hansen, D.J., Middaugh, D.P., Cripe, G.M., Moore, J.C., 1985. Method for early life-stage toxicity tests using three atherinid fishes and results with chlorpyrifos. In: Cardwell, R.D., Purdy, R., Bahner, R.C. (Eds.), Aquatic Toxicology and Hazard Evaluation. ASTM STP 854. Americant Society for Testing and Materials, Philadelphia, PA, USA.

Gries, T.H., 1998. Larval Bioassay Workshop Summary Workshop Proceedings Sponsored by the Puget Sound Dredged Materials Management Program. Washington Department of Ecology.

SETAC Pellston Workshop on Whole Effluent Toxicity. In: Grothe, D.R., Dickson, K.L., Reed-Judkins, D.K. (Eds.), 1996. Whole Effluent Toxicity Testing: An Evaluation of Methods and Prediction of Receiving System Impacts. SETAC Press, Pensacola, FL.

Gully, J.R., Bottomley, J.P., Baird, R.B., 1999. Effects of sporophyll storage on giant kelp *Macrocystis pyrifera* (Agardh) bioassay. Environ. Toxicol. Chem. 18, 1474–1481.

Hall, L.W., Alden, R.W., 1997. A review of concurrent ambient water column and sediment toxicity testing in the Chesapeake Bay watershed: 1990–1994. Environ. Toxicol. Chem.

Harmon, V.L., Langdon, C.J., 1996. A 7-D toxicity test for marine pollutants using the Pacific mysid Mysidopsis intii. 2. Protocol evaluation. Environ. Toxicol. Chem. 15, 1824–1830.

Hepburn, C.D., Pritchard, D.W., Cornwall, C.E., McLeod, R.J., Beardall, J., Raven, J.A., Hurd, C.L., 2011. Diversity of carbon use strategies in a kelp forest community: implications for a high CO_2 ocean. Glob. Change Biol. 17, 2488–2497.

Hettinger, A., Sanford, E., Hill, T.M., Russell, A.D., Sato, K.N., Hoey, J., Forsch, M., Page, H.N., Gaylord, B., 2012. Persistent carry-over effects of planktonic exposure to ocean acidification in the Olympia oyster. Ecology 93, 2758–2768.

Hicken, C.E., Linbo, T.L., Baldwin, D.H., Willis, M.L., Meyers, M.S., Holland, L., Larsen, M., Stekoll, M.S., Rice, S.D., Collier, T.K., Scholz, N.L., Incardona, J.P., 2011. Sublethal exposure to crude oil during embryonic development alters cardiac morphology and reduces aerobic capacity in adult fish. Proc. Natl. Acad. Sci. USA 108, 7086–7090.

Ho, K., Caudle, D., 1997. Ion toxicity and produced water. Letter to the Editor Environ. Toxicol. Chem. 16, 1993–1995.

Ho, K.T., McKinney, R.A., Kuhn, A., Pelletier, M.C., Burgess, R.M., 1997. Identification of acute toxicants in New Bedford Harbor sediments. Environ. Toxicol. Chem. 16, 551–558.

Hofmann, G.E., Evans, T.G., Kelly, M.W., Padilla-Gamino, J.L., Blanchette, C.A., Washburn, L., Chan, F., McManus, M.A., Menge, B.A., Gaylord, B., Hill, T.M., Sanford, E., LaVigne, M., Rose, J.M., Kapsenberg, L., Dutton, J.M., 2014. Exploring local adaptation and the ocean acidification seascape studies – in the California current large marine ecosystem. Biogeosciences 11, 1053–1064.

Hook, S.E., Lee, R.F., 2004. Genotoxicant induced DNA damage and repair in early and late developmental stages of the grass shrimp *Paleomonetes pugio* embryo as measured by the comet assay. Aquat. Toxicol. 66, 1–14.

Hooten, R.L., Carr, R.S., 1998. Development and application of a marine sediment pre-water toxicity test using *Ulva fasciata* zoopsores. Environ. Toxicol. Chem. 17, 932–940.

Hunt, J.W., Anderson, B.S., 1989. Sublethal effects o zinc and municipal effluents on larvae of the red abalone *Haliotis rufescens*. Mar. Biol. 101, 545–552.

Hunt, J.W., Anderson, B.S., 1993. From research to routine: a review of toxicity testing with marine molluscs. In: Landis, W.G., Hughes, J.S., Lewis, M.A. (Eds.), Environmental Toxicology and Risk Assessment, ASTM STP 1179. American Society for Testing and Materials, Philadelphia, PA, USA, pp. 320–339.

Hunt, J.W., Anderson, B.S., Turpen, S.L., Englund, M.A., Piekarski, W., 1997. Precision and sensitivity of a seven-day growth and survival toxicity test using the west coast marine mysid crustacean *Holmesimysis costata*. Environ. Toxicol. Chem. 16, 824–834.

Hunt, J.W., Anderson, B.S., Phillips, B.M., Tjeerdema, R.S., Puckett, H.M., deVlaming, V., 1999. Patterns of aquatic toxicity in an agriculturally dominated coastal watershed in California. Agric. Ecosystems Environ. 75, 75–91.

Hunt, J.W., Anderson, B.S., Phillips, B.M., Tjeerdema, R.S., Puckett, H.M., Stephenson, M., Tucker, D.W., Watson, D., 2002. Acute and chronic toxicity of nickel to marine organisms: implications for water quality criteria. Environ. Toxicol. Chem. 21, 2423–2430.

Incardona, J., Collier, T., Scholz, N., 2004. Defects in cardiac function precede morphological abnormalities in fish exposed to polycyclic aromatic hydrocarbons. Toxicol. Appl. Pharmacol. 196, 191–205.

Incardona, J.P., Carls, M.G., Teraoka, H., Sloan, C.A., Collier, T.K., Scholz, N.L., 2005. Aryl hydrocarbon receptor-independent toxicity of weathered crude oil during fish development. Environ. Health Perspect. 113, 1755–1762.

Jayasinghe, S., Kroll, K., Adeyemo, O., Lavelle, C., Denslow, N., Mehinto, A., Bay, S., Maruya, K., 2014. Linkage of *In Vitro* Assay Results with *In Vivo* End Points Final Report – Phase 1. San Francisco Estuary Institute, Richmond. CA. Contribution #734.

Jones, R.J., Muller, J., Haynes, D., Schreiber, U., 2003. Effects of herbicides diuron and atrazine on corals of the Great Barrier Reef, Australia. Mar. Ecol. Prog. Ser. 251, 153–167.

Kahn, A.A., Barbieri, J., Khan, S.A., Sweeney, F.P., 1993. Toxicity of ambient waters to the estuarine mysid *Mysidposis bahia*. In: Landis, W.G., Hughes, J.S., Lewis, M.A. (Eds.), Environmental Toxicology and Risk Assessment, ASTM STP 1179. American Society for Testing and Materials, Philadelphia, PA, USA, pp. 405–412.

Karbe, L., Borchardt, T., Dannenberg, R., Meyer, E., 1984. Ten years of experience using marine and freshwater hydroid bioassays. In: Persoone, G., Jaspers, E., Claus, C. (Eds.), Ecotoxicological Testing for the Marine Environment. State University of Ghent and Institute for Marine Scientific Research, Bredene, Belgium.

Kelly, M.W., Grosberg, R.K., Sanford, E., 2013. Trade-offs, geography, and limits to thermal adaptation in a tide pool copepod. Am. Nat. 181, 846–854.

Kobayashi, N., 1972. Marine pollution bioassay by using sea urchin eggs in the inland Sea of Japan. Publ. Seto Mar. Biol. Lab. 19, 359–381.

Kocan, R.M., 1996. Fish embryos as in situ monitors of aquatic pollution. In: Ostrander, G.K. (Ed.), Techniques in Aquatic Toxicology. CRC Press, Boca Raton, FL, USA, pp. 73–92.

Kocan, R.M., Hose, J.E., Brown, E.D., Baker, T.T., 1996. Pacific herring (*Clupea pallasi*) embryo sensitivity to Prudhoe Bay petroleum hydrocarbons: laboratory evaluation and in situ exposure at oiled and unoiled sites in Prince William Sound. Can. J. Fish. Aquat. Sci. 53, 2366–2375.

Kusk, K.O., Petersen, S., 1997. Acute and chronic toxicity of tributyltin and linear alkylbenzene sulfonate to the marine copepod *Acartia tonsa*. Environ. Toxicol. Chem. 16.

Langdon, C.J., Harmon, V.L., Vance, M.M., Kreeger, K.E., Kreeger, D.A., Chapman, G.A., 1996. A 7-d toxicity test for marine pollutants using the Pacific mysid *Mysidopsis intii*. 1. Culture and protocol development. Environ. Toxicol. Chem. 15, 1815–1823.

Lassus, P., Le Baut, C., Le Dean, L., Bardouil, M., Truquet, P., Bocquene, G., 1984. Marine ecotoxicological tests with zooplankton. In: Persoone, G., Jaspers, E., Claus, C. (Eds.), Ecotoxicological Testing for the Marine Environment, vol. 2. State University of Ghent and Institute for Marine Scientific Research, Bredene, Belgium.

Lewis, M.A., 1995. Algae and vascular plant tests. In: Rand, G.A. (Ed.), Fundamentals of Aquatic Toxicology, second ed. Taylor and Francis, Washington, DC, pp. 135–169.

Lussier, S.M., Kuhn, A., Chammas, M.J., Sewell, J., 1988. Techniques for the laboratory culture of *Mysidopsis* spp. Crustacea Mysidacea. Environ. Toxicol. Chem. 7, 969–978.

Lussier, S.M., Kuhn, A., Comeleo, R., 1999. An evaluation of the seven-day toxicity test with *Americamysis bahia* (formerly *Mysidopsis bahia*). Environ. Toxicol. Chem. 18, 2888–2893.

McIntosh, S., King, T., Wu, D., Hodso, P., 2010. Toxicity of dispersed weathered crude oil to early life stages of Atlantic herring (*Clupea harengus*). Environ. Toxicol. Chem. 29, 1160–1167.

McKenney Jr., C.L., 1998. Physiological dysfunction in estuarine mysids and larval decapods with chronic pesticide exposure. In: Wells, P.G., Lee, K., Blaise, C., Gauthier, J. (Eds.), Microscale Tesing in Aquatic Toxicology, Advances, Techniques, and Practice. CRC Press, Boca Raton, FL, pp. 465–478.

McNulty, H.R., Anderson, B.S., Hunt, J.W., Turpen, S.L., Singer, M.M., 1994. Age-specific toxicity of copper to larval topsmelt *Atherinops affinis*. Environ. Toxicol. Chem. 3, 487–492.

Middaugh, D.P., Anderson, B.S., 1993. Utilization of topsmelt, *Atherinops affinis*, in environmental toxicology studies along the Pacific coast of the United States. Rev. Environ. Toxicol. 5, 1–49.

Miller, S.H., Zarate, S., Smith, E.H., Gaylord, B., Hosfelt, J.D., Hill, T.M., 2014. Effect of elevated pCO(2) on metabolic responses of porcelain crab (*Petrolisthes cinctipes*) larvae exposed to subsequent salinity stress. PLoS One 9, e109167.

Morrison, G., Petrocelli, E., 1990. Short-Term Methods for Estimating the Chronic Toxicity of Effluents and Receiving Waters to Marine and Estuarine Organisms: Supplement: Test Method for Coot Clam, *Mulinia lateralis*, Embryo/larval Test. Draft Report. U.S. Environmental Protection Agency, Narragansett, RI.

Moulin, L., Grosjean, P., Leblud, J., Batigny, A., Collard, M., Dubois, P., 2015. Long-term mesocosms study of the effects of ocean acidification on growth and physiology of the sea urchin *Echinometra mathaei*. Mar. Environ. Res. 103, 103–114.

Nacci, D., Comeleo, P., Petrocelli, E., Kuhn-Hines, A., Modica, G., Morrison, G., 1991. Performance evaluation of sperm cell toxicity test using the sea urchin, *Arbacia punctulata*. In: Mayes, M.A., Barron, M.G. (Eds.), Aquatic Toxicology and Risk Assessment, vol. 14. American Society for Testing and Materials, Philadelphia, PA, USA, pp. 324–336.

Nacci, D., Morrison, G.E., 1993. Standard Operating Procedures for Conducting a Sperm Toxicity Test Using the Hawaiian Sea Urchin *Tripneustes gratilla*. Environmental Research Laboratory – Narragansett Contribution 1516. U.S. Environmental Protection Agency, Narragansett, RI.

Nacci, D., Coiro, L., Kuhn, A., Champlin, D., Munn Jr., W., Specker, J., Cooper, K., 1998. Non-destructive indicator of ethoxyresorufin-O-Deethylase activity in embryonic fish. Environ. Toxicol. Chem. 17, 2481–2486.

Negri, A.P., Heyward, A.J., 2000. Inhibition of fertilization and larval metamorphosis of the coral *Acropora millepora* (Ehrenberg, 1834) by petroleum products. Mar. Poll. Bull. 41, 420–427.

Nimmo, D.R., Bahner, L.H., Rigby, R.A., Sheppard, J.M., Wilson, A.J.J., 1977. *Mydisopsis bahia*, an estuarine species suitable for life cycle toxicity tests to determine the effects of a pollutant. Special technical publication, no. 634. In: Mayer, F.L., Hamelink, J.L. (Eds.), Aquatic Toxicology and Hazard Evaluation. Proceedings of the First Annual Sumposium American Society for Testing and Materials, Memphis, TN, pp. 109–116.

Norberg-King, T.J., Ausley, L.W., Burton, D.T., Goodfellow, W.L., Miller, J.L., Waller, W.T. (Eds.), 2005. Toxicity Reduction and Toxicity Identification Evaluations for Effluents, Ambient Waters, and Other Aqueous Media. Society of Environmental Toxicology and Chemistry (SETAC), Pensacola, FL.

Oberdorster, E., Brouwer, M., Hoexum-Brouwer, T., Manning, S., McLachlan, 2000. Long-term pyrene exposure of grass shrimp *Palaemonetes pugio*, affects molting and reproduction of exposed males and offspring of exposed females. Environ. Health Perspect. 108, 1–9.

Padilla-Gamino, J.L., Kelly, M.W., Evans, J.M., Hofmann, G.E., 2013. Temperature and CO_2 additively regulate physiology, morphology and genomic responses of larval sea urchins, *Strongylocentrotus purpuratus*. Proc. R. Soc. B 280 (1759), 20130155.

Phillips, B.M., Nicely, P.A., Anderson, B.S., Hunt, J.W., Tjeerdema, R.S., 2003a. Marine Bioassay Project Eleventh Report. State Water Resources Control Board, Sacramento, CA.

Phillips, B.M., Nicely, P.A., Hunt, J.W., Anderson, B.S., Tjeerdema, R.S., Palmer, S.E., Palmer, F.H., Puckett, H.M., 2003b. Toxicity of cadmium-copper-nickel-zinc mixtures to larval purple sea urchins (*Strongylocentrotus purpuratus*). Bull. Environ. Contam. Toxicol. 70, 592–599.

Phillips, B.M., Anderson, B.S., Hunt, J.W., Tjeerdema, R.S., Beegan, C., Palmer, F.H., 2004. Marine Bioassay Project Twelfth Report. State Water Resources Control Board, Sacramento, CA.

Phillips, B.M., Nicely, P.A., Hunt, J.W., Anderson, B.S., Tjeerdema, R.S., Palmer, F.H., 2005. Tolerance of five west coast marine toxicity test organisms to ammonia. Bull. Environ. Contam. Toxicol. 75, 23–27.

Pillard, D.A., DuFresne, D.L., Caudle, D.D., Tietge, J.E., Evans, J.M., 2000. Predicting the toxicity of major ions in seawater to mysid shrimp (*Mysidopsis bahia*), sheepshead minnow (*Cyprinodon variegatus*), and inland silverside minnow (*Menidia beryllina*). Environ. Toxicol. Chem. 19, 183–191.

Prosperi, V.A., Bertolettia, E., Buratini, S.V., 1998. Toxicity tests with different age groups of *Mysidopsis juniae*. In: Conference Proceeding. Society of Environmental Toxicology and Chemistry (SETAC) 19th Annual Meeting. Charlotte, NC.

Rand, G.M., Petrocelli, S.R., 1985. Fundamentals of Aquatic Toxicology: Methods and Applications. Taylor and Francis, Washington, DC.

Rodgers, J.H., Dorn, P.B., Duke, T., Parrish, R., Venables, B., 1986. Mysidopsis sp.: Life History and Culture. Workshop Report. Gulf Breeze, FL. American Petroleum Institute, Washington, DC.

Rose, W.L., Hobbs, J.A., Nisbet, R.M., Green, P.G., Cherr, G.N., Anderson, S.L., 2005. Validation of otolith growth rate analysis using cadmium-exposed larval topsmelt (*Atherinops affinis*). Environ. Toxicol. Chem. 24, 2612–2620.

Sanford, E., Gaylord, B., Hettinger, A., Lenz, E.A., Meyer, K., Hill, T.M., 2014. Ocean acidification increases the vulnerability of native oysters to predation by invasive snails. Proc. R. Soc. Biol. Sci. Ser. B 281, 20132681.

Schiff, K.C., Greenstein, D.J., Anderson, J.W., Bay, S.M., 1992. A comparative evaluation of produced water toxicity. In: Ray, J.P., Engelhart, F.R. (Eds.), Produced Water. Plenum Press, New York, NY, pp. 199–207.

Schiff, K., Bay, S.M., Diehl, D.W., 2003. Stormwater toxicity in Chollas creek and san Diego Bay, California. Environ. Monit. Assess. 81, 119–132.

Schimmel, S.C., Morrison, G.E., Heber, M.A., 1989. Marine complex effluent toxicity test program: test sensitivity, repeatability and relevance to receiving water toxicity. Environ. Toxicol. Chem. 8, 739–746.

Schimmel, S.C., Thursby, G.B., 1996. Predicting receiving system impacts from effluent toxicity: a marine perspective. In: Grothe, D.R., Dickson, K.L., Reed-Judkins, D.K. (Eds.), Whole Effluent Toxicity Testing: An Evaluation of Methods and Prediction of Receiving Water Impacts. SETAC Pellston Workshop on Whole Effluent Toxicity. SETAC Press, Pensacola, FL, USA, pp. 322–330.

Scholz, N.L., Incardona, J.P., 2015. In response: scaling polyaromatic hydrocarbon toxicity to fish early life stages: a governmental perspective. Environ. Toxicol. Chem. 34, 459–461.

Shugart, L., Bickham, J., Jakim, G., McMahon, G., Ridley, W., Stein, J., Stenert, S.M., 1992. DNA alterations. In: Huggett, R.J., Kimerle, R.A., Mehrle, P.M., Berman, H.L. (Eds.), Biomarkers: Biochemical, Physiological, and Histological Markers of Anthropogenic Stress. CRC Press, Boca Raton, FL, pp. 125–153.

Singer, M.M., George, S., Lee, I., Jacobson, S., Weetman, L.L., Blondina, G., Tjeerdema, R.S., 1998. Effects of dispersant treatment on the acute toxicity of petroleum hydrocarbons. Arch. Environ. Contam. Toxicol. 34, 177–187.

Skinner, L., dePeyster, A., Schiff, K., 1998. Developmental effects of urban stormwater in San Diego County, California in medaka (*Oryzias latipes*) and inland silverside (*Menidia beryllina*). Arch. Environ. Contam. Toxicol. 37, 227–235.

Snell, T.W., Janssen, C.R., 1998. Microscale toxicity testing with rotifers. In: Wells, P.G., Lee, K., Blaise, C., Gauthier, J. (Eds.), Microscale Testing in Aquatic Toxicology, Advances, Techniques, and Practice. CRC Press, Boca Raton, FL, pp. 409–422.

Stebbing, A.R.D., Brown, B.E., 1984. Marine ecotoxicological test with coelenterates. In: Persoone, G., Jaspers, E., Claus, C. (Eds.), Ecotoxicological Testing for the Marine Environment. State University of Ghent and Institute for Marine Scientific Research, Bredene, Belgium.

Stransky, C., Rosen, G., Colvin, M., Dolecal, R., Cibor, A., Tait, K., 2014. In situ storm water impact assessment in San Diego Bay, CA, USA. In: Proceedings, 35th Annual Meeting of the Society of Environmental Toxicology and Chemistry (SETAC). British Columbia, Vancouver.

Tait, K.J., Stransky, C., Wells, D., Kolb, R., Sonksen, A., Cibor, A., 2014. Wet weather receiving water evaluation of a rocky intertidal area of special biological significance in La Jolla, CA, USA. In: Proceedings, 35th Annual Meeting of the Society of Environmental Toxicology and Chemistry (SETAC). British Columbia, Vancouver.

Thursby, G.B., Anderson, B.S., Walsh, G.E., Steele, R.L., 1993. A review of the current status of marine algal toxicity testing in the United States. In: Landis, W.G., Hughes, J.S., Lewis, M.A. (Eds.), Environmental Toxicology and Risk Assessment, ASTM STP 1179. American Society for Testing and Materials, Philadelphia, PA, USA, pp. 362–377.

Tsvetnenko, Y.B., Evans, L.H., Gorrie, J., 1996. Toxicity of the Produced Formation Water to Three Marine Species. Final Technical Report Prepared for Ampolex Ltd. Curtin University of Technology.

Tyler-Schroeder, D.B., 1978a. Entire life-cycle toxicity test using grass shrimp (*Palaemonetes pugio* Holthuis). In: EPA, U.S. (Ed.), Bioassay Procedures for the Ocean Disposal Permit Program. EPA 600/9–78/010. United States Environmental Protection Agency, Gulf Breeze, FL.

Tyler-Schroeder, D.B., 1978b. Static bioassay procedure using grass shrimp (*Palaemonetes* sp.) larvae. In: EPA, U.S. (Ed.), Bioassay Procedures for the Ocean Disposal Permit Program. EPA 600/9–78/010. United States Environmental Protection Agency, Gulf Breeze, FL.

Tyler-Schroeder, D.B., 1979. Use of the grass shrimp (*Palaemonetes pugio*) in a life-cycle toxicity test. In: Marking, L.L., Kimerle, R.A. (Eds.), Aquatic Toxicology. ASTM STP 667. American Society for Testing and Materials, Philadelphia, PA, USA, pp. 153–170.

U.S. EPA, 1985a. Standard Evaluation Procedure: Acute Toxicity Test for Estuarine and Marine Organisms (Estuarine Fish 96-hour Acute Toxicity). EPA-540/9-85-009.

U.S. EPA, 1985b. Standard Evaluation Procedure: Acute Toxicity Test for Estuarine and Marine Organisms (Mollusk 48-hour Embryo Larvae Study). EPA-540/9-85-012.

U.S. EPA, 1985c. Standard Evaluation Procedure: Acute Toxicity Test for Estuarine and Marine Organisms (Mollusk 96-hour Flow-Through Shell Deposition Study). EPA-540/9-85-011.

U.S. EPA, 1985d. Standard Evaluation Procedure: Acute Toxicity Test for Estuarine and Marine Organisms (Shrimp 96-hour Acute Toxicity). EPA-540/9-85-010.

U.S. EPA, 1986a. Non-target Plants: Growth and Reproduction of Aquatic Plants – Tiers 1 and 2. EPA-540/9-86-134. Hazard Evaluation Division.

U.S. EPA, 1986b. Standard Evaluation Procedure: Fish Early Life-Stage. EPA-540/9-86-138. Hazard Evaluation Division.

U.S. EPA, 1986c. Standard Evaluation Procedure: Fish Life-Cycle Toxicity Tests. EPA-540/9-86-137. Hazard Evaluation Division.

U.S. EPA, 1991. Technical Support Document for Water Quality-Based Toxics Control. Office of Water, Washington, DC. EPA/505/2–90/001.

U.S. EPA, 1995. Short-Term Methods for Estimating the Chronic Toxicity of Effluents and Receiving Waters to West Coast Marine and Estuarine Organisms. EPA/600/R-95/136. Office of Research and Development, Washington DC, USA.

U.S. EPA, 1996. Marine Toxicity Identification Evaluation (TIE), Phase I Guidance Document. EPA/600/R-95/054. Office of Research and Development, Washington, DC.

U.S. EPA, 2002a. Methods for Measuring Acute Toxicity of Effluents and Receiving Water to Freshwater and Marine Organisms. EPA-821-R-02–012. Office of Research and Development, Washington, DC.

U.S. EPA, 2002b. Short-Term Methods for Estimating the Chronic Toxicity of Effluents and Receiving Waters to Marine and Estuarine Organisms. EPA-821-R-02–014. Office of Water, Washington, DC, USA.

U.S. EPA, 2012. Tropical Collector Urchin, *Tripneustes gratilla*, Fertilization Test Method. EPA/600/R-12/022. Office of Research and Development, Washington, DC, USA.

Vazquez, L.C., 2003. Effect of sperm cell density on measured toxicity from the sea urchin *Tripneustes gratilla* fertilization bioassay. Environ. Toxicol. Chem. 22, 2191–2194.

Voorhees, J.P., Phillips, B.M., Anderson, B.S., Siegler, K., Katz, S., Jennings, L.L., Tjeerdema, R.S., 2013. Hypersalinity toxicity thresholds for nine California ocean plan toxicity test protocols. Arch. Environ. Contam. Toxicol. 65, 665–670.

Ward, G.S., 1995. Saltwater tests. In: Rand, G.M. (Ed.), Fundamentals of Aquatic Toxicology: Effects, Environmental Fate, and Risk Assessment, second ed. Taylor and Francis, Washington, DC.

Ward, T.J., Rider, E.D., Drozdowski, D.A., 1979. A chronic toxicity test with the marine copepod *Acartia tonsa*. In: Marking, L.L., Kimerle, R.A. (Eds.), Aquatic Toxicology. ASTM STP 667. American Society for Testing and Materials, Philadelphia, PA, pp. 148–158.

Weis, P., Weis, J.S., 1982. Toxicity of methylmercury, mercuric chloride, and lead in killifish (*Fundulus heteroclitus*) from Southampton, New York. Environ. Res. 28, 364–374.

Werner, I., Deanovic, L.A., Markewicz, D., Khamphanh, M., Reece, C.K., Stillway, M., Reece, C., 2010. Monitoring acute and chronic water column toxicity in the northern Sacramento-San Joaquin Estuary, California, USA, using the euryhaline amphipod, *Hyalella azteca*: 2006 to 2007. Environ. Toxicol. Chem. 29, 2190–2199.

Woelke, C.E., 1967. Measurement of water quality with the Pacific Oyster embryo bioassay. Special Technical Publication 416. American Society for Testing and Materials, Philadelphia, PA, pp. 112–120.

Woodworth, J.G., King, C., Miskiewicz, A.G., Laginestra, E., Simon, J., 1999. Assessment of the comparative toxicity of sewage effluent from 10 sewage treatment plants in the area of Sydney, Australia using an amphipod and two sea urchin bioassays. Mar. Pollut. Bull. 39, 174–178.

CHAPTER

7

Sediment Toxicity Testing

S.L. Simpson[1], O. Campana[2], K.T. Ho[3]

[1]CSIRO Land and Water, Sydney, NSW, Australia; [2]University of York, York, United Kingdom; [3]U.S. Environmental Protection Agency, Narragansett, RI, United States

7.1 INTRODUCTION

Sediments are the ultimate repository for many contaminants that enter water bodies from urban runoff, agriculture, and industries. Sediments in estuaries and coastal marine environments are a highly valued component of aquatic ecosystems, providing critical habitat for benthic species and early life stages of pelagic species that support the broader marine food chain. This creates a need to understand the forms, fate, and effects of contaminants in sediments.

Sediment toxicity tests that expose organisms to sediments under controlled conditions give an estimation of the level of toxicity that contaminated sediments pose to organisms in the field (ASTM, 2008a; Greenstein et al., 2008; Kennedy et al., 2009; Simpson and Spadaro, 2011; Rodríguez-Romero et al., 2013). Levels of toxicity can only be inferred from chemical analyses, where the measured contaminant concentrations may be compared to sediment quality guidelines to classify sediments as having a low or high likelihood of causing adverse effects. However, chemical analyses do not assess the potential effects from unmeasured contaminants, contaminant mixtures, and noncontaminants (grain size or other sediment characteristics) that may interact to contribute to effects. While chemical analyses may provide information on the potential bioavailability of contaminants, sediment toxicity test responses reflect the bioavailable fraction of contaminants, the cumulative effects of contaminant mixtures, and the interaction of noncontaminant stressors because they directly affect the organisms tested. Sediment toxicity tests also frequently provide more quantifiable evidence of effects of contaminants than benthic ecology assessments (eg, community analyses, species diversity, abundance, and function) because of difficulty in interpreting community endpoints (Johnston and Roberts, 2009; Burton and Johnston, 2010; Dafforn et al., 2012; Schleckat et al., 2015).

Sediment toxicity testing is undertaken for a range of assessment purposes, and most often

Marine Ecotoxicology
http://dx.doi.org/10.1016/B978-0-12-803371-5.00007-2

these are regulatory in nature. Common applications include:

1. the setting of benchmarks (eg, guidelines, criteria, standards) for existing chemicals using toxicity databases with matching concentration data and for new chemicals that are proposed for use under frameworks such as Regulation on Registration, Evaluation, Authorization and Restriction of Chemicals, EU;
2. investigating interactions between contaminants and environmental variables, for example elevated ammonia concentration arising from excessive nutrients, and predicted changes in climatic conditions (eg, temperature);
3. comparing the sensitivities and relative routes of exposure of different organisms to contaminants;
4. determining relationships between contaminant concentrations, exposure, bioavailability, and toxicity;
5. determining the spatial distribution and ranking of toxic sediments at contaminated field sites for the purpose of management, including dredged sediment disposal;
6. developing site-specific management limits for sediment contaminants or potential remedial actions and assessing their effectiveness; and
7. input into ecological risk assessments to focus assessment outcomes.

Sediment toxicity testing has advanced considerably in the past 20 years, with tests now providing greater environmental relevance. There is better understanding of species sensitivity to contaminants, organism behavior, and exposure pathways, and design considerations that provide more relevant exposure conditions. There has also been a gradual transition from use of mostly acute lethality tests to methods that consider a wide range of possible sublethal and chronic responses (Scarlett et al., 2007a; Kennedy et al., 2009; Simpson and Spadaro, 2011; Fox et al., 2014; Simpson et al., 2016). In addition, there is a greater range of approaches used to assess toxicity, for example cellular responses measured using biomarkers or genomic endpoints. These cellular responses may be used independently or in concert with traditional sediment toxicity test assessments to better understand organism survival, development, and reproduction. As toxicity test procedures can have a major influence on the sediment quality assessment outcome, to achieve sound test outcomes there is a need to ensure both good test design, measurements, and reporting (Harris et al., 2014). To apply sediment toxicity tests effectively, there is a need for rigor, an understanding of the limitations of various methods (confounding factors and operationally defined boundaries), and how laboratory-based exposures of surrogate organisms may not always adequately reflect the exposure conditions of species that exist in the natural environment. The use of sediment toxicity tests to discriminate or identify effects of individual chemicals within sediments is still challenging.

This chapter describes sediment toxicity testing methods that primarily assess effects to whole organisms, including survival, reproduction, development, and behavior. However, for many whole-organism tests, biomarker endpoints may also be assessed and are discussed in Chapter 5. Methods that use molecular and biochemical endpoints in cell-line bioassays (eg, cultured bacteria and eukaryotic cell-based assays and other surrogate organisms) are not described here. These suborganismic-level responses are increasingly important for high-throughput testing of individual chemicals; however, they do not attempt to create the chemical exposure conditions and bioavailability similar to those of organisms present in sediments. Well-designed organism-level sediment toxicity tests provide the science to anchor these endpoints to organism and population-level responses (Ankley et al., 2010). The chapter focuses on laboratory-based whole-sediment toxicity tests, with mesocosm and field-based toxicity testing methods described in Chapter 8.

7.1.1 Principal Considerations for Sound Ecotoxicology

To enable a sound ecotoxicological assessment to be achieved from sediment toxicity testing, there are a wide range of principles that should be considered, starting from the essential aspects of experimental design and measurements, to demonstrate that the desired exposure conditions have been achieved, to the appropriate choice of control and reference sediments and endpoints, through to achieving an unbiased analysis of the results. While different principles may apply for different assessment needs, there are basic principles that are essential for achieving a realistic assessment. Harris et al. (2014) describe a number of principles that are all essential for ecotoxicological assessments; these are here modified to align more closely with the considerations relevant to sediment toxicity assessments:

7.1.1.1 Ensure Adequate Planning and Good Design

As the purposes of tests differ, so do the design requirements, and they are influenced by the sediment properties, contaminants of concern, desired exposure conditions, and the proposed data analyses (eg, establishing presence, causes of toxicity in sediments, or effect concentrations (such as EC10, EC50) for a specific measured chemical or endpoint). The development of data quality objectives for the study is recommended during the initial planning process, including test performance and acceptance criteria (USEPA, 2006a,b) (see Section 2.5). The design will need to consider:

- the species relevance: number and types of species to be tested if selecting a battery of tests.
- the endpoints to be analyzed including expected sensitivity and test duration necessary to allow the endpoint to be fully expressed. The choice of endpoint can influence how useful the results are for extrapolating from the individual-level to population-level effects.
- the number of replicates necessary to assess whether a specific sediment causes toxicity, or the number of exposure concentrations and replicates of each concentration necessary to develop a concentration–response relationship. The number of replicates may vary, for instance if the test is a range finder or if it is a definitive test.
- the exposure pathways being assessed, how a desired exposure may change during the test, and steps necessary to maintain a specific exposure; which exposure parameters to measure, and the frequency of measurements, to analyze the effect endpoints in relation to the concentration in sediment porewater, or overlying water, or how exposure pathways may be modified by factors such as pH, acid-volatile sulfide (AVS), and organic carbon (TOC).
- the need for food addition, and how this may influence organism behavior and contaminant uptake.
- whether the test is static, semi-static, or flow-through, as each of these designs will influence the overall results. Static and semi-static tests may overestimate the exposure concentration in the overlying water, whereas flow-through tests with clean water may underestimate exposure.
- when creating and assessing an exposure for a specific chemical, what measurement(s) are necessary to assess whether the spiked (added) chemical has equilibrated sufficiently, or whether the concentration of the bioavailable forms change during time (eg, volatilization of organic contaminants, oxidation of AVS in surface sediments increasing bioavailability of metals).

- monitoring to assist in the identification of confounding factors, potentially relating to organism behavior or the presence of undesired stressors or uncharacterized contaminants, which may lead to the need to consider toxicity identification and evaluation (TIE) manipulations.

7.1.1.2 *Understand and Define the Baseline for the Response Endpoints*

There is a need to define a normal endpoint response and variability in responses for unexposed organisms in a range of sediment types. Field-collected organisms may be more variable in response due to greater genetic variability than laboratory-reared organisms. For tests where greater variability exists, there may be a need for greater replication to assess whether toxicity is observed. The baseline considerations should include:

- knowledge of how the properties of control sediments influence responses (eg, variations in grain size) and use of reference sediments to bracket the physical and chemical characteristics of test sediments. These characteristics should be reported along with the results.
- the influence on the test endpoints of organism size, life stage, sex (eg, behavior of males may differ considerably from females), and the density of individuals. It is important to appropriately describe the source, life stage, and history of the individuals being used.
- knowledge of the feeding behavior and the organism nutrition requirements, as providing too much or too little food will modify behaviors, contaminant exposure, and responses (particularly growth).
- an understanding of how density and cannibalization may influence outcomes, and how organism health may be influenced by untested factors (eg, parasites).
- an understanding of responses to noncontaminant stressors (eg, grain size).

7.1.1.3 *Use Appropriate Exposure Routes and Concentrations (Relevant to the Environment Being Assessed) and Make Measurements to Define the Exposure*

The environmental relevance of the exposure is a critical consideration, both in relation to the expected behavior of the test organisms in their natural environment and the exposure created by the test design. Multiple exposure routes may contribute to effects to organisms in sediments. Exposure considerations should include:

- the environmental relevance of the exposure concentrations, eg, the concentrations should not be too high compared to the environmental occurrence to be considered environmentally relevant.
- a design needed to create and maintain a desired exposure, eg, would sediment resuspension occur naturally at the assessment location and should the design replicate this?
- the measurements necessary to define the exposure and achieve the data requirements for the assessment, including the measured concentrations in the different water–sediment phases present during the tests and estimation of the potential bioavailable fraction(s) of contaminants of concern.
- the water-renewal rates for semi-static and flow-through tests necessary to replicate the exposure conditions of an assessment location.
- whether the dominant exposure routes may change during the test, due to changes in organism life stage, behavior, or the contaminant forms, and how this may influence reporting.

7.1.1.4 *Statistical Analyses and Repeatability of the Results*

The numbers of replicates per test, and number of organisms per treatment or replicate, will influence whether the results have sufficient statistical power to determine whether toxicity has occurred (ie, whether hypotheses are statistically supported or rejected). In addition, the number of exposures tested will influence whether effect concentrations such as EC10 and EC50 can be determined and their precision. It may be useful to undertake range-finder tests and avoid potentially repeating large tests. It is necessary to consider statistical significance separately from biological or environmental significance. If the variance or standard deviation among replicates is very small, the results may be statistically significantly, but if those differences are less than 10% or 15% (relative to controls), then the difference may not be biologically significant. For example, if the control survival is 100% with a standard deviation of 0, and a test concentration survival is 90% with a standard deviation of 5%, the two may be statistically different, but not biologically significant. Statistical design considerations include:

- whether tests are designed to identify if significant toxicity effects occur for a specific exposure or concentration.
- the need to develop dose–response relationships for the purpose of empirical regression modeling (eg, EC10, EC20, EC50 estimates).
- data use for more complex exposure-effects models that consider effects through time (eg, toxicokinetic/toxicodynamic models).
- requirements for greater replication of tests and potentially greater numbers of individuals within a test if the effect is represented by a small change in a biological response, eg, small change in growth relative to controls. Note, the repeatability should be considered in relation to both the exposure and the effects and may be more easily achieved in the laboratory (more controlled exposure) than the field.
- whether range-finder studies may be useful when seeking to derive effect thresholds, especially if many measurements are necessary to characterize the actual exposure and there is an inability to repeat the test if a suitable range of test concentrations is not initially achieved.

More specific considerations regarding statistics are discussed in Chapter 2 of this book.

7.1.1.5 Consider Confounding Factors

A large range of factors will influence the organism response and test outcomes, and while some of these are desired by design (eg, choice of controlled temperature or salinity), the uncontrolled factors (eg, disease, parasites) are undesirable and need to be identified and accounted for if possible. There also exist test sample-related factors that may confound the interpretation of any concentration–response relationships (eg, exposure to multiple unmeasured substances/stressors). Confounding factors to consider include:

- factors specific to the test organism, or from other organisms, such as competition, predation, disease, and parasites (competition and predation interactions are designed into community tests but should be controlled for or eliminated from single organism laboratory-based whole-sediment toxicity tests).
- factors influenced by the sediment, such as biofilm or fungi growth.
- unmeasured or varying exposure to dissolved or particulate chemical forms, which are particularly important when calculating and reporting effect thresholds with respect to a single exposure route (eg, EC50 in mg/kg (specify wet or dry wt) or μg/L).
- unmeasured stressors (eg, ammonia, sulfide) or other chemicals (thousands of chemical compounds may exist in a sediment from anthropogenic or natural origins, and only major contaminants are usually quantified).
- mismatches between properties of control and reference sediment ranges (eg, extremes in particle size).

- variations in test conditions (eg, nutrition provided as food, light, dissolved oxygen, temperature, hardness, salinity).

7.1.1.6 Analyzing and Reporting Results in an Unbiased Manner

A good experiment design should have eliminated most potential biases, with factors such as the random allocation of organisms between treatments and use of treatment and sample names that facilitate a "blind" unbiased assessment of endpoints and related analyses. There is a need to ensure that there are no expectations, or pressures for tests to achieve ideal dose–response relationships, or that an assessment have a specific outcome. When analyzing, reporting, and discussing the results, additional considerations include:

- a systematic discussion of uncertainties relating to variability and repeatability.
- discussing the exposure conditions achieved and how representative they may be of those expected for the sediments in their proposed field environment.
- whether the dose–response observed in the laboratory would likely exist in the field. For example, could the potentially greater rates of water exchange or dynamics expected in the field result in different outcomes?
- whether nonmonotonic dose–responses (eg, those not creating the ideal sigmoidal curve) may be a result of confounding factors, including poor experimental design or technique.
- discussing whether the exposure concentration–response was plausible based on all the lines of evidence (exposure, conditions, observations, measurements, responses).
- considering whether effects being attributed to the sediments are due to a particulate (EC50 reported in mg/kg [wet or dry wt specified]) or dissolved (EC50 reported in μg/L) exposure route and whether the described exposures may realistically occur in the field. This is particularly important when dissolved contaminant concentrations are elevated in overlying waters compared to contaminated field settings.

Care should be taken to not overextrapolate beyond the test or environmental exposure.

7.1.2 Considerations for Selection of Test Organisms

Benthic organisms exist in virtually all natural sediment environments and exhibit a wide range of behaviors that result in differing contaminant exposure pathways and sensitivities. Due to this diversity, no one organism is best suited for the ecotoxicological assessment of all types of contaminated sediments; a range of organisms, with differing exposure pathways and from different phyla, should ideally be used. However, as the choice of test organisms and endpoints, along with the test design, may have a major influence on the outcome of assessment programs, the selection of test organisms needs to be carefully considered. The test organism(s) chosen for the assessment program will act as surrogate(s) for other organisms within the sediment ecosystem targeted for protection. The basis for the use of surrogate species for environmental risk assessment is all organisms share similar DNA; therefore, they should also have similar response systems for contaminants. However, test aspects relating to sensitivity of life stages, behavior, and exposure pathways will influence how representative the surrogates are of the overall ecosystem. As general guidance on the selection of organisms for biological tests with sediments (ASTM, 2008b), ideally, the test organism should:

1. be one that has direct contact with sediment during normal behavior, to enable assessments of potential effects of contaminants via water and sediment exposure routes, as opposed to use of organisms or life stages of organisms that are not usually in contact with sediments;
2. have a demonstrated sensitivity to a range of contaminants of interest in sediments, if those

contaminants are known, or at least not be known to be insensitive to known sediment contaminants of potential concern;
3. be tolerant of a broad range of sediment physicochemical characteristics (eg, grain size, salinity);
4. be compatible with selected exposure methods and endpoints;
5. have standardized application and quality assurance procedures, to provide a means for interlaboratory comparisons;
6. be readily available from culture or through field collection;
7. be easily maintained in the laboratory;
8. be easily identified (especially when collected from the field); and
9. have short to moderate life cycles (days to weeks) to enable effects to reproduction to be assessed without the need for long (months) test durations.

Few test species may ever meet all of these criteria; however, they are important considerations when designing sediment ecotoxicology assessments. The first two points, regarding exposure directly to sediments and sensitivity of the effect endpoints to contaminants of potential concern may have the greatest influence on the assessment outcomes. As the organism used will act as a surrogate, it may be desirable to develop tests that use an organism that is indigenous (either present or historical) to the site being evaluated, or have a similar niche (eg, behavior and feeding guilds) to the indigenous organisms or other organisms that may be regarded as high conservation value ecologically or economically.

7.1.2.1 Contaminant Exposure Pathways and Sensitivity

Benthic organisms interact with, and are exposed to, contaminants present in both dissolved and particulate forms, with contaminant uptake occurring from porewater and overlying water, the direct ingestion of prey, and direct or inadvertent ingestion of sediments (Rainbow, 2007; Simpson and Batley, 2007). The concentrations of contaminants associated with sediment particles are often 100s to 10,000s of time greater than those in the porewaters or overlying waters (Hassan et al., 1996; Simpson and Batley, 2007). Due to the greater surface area, the concentrations of contaminants associated with fine sediment particles are frequently greater than those on the coarse particles (Chariton et al., 2010). While dissolved contaminants may be expected to represent the more bioavailable form of contaminants, and porewaters are an important contaminant exposure route to many infaunal species, the exposure to contaminants from ingestion of sediment particles may often be greater (Rainbow, 2007; Casado-Martinez et al., 2010; Strom et al., 2011; Campana et al., 2012). When isolated from sediments, contaminant concentrations in porewaters generally decrease due to volatilization, precipitation, or oxidation, making it difficult to maintain environmentally relevant exposure from isolated porewater exposures. For these reasons, whole-sediment toxicity tests using sediment-dwelling organisms or biota are the most appropriate to capture all exposure pathways.

Benthic organisms are often deposit feeders, obtaining nutrition from ingesting biota, organic and inorganic particles from the sediment surface or within the sediments. Many species, particularly polychaete worms, ingest subsurface sediments and convey them to the sediment–water interface as fecal pellets. These differing behaviors are notable both in terms of the likely differing exposure routes during feeding and also the influence they have on the sediment chemistry and the exposure that may occur during a test. These "bioturbation" activities modify the partitioning of contaminants between dissolved and particulate forms, frequently increasing the release of contaminants from sediment to the water column for both dissolved (from porewaters) and particulate forms (fine suspended solids) (Aller et al., 2001; Ciutat and Boudou, 2003; Belzunce et al., 2015). Bioturbation increases the interaction of sediment with the overlying water and maintains contaminants in oxidized forms within surface sediment layers and burrow walls (Peterson et al., 1996; Gerould and Gloss, 1986; Simpson et al., 2012; Volkenborn et al., 2010). Changes in partitioning dissolved and particulate fluxes that occur during toxicity tests may modify the contaminant exposure and outcomes of the tests, and therefore need to be considered during test design and monitored where possible.

The sensitivities of benthic organisms to contaminants in sediments differ considerably between species as well as organism life stage. The organism's burrowing behavior (eg, whether tube builders or free-living, bioturbation activity, bioirrigation rates) and feeding habits (filter or deposit feeding, and their food selectivity) considerably influence their exposure pathways and sensitivity. Sediment toxicity assessments should therefore, where possible, use a suite of tests with organisms that have different behaviors and feeding habits, to cover a wide range of potential routes of contaminant exposure. The use of local species may sometimes be desirable for assessment programs; however, developing accepted standardized test procedures using local species is often time-consuming and frequently cannot be performed in many testing programs. Local species may be more, or less, sensitive to the contaminants of potential concern than surrogate species.

In general, there is no one species that is sensitive to all contaminants, which is why it is recommended to test a variety of different phyla, feeding and exposure regimes. That said, there are species that are known to be particularly insensitive to many contaminants, for example, the marine worm (*Nereis virens*) and the brine shrimp (*Artemia* sp.). These organisms can be useful in other types of tests, for example *N. virens* is often used in bioaccumulation tests (USEPA, 1993) because of its large size and ability to live in contaminated sediments where more sensitive organisms would not survive. *Artemia* sp. is an easily cultured food source used in many toxicity and bioaccumulation tests because of its ability to survive in contaminated systems which ensures that test organisms have adequate nutrition.

7.1.2.2 Potential Test Organisms

Most phyla of benthic organisms contain species that have been used for sediment toxicity testing, including bacteria, algae, crustaceans (eg, amphipods, copepods, mysids), mollusks (bivalves/clams), annelids (polychaetes), gastropods (snails), nematodes, and echinoderms (sea urchins, sea cucumbers, sea stars) (Fig. 7.1). All play important roles in benthic communities, with food sources comprising bacteria, algae, diatoms, other fauna and plant detritus, and/or in turn they are important food sources for larger invertebrates, juvenile fish, and water birds. Many directly ingest sediment particles and are therefore directly exposed to sediment-bound contaminants along with contaminants in porewaters and overlying waters. ASTM (2012a) discusses many of the important considerations for the selection of resident species as test organisms. Examples of benthic species frequently used for conducting sediment toxicity tests are shown in Table 7.1.

Bacteria and algae are important components of all sediments, acting as a food source for many benthic organisms and facilitating transfer of some sediment contaminants to higher trophic levels. Both influence organic matter degradation, nutrient cycling, organic compound degradation, and remobilization of metals at the sediment–water interface. Bacteria form biofilms (with alga), and both phyla stabilize surface sediments and form microhabitats. Bacteria or algae do not have a means of directly ingesting particles, so toxicity tests assess only exposure to contaminants in porewater or overlying water.

For estuarine and marine sediments, the commercially available Microtox test kit (Azur Environmental, 1998) provides the best example of a bacterial toxicity test. This uses the marine luminescent bacterium (*Vibrio fischeri*) to assess

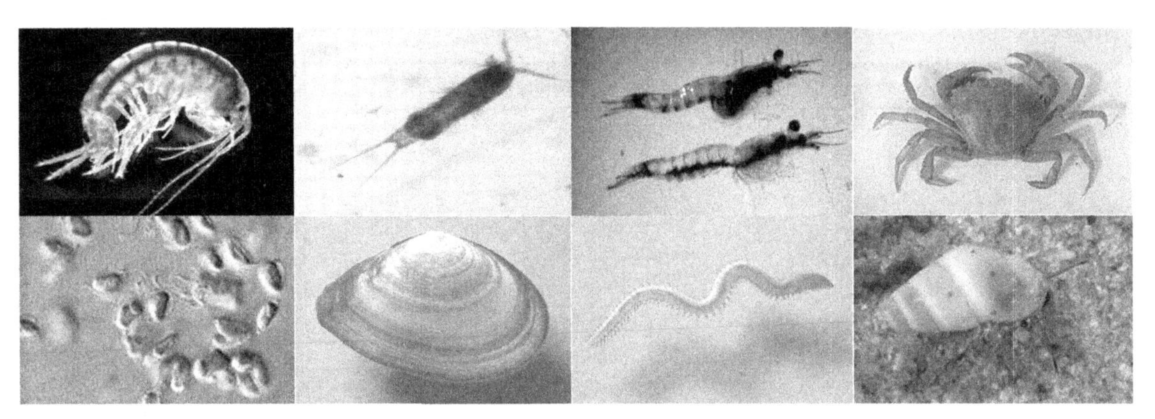

FIGURE 7.1 Whole-sediment toxicity test species: *(top, left to right)* amphipod, copepod, mysid, and crab (crustaceans); *(bottom, left to right)* benthic algae (plant), bivalve/clam (mollusk), polychaete worm (annelid), and snail (gastropod).

a decrease in light output after a 20-min exposure to sediments compared to controls. The method has been found to be a useful tool for screening and providing ecotoxicological information to assist in ranking of large numbers of sediments to quickly identify areas of potential concern. There are issues regarding sediment turbidity interfering with measured light output from the luminescent bacteria, but this can generally be controlled with appropriate controls. Most algal bioassays for sediments have used endpoints that assess effects to chlorophyll production (affecting photosynthesis) and enzyme inhibition, rather than measurements of inhibition of algal growth, as growth is strongly influenced by ammonia or nutrient concentrations in sediments (Adams and Stauber, 2004). Plants, such as mangroves species that are common in tropical estuaries, are underrepresented in toxicity tests due to difficulty in testing and their generally low sensitivity to many common contaminants.

A wide range of crustaceans have been successfully used for sediment toxicity assessments, most notably amphipods, meiofaunal harpacticoid copepods, and mysids. They meet many of the criteria used for selecting test organisms for whole-sediment toxicity tests; a range of standardized or in peer-reviewed methods are available (ASTM, 2014; Greenstein et al., 2008; Perez-Landa and Simpson, 2011). As the life cycles of small crustaceans are frequently short (eg, 20–30 days for many copepod species), and copepods go through multiple larval stages and mid-life stages before reaching full form (eg, nauplii to copepodites to reproducing adult copepod), they are frequently amenable to the development of short subchronic tests or life-cycle whole-sediment toxicity tests (Chandler and Green, 1996; Hack et al., 2008; Simpson and Spadaro, 2011; Araujo et al., 2013). Test endpoints typically include survival, reproduction, and development/growth. Crabs have not yet been used extensively for sediment tests despite their close association with sediments; however, many species are well studied with respect to effects of contaminants in water-only exposures (Rodrigues and Pardal, 2014).

Sediment toxicity tests using mollusk, annelid, gastropod, and echinoderm phyla are less common than those using crustaceans, in part due to their either longer or more complex life cycles. Bivalve mollusks often have wide distribution, high abundance, are generally easy to handle, and many species deposit-feed at the sediment–water interface (Ringwood and Keppler, 1998; Keppler and Ringwood,

TABLE 7.1 Estuarine and Marine Whole-Sediment Toxicity Tests

Organism	Test species	Duration/Endpoint	Acute/Chronic	References
Bacteria	*Vibrio fischeri* (Microtox)	20-min luminescence	Acute	Azur Environmental (1998), Environment Canada (2002)
Protozoa	*Euplotes crassus*	Elutriate tests: 8-h cell viability; 24-h replication	Acute	Gomiero et al. (2013)
Miroalga	*Entomoneis* cf *punctulata*	24-h enzyme (esterase) inhibition	Acute	Adams and Stauber (2004), Adams (2016)
Microalga	*Cylindrotheca closterium* (formerly *Nitzschia closterium*)	72-h growth rate	Chronic	Moreno-Garrido et al. (2003a,b, 2007), Araujo et al. (2010)
Copepod	*Amphiascus tenuiremis* and *Microarthridion littorale*	14-day survival and reproduction; 21-day full life cycle	Chronic	Chandler and Green (1996), Kovatch et al. (1999), Kennedy et al. (2009)
Copepod	*Nitocra spinipes*	10-day reproduction	Chronic	Perez-Landa and Simpson (2011), Simpson and Spadaro (2011), Krull et al. (2014), Araujo et al. (2013), Spadaro and Simpson (2016a)
Copepod	*Robertsonia propinqua*	24-day life-cycle test	Chronic	Hack et al. (2008)
Copepod	*Tisbe biminiensis*	7-day reproduction	Chronic	Araujo et al. (2013)
Mysid	*Americamysis bahia*	10-day survival	Acute	Kennedy et al. (2009)
Amphipod	*Ampelisca brevicornis*	28-day survival, fecundity, and growth	Chronic	Costa et al. (1998, 2005)
Amphipod	*Corophium multisetosum*	10-day survival; 21-day fecundity and growth	Acute Chronic	Casado-Martinez et al. (2006), Castro et al. (2006)
Amphipod	*Corophium volutator*	28-day survival and growth; 28-day and 76-day survival, growth, and reproduction	Chronic	Scarlett et al. (2007a), Fox et al. (2014)
Amphipod	*Ampelisca brevicornis, Corophium volutator, Eohaustorius estuarius, Leptocheirus plumulosus, Rhepoxynius abronius*	10-day survival	Acute	Rodrıguez-Romero et al. (2013), ASTM (2014), Greenstein et al. (2008)
Amphipod	*Gammarus locusta*	28-day survival, fecundity, and growth	Chronic	Costa et al. (1998, 2005)

TABLE 7.1 Estuarine and Marine Whole-Sediment Toxicity Tests—cont'd

Organism	Test species	Duration/Endpoint	Acute/Chronic	References
Amphipod	*Hyalella azteca* (up to 15 ppt)	10-day and 28-day survival and growth; 42-day survival, growth, and reproduction	Chronic	ASTM (2010)
Amphipod	*Leptocheirus plumulosus*	28-day reproduction and growth	Chronic	ASTM (2008a), Kennedy et al. (2009)
Amphipod	*Melita plumulosa*	10-day juvenile survival	Acute	Spadaro et al. (2008), Strom et al. (2011)
Amphipod	*Melita plumulosa*	10-day reproduction	Chronic	Mann et al. (2009), Simpson and Spadaro (2011), Spadaro and Simpson (2016b)
Bivalve	*Mercenaria mercenaria*	7-day juvenile growth	Sublethal	Ringwood and Keppler (1998), Keppler and Ringwood (2002)
Bivalve	*Tellina deltoidalis*	10-day survival	Acute	King et al. (2010)
Bivalve	*Tellina deltoidalis*	30-day survival and growth	Chronic	Campana et al. (2013), Spadaro and Simpson (2016c)
Polychaete worm	*Arenicola marina*	10-day, 21-day survival	Acute	Bat and Raffaelli (1998), Morales-Caselles et al. (2008)
Polychaete worm	*Neanthes arenaceodentata*	20-day survival, 28-day growth	Chronic	Bridges and Farrar (1997), Farrar and Bridges (2011)
Polychaete worm	*Nereis virens*	7-day avoidance and damage to body condition	Acute	Van Geest et al. (2014a,b)
Mussel	*Mytilus galloprovincialis*	2-day embryo development at sediment–water interface	Sublethal	Anderson et al. (1996), Greenstein et al. (2008)
Snail	*Hydrobia ulvae*	48-h postexposure feeding 24-h avoidance	Sublethal	Krell et al. (2011), Araujo et al. (2012)

2002; King et al., 2010; Campana et al., 2013). Thus contaminants bound to particles (food) and in solutes within the water column are important exposure pathways (Griscom and Fisher, 2004; King et al., 2010; Campana et al., 2013). Adult oysters have not been used for whole-sediment toxicity tests, but the larval and juvenile stages are used for porewater testing (Ringwood, 1992) and are applicable to studying effects of contaminant release to the water column and from suspended solids (Edge et al., 2012, 2014). Polychaetes (marine worms) may reside in constructed tubes or be free-living within the sediment and are generally subjected to high exposure to sediment-bound contaminants, via ingested sediments, porewater, and overlying water, and via dermal contact by direct diffusion of some organic contaminants (Bat and Raffaelli, 1998; Morales-Caselles et al., 2008; Ramos-Gómez et al., 2011; Farrar and Bridges, 2011). Gastropods, nematodes, and echinoderms are underrepresented in sediment toxicity tests, due to their often complex behaviors or life cycles and difficulties in assessing endpoints more sensitive than survival (Ringwood, 1992).

7.1.3 Possible Test Endpoints

To maximize the benefit of sediment toxicity test results for assisting in management decision-making regarding options for contaminated sediments, it is necessary to consider the biological and ecological relevance of the effects being assessed to select the most appropriate test endpoint(s). Sediment toxicity tests are most frequently used to provide information about the effects of contaminants at the organism level. These effects include survival, growth, development, reproduction, and behavior. Many of these test endpoints can be used to make predictions about the risks that chemicals may pose at the population level by incorporating them into models (Kuhn et al., 2000, 2002). Increasingly, suborganism-level endpoints are being developed that can be used to predict organism-level effects (eg, biomarker responses or assays for cellular toxicity, genotoxicity, immunotoxicity, and endocrine effects). However, a greater level of mechanistic understanding of toxicity is necessary to anchor effects through suborganism to organism and through to the population level; one approach to do so is using Adverse Outcome Pathway (AOP) models (Ankley et al., 2010). AOPs are a formalization of the link between toxicants and changes in populations. AOPs strive to elucidate and attempt to link key events via relationships until an adverse outcome is reached. Key events can occur at critical levels of organization including molecular, cellular, organ, organismal, and population. Documentation and elucidation of all key events along the AOP can be very complex and difficult to complete. In this chapter we deal primarily with organism-level endpoints, with Chapter 5 discussing the use of biomarkers and Chapter 6 cellular toxicity endpoints.

Test endpoints are often referred to as acute or chronic, although the terms subchronic, sublethal, and lethal are also frequently used. Whether a test is considered acute or chronic depends on both the species and endpoint being considered; the definition is not rigid. A chronic toxicity test may be defined as one in which the species is exposed to the contaminated sediment for at least one full life cycle, or the species is exposed during one or more critical and sensitive life stages. However, the latter part of this definition better describes a subchronic toxicity test, in which the exposure duration would not cover a full life cycle but occurs during at least one critical stage in the organisms life cycle (eg, reproductive output: Mann et al., 2009; Perez-Landa and Simpson, 2011). As many organisms have relatively long life cycles (eg, greater than 4 months), for practical purposes an exposure period greater than 10% of the organism's life cycle is typically considered substantial enough to describe a chronic test (Newman, 2010). Consequently, an exposure duration of ≥10 days is often considered chronic for most organisms other than bacteria (24 h or less) and algae (48–72 h for growth rate endpoints). Durations of 28–60 days are commonly used for assessing chronic effects to survival, growth, and reproduction of amphipod, bivalve, and worm species (ASTM, 2008a). For amphipod and copepod species with shorter life cycles (eg, 20–60 days), it is possible to assess effect to their reproduction and development using subchronic tests with durations of 10 days (Mann et al., 2009; Perez-Landa and Simpson, 2011). Acute toxicity typically refers to an adverse effect that occurs as the result of an exposure duration that is short relative to the organism's full life cycle; acute lethality tests represent the most widely used test endpoint. Acute sediment toxicity tests most frequently assess survival following exposure periods of 4–10 days, although assessment periods for behavioral endpoints such as avoidance and postexposure feeding may be as short as 48 h.

The acute survival of juvenile or young adult organisms following exposure periods of $\leq$10 days have been the most used sediment toxicity test endpoints for sediment toxicity assessment programs (USEPA, 1994). However, it is now widely recognized that these tests may not detect potential impacts that may be observed from longer exposures, or by other more sensitive life stages (Simpson and Spadaro, 2011). Adult organisms are often far less sensitive to contaminants than embryos or very early life stages (Williams et al., 1986; Hutchinson et al., 1998). Sublethal tests that assess chronic or subchronic effects to organisms such as changes in biomass (eg, average weight of individual), growth (eg, change in mean weight, or sometimes body length or surface area), development (eg, larval-juvenile-adult life stages), and reproduction (eg, fertility, fecundity, or number of offspring produced) are generally more sensitive than acute survival tests. Such sublethal test endpoints are also more relevant for risk assessments as they provide greater information for predicting long-term effects at the population level (Kuhn et al., 2000, 2002). Test endpoints that assess suborganism-level effects (eg, biomarkers for physiological or biochemical responses) and behavioral responses (eg, avoidance and feeding) provide useful information for identifying potential for impairment to the health of individuals, but for risk assessment purposes the actual information provided by these endpoints is often considered more difficult to link to potential population-level effects. A number of authors have compared the sensitivity of lethal and sublethal test endpoints (Scarlett et al., 2007a; Greenstein et al., 2008; Kennedy et al., 2009; Simpson and Spadaro, 2011).

7.1.3.1 Behavioral Test Endpoints

Behavioral changes may be exhibited by organisms in response to contaminant and non-contaminant stressors that are suitable as endpoints for sediment toxicity tests (Scarlett et al., 2007b; Hellou, 2011). Behavioral endpoints are often considered to be "early-warning" responses, as rapid responses may provide immediate protection from the stressor, and may represent an early reaction (eg, ceasing filtration of water or feeding when contaminants are detected), whereas some longer-term behavioral responses may have adverse consequences, eg, increased predation due to failure of benthic organisms to burrow (amphipods or bivalves), impairment of movement (eg, righting of brittle stars), or failure to mate and reproduce. Conversely, failure to avoid contamination can increase the likelihood of adverse effects (Ward et al., 2013a). Decreases in postexposure feeding have been the most widely used behavioral response endpoint for sediment toxicity tests (Moreira et al., 2005, 2006; Krell et al., 2011; Rosen and Miller, 2011) and avoidance behavior responses the next most used (Araujo et al., 2012; Ward et al., 2013a,b). Behavioral responses that concern feeding may be more easily linked to potential organism-level effects such as organism growth, development, and reproduction than avoidance behavior.

7.1.3.2 Biomarker, Gene Expression, and Genotoxic and Cell-Based Bioassay Test Endpoints

Suborganism-level test endpoints that assess effects to physiological and biochemical responses (eg, changes in biomarkers or gene expression) are discussed in greater detail in Chapter 5. While changes that occur at the molecular level could theoretically be more sensitive and more specific than responses at higher organism levels, it has not yet been demonstrated that these molecular initiating events can provide greater sensitivity or reduced variability compared to the well-established sublethal endpoints in whole-sediment tests (Martín-Díaz et al., 2004; Simpson and Spadaro, 2011; Edge et al., 2014). The most commonly assessed biomarkers are those relating to activity of biotransformation and antioxidant enzymes, and biochemical indications of oxidative damage to cells, eg, lysosomal instability (Monserrat et al., 2007; Edge et al., 2012; Martins et al., 2012; Taylor and Maher, 2016). A greater range of biomarkers measured in whole organisms provide evidence of exposure to contaminants, rather than direct evidence of effects, although evidence for the latter is increasing (Boldina-Cosqueric et al., 2010; Taylor and Maher, 2010; Hook et al., 2014a; Regoli and Giuliani, 2014).

Molecular-based biomarkers are increasingly being proposed as test endpoints, including those referred to as ecotoxicogenomics and transcriptomics when assessing changes in gene expression (RNA extractions), for the purpose of identifying gene classes that are switched on or off, or proteomics and metabolomics for identifying changes in protein or metabolite production rates (Biales et al., 2013; Hook et al., 2014a). Such endpoints have the potential to provide molecular fingerprints specific to the bioavailable fraction of the chemical; however, these endpoints need to anchored, eg, through AOPs (Ankley et al., 2010; Lee et al., 2015) to organism-level (eg, reproductive output) and/or population-level responses (eg, population growth). Development of detailed and useful AOPs are not yet commonplace as discussed later in this chapter.

The variability of test endpoints based on biomarker and molecular responses is frequently high and may also be more sensitive to many noncontaminant factors, such as changes in salinity and light conditions. However, further advances in these techniques may identify more specific responses to both individual contaminant and noncontaminant stressors.

Interactions of contaminants with the genetic material within the cells of organisms may cause a range of genotoxic/mutagenic responses that can potentially result in effects such as developmental abnormalities (teratogenicity) and cancer formation (carcinogenicity). Cell-based bioassays can be useful for high-throughput screening of endocrine disrupting potential of chemicals (eg, CALUX; USEPA (2014) Method 4435-59). However, for sediment toxicity assessments such as those that are used with chemical extracts of sediment (Li et al., 2013; Gao et al., 2015), the relevance of the endpoints for many assessments is greatly reduced as the bioavailability of contaminants is not well represented in the observed responses.

7.1.3.3 Time-Dependent Endpoints

The response of organisms to toxicants depends not only on the dose but also on the duration of the exposure. Toxicity occurs when a series of biological perturbations inside the organism produce an adverse response. When this outcome needs to be observable will depend on the type of toxicant, the environmental conditions, the species, and the endpoint selected.

Endpoints such as survival, avoidance, or immobility are quantal, which means that we can't see a gradual effect but only observe the presence/absence of the response (eg, the fraction of the individuals in which the response is observed), which depends on the duration of the exposure. Survival is obviously irreversible but mobility might be recovered if the stress is removed. Recovery time is a valuable response, especially in the context of ecological risk assessment actions, which can be measured for many endpoints to investigate the "resilience" of organisms/populations. Resilience represents the capacity to return to the original state (before the disturbing event) following the disappearance of the stress, eg, a concentration of the toxicant at which it no longer has adverse effect on the measurement endpoint.

Graded endpoints refer to effects that can be quantified over time at different degrees, eg, growth, reproduction, feeding rate, and biochemical or genetic biomarkers. When choosing the endpoint, it is critical to consider the time scale in which the response to stress is measured. Time scales vary from minutes to days. No single time scale is appropriate for all endpoints because some endpoints will need longer exposure duration than others for the effect to be observed. Sensitivity is a characteristic specific for each endpoint that relies on a temporal component. If the time scale chosen is too short, the response would be undetectable because the signal is undiscernible from natural variability, eg, activation/inhibition of enzymes, DNA alteration (Regoli et al., 2004). Likewise, latent responses could be missed or become less apparent in organisms if the endpoint is measured before the effect has enough time to be realized. For example, in assessing the reproduction endpoint, the response of the offspring is not routinely observed in toxicity tests nor is exposure time considered in cases of pulsed exposure typical of many spills, episodic runoff events, or periodic pesticide treatment (Zhao and Newman, 2007).

7.1.4 Confounding Factors

A wide range of abiotic and biotic factors can influence the responses of organisms in sediments; it is important to consider these when designing tests and interpreting the results. Confounding factors may result in both synergistic and antagonistic influences that may potentially affect attribution of causes of effects to certain sediment contaminants, potentially resulting in significant toxicity (negative effects relative to control responses) when contamination is low, greater toxicity than expected, or no (or less) observed toxicity when contaminants are present at concentrations expected to cause great toxicity (Bridges et al., 1997; Spadaro et al., 2008).

Developing an understanding of potential confounding factors and tolerance limits for test species is an important element in toxicity test development and interpretation. Abiotic factors include the influence of sediment properties (eg, particle size) and test conditions (eg, temperature, salinity, light) and their variability (eg, sediment heterogeneity), along with the potential effects from naturally occurring stressors such as ammonia and sulfide. Biotic factors include disease, predation, cannibalization, and nutritional requirements that may be influenced by the quantity and quality of added food, organic matter, bacteria, algae, and/or organisms such as meiofauna in the sediments.

Confounding factors, such as predation or cannibalism in experimental systems can often be avoided by appropriate test design including providing adequate and preferred prey items and avoiding overcrowding. The number of organisms per test container should be optimized to provide the maximum statistical power, yet ensure organisms are not stressed which could be manifested in predation and negative responses that are not attributable to chemical contamination. Disease in either cultured or field-collected animals may be detected by adequate controls, which generally range between 80% and 100% survival or 70% and 100% growth or reproduction. Test organisms, particularly cultured test organisms, should also be subjected to quarterly control tests to ensure they are healthy. A reference chart of quarterly culture responses should be maintained to ensure that organisms react consistently to a chemical challenge.

The physicochemical conditions, including sediment particle size distribution, should be within the known tolerance limits of the selected test organism (Moore et al., 1997). Some organisms have narrow tolerance limits for sediment grain size distribution, with thresholds often existing toward either the ends of the silty or sandy sediment ranges. These tolerances may be influenced by the organism's burrowing behavior and may be modified by the provision of food (which may increase tolerance in very sandy sediments that have low baseline nutrients). Sediments that contain compact clays may cause stress to organisms due to the difficulty for the organisms to burrow. Control and reference sediments should be used that cover the particle size and salinity ranges of the test sediments.

Dissolved ammonia and sulfide occur naturally in most sediments, but concentrations can be greatly exacerbated by anthropogenic inputs such as nutrients and organic carbon. Many tube-building benthic invertebrates circulate overlying water through their burrows to provide oxygen and lower their exposure to porewater contaminants including ammonia and hydrogen sulfide (Knezovich et al., 1996; Wang and Chapman, 1999). However, ammonia and sulfide concentrations in porewaters may result in toxic exposures in porewater tests that use species life stages that do not inhabit porewaters (eg, bivalve and sea urchin larval tests) (ASTM, 2012a). For both ammonia and sulfide, the tolerance will depend on organism sensitivity and behavior; tolerance limits should be specified within the test acceptability criteria. Differences in the overlying water pH, temperature, and method of aeration of waters during tests will affect the toxicity of ammonia and hydrogen sulfide, where the relative proportion of the more toxic unionized forms increase for NH_3 and decrease for H_2S with increasing pH, and increase with temperature (Miller et al., 1990; Wang and Chapman, 1999).

Ammonia and sulfide concentrations in porewaters and overlying waters should be measured and the exposure in the overlying waters of laboratory-based toxicity tests considered during the test design. Modification of test design may be appropriate for achieving the desired condition, for example the rate of renewal of overlying water may be increased to lower ammonia concentrations (Word et al., 2005). However, in a review of 30 sediments that underwent whole-sediment TIE analysis, no whole sediments tested indicated sulfide toxicity and only two sediments indicated toxicity from ammonia (Ho and Burgess, 2013). The emphasis of earlier papers on ammonia as a significant toxicant in sediments (Ankley et al., 1990) may be an artifactual result of porewater testing. Any manipulation of the sediments to reduce the influence of modifying factors such as ammonia should also be considered in relation to how such may also modify the concentration or bioavailability of other water-soluble contaminants. Because of the volatile nature of sulfides, and the tendency of sulfides to become less toxic sulfates upon exposure to oxygen, normal handling of sediments (mixing, distribution, and a 24-h equilibration of the sediment in a sediment–water test system with aeration) may decrease moderate sulfide levels to below toxic concentrations. Changes in pH and salinity may also modify the bioavailability and toxicity of other contaminants such as metals (Ho et al., 1999a; Riba et al., 2004). TIE procedures can be used to identify and remove ammonia if it is a major toxicant (see Section 7.3).

Food (nutrition) availability may frequently influence test outcomes by the modification of feeding behaviors, with the potential for selective feeding on added clean food rather than natural foraging and sourcing of food from the test sediments. But, particularly for sandy sediments that contain little organic matter, the addition of food may be necessary for organisms to maintain their growth and development within normal ranges during long-term exposures. However, feeding with uncontaminated food may reduce the exposure of the test organisms to sediment-bound contaminants due to selective feeding, and this needs to be evaluated (McGee et al.,

2004; Spadaro et al., 2008). McGee et al. (2004) observed differences in acute and chronic toxicity that were possibly related to the test diet used, where the presence of supplemental food was likely ameliorating toxic effects in chronic exposures. Conversely, contaminated or spiked food may be added to the system if the contaminant is known and the spiking method for the food is characterized and reliable. However, ensuring a supply of well-characterized spiked food adds a complicated step to conducting a toxicity test. Where it is possible to achieve acceptable endpoint responses for the range of sediment types being tested, it is generally recommended that additional food not be provided; however, enough nutrition should be available so as to not jeopardize test organism health. The presence of uncharacterized meiofauna can influence food availability to the test organisms, eg, nematode numbers can increase very rapidly when provided additional food and potentially result in competitive stress on test species. Also, excess food may result in elevated ammonia concentrations or algal or biofilm growth.

Photoinduced toxicity has been observed to be an important factor in controlling the toxicity of some contaminants associated with sediment (eg, certain PAHs) (Ankley et al., 1995; Swartz et al., 1997; Fathallah et al., 2012). As this may occur naturally in the field, the light conditions used during laboratory tests may need to reflect conditions at the field location including water depth and turbidity. Even though sunlight may not penetrate through turbid waters, water column larvae of sediment organisms may be affected (Pelletier et al., 1997).

7.2 HOW TO CONDUCT A SEDIMENT TOXICITY TEST

The intent of the following section is to sequentially describe each of the steps taken and the information recorded when undertaking sediment toxicity tests. This section assumes that the test species and endpoints have already been selected, following the considerations outlined above. It refers to many of the considerations for sound ecotoxicology discussed in the introduction, with emphasis on design considerations that can be applied consistently to allow a statistically robust evaluation of organism responses to the contaminated sediments using environmentally realistic exposure conditions.

7.2.1 Sediment Collection and Preparation

7.2.1.1 Collection and Storage

The number of samples collected and tested for sediment toxicity assessments can be determined by a broader sediment quality assessment program that will specify much of the quality assurance necessary to achieve the data quality objectives (see Section 7.2.5). This will also determine the numbers and types of control and reference sediments used to assess the performance of tests. The methods used in sample collection and all forms of handling (transport, storage, and manipulation) before testing will alter the physicochemical properties, potentially change total contaminant concentrations, and/or bioavailability of contaminants and noncontaminant stressors such as ammonia, and therefore influence the results of toxicological analyses. As a consequence, there is a need to provide various levels of quantitative information relating how the tested sediments resemble the sediments collected from the field. Field records should include the location, depth, water quality parameters, and other relevant observations about the site (ASTM, 2008c; Batley and Simpson, 2016).

Depending upon the goals of the assessment, the upper few centimeters of sediment are often the horizon of interest (eg, 0–2 cm or 0–10 cm surface sediments). This sediment horizon generally represents the zone of contamination available to many benthic invertebrates and also is the sediment depth that can be easily resuspended and made bioavailable to epibenthic and water column organisms. The devices and techniques available for collecting sediments are well described elsewhere (Mudroch and Azcue, 1995; USEPA, 2001). Grab-sampling devices are typically used for many sampling programs. Where historical contamination is of interest, sediment cores can be collected and either homogenized or sectioned to provide sediment from a particular depth for toxicity testing. The quantities of sediment required will depend on the particular test being undertaken (species and procedure), but generally 3 kg of wet sediment would be sufficient for most replicated toxicity tests. If range-finding tests, chemistry, porewater testing, or TIEs are to be undertaken, more sediment would need to be collected. If sediment chemistry analyses are also to be undertaken on the same sediments, different handling and storage needs should be considered. Procedures for spiking of sediments with contaminants to achieve desired exposure concentrations are described in detail elsewhere (Northcott and Jones, 2000; Simpson et al., 2004; Hutchins et al., 2008).

Sediments for toxicity testing should be stored cold (not frozen) (USEPA, 1993; ASTM, 2008c); however, there are different opinions about appropriate storage times. Practicalities can prevent tests from commencing immediately; however, it appears reasonable that if the sediment toxicant is known to be volatile or labile, testing should be undertaken generally within two weeks (ASTM, 2008c). If the toxicant is known to be relatively stable (eg, PCBs, PAHs), it appears that moderate or even long-term storage may not alter the toxicity (DeFoe and Ankley, 1998). It is reasonable to test sediments as soon as practicable, but it seems that even long-term storage may not invalidate sediment toxicity. To minimize factors modifying sediment toxicity, which result from sediment disturbances, it is recommended that sediments not be finely sieved unless it is necessary to remove predatory or indigenous species that would interfere with test endpoints.

Methods for collection of porewater from sediments are well described elsewhere as well as discussions of precautions and artifacts (Carr and Nipper, 2003; Chapman et al., 2002). Reporting of procedures and measurements, as well as interpretation of results, are critical for porewater testing, as once porewaters are isolated from sediments, dissolved concentrations of contaminants of interest may change considerably before and during tests (Chapman et al., 2002; Simpson and Batley, 2003) (see Section 7.3 for more discussion of porewater testing).

7.2.1.2 Control and Reference Sediments

Few organisms will have responses that are not influenced by sediment properties (eg, particle size). Control sediment tests provide experimental conditions which are conducive to achieving the near-optimum response of the test organism. Control sediments may be comprised of natural field-collected sediments (collected from organism collection sites) or formulated sediments that are prepared by mixing sediment substrates such as sand, silts, and clays to achieve the desired conditions. When sediment properties (eg, grain size, organic carbon) influence the test response, control sediments may be used to match these conditions (eg, silty and sandy grain-size controls). These controls are frequently referred to as "reference sediments" and are usually collected from sites near the contaminated assessment site but outside the sphere of influence of contaminant sources or at locations known to not contain the contaminants of interest that are present at the contaminated site. Parallel tests with reference sediments provide information on how noncontaminant factors that vary within the assessment area, including natural stressors and other anthropogenic toxicants beyond the chemical(s) of interest, may influence the test endpoints. Depending on the range of properties of the sediments being tested, the statistical comparisons of significance of biological responses may be made against the control or reference sediments. As these sediments act as "negative controls," the sediments should be relatively free of contaminants and nontoxic. Control charts are useful for monitoring the performance of test endpoints within and between testing programs (ASTM, 2013).

7.2.1.3 Characterization

As the bioavailability of contaminants influences their toxicity in sediments, it is important to consider how the collection, handling, and storage of the collected sediment alter the chemical and physical characteristics of the sediments (Bull and Williams, 2002; Simpson and Batley, 2003). Before toxicity testing many assessment programs will analyze sediments for contaminants of interest, as well as pH, redox potential, moisture content, grain size, organic carbon, AVS, particulate iron and manganese, and porewater constituents (eg, ammonia, sulfide, metals) (Vandegehuchte et al., 2013). For most sediment toxicity test procedures, the sediment will be thoroughly homogenized and distributed between replicate treatments before commencing the test, thus modifying the bioavailability and possibly the total concentrations of some contaminants. For test procedures that require sediments to be sieved to remove large particles and debris, or the separation of indigenous biota, or to be considerably disturbed in other ways, it is recommended that a description indicating how the sediment disturbance may have potentially modified the concentrations or bioavailability of the contaminants within the test sediments is provided together with the test results.

Following the placement of sediment and overlying water within test containers, it may be desirable to allow a period of days to weeks for the sediment to reequilibrate and redox stratification to reestablish. It is recommended that some assessment of how the treatment of the sediment before testing may have influenced contaminant concentrations (eg, loss of volatiles), bioavailability (eg, porewater and AVS), or other factors that may influence their toxicity.

7.2.2 Test Organisms

Organisms may be sourced from cultures (laboratory or aquaculture facilities) or collected from the field. Methods for culturing will be specific to the organism, and variations in culturing procedures may influence the sensitivity of the organisms used in the tests. The sex (male/female), life stage, or state (eg, juvenile/adult, gravid, size) should be specified, along with feeding regime and environmental exposure conditions (eg, salinity, temperature, light) in the days to weeks before tests. Along with the other factors that may influence the condition of the organisms, these should be consistent between replicates and sediments tested. Organisms that die, appear unhealthy, or behave atypically during the test preparation stages should be removed and discarded. Signs of unhealthy organisms include open bivalves, abnormal color, organisms that are "too relaxed" or don't respond to touch, or organisms that have no food in their gut.

Lower variability between test organisms may be more easily achieved using laboratory-cultured organisms than field-collected organisms, due to the greater difficulty in knowing the history of field organisms. To minimize stress to the organisms during transport from field collection sites or culturing facilities to the laboratory, and during pretest holding stages, the organisms should be provided a substrate to burrow in during transport to and in the laboratory (eg, 2–4 cm depth of field sediment with overlying water). The holding densities should be similar to those being used in the tests, and an acclimation period should be provided for adjusting the organisms to laboratory testing conditions (eg, 2–10 days) before test commencement (ASTM, 2012b). Where test conditions differ significantly from the field, eg, salinity, the acclimation to test conditions should be undertaken slowly, eg, 5‰ salinity and 3°C/24 h (ASTM, 2012b). Likewise, cultured organisms should be gradually adapted to the test pH, salinity, and temperature if test conditions are different than culture conditions.

For tests that assess effects specific to certain life stages, eg, reproductive endpoints, males and females need to be isolated to separate holding trays and gravid females or juveniles prepared in the days to weeks before test commencement.

7.2.3 Test Exposure Conditions and Setup

The test design will influence the exposure conditions that are achieved and the ability to detect the desired endpoint response. The exposure chamber dimensions and configuration are the first consideration, including volume of sediment and overlying water, and the water exchange rate necessary to achieve the desired exposure. To allow the development of an oxic sediment layer, which occurs in situ in most surficial sediments, and some equilibration of deeper sediments, it is recommended that sediments be added to toxicity test chambers at least 24 h before the addition of test organisms (USEPA, 2000; ASTM, 2014).

Static or semi-static designs, with water renewal occurring daily or less frequently, are often easier to configure than flow-through designs; the advantages or disadvantages often depend on the test species and duration. Owing to the flux of contaminants from sediments, static tests may frequently result in an overestimation of the exposure of the organisms to dissolved contaminants from the overlying water due to the lack of exchange/replacement of the overlying water. Flow-through designs will generally allow better simulation of overlying water conditions that may exist in the field, but often require larger volumes of clean overlying water to maintain and may not be practical if life stages being assessed are very small and frequently present in the water column. If field conditions are such that overlying water is the source of sediment contamination, then clean overlying water in a flow-through test may underestimate toxicity in a flow-through test. The frequency of water renewal should be specified for both static-renewal and flow-through procedures.

Ammonia concentrations in the porewaters should be characterized before test designs are finalized. However, any manipulation of sediments to reduce the influence of modifying factors such as ammonia should also be considered in relation to how they may modify the concentration or bioavailability of the contaminant exposure as well as other soluble contaminants (discussed in section on "Confounding Factors"). While it is ideal to have test organisms that can tolerate wide variations in environmental conditions so that the test can be applied to sediments with varying physicochemical properties, most organisms have narrow tolerance limits for some variables. For most organisms, temperature (eg, 21 ± 1°C), light (eg, 12 h dark/12 h light photoperiod and specified light intensity), pH (eg, 8.0 ± 0.2), salinity (eg, 30 ± 2‰), and dissolved oxygen concentrations (eg, 80–110%-saturation of DO) must be maintained within specific ranges. The "ruggedness" of a test is a term frequently used to describe the insensitivity of a test method to departures from specified test or environmental conditions. The desired ranges and limits for these conditions should be specified and reported to evaluate test performance and acceptability of the endpoint results.

7.2.4 Monitoring of Exposure Conditions

The concentrations and bioavailability of the contaminants of interest may change due to test manipulation, time required for test setup, and changes in physicochemical properties due to organism interactions during the tests. To relate any organism effects to exposure to a particular contaminant, it is useful to have sediment samples taken for analysis at the start and completion of tests, and potentially from surface sediments (eg, 0–0.5 cm depth) and deeper sediments to provide a more detailed analysis of exposure. The analyses may include contaminant concentrations, along with measurements that provide information on contaminant bioavailability (eg, pH, redox potential, grain size, TOC, AVS, porewater contaminant concentrations, and ammonia and sulfide) (Simpson et al., 2016; Vandegehuchte et al., 2013).

During the exposure, the standard water quality parameters (temperature, pH, salinity, dissolved oxygen) need to be monitored to check that they remain within the acceptable test range. Particularly for static and static-renewal tests, it is useful to monitor dissolved ammonia and target contaminants within the overlying water. Passive sampling techniques can be useful for monitoring concentrations of contaminants in the overlying waters and porewaters if the test chambers are large enough to accommodate them (USEPA, 2012; Perron et al., 2013; Lydy et al., 2014; Peijnenburg et al., 2014; Amato et al., 2014).

In addition to physicochemical measurements, biological behavioral observations should be recorded. These include sediment avoidance, failure to make burrows, cessation of eating or filtering, or any behavior that differs from the control/reference.

7.2.5 Test Acceptability, Data Analyses, and Interpretation

The quality of the experiments and data generated from the responses of the sediment toxicity tests should be considered from the stage of sampling collection to the final statistical analyses. This will include consideration of any general data quality objectives and principles outlined in Section 7.1.1. While the objective of the analyses of the test data is to quantify the changes in effects relative to control/reference sediments, the use of the data in risk assessment programs will also consider how the test conditions may resemble the exposure that may occur to organisms in the field.

7.2.5.1 Test Acceptability Criteria

Test acceptability criteria need to be developed and reported to confirm that desired conditions and accurate responses for all exposures are achieved. The criteria for determining test acceptability are specific to the test species, test conditions, procedures, and endpoint and may include both observations and measurements. Measured criteria including sediment and water properties and other confounding factors were discussed earlier in this chapter.

The use of positive and negative controls can assist in determining whether the response of the test organism is consistent with the test conditions and their inter- or intralaboratory performance. The control and reference sediments are intended to act as negative controls. Negative controls are usually run for every batch of tests and are often used for the hypothesis testing from which the observation of effects is determined. Positive controls are intended to identify whether the pool of test organisms used is behaving or responding in an abnormal manner. Reference toxicant tests, which expose the test organisms to a toxicant for which the endpoint response is known (within a specific range), are the most common form of positive control. These are used periodically to determine the health or responsiveness of cultured or field-collected organisms. Positive control tests should give EC50/LC50 values within ± 2 standard deviations (SD) of the mean value from quality control charts specific to that test organism and test conditions. Reference toxicants are intended to provide a general measure of the reproducibility (precision) of a toxicity test method or changes in cultured or collected test organisms over time. If results fall outside the criteria range, this may trigger an investigation. Control charts are recommended for tracking the performance

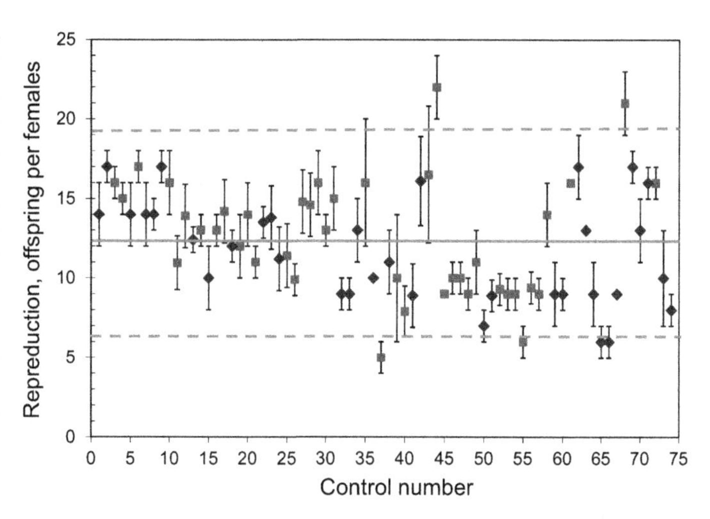

FIGURE 7.2 Example of control chart for tracking the performance of both negative and positive controls over time: total reproductive output for the amphipod (*Melita plumulosa*) (*Modified, with additional data, from Spadaro, D.A., Simpson, S.L., 2016b. Appendix E. Protocol for 10-day whole-sediment sub-lethal (reproduction) and acute toxicity tests using the epibenthic amphipod Melita plumulosa. In: Simpson, S.L., Batley, G.E. (Eds.), Sediment Quality Assessment: A Practical Handbook. CSIRO Publishing, Canberra, Australia, pp. 265–275.*) The *lines* show the mean (*solid line*) and two standard deviations of the mean (*dashed lines*). The *squares* indicate the more silty sediments (50% < 63 μm) and *diamonds* the more sandy sediments.

of both negative and positive controls over time (Fig. 7.2).

Reference toxicant tests are typically conducted as water only, short-term, eg, 96-h to 7-day static tests, and generally with exposure to a single concentration or dilution series of a single chemical, although spiked-sediment exposures may also be used. Ideally, these should be initiated using the same batch of field-collected test organisms and within a few days of commencing the whole-sediment tests. For organisms from cultures, reference toxicants tests may be undertaken on a routine basis, rather than for each batch of test organisms used in tests. If a sediment is spiked with a reference toxicant, careful consideration should be given to the spiking and equilibration period (Simpson et al., 2004).

7.2.5.2 Determination of Toxicity

The statistical analyses to determine whether toxicity has occurred are usually straightforward if the test acceptability criteria are met. These analyses will determine whether the effects to replicate groups of organisms in the test sediments are significantly different from those occurring in the replicate groups of control or reference sediments (ASTM, 2013). The test design will specify the degree of replication necessary to provide the statistical power to correctly detect toxic effects and evaluate the magnitude of response. Basic statistical endpoints such as percentage survival (mean ± SD) or the percentage of impairment (eg, growth, reproduction, behavior) should be calculated for each treatment and compared to the control and reference sediment results.

Initially data normality and homogeneity of variance should be tested using pairwise comparisons of treatment versus control (or reference) data (eg, by student's *t*-test or analysis of variance (ANOVA) followed by Dunnett's or Tukey's tests). If the requirements for normality and homogeneity of variance are not met, the data can be transformed and retested or, if the data still fail, a nonparametric test such as the Wilcoxon Rank Sum test can be used for the statistical comparison. The analysis of statistical differences between the responses of test organisms to control and test sediments is achieved by hypothesis testing usually with the level of significance of $\alpha = 0.05$ (ie, resulting in a 5% probability of a "false positive"). Note that both the likelihood for a "false positive"

(Type I error—detecting an effect that is not present) and "false negative" (Type II error—failing to detect an effect that is present) can generally be minimized during the design of a biological test through increased replication. However, the need for statistical replication needs to be balanced with laboratory feasibility. More detailed descriptions of statistical methods for toxicity test data analyses are provided elsewhere (ASTM, 2008a, 2013; OECD, 2006) and also described in Chapter 2 of this book.

Beyond the statistical analyses, criteria for determining whether a test sediment is toxic usually specify a magnitude of effect, eg, $\geq$20% lower survival or reproduction than the control or reference sediment. This magnitude will generally consider the past performance of the test (both inter- and intralaboratory) in relation to the variability in responses that are typical of the type of sediments being tested (Simpson and Spadaro, 2011). For example, the variability in response may allow the detection of significant differences but be greater for sandy sediments rather than silty sediments, and the differences in variability may need to be considered if classifying the sediments as slightly or highly toxic. Beyond that, additional magnitudes of toxicity may be set on an arbitrary basis (eg, 20–50% difference = moderately toxic and >50% differences = highly toxic). Magnitudes of toxicity can become important when ranking series of toxicity test results for multiple bioassays.

7.2.5.3 *Calculation of Effect Thresholds (LC50, EC/IC10)*

Where studies have been designed to provide a concentration gradient of a contaminant, it may be possible to use the effects data to calculate effect thresholds for the contaminant, eg, EC/IC10 and EC/IC50 values. Effect thresholds and the associated uncertainty should be calculated based on the measured concentration of the chemical, eg, EC50 = 50 (40–60) mg copper/kg dry weight (where the bracketed values are confidence limits). Nominal concentrations of the chemical should only be used where measurements are not possible, and uncertainties due to variation from nominal concentration must be described. In addition to reporting the effect threshold, the partitioning of the chemical between the dissolved and particulate phases should be described (eg, the partition coefficient, Kd) to provide information on the exposure route contributing to the effect.

When the intention of the effects data is for use in the construction of species sensitivity distributions (SSDs) for sediment contaminants (Simpson et al., 2011; Vangheluwe et al., 2013) or to establish site-specific management limits (Simpson et al., 2013), information on contaminant speciation and exposure routes becomes even more important. For sediment contaminants, the desired output of SSDs will usually be % hazard or % species-protection concentrations in units of mg/kg dry weight. The no-effect threshold for each species in the SSD will be specific to the sediment properties, as they influence contaminant speciation and relative contributions of dissolved and particulate contaminant exposure. This remains a significant challenge for the derivation of guideline values that can be applied to field sediments with properties that differ significantly from the properties of the sediments used to derive effect thresholds (eg, different particle size, AVS, TOC).

7.3 TOXICITY IDENTIFICATION EVALUATION FOR WHOLE SEDIMENTS

Use of TIE procedures for dissolved toxicants in natural waters is now well established (eg, USEPA, 1996). TIE procedures for sediments (either whole sediments or porewaters) (USEPA, 2007) are expanding and are being considered for regulatory purposes. This is largely because identification of toxicant classes affecting aquatic ecosystems, in particular the health of benthic organisms, is becoming an increasingly important part of sediment quality assessment programs (NFESC, 2003; USEPA, 2007; Ho and Burgess, 2009; Ho et al., 2009, 2013; Burgess et al., 2011; Araujo et al., 2013; Camargo et al., 2015). TIEs involve the manipulation of sediments, or sediment components (eg, porewaters) to remove or alter bioavailability of an individual toxicant (eg, ammonia) or class of contaminants (eg, hydrophobic organics, metals) from the porewaters or whole sediments, thus allowing identification of an individual toxicant or class of toxicants responsible for the observed toxicity (Ankley and Schubauer-Berigan, 1995; Burgess et al., 2003, 2004, 2007, 2011; Ho et al., 2002, 2004; USEPA, 2007). In the USEPA (2007) framework, TIE methods are divided into three phases: characterization, identification, and confirmation.

The decision to use porewater or whole-sediment TIEs is based on many of the same factors used to decide between porewater or whole-sediment toxicity tests. Porewater testing for TIEs has the advantage of being able to use many of the established manipulations developed for effluent TIEs. Other advantages of using porewater TIEs include the ability to use test organisms that are not compatible with a sediment matrix, assuming that porewaters are considered a major route of exposure of many toxicants, and there is a good understanding of how sample manipulation affects water chemistry.

However, there are a number of factors that disadvantage porewater TIE procedures, including many of the sample manipulation artifacts that affect contaminant speciation and bioavailability of porewater toxicity testing. Isolation of porewater for testing or TIEs results in changes in the equilibrium of many chemicals, eg, metals may become oxidized, organic contaminants and organic carbon in the porewaters may precipitate, the equilibrium of contaminants between the porewaters and the sediment particles is disrupted and cannot be reestablished because the porewater has been removed from the system. In addition, organisms whose exposure routes are usually a mixture of porewater and overlying water (burrowing amphipods or worms) are now exposed to 100% porewater, increasing their exposure to water-soluble contaminants such as ammonia, metals, and sulfides. This may be the reason why early porewater TIE studies implicated ammonia as a toxicant much more often than later whole-sediment TIEs. In general, whole-sediment TIEs are the preferred method if they can be successfully performed. Reasons why they cannot be performed include toxicity is found only in the porewaters and not the whole-sediment phase, or methods for identification of toxicants are more advanced with the aqueous phase relative to the whole sediments.

The desire to incorporate ecological realism and procedural differences between porewater and whole-sediment toxicity test methods (including sample preparation) make it desirable to have TIE procedures that can be applied to whole sediments. Whole-sediment TIEs are expected to be more accurate and provide more realistic exposure pathways for organisms. To date three major TIE separations have been applied:

1. The addition of a metal-chelating resin to sediments has been found to be a useful whole-sediment TIE method that reduced the concentration and toxicity of metals but had only minor effects on the toxicity of ammonia and a nonpolar toxicant present in the sediments (Burgess et al., 2000). The resin and accumulated metals were able to be isolated from the test system following exposures, allowing for the initiation of the identification stage of the TIE procedure.
2. Removal of ammonia toxicity was achieved in whole-sediment toxicity tests by addition of the marine algae (*Ulva lactuca)* or zeolite (Besser et al., 1998; Ho et al., 1999b; Burgess et al., 2004).

3. Powdered coconut charcoal addition has been found to effectively remove the toxicity of organic contaminants such as PAHs, PCBs, and pesticides (Ho et al., 2004).

The incorporation of knowledge of contaminant exposure pathways sensitivity to selected contaminants can further improve future TIE methodologies. Further, recent advances in genomic biomarkers create potential for TIEs to identify specific toxicants and pathways, not just toxicant classes (Biales et al., 2013; Hook et al., 2014b).

7.4 MORE FROM LESS (FUTURE POSSIBILITIES AND PROBABILITIES)

The key challenge in ecological risk assessment of chemicals is to determine the probability and extent of an adverse effect occurring in an ecological system with the ultimate goal of protecting the long-term viability of populations, communities, and ecosystems. Currently, risk assessment schemes are based on proxies: standard tests under constant and typically favorable laboratory conditions, which provide data only on organism-level endpoints over a relatively small number of species. For chemical risk assessment, tests are for one chemical at a time, and based on United Nations Environment Programme data there are over 100,000 different chemical substances in use today (UNEP, 2010). For assessment of sediment quality at sites impacted by many chemicals, considerable challenges remain in predicting the effects of mixtures and how bioavailability modifying factors and noncontaminant stressors modify the risks posed by chemical contamination. The total number of marine species known to us is 212,000 species; however, it is estimated that there are 1.4–1.6 million marine species on earth (Bouchet, 2006). Increasingly, humans are looking to access resources from more remote and less well-characterized environments, for example, mining operations in the deep-sea (Collins et al., 2013) or polar regions. These activities will expose entirely new species to chemicals and may require the development of new forms of sediment ecotoxicology methods for assessing risks in these environments. For example, methods appropriate in water depths of greater than 2000 m (eg, organisms that live without light and under higher pressure), or in near-frozen waters with short summers, will need to be developed. No calculation is needed to realize that testing all species and chemical combinations under the myriad of possible environmental conditions is impractical.

7.4.1 Derivation of Bioavailability-Based Sediment Quality Guidelines

Since the introduction of whole-sediment toxicity testing to ecological risk assessment frameworks, challenges have arisen regarding the differences in contaminant bioavailability for different sediment types (Chapman et al., 1998; Simpson and Batley, 2007; Maruya et al., 2012). The most well-recognized bioavailability considerations for contaminants are the normalization of nonionic hydrophobic organics contaminant concentrations (HOCs) to the concentration of sediment organic carbon (Di Toro and McGrath, 2000; USEPA, 2003, 2012) and the acid-volatile sulfide (AVS)—simultaneously extracted metals (SEM) theory to predict the absence of toxicity when the molar concentration of available sulfide (AVS) exceeds that of SEM (USEPA, 2005). However, both approaches come with limitations; the bioavailability predictions for HOCs are complicated by presence of black carbon (eg, pyrogenic carbon such as soot, coal tars, coal, and residues of incomplete combustion such as charcoal) (USEPA, 2003, 2012), and AVS-SEM theory generally predicts nontoxicity rather than toxicity as it does not account for the additional influence on metal bioavailability from the portion of clay/silt, organic carbon, and iron and manganese oxyhydroxide phases that are not accounted for in the AVS calculation (Strom et al., 2011; Simpson et al., 2011; Campana et al., 2012, 2013; Besser et al., 2013). Improved understanding of bioavailability modifying factors is now facilitating the development of bioavailability-based sediment quality guidelines. Using a range of standardized chronic ecotoxicity tests predicted no effects concentrations (PNECsediment) may be derived from SSDs, and bioavailability models provide the means to tailor the PNECsediment for all sediment types and conditions.

The establishment of PNECsediment values for sediment contaminants may initially consider chronic effects for single species (eg, Fox et al., 2014). For metals in particular, it is now considered necessary to incorporate bioavailability into PNECsediment derivations (Simpson et al., 2011; Campana et al., 2013; Schlekat et al., 2015). The best current example of this is for nickel and freshwater sediments (Schlekat et al., 2015). This study generated chronic ecotoxicity data for nine benthic species, and by examining bioavailability relationships for many of these species allows an SSD to be developed for benthic organisms in various sediment types and a reasonable worst case PNECsediment for nickel. For marine ecotoxicology, there may not yet be the necessary number of standardized chronic tests available; however, provided that the principles of sound ecotoxicology are adhered to, many of the methods outlined in Table 7.1 will be capable of providing effect data of a suitable level of environmental and statistical quality as standardized methods. Using an example of copper in marine sediments with varying properties, combinations of SSDs for a single sediment type (Simpson et al., 2011) accompanied by models (Simpson and King, 2005; Simpson, 2005) or detailed studies of bioavailability modifying factors (Strom et al., 2011; Campana et al., 2012) may also be suitable for deriving bioavailability-based PNECs that can be applied with risk assessment programs (Simpson et al., 2013). The use of whole-sediment tests with spiked sediments is a fundamental part of developing the necessary bioavailability relationships (Campana et al., 2012; Besser et al., 2013), with care to ensure the contaminant exposure created through spiking and equilibration provides the desired partitioning between sediments and porewater to allow concentration–toxicity response relationships to be established for various benthic invertebrate species (Simpson et al., 2004; Hutchins et al., 2008; Brumbaugh et al., 2013).

7.4.2 Novel Sediment Exposure Methods to Assess Ecological–Ecotoxicology Effects

Following the principles for sound ecotoxicology (Section 7.1.1), standard methods for whole-sediment ecotoxicology have been, or can now be, developed for a wide range of organisms from a wide range of environments. However, a major dilemma for assessments remains that, for most assessment environments, we have no effects data for the many other species, eg, >99% of species present. How do we ensure that we are protecting these other species when deriving PNECsediment values?

The control of exposure conditions provided by laboratory-based sediment toxicity test procedures is a major reason why effects threshold for contaminants can be derived more easily than field-based benthic ecology assessments. The dilemma is that laboratory-based sediment toxicity tests do not provide information on the many benthic organisms that make up a community nor do they allow for organism interactions that may influence both contaminant exposure and well-being. Conversely, field-based experiments (eg, in mesocosms with natural organism assemblages or in field recolonization experiments, Olsgard, 1999; Chariton et al., 2011; Hill et al., 2013) may provide information on the sensitivity of many benthic organisms and allow for community interactions; however, as the exposures of many contaminants are difficult to control and measure within the field exposure chambers, the resulting effect relationships are difficult to interpret.

Ho et al. (2013) demonstrated that some of the difficulty in interpreting community endpoints may be resolved through the use a novel hybrid exposure method that brings intact sediment cores into the laboratory and then exposes the existing benthic communities by addition of toxicant-spiked sediments to the core surface. The approach was an adaption of that described by Chandler et al. (1997), used a sediment core to bring intact benthic meio- and macrofaunal communities into the laboratory, and exposed them to a 2-cm layer of contaminant-spiked sediment through which the "community" must migrate vertically into to reach the oxygenated layer and survive. After a two-week exposure to the contaminant layer, a 2-cm layer of clean "DNA-free" sediment is added and those organisms that survive the contaminant exposure again must migrate vertically to the oxidized surface layer. The surviving community is assessed after another week. The use of DNA-free sediments to create the layers into which only live organisms can migrate allows genomic endpoints to be included, as these endpoints cannot distinguish between live and dead DNA (Chariton et al., 2014). The approach has been applied successfully to examine the effect on estuarine meio- and macrobenthic communities of triclosan-spiked sediments (Ho et al., 2013; Chariton et al., 2014), bifenthrin, and nano copper (unpublished).

7.4.3 Toxicity Is Dynamic (Modeling Our Way Out of the Black Box)

Further impacting our current risk assessment approaches is the lack of environmental realism inherent in many methods, which generates a high level of uncertainty for the actual consequences of environmental contamination of ecosystems. At the origins of this uncertainty is the high variability in time and space of ecological factors (eg, competition, predation, resource limitations), environmental conditions (eg, habitat quality, physical stressors), and the chemical stress itself (eg, mixtures of chemicals). Therefore, the likelihood and degree of ecological effects of many chemicals often remains uncertain when extrapolating beyond the testing conditions.

Given the impracticality of achieving assessments for many chemicals in all the necessary environments, concurrently with improving the range of methods available for robust assessment of chronic effects of chemicals in sediments, there is a need for models that allow accurate extrapolation of effects (and risks) to other environments and conditions. This may only be achieved through developing the ability to predict toxicity through a stronger understanding of the factors influencing the routes of exposure and mechanisms through which chemicals cause toxicity to benthic organisms.

A current weakness in toxicity testing methods (both aqueous and sediment) relates to our assessment of endpoints for single exposure periods. There is a need to think about organisms as dynamic systems and move away from the traditional "black box" animal-based paradigm of dosing an organism with a chemical and looking at the end result. Conventional tests focus on collecting knowledge about few final adverse outcomes such as survival, growth, or reproduction that provide valuable, but mostly descriptive, information without any insight into the mechanisms that regulate such outcome. The pathways to toxicity are dynamic. At low concentrations of chemicals, toxicity might be reversible and organisms might recover through adaptive responses; at higher concentrations the changes might be irreversible and eventually lead to impaired biological function or death. Whichever pathway occurs, the generation and analysis of quality data describing life-history traits (survival, growth, reproduction) over time is critical to accurately represent individual processes relevant for population dynamics and the effects of chemical stress on those processes.

Advances in science have led to a new vision for predicting toxicity, based on strong scientific knowledge of the underpinning pathways of toxicity. The following paragraphs introduce the reader to different emerging technologies and concepts that have recently created amazing opportunities for a more modern approach to ecotoxicology.

7.4.3.1 Mechanistic Effect Models

Ecological and mechanistic effect modeling may help to implement the above vision (EC Report, 2012). Mechanistic effect models such as toxicokinetic-toxicodynamic (TK-TD), Dynamic Energy Budget (DEB) or Individual-based models (IBMs) of populations, described further below, are valuable tools that explicitly represent biological and chemical processes over time and translate effects on individuals into effects on ecological systems such as population and communities.

The following overview describes some mechanistic effect models; a complete insight into modeling approaches in ecotoxicology is presented in Chapter 3 of this book.

TK-TD models simulate the time course of external concentrations of a toxicant (in water or sediment) leading to the effects on life-history traits of the exposed organism over time. Double-step TK-TD modeling applies toxicokinetics and toxicodynamics concepts. Toxicokinetics includes the processes of absorption (uptake, bioaccumulation), distribution, biotransformation, and elimination of a toxicant within the organism (what the organism does with the chemical). The toxicodynamic step quantitatively links the internal concentration to the effect at the level of individual organisms over time (what the chemical does to the organism).

Rather than designing a new model for each species, generic models are constructed to capture the processes animals share in common. Strong evidence suggests that the metabolic processes that regulate individual characteristics relevant for understanding population dynamics are conserved among species. For example, despite the vast morphological, developmental, and behavioral differences among species, nearly all follow the characteristic Von Bertalanffy growth pattern (Von Bertalanffy, 1957). DEB theory (Kooijman et al., 1989) is built on this premise and tries to explain the diversity of key physiological life traits (ie, growth and reproductive output over time) via a shared set of processes based on first principles, such as conservation of mass energy. Focusing on commonalities rather than differences allows generic models, as DEB-based models, to conserve model structure from one species to the next because differences among species are characterized through variation in the model's parameters, which regulate energy acquisition and allocation. The change in energy allocation patterns in response to a toxicant reveals the "physiological modes of action." The ability to understand the details of the physiological modes of action of toxicants highlights another important potential use of DEB-based models: extrapolation to population-level effects. For example, an observed reduction in cumulative reproduction might be due to either mortality of embryos during the embryonic stage or to a reduction in feeding of the mother (ie, less energy available to allocate reproduction), but these two different physiological modes of action might have very different consequences for population structure and dynamics.

IBMs (Martin et al., 2012, 2013) are based on the premise that population-level processes emerge from what individuals do. These models represent individual organisms as unique entities that differ from each other and change over their life cycle. The population dynamics emerge via interactions of individuals with each other or their abiotic environment.

Routinely used, mechanistic effect models can help close the gap between laboratory tests on individuals and ecological systems in real landscapes because they can integrate ecological factors and environmental conditions with chemical effects.

7.4.4 New Technologies and Visions for Ecotoxicology

A new vision for toxicity testing moves toward more predictive, higher throughput, and lower cost methods while reducing organisms use and time required for chemical testing. So far only minimal levels of automation have been adopted in ecotoxicology.

7.4.4.1 Lab-On-A-Chip Technology

Microfluidic chips, also called lab-on-a-chip or micro total analysis systems (μTAS), integrate many biological and chemical operations on a single microchip. In combination with microscopy-based readouts, as automated imaging systems, these devices provide a new tool to miniaturize, automatize, and parallelize assays, simultaneously increasing accuracy and resolution. For toxicity testing, these integrated microsystems allow reducing experimental costs and analysis time. Fabricated using nontoxic, transparent, and inexpensive polymers (eg, polydimethylsiloxane), they can be molded to obtain a microstructure that mimics the environment for microorganism maintenance so researchers can investigate physiological functions (Zheng et al., 2014). Using channels or chamber arrays, valves, mixers, and other building blocks with typical size on the order of tens of microliters, in a single microfluidic device, it is possible to generate multiple toxicant concentration gradients and conduct simultaneously high-throughput screening for different compounds with the advantage of maintaining well-controlled microenvironment parameters (eg, dissolved oxygen, temperature, toxicant concentration).

Microfluidic platforms have been created to obtain high definition imaging of morphological features of zebrafish larvae (*Dario rerio*) using 3D environmental scanning electron microscopy without any need for staining protocols (Akagi et al., 2014) or to detect sublethal changes in swimming activity of the marine amphipod (*Allorchestes compressa)* using miniaturized video cameras with a complete automation analysis approach (Cartlidge et al., 2015). Zheng et al. (2013) demonstrated that, using integrated microfluidic devices, marine microalgae can be successfully cultured and exposed on-chip, providing online measurements of multiple biological responses (eg, cell division rate, autofluorescence, esterase activity).

7.4.4.2 Adverse Outcome Pathways

As discussed earlier, the AOP is a conceptual framework for organizing and describing existing knowledge on the toxicity mechanisms and outcomes across levels of biological organization. An AOP represents the progression from a molecular initiating event (MIE; a direct interaction of a toxicant with its molecular target) that triggers a biological perturbation and occurrence of an adverse outcome (AO) considered relevant to regulation decision making (Ankley et al., 2010). In ecotoxicology, an AO might describe an impact on survival, growth, or reproduction at the individual level, or a decline at a higher level of biological organization, eg, at the population level. The sequence of events necessary to cause an AO, initiated by MIE, are a series of measurable/observable biological changes at the cellular, tissue, organ, or individual level that are essential (defined key events; KEs), while key event relationships (KERs) provide the description of linkages (qualitative and quantitative) between MIEs, KEs, and AOs (Groh et al., 2015).

The basic principle that underlies AOP developments focuses on AOPs as generalizable and not chemical-specific. KEs and KERs are described independently and are generally linked in a nonbranching sequence to define an individual AOP. AOPs can also be linked in a broader network context that considers potential interactions, cumulative impacts of multiple perturbations, similarities, and differences in response as a function of life stage, sex, taxa, etc. (Villeneuve et al., 2014a,b). Validating an AOP through the identification and quantitative definition of KERs would allow prediction of an AO based on measurements of KEs at the beginning of the pathway without the need of performing whole-organism toxicity tests or field studies to directly observe the AO, which can be expensive or even impossible to test (Groh et al., 2015).

Documentation and elucidation of all key events along the AOP pathway can be very complex and difficult to complete. In 2013, a *Guidance Document on Developing and Assessing Adverse Outcome Pathways* was published (OECD, 2013), whose purpose was to provide the best practices for the AOP development and description. It is expected that measurements and scientific support that underlie AOP development will evolve over time, increasing the AOP knowledge base.

The transition to an AOP-based paradigm for chemical safety assessment also focuses on the integration of existing in vivo data with in vitro and in silico approaches. In vitro methods are widely used to investigate toxicity of chemicals based on the study of factors that control the time course for adsorption, distribution, metabolism, and excretion (ADME) in target tissues. However, one of the major drawbacks of in vitro tests is their limited power to predict toxicity in vivo because the microenvironment generated during cell culture is very different from the in vivo environment. Microfluidic technologies might finally enable the fabrication of advanced in vitro systems allowing the culture of in vitro preparations (tissue or cells) that preserve all the properties of their in vivo original source for prolonged periods of time, providing means of reflecting in vivo toxicokinetics in vitro and establishing clear relationship between in vitro endpoints and adverse effects in vivo. For example, in an effort to bypass animal testing and take into account the complex interactions between different organs and tissues, multitissue-based microfluidic devices mimic, on a single chip, some of the physiologically relevant processes (ADME) in the organism that describe the toxicokinetic models (Baker, 2011; van Midwoud et al., 2011). Combining these processes with environmental bioavailability factors is a future challenge, ie, to link these assays to whole organism sediment toxicity assessment.

7.4.5 Conclusions

The application of these new technologies and approaches to ecotoxicology in general and to sediment toxicity testing in particular, presents obvious limitations and challenges. The complexity in the application of mechanistic models and the resources needed to perform time-intensive sediment toxicity tests necessary for model input have presented a serious hurdle for their use in ecotoxicology. However, for example, mechanistic modeling approach has already been applied to different studies using sediment toxicity tests to analyze the effect of toxicants on mollusks (Ducrot et al., 2007), insects (Beaudouin et al., 2012), polychaetes (Jager and Selck, 2011), and marine microalgae (Miller et al., 2010). Regarding AOPs, an important limitation is represented by their own definition as nonchemical-specific entities that do not require, as of yet, knowledge of external or internal exposure. This prevents the use of bioavailability data or toxicokinetic processes that give critical information, eg, what are the external or internal concentrations of the toxicant that trigger MIE or how the frequency and duration of the exposure affect the activation of MIE. Clearly, the application of lab-on-a-chip technology for sediment toxicity testing is still in the future. The problem of mimicking bioavailability data that are modified by sediment properties or the organism itself is a challenge that microfluidic technologies are not probably able to solve in the short term. However, their innovative ability to spatiotemporally control microenvironments with higher resolution and accuracy could be used to extrapolate toxicokinetic data from in vitro systems to predicted in vivo outcomes and, in combination with well-established conventional tests for assessing external exposure, limit the need for animal testing.

References

Adams, M.S., Stauber, J.L., 2004. Development of a whole-sediment toxicity test using a benthic marine microalga. Environ. Toxicol. Chem. 23, 1957–1968.

Adams, M.S., 2016. Appendix D. Protocol for whole-sediment bioassay using the marine microalga *Entomoneis* cf *punctulata*. In: Simpson, S.L., Batley, G.E. (Eds.), Sediment Quality Assessment: A Practical Handbook. CSIRO Publishing, Canberra, Australia, pp. 255–264.

Akagi, J., Zhu, F., Hall, C.J., Crosier, K.E., Crosier, P.S., Wlodkowic, D., 2014. Integrated chip-based physiometer for automated fish embryo toxicity biotests in pharmaceutical screening and ecotoxicology. Cytometry 85A, 537–547.

Aller, J.Y., Woodin, S.A., Aller, R.C., 2001. Organism-sediment Interactions. University of South Carolina Press, Columbia, SC, USA.

Amato, E.D., Simpson, S.L., Jarolimek, C., Jolley, D.F., 2014. Diffusive gradients in thin films technique provide robust prediction of metal bioavailability and toxicity in estuarine sediments. Environ. Sci. Technol. 48, 4485–4494.

Anderson, B.S., Hunt, J.W., Hester, M., Phillips, B.M., 1996. Assessment of sediment toxicity at the sediment–water interface. In: Ostrander, G.K. (Ed.), Techniques in Aquatic Toxicology. CRC Press, Boca Raton, FL, USA, pp. 609–624.

Ankley, G.T., Schubauer-Berigan, M.K., 1995. Background and overview of current sediment toxicity identification evaluation procedures. J. Aquat. Ecosyst. Health 4, 133–149.

Ankley, G.T., Katko, A., Arthur, J.W., 1990. Identification of ammonia as an important sediment-associated toxicant in the lower Fox River and Green Bay, Wisconsin. Environ. Toxicol. Chem. 9, 312–322.

Ankley, G.T., Erickson, R.J., Phipps, G.J., Mattson, V.R., Kosian, P.A., Sheedy, B.R., Cox, J.S., 1995. Effects of light intensity on the phototoxicity of flouranthene to a benthic invertebrate. Environ. Sci. Technol. 29, 2828–2833.

Ankley, G.T., Bennett, R.S., Erickson, R.J., Hoff, D.J., Hornung, M.W., Johnson, R.D., Mount, D.R., Nichols, J.W., Russom, C.L., Schmieder, P.K., Serrrano, J.A., Tietge, J.E., Villeneuve, D.L., 2010. Adverse outcome pathways: a conceptual framework to support ecotoxicology research and risk assessment. Environ. Toxicol. Chem. 29, 730–741.

Araujo, C.V.M., Tornero, V., Lubian, L.M., Blasco, J., van Bergeijk, S.A., Canavate, P., Cid, A., Franco, D., Prado, R., Bartual, A., Lopez, M.G., Ribeiro, R., Moreira-Santos, M., Torreblanca, A., Jurado, B., Moreno-Garrido, I., 2010. Ring test for whole-sediment toxicity assay with -a- benthic marine diatom. Sci. Tot. Environ. 408, 822–828.

Araujo, C.V.M., Blasco, J., Moreno-Garrido, I., 2012. Measuring the avoidance behaviour shown by the snail *Hydrobia ulvae* exposed to sediment with a known contamination gradient. Ecotoxicology 21, 750–758.

Araujo, G.S., Moreira, L.B., Morais, R.D., Davanso, M.B., Garcia, T.F., Cruz, A.C.F., Abessa, D.M.S., 2013. Ecotoxicological assessment of sediments from an urban marine protected area (Xixova-Japui State Park, SP, Brazil). Mar. Pollut. Bull. 75, 62–68.

ASTM (American Society for Testing and Materials), 2008a. Standard Guide for Conducting 10-day Static Sediment Toxicity Tests with Marine and Estuarine Amphipods (E1367-03(2008)). Annual Book of ASTM Standards, Vol 11.06. West Conshohocken, PA, USA. http://www.astm.org/Standards/E1367.htm.

ASTM, 2008b. Standard Guide for Designing Biological Tests with Sediments (E1525-02(2008)). Annual Book of ASTM Standards, Vol 11.06, West Conshohocken, PA, USA. http://www.astm.org/Standards/E1525.htm.

ASTM, 2008c. Standard Guide for Collection, Storage, Characterization, and Manipulation of Sediments for Toxicological Testing and for Selection of Samplers Used to Collect Benthic Invertebrates (E1391–03(2008)). Annual Book of ASTM Standards, Vol 11.06, West Conshohocken, PA, USA. http://www.astm.org/Standards/E1391.htm.

ASTM, 2010. Standard Test Method for Measuring the Toxicity of Sediment-associated Contaminants with Freshwater Invertebrates (E1706–05(2010)). Annual Book of ASTM Standards, Vol 11.06, West Conshohocken, PA, USA. http://www.astm.org/Standards/E1706.htm.

ASTM, 2012a. Standard Guide for Conducting Static Acute Toxicity Tests Starting with Embryos of Four Species of Saltwater Bivalve Molluscs (E724-98(2012)). Annual Book of ASTM Standards, Vol 11.06. West Conshohocken, PA, USA. http://www.astm.org/Standards/E724.htm.

ASTM, 2012b. Standard Guide for Selection of Resident Species as Test Organisms for Aquatic and Sediment Toxicity Tests (E1850-04(2012)). Annual Book of ASTM Standards, Vol 11.06, West Conshohocken, PA, USA. http://www.astm.org/Standards/E1850.htm.

ASTM, 2013. Standard Practice for Statistical Analysis of Toxicity Tests Conducted Under ASTM Guidelines (E1847-06(2013)). Annual Book of ASTM Standards, Vol 11.06, West Conshohocken, PA, USA. http://www.astm.org/Standards/E1847.htm.

ASTM, 2014. Standard Test Method for Measuring the Toxicity of Sediment-associated Contaminants with Estuarine and Marine Invertebrates (E1367–03(2014)). Annual Book of ASTM Standards, Vol 11.06. West Conshohocken, PA, USA. http://www.astm.org/Standards/E1367.htm.

Azur Environmental, 1998. Microtox Acute Toxicity Solid Phase Test. Microtox® Manual, Carlsbad, CA, USA.

Baker, M., 2011. Tissue model: a living system on a chip. Nature 471, 661–665.

Bat, L., Raffaelli, D., 1998. Sediment toxicity testing: a bioassay approach using the amphipod *Corophium volutator* and the polychaete *Arenicola marina*. J. Exp. Mar. Biol. Ecol. 226, 217–239.

Batley, G.E., Simpson, S.L., 2016. Sediment sampling, sample preparation and general analysis. In: Simpson, S.L., Batley, G.E. (Eds.), Sediment Quality Assessment: A Practical Handbook. CSIRO Publishing, Canberra, Australia, pp. 15–46.

Beaudouin, R., Dias, V., Bonzom, J.M., 2012. Individual-based model of *Chironomus riparius* population dynamics over several generation to explore adaptation following exposure to uranium-spiked sediments. Ecotoxicology 21, 1225–1239.

Belzunce-Segarra, M.J., Simpson, S.L., Amato, E.D., Spadaro, D.A., Hamilton, I., Jarolimek, C., Jolley, D.F., 2015. Interpreting the mismatch between bio-accumulation occurring in identical sediments deployed in field and laboratory environments. Environ. Pollut. 204, 48–57.

Besser, J.M., Ingersoll, C.G., Leonard, E.N., Mount, D.R., 1998. Effect of zeolite on toxicity of ammonia in freshwater sediments: implications for toxicity identification evaluation procedures. Environ. Toxicol. Chem. 17, 2310–2317.

Besser, J.M., Brumbaugh, W.G., Ingersoll, C.G., Ivey, C.D., Kunz, J.L., Kemble, N.E., Schlekat, C.E., Garman, E.R., 2013. Chronic toxicity of nickel-spiked freshwater sediments: variation in toxicity among eight invertebrate taxa and eight sediments. Environ. Toxicol. Chem. 32, 2495–2506.

Biales, A.D., Kostich, M., Burgess, R.M., Ho, K.T., Bencic, D.C., Flick, R.L., Portis, L.M., Pelletier, M.C., Perron, M.M., Reiss, M., 2013. Linkage of genomic biomarkers to whole organism end points in a toxicity identification evaluation (TIE). Environ. Sci. Technol. 47, 1306–1312.

Boldina-Cosqueric, I., Amiard, J.-C., Amiard-Triquet, C., Dedourge-Geffard, O., Métais, I., Mouneyrac, C., Moutel, B., Berthet, B., 2010. Biochemical, physiological and behavioural markers in the endobenthic bivalve *Scrobicularia plana* as tools for the assessment of estuarine sediment quality. Ecotoxicol. Environ. Saf. 73, 1733–1741.

Bouchet, P., 2006. The magnitude of marine biodiversity. In: Duarte, C. (Ed.), The Exploration of Marine Biodiversity: Scientific and Technological Challenges. Fundación BBVA, Bilbao, Spain, pp. 31–62.

Bridges, T.S., Farrar, J.D., 1997. The influence of worm age, duration of exposure and endpoint selection on bioassay sensitivity for *Neanthes arenaceodentata* (Annelida: Polychaeta). Environ. Toxicol. Chem. 16, 1650–1658.

Bridges, T.S., Farrar, J.D., Duke, B.M., 1997. The influence of food ration on sediment toxicity in *Neanthes arenaceodentata* (*Annelida*: Polychaeta). Environ. Toxicol. Chem. 16, 1659–1665.

Brumbaugh, W.G., Besser, J.M., Ingersoll, C.G., May, T.W., Ivey, C.D., Schlekat, C.E., Rogevich-Garman, E., 2013. Preparation and characterization of nickel spiked freshwater sediments for toxicity tests: toward more environmentally realistic nickel partitioning. Environ. Toxicol. Chem. 32, 2482–2494.

Bull, D.C., Williams, E.K., 2002. Chemical changes in an estuarine sediment during laboratory manipulation. Bull. Environ. Contam. Toxicol. 68, 852–861.

Burgess, R.M., Cantwell, M.G., Pelletier, M.C., Ho, K.T., Serbst, J.R., Cook, H.F., Kuhn, A., 2000. Development of a toxicity identification evaluation procedure for characterizing metal toxicity in marine sediments. Environ. Toxicol. Chem. 19, 982–991.

Burgess, R.M., Pelletier, M.C., Ho, K.T., Serbst, J.R., Ryba, S.A., Kuhn, A., Perron, M.M., Raczelowski, P., Cantwell, M.G., 2003. Removal of ammonia toxicity in marine sediment TIEs: a comparison of *Ulva lactuca*, zeolite and aeration methods. Mar. Pollut. Bull. 46, 607–618.

Burgess, R.M., Perron, M.M., Cantwell, M.G., Ho, K.T., Serbst, J.R., Pelletier, M.C., 2004. Use of zeolite for removing ammonia and ammonia-caused toxicity in marine toxicity identification evaluations. Arch. Environ. Contam. Toxicol. 47, 440–447.

Burgess, R.M., Perron, M.M., Cantwell, M.G., Ho, K.T., Pelletier, M.C., Serbst, J.R., Ryba, S.A., 2007. Marine sediment toxicity identification evaluation methods for the anionic metals arsenic and chromium. Environ. Toxicol. Chem. 26, 61–67.

Burgess, R., Ho, K., Biales, A., Brack, W., 2011. Recent developments in whole sediment toxicity identification evaluations: innovations in manipulations and endpoints. In: Brack, W. (Ed.), Effect-directed Analysis of Complex Environmental Contamination. Springer, Berlin, Germany, pp. 19–40.

Burton, G.A., Johnston, E.L., 2010. Assessing contaminated sediments in the context of multiple stressors. Environ. Toxicol. Chem. 29, 2625–2643.

Camargo, J.B.D.A., Cruz, A.C.F., Campos, B.G., Araújo, G.S., Fonseca, T.G., Abessa, D.M.S., 2015. Use, development and improvements in the protocol of whole-sediment toxicity identification evaluation using benthic copepods. Mar. Pollut. Bull. 91, 511–517.

Campana, O., Spadaro, D.A., Blasco, J., Simpson, S.L., 2012. Sublethal effects of copper to benthic invertebrates explained by changes in sediment properties and dietary exposure. Environ. Sci. Technol. 46, 6835–6842.

Campana, O., Blasco, J., Simpson, S.L., 2013. Demonstrating the appropriateness of developing sediment quality guidelines based on sediment geochemical properties. Environ. Sci. Technol. 47, 7483–7489.

Carr, R.S., Nipper, M.J., 2003. Porewater Toxicity Testing. Society of Environmental Toxicity and Chemistry (SETAC), Pensacola, FL, USA.

Cartlidge, R., Nugegoda, D., Wlodkowic, D., 2015. Gammarus Chip: innovative lab-on-a-chip technology for ecotoxicological testing using the marine amphipod *Allorchestes compressa*. Proc. SPIE 9518, Bio-MEMS and Medical Microdevices II 951812. http://dx.doi.org/10.1016/B978-0-12-803371-5.00007-2.

Casado-Martinez, M.C., Beiras, R., Belzunce, M.J., Gonzalez-Castromil, M.A., Marin-Guirao, L., Postma, J.F., Riba, I., DelValls, T.A., 2006. Interlaboratory assessment of marine bioassays to evaluate the environmental quality of coastal sediments in Spain. IV. Whole sediment toxicity test using crustacean amphipods. Ciencias Mar. 32, 149–157.

Casado-Martinez, M.C., Smith, B.D., Luoma, S.N., Rainbow, P.S., 2010. Metal toxicity in a sediment-dwelling polychaete: threshold body concentrations or overwhelming accumulation rates? Environ. Pollut. 158, 3071–3076.

Castro, H., Ramalheira, F., Quintino, V., Rodrigues, A.M., 2006. Amphipod acute and chronic sediment toxicity assessment in estuarine environmental monitoring: an example from Ria de Aveiro, NW Portugal. Mar. Pollut. Bull. 53, 91–99.

Chandler, G.T., Green, A.S., 1996. A 14-day harpacticoid copepod reproduction bioassay for laboratory and field contaminated muddy sediments. In: Ostrander, G.K. (Ed.), Techniques in Aquatic Toxicology. CRC, Boca Raton, FL, USA, pp. 23–39.

Chandler, G.T., Coull, B.C., Schizas, N.V., Donelan, T.L., 1997. A culture-based assessment of the effects of chlorpyrifos on multiple meiobenthic copepods using microcosms of intact sediments. Environ. Toxicol. Chem. 16, 2339–2346.

Chapman, P.M., Wang, F., Janssen, C., Persoone, G., Allen, H.E., 1998. Ecotoxicology of metals in aquatic sediments: binding and release, bioavailability, risk assessment, and remediation. Can. J. Fish Aquat. Sci. 55, 2221–2243.

Chapman, P.M., Wang, F., Germano, J.D., Batley, G.E., 2002. Porewater testing and analysis: the good, the bad, and the ugly. Mar. Pollut. Bull. 44, 359–366.

Chariton, A.A., Roach, A.C., Simpson, S.L., Batley, G.E., 2010. The influence of the choice of physical and chemistry variables on interpreting the spatial patterns of sediment contaminants and their relationships with benthic communities. Mar. Freshwater Res. 61, 1109–1122.

Chariton, A.A., Maher, W.A., Roach, A.C., 2011. Recolonisation of translocated metal-contaminated sediments by estuarine macrobenthic assemblages. Ecotoxicology 20, 706–718.

Chariton, A.A., Ho, K.T., Proestou, D., Bik, H., Simpson, S.L., Portis, L.M., Cantwell, M., Baguley, J.G., Burgess, R.M., Pelletier, M., 2014. A molecular-based approach for examining responses of microcosm-contained eukaryotes to contaminant-spiked estuarine sediments. Environ. Toxicol. Chem. 33, 359–369.

Ciutat, A., Boudou, A., 2003. Bioturbation effects on cadmium and zinc transfers from a contaminated sediment and on metal bioavailability to benthic bivalves. Environ. Toxicol. Chem. 22, 1574–1581.

Collins, P.C., Croot, P., Carlsson, J., Colaco, J., Grehan, A., Hyeong, K., Kennedy, R., Mohn, C., Smith, S., Yamamoto, H., Rowden, A., 2013. A primer for the Environmental Impact Assessment of mining at seafloor massive sulfide deposits. Mar. Policy 42, 198–209.

Costa, F.O., Correia, A.D., Costa, M.H., 1998. Acute marine sediment toxicity: a potential new test with the amphipod *Gammarus locusta*. Ecotoxicol. Environ. Saf. 40, 81–87.

Costa, F.O., Neuparth, T., Correia, A.D., Costa, M.H., 2005. Multi-level assessment of chronic toxicity of estuarine sediments with the amphipod *Gammarus locusta*: II. Organism and population-level endpoints. Mar. Environ. Res. 60, 93–110.

Dafforn, K.A., Simpson, S.L., Kelaher, B.P., Clark, G.F., Komyakova, V., Wong, C.K.C., Johnston, E.L., 2012. The challenge of choosing environmental indicators of anthropogenic impacts in estuaries. Environ. Pollut. 163, 207–217.

DeFoe, D.L., Ankley, G.T., 1998. Influence of storage time on the toxicity of freshwater sediments to benthic macroinvertebrates. Environ. Pollut. 99, 123–131.

Di Toro, D.M., McGrath, J.A., 2000. Technical basis for narcotic chemicals and polycyclic aromatic hydrocarbon criteria. II. Mixtures and sediments. Environ. Toxicol. Chem. 19, 1971–1982.

Ducrot, V., Pery, A.R.R., Mons, R., Queau, H., Charles, S., Garric, J., 2007. Dynamic energy budget as a basis to model population-level effects of zinc-spiked sediments in the gastropod *Valvata piscinalis*. Environ. Toxicol. Chem. 26, 1774–1783.

EC (European Commission) Report, 2012. Addressing the New Challenges for Risk Assessment. SCENIHR (Scientific Committee on Emerging and Newly Identified Health Risks), SCHER (Scientific Committee on Health and Environmental Risks), SCCS (Scientific Committee on Consumer Safety). Available at: http://ec.europa.eu/health/scientific_committees/emerging/docs/scenihr_o_037.pdf.

Edge, K., Johnston, E., Roach, A., Ringwood, A., 2012. Indicators of environmental stress: cellular biomarkers and reproductive responses in the Sydney rock oyster (*Saccostrea glomerata*). Ecotoxicology 21, 1–11.

Edge, K., Dafforn, K.A., Roach, A.C., Simpson, S.L., Johnston, E.L., 2014. A biomarker of contaminant exposure is effective in large scale study of ten estuaries. Chemosphere 100, 16–26.

Environment Canada, 2002. Biological Test Method: Reference Method for Determining the Toxicity of Sediment Using Luminescent Bacteria in a Solid-Phase Test. Report EPS 1/RM/42. Ottawa, ON, Canada.

Farrar, J.D., Bridges, T.S., 2011. 28-Day Chronic Sublethal Test Method for Evaluating Whole Sediments Using an Early Life Stage of the Marine Polychaete *Neanthes arenaceodentata*. US Army Corps of Engineers, Vicksburg, MS, USA. Report ERDC TN-DOER-R14.

Fathallah, S., Medhioub, M.N., Kraiem, M.M., 2012. Photo-induced toxicity of four polycyclic aromatic hydrocarbons (PAHs) to embryos and larvae of the carpet shell clam *Ruditapes decussatus*. Bull. Environ. Contam. Toxicol. 88, 1001–1008.

Fox, M., Ohlauson, C., Sharpe, A.D., Brown, R.J., 2014. The use of a *Corophium volutator* chronic sediment study to support the risk assessment of medetomidine for marine environments. Environ. Toxicol. Chem. 33, 937–942.

Gao, J.J., Shi, H.H., Dai, Z.J., Mei, X.F., 2015. Variations of sediment toxicity in a tidal estuary: a case study of the South Passage, Changjiang (Yangtze) Estuary. Chemosphere 128, 7–13.

Gerould, S., Gloss, S.P., 1986. Mayfly-mediated sorption of toxicants into sediments. Environ. Toxicol. Chem. 5, 667–673.

Gomiero, A., Dagnino, A., Nasci, C., Viarengo, A., 2013. The use of protozoa in ecotoxicology: application of multiple endpoint tests of the ciliate *E. crassus* for the evaluation of sediment quality in coastal marine ecosystems. Sci. Tot. Environ. 442, 534–544.

Greenstein, D., Bay, S., Anderson, B., Chandler, G.T., Farrar, J.D., Keppler, C., Phillips, B., Ringwood, A., Young, D., 2008. Comparison of methods for evaluating acute and chronic toxicity in marine sediments. Environ. Toxicol. Chem. 27, 933–944.

Griscom, S.B., Fisher, N.S., 2004. Bioavailability of sediment-bound metals to marine bivalve molluscs: an overview. Estuaries 27, 826–838.

Groh, K.J., Carvalho, R.N., Chipman, J.K., Denslow, N.D., Halder, M., Murphy, C.A., Roelofs, D., Rolaki, A., Schirmer, K., Watanabe, K.H., 2015. Development and application of the adverse outcome pathway framework for understanding and predicting chronic toxicity: challenges and research needs in ecotoxicology. Chemosphere 120, 764–777.

Hack, L.A., Tremblay, L.A., Wratten, S.D., Forrester, G., Keesing, V., 2008. Toxicity of estuarine sediments using a full life-cycle bioassay with the marine copepod *Robertsonia propinqua*. Ecotox. Environ. Saf. 70, 469–474.

Harris, C.A., Scott, A.P., Johnson, A.C., Panter, G.H., Sheahan, D., Roberts, M., Sumpter, J.P., 2014. Principles of sound ecotoxicology. Environ. Sci. Technol. 48, 3100–3111.

Hassan, S.M., Garrison, A.W., Allen, H.E., Di Toro, D.M., Ankley, G.T., 1996. Estimation of partition coefficients for five trace metals in sandy sediments and application to sediment quality criteria. Environ. Toxicol. Chem. 15, 2198–2208.

Hellou, J., 2011. Behavioural ecotoxicology, an "early warning" signal to assess environmental quality. Environ. Sci. Pollut. Res. 18, 1–11.

Hill, N.A., Simpson, S.L., Johnston, E.L., 2013. Beyond the bed: effects of metal contamination on recruitment to bedded sediments and overlying substrata. Environ. Pollut. 173, 182–191.

Ho, K.T., Burgess, R.M., 2009. Marine sediment toxicity identification evaluations (TIEs): history, principles, methods, and future research. In: Kassin, T.A., Barcelo, D. (Eds.), Contaminated Sediments. Springer, Berlin, Germany, pp. 75–95.

Ho, K.T., Burgess, R.M., 2013. What's causing toxicity in sediments? Results of 20 years of toxicity identification and evaluations. Environ. Toxicol. Chem. 32, 2424–2432.

Ho, K.T., Kuhn, A., Pelletier, M.C., Hendricks, T.L., Helmstetter, A., 1999a. pH dependent toxicity of five metals to three marine organisms. Environ. Toxicol. Chem. 14, 235–240.

Ho, K.T., Kuhn, A., Pelletier, M.C., Burgess, R.M., Helmstetter, A., 1999b. Use of *Ulva lactuca* to distinguish pH-dependent toxicants in marine waters and sediments. Environ. Toxicol. Chem. 18, 207–212.

Ho, K.T., Burgess, R.M., Pelletier, M.C., Serbst, J.R., Ryba, S.A., Cantwell, M.G., Kuhn, A., Raczelowski, P., 2002. An overview of toxicant identification in sediments and dredged materials. Mar. Pollut. Bull. 44, 286–293.

Ho, K.T., Burgess, R.M., Pelletier, M.C., Serbst, J.R., Cook, H., Cantwell, M.G., Ryba, S.A., Perron, M.M., Lebo, J., Huckins, J., Petty, J., 2004. Use of powdered coconut charcoal as a toxicity identification and evaluation manipulation for organic toxicants in marine sediments. Environ. Toxicol. Chem. 23, 2124–2131.

Ho, K.T., Gielazyn, M.L., Pelletier, M.C., Burgess, R.M., Cantwell, M.C., Perron, M.M., Serbst, J.R., Johnson, R.L., 2009. Do toxicity identification and evaluation laboratory-based methods reflect causes of field impairment? Environ. Sci. Technol. 43, 6857–6863.

Ho, K.T., Chariton, A.A., Portis, L.M., Proestou, D., Cantwell, M., Baguley, J.G., Burgess, R.M., Simpson, S.L., Pelletier, M., Perron, M., Gunsch, C., Bik, H., Katz, D., Kamikawa, A., 2013. Use of a novel sediment exposure to determine the effects of triclosan on estuarine benthic communities. Environ. Toxicol. Chem. 32, 384–392.

Hook, S.E., Gallagher, E.P., Batley, G.E., 2014a. The role of biomarkers in the assessment of aquatic ecosystem health. Integr. Environ. Assess. Manag. 10, 327–341.

Hook, S.E., Osborn, H.L., Golding, L.A., Spadaro, D.A., Simpson, S.L., 2014b. Dissolved and particulate copper exposure induce differing gene expression profiles and mechanisms of toxicity in a deposit feeding amphipod. Environ. Sci. Technol. 48, 3504–3512.

Hutchins, C.M., Teasdale, P.R., Lee, S.Y., Simpson, S.L., 2008. Cu and Zn concentration gradients created by dilution of pH neutral metal-spiked marine sediment: a comparison of sediment geochemistry with direct methods of metal addition. Environ. Sci. Technol. 42, 2912–2918.

Hutchinson, T.H., Solbe, J., Kloepper-Sams, P.J., 1998. Analysis of the Ecetox Aquatic Toxicity (EAT) database III—Comparative toxicity of chemical substances to different life stages of aquatic organisms. Chemosphere 36, 129–142.

Jager, T., Selck, H., 2011. Interpreting toxicity data in a DEB framework: a case study for nonylphenol in the marine polychaete *Capitella teleta*. J. Sea Res. 66, 456–462.

Johnston, E.L., Roberts, D.A., 2009. Contaminants reduce the richness and evenness of marine communities: a review and meta-analysis. Environ. Pollut. 157, 1745–1752.

Kennedy, A.J., Steevens, J.A., Lotufo, G.R., Farrar, J.D., Reiss, M.R., Kropp, R.K., Doi, J., Bridges, T.S., 2009. A comparison of acute and chronic toxicity methods for marine sediments. Mar. Environ. Res. 68, 118–127.

Keppler, C.J., Ringwood, A.H., 2002. Effects of metal exposures on juvenile clams, *Mercenaria mercenaria*. Bull. Environ. Contam. Toxicol. 68, 43–48.

King, C.K., Dowse, M.C., Simpson, S.L., 2010. Toxicity of metals to the bivalve *Tellina deltoidalis* and relationships between metal bioaccumulation and metal partitioning between seawater and marine sediments. Arch. Environ. Contam. Toxicol. 58, 657–665.

Knezovich, J.P., Steichen, D.J., Jelinski, J.A., Anderson, S.L., 1996. Sulfide tolerance of four marine species used to evaluate sediment and porewater toxicity. Bull. Environ. Contam. Toxicol. 57, 450–457.

Kooijman, S.A.L.M., Van Der Hoeven, N., Van Der Werf, D.C., 1989. Population consequences of a physiological model for individuals. Funct. Ecol. 3, 325–336.

Kovatch, C.E., Chandler, G.T., Coull, B.C., 1999. Utility of a full life-cycle copepod bioassay approach for assessment of sediment-associated contaminant mixtures. Mar. Pollut. Bull. 38, 692–701.

Krell, B., Moreira-Santos, M., Ribeiro, R., 2011. An estuarine mud snail in situ toxicity assay based on postexposure feeding. Environ. Toxicol. Chem. 30, 1935–1942.

Krull, M., Abessa, D.M.S., Hatje, V., Barros, F., 2014. Integrated assessment of metal contamination in sediments from two tropical estuaries. Ecotoxicol. Environ. Saf. 106, 195–203.

Kuhn, A., Munns, W.R.J., Poucher, S., Champlain, D., Lussier, S., 2000. Prediction of population-level response from mysid toxicity test data using population modeling techniques. Environ. Toxicol. Chem. 19, 2364–2371.

Kuhn, A., Munns, W.R.J., Serbst, J., Edwards, P., Cantwell, M.G., Gleason, T., Pelletier, M.C., Berry, W.J., 2002. Evaluating the ecological significance of laboratory response data to predict population-level effects for the estuarine amphipod *Ampelisca abdita*. Environ. Toxicol. Chem. 21, 865–874.

Lee, J.W., Won, E.-J., Raisuddin, S., Lee, J.-S., 2015. Significance of adverse outcome pathways in biomarker-based environmental risk assessment in aquatic organisms. J. Environ. Sci. 35, 115–127.

Li, J.Y., Tang, J.Y.M., Jin, L., Escher, B.I., 2013. Understanding bioavailability and toxicity of sediment-associated contaminants by combining passive sampling with in vitro bioassays in an urban river catchment. Environ. Toxicol. Chem. 32, 2888–2896.

Lydy, M.J., Landrum, P.F., Oen, A., Allinson, M., Smedes, F., Harwood, A., Li, H., Maruya, K., Liu, J.F., 2014. Passive sampling methods for contaminated sediments: state of the science for organic contaminants. Integr. Environ. Assess. Manag. 10, 167–178.

Mann, R.M., Hyne, R.V., Spadaro, D.A., Simpson, S.L., 2009. Development and application of a rapid amphipod reproduction test for sediment quality assessment. Environ. Toxicol. Chem. 28, 1244–1254.

Martin, B.T., Zimmer, E.I., Grimm, V., Jager, T., 2012. Dynamic Energy Budget theory meets individual-based modelling: a generic and accessible implementation. Meth. Ecol. Evol. 3, 445–449.

Martin, B.T., Jager, T., Nisbet, R.M., Preuss, T.G., Hammers-Wirtz, M., Grimm, V., 2013. Extrapolating ecotoxicological effects from individuals to populations: a generic approach based on dynamic energy budget theory and individual-based modeling. Ecotoxicology 22, 574–583.

Martins, M., Costa, P.M., Raimundo, J., Vale, C., Ferreira, A.M., Costa, M.H., 2012. Impact of remobilized contaminants in *Mytilus edulis* during dredging operations in a harbour area: bioaccumulation and biomarker responses. Ecotoxicol. Environ. Saf. 85, 96–103.

Martín-Díaz, M.L., Blasco, J., Sales, D., DelValls, T.Á., 2004. Biomarkers as tools to assess sediment quality: laboratory and field surveys. Trends Anal. Chem. 23, 807–818.

Maruya, K.A., Landrum, P.F., Burgess, R.M., Shine, J.P., 2012. Incorporating contaminant bioavailability into sediment quality assessment frameworks. Integr. Environ. Assess. Manag. 8, 659–673.

McGee, B.L., Fisher, D.J., Wright, D.A., Yonkos, L.T., Ziegler, G.P., Turley, S.D., Farrar, J.D., Moore, D.W., Bridges, T.S., 2004. A field test and comparison of acute and chronic sediment toxicity with the marine amphipod *Leptocheirus plumulosus* in Chesapeake Bay, USA. Environ. Toxicol. Chem. 23, 1751–1761.

Miller, D.C., Poucher, S., Cardin, J.A., Hansen, D., 1990. The acute and chronic toxicity of ammonia to marine fish and a mysid. Arch. Environ. Contam. Toxicol. 19, 40–48.

Miller, R.J., Lenihan, H.S., Muller, E.B., Tseng, N., Hanna, S.K., Keller, A.A., 2010. Impacts of metal oxide nanoparticles on marine phytoplankton. Environ. Sci. Technol. 44, 7329–7334.

Monserrat, J.M., Martinez, P.E., Geracitano, L.A., Amado, L.L., Martins, C.M.G., Pinho, G.L.L., Chaves, I.S., Ferreira-Cravo, M., Ventura-Lima, J., Bianchini, A., 2007. Pollution biomarkers in estuarine animals: critical review and new perspectives. Comp. Biochem. Physiol. C Toxicol. Pharmacol. 146, 221–234.

Moore, D.W., Bridges, T.S., Gray, B.R., Duke, B.M., 1997. Risk of ammonia toxicity during sediment bioassays with the estuarine amphipod *Leptocheirus plumulosus*. Environ. Toxicol. Chem. 16, 1020–1027.

Morales-Caselles, C., Ramos, J., Riba, I., DelValls, T.A., 2008. Using the polychaete *Arenicola marina* to determine toxicity and bioaccumulation of PAHS bound to sediments. Environ. Monit. Assess. 142, 219–226.

Moreira, S.M., Moreira-Santos, M., Guilhermino, L., Ribeiro, R., 2005. Short-term sublethal in situ toxicity assay with *Hediste diversicolor* (Polychaeta) for estuarine sediments based on postexposure feeding. Environ. Toxicol. Chem. 24, 2010–2018.

Moreira, S.M., Lima, I., Ribeiro, R., Guilhermino, L., 2006. Effects of estuarine sediment contamination on feeding and on key physiological functions of the Polychaete *Hediste diversicolor*: laboratory and in situ assays. Aquat. Toxicol. 78, 186–201.

Moreno-Garrido, I., Hampel, M., Lubian, L.M., Blasco, J., 2003a. Marine benthic microalgae *Cylindrotheca closterium* (Ehremberg) Lewin and Reimann (*Bacillariophyceae*) as a tool for measuring toxicity of linear alkylbenzene sulfonate in sediments. Bull. Environ. Contam. Toxicol. 70, 242–247.

Moreno-Garrido, I., Hampel, M., Lubian, L.M., Blasco, J., 2003b. Sediment toxicity tests using benthic marine microalgae *Cylindrotheca closterium* (Ehremberg) Lewin and Reimann (*Bacillariophyceae*). Ecotoxicol. Environ. Saf. 54, 290–295.

Moreno-Garrido, I., Lubian, L.M., Jimenez, B., Soares, A., Blasco, J., 2007. Estuarine sediment toxicity tests on diatoms: sensitivity comparison for three species. Estuar. Coastal Shelf Sci. 71, 278–286.

Mudroch, A., Azcue, J.M., 1995. Manual of Aquatic Sediment Sampling. CRC Press, Boca Raton, FL, USA.

Newman, C.M., 2010. Fundamentals of Ecotoxicology, third ed. CRC Press, Boca Raton, FL, USA.

NFESC (Naval Facility Engineering Services Center), 2003. Using Sediment Toxicity Identification Evaluations to Improve the Development of Remedial Goals for Aquatic Habitats. Special Publication SP-2132-ENV, Port Hueneme, CA, USA.

Northcott, G.L., Jones, K.C., 2000. Spiking hydrophobic organic compounds into soil and sediment: a review and critique of adopted procedures. Environ. Toxicol. Chem. 19, 2418–2430.

OECD (Organisation for Economic Cooperation and Development), 2006. Current Approaches in the Statistical Analysis of Ecotoxicity Data: A Guidance to Application. ENV/JM/MONO(2006)18. OECD Series on Testing and Assessment No. 54. Environment Directorate, OECD, Paris, France.

OECD, 2013. Guidance Document on Developing and Assessing Adverse Outcome Pathways. In: Series on Testing and Assessment, No 184, Vol. ENV/JM/MONO(2013)6. Environment Directorate, Paris, France.

Olsgard, F., 1999. Effects of copper contamination on recolonisation of subtidal marine soft sediments—an experimental field study. Mar. Pollut. Bull. 38, 448–462.

Peijnenburg, W.J., Teasdale, P.R., Reible, D., Mondon, J., Bennett, W.W., Campbell, P.G., 2014. Passive sampling methods for contaminated sediments: state of the science for metals. Integr. Environ. Assess. Manag. 10, 179–196.

Pelletier, M.C., Burguess, R.M., Ho, K.T., Kuhn, A., McKinney, R.A., Ryba, S.A., 1997. Phototoxicity of individual polycyclic aromatic hydrocarbons and petroleum to marine invertebrate larvae and juveniles. Environ. Toxicol. Chem. 16, 2190–2199.

Perez-Landa, V., Simpson, S.L., 2011. A short life-cycle test with the epibenthic copepod *Nitocra spinipes* for sediment toxicity assessment. Environ. Toxicol. Chem. 30, 1430–1439.

Perron, M.M., Burgess, R.M., Suuberg, E.M., Cantwell, M.G., Pennell, K.G., 2013. Performance of passive samplers for monitoring estuarine water column concentrations: 1. Contaminants of concern. Environ. Toxicol. Chem. 32, 2182–2189.

Peterson, G.S., Ankley, G.T., Leonard, E.N., 1996. Effect of bioturbation on metal-sulfide oxidation in surficial freshwater sediments. Environ. Toxicol. Chem. 15, 2147–2155.

Rainbow, P.S., 2007. Trace metal bioaccumulation: models, metabolic availability and toxicity. Environ. Int. 33, 576–582.

Ramos-Gómez, J., Martins, M., Raimundo, J., Vale, C., Martín-Díaz, M.L., DelValls, T.Á., 2011. Validation of *Arenicola marina* in field toxicity biomass using benthic cages biomarkers as tools for assessing sediment quality. Mar. Pollut. Bull. 62, 1538–1549.

Regoli, F., Giuliani, M.E., 2014. Oxidative pathways of chemical toxicity and oxidative stress biomarkers in marine organisms. Mar. Environ. Res. 93, 106–117.

Regoli, F., Frenzilli, G., Bocchetti, R., Annarumma, F., Scarcelli, V., Fattorini, D., Nigro, M., 2004. Time-course variation of oxyradical metabolism, DNA integrity and lysosomal stability in mussels, *Mytilus galloprovincialis*, during a field translocation experiment. Aquat. Toxicol. 68, 167–178.

Riba, I., DelValls, T.A., Forja, J.M., Gómez-Parra, A., 2004. The influence of pH and salinity values in the toxicity of heavy metals in sediments to the estuarine clam *Ruditapes phillipinarum*. Environ. Toxicol. Chem. 23, 1100–1107.

Ringwood, A.H., Keppler, C.J., 1998. Seed clam growth: an alternative sediment bioassay developed during EMAP in the Carolinian Province. Environ. Monit. Assess. 51, 247–257.

Ringwood, A., 1992. Comparative sensitivity of gametes and early developmental stages of a sea urchin species (*Echinometra mathaei*) and a bivalve species (*Isognomon californicum*) during metal exposures. Arch. Environ. Contam. Toxicol. 22, 288–295.

Rodrigues, E.T., Pardal, M.Â., 2014. The crab *Carcinus maenas* as a suitable experimental model in ecotoxicology. Environ. Int. 70, 158–182.

Rodriguez-Romero, A., Khosrovyan, A., Del Valls, T.A., Obispo, R., Serrano, F., Conradi, M., Riba, I., 2013. Several benthic species can be used interchangeably in integrated sediment quality assessment. Ecotoxicol. Environ. Saf. 92, 281–288.

Rosen, G., Miller, K., 2011. A postexposure feeding assay using the marine polychaete *Neanthes arenaceodentata* suitable for laboratory and in situ exposures. Environ. Toxicol. Chem. 30, 730–737.

Scarlett, A., Rowland, S.J., Canty, M., Smith, E.L., Galloway, T.S., 2007a. Method for assessing the chronic toxicity of marine and estuarine sediment-associated contaminants using the amphipod Corophium volutator. Mar. Environ. Res. 63, 457–470.

Scarlett, A., Canty, M.N., Smith, E.L., Rowland, S.J., Galloway, T.S., 2007b. Can amphipod behaviour help to predict chronic toxicity of sediments? Hum. Ecol. Risk Assess. 13, 506–518.

Schlekat, C.E., Garman, E.R., Vangheluwe, M.L.U., Burton Jr., G.A., 2015. Development of a bioavailability-based risk assessment approach for nickel in freshwater sediments. Integr. Environ. Assess. Monit. http://dx.doi.org/10.1016/B978-0-12-803371-5.00007-2.

Simpson, S.L., Batley, G.E., 2003. Disturbances to metal partitioning during toxicity testing of iron(II)-rich estuarine porewaters and whole-sediments. Environ. Toxicol. Chem. 22, 424–432.

Simpson, S.L., Batley, G.E., 2007. Predicting metal toxicity in sediments: a critique of current approaches. Integr. Environ. Assess. Manag. 3, 18–31.

Simpson, S.L., King, C.K., 2005. Exposure-pathway models explain causality in whole-sediment toxicity tests. Environ. Sci. Technol. 39, 837–843.

Simpson, S.L., Kumar, A., 2016. Sediment ecotoxicology. In: Simpson, S.L., Batley, G.E. (Eds.), Sediment Quality Assessment: A Practical Handbook. CSIRO Publishing, Canberra, Australia, pp. 77–122.

Simpson, S.L., Spadaro, D.A., 2011. Performance and sensitivity of rapid sublethal sediment toxicity tests with the amphipod *Melita plumulosa* and copepod *Nitocra spinipes*. Environ. Toxicol. Chem. 30, 2326–2334.

Simpson, S.L., Angel, B.M., Jolley, D.F., 2004. Metal equilibration in laboratory-contaminated (spiked) sediments used for the development of whole-sediment toxicity tests. Chemosphere 54, 597–609.

Simpson, S.L., Batley, G.E., Hamilton, I., Spadaro, D.A., 2011. Guidelines for copper in sediments with varying properties. Chemosphere 85, 1487–1495.

Simpson, S.L., Ward, D., Strom, D., Jolley, D.F., 2012. Oxidation of acid-volatile sulfide in surface sediments increases the release and toxicity of copper to the benthic amphipod *Melita plumulosa*. Chemosphere 88, 953–961.

Simpson, S.L., Spadaro, D.A., O'Brien, D., 2013. Incorporating bioavailability into management limits for copper and zinc in sediments contaminated by antifouling paint and aquaculture. Chemosphere 93, 2499–2506.

Simpson, S.L., Batley, G.E., Maher, W.A., 2016. Chemistry of sediment contaminants. In: Simpson, S.L., Batley, G.E. (Eds.), Sediment Quality Assessment: A Practical Handbook. CSIRO Publishing, Canberra, Australia, pp. 47–75.

Simpson, S.L., 2005. An exposure-effect model for calculating copper effect concentrations in sediments with varying copper binding properties: a synthesis. Environ. Sci. Technol. 39, 7089–7096.

Spadaro, D.A., Simpson, S.L., 2016a. Appendix F. Protocol for whole-sediment sub-lethal (reproduction) toxicity tests using the copepod *Nitocra spinipes* (harpacticoid). In: Simpson, S.L., Batley, G.E. (Eds.), Sediment Quality Assessment: A Practical Handbook. CSIRO Publishing, Canberra, Australia, pp. 276–284.

Spadaro, D.A., Simpson, S.L., 2016b. Appendix E. Protocol for 10-day whole-sediment sub-lethal (reproduction) and acute toxicity tests using the epibenthic amphipod *Melita plumulosa*. In: Simpson, S.L., Batley, G.E. (Eds.), Sediment Quality Assessment: A Practical Handbook. CSIRO Publishing, Canberra, Australia, pp. 265–275.

Spadaro, D.A., Simpson, S.L., 2016c. Appendix G. Protocols for 10-day whole-sediment lethality toxicity tests and 30-day bioaccumulation tests using the deposit-feeding benthic bivalve *Tellina deltoidalis*. In: Simpson, S.L., Batley, G.E. (Eds.), Sediment Quality Assessment: A Practical Handbook. CSIRO Publishing, Canberra, Australia, pp. 285–293.

Spadaro, D.A., Micevska, T., Simpson, S.L., 2008. Effect of nutrition on toxicity of contaminants to the epibenthic amphipod, *Melita plumulosa*. Arch. Environ. Contam. Toxicol. 55, 593–602.

Strom, D., Simpson, S.L., Batley, G.E., Jolley, D.F., 2011. The influence of sediment particle size and organic carbon on toxicity of copper to benthic invertebrates in oxic/sub-oxic surface sediments. Environ. Toxicol. Chem. 30, 1599–1610.

Swartz, R.C., Ferraro, S.P., Lamberson, J.O., Cole, F.A., Ozretich, R.J., Boese, B.L., Schults, D.W., Behrenfeld, M., Ankley, G.T., 1997. Photoactivation and toxicity of mixtures of polycyclic aromatic hydrocarbon compounds in marine sediment. Environ. Toxicol. Chem. 16, 2151–2157.

Taylor, A.M., Maher, W.A., 2010. Establishing metal exposure–dose–response relationships in marine organisms: illustrated with a case study of cadmium toxicity in *Tellina deltoidalis*. In: Puopolo, K., Martorino, L. (Eds.), New Oceanography Research Developments: Marine Chemistry, Ocean Floor Analyses and Marine Phytoplankton. Nova Science, New York, USA, pp. 1–57.

Taylor, A.M., Maher, W.A., 2016. Biomarkers. In: Simpson, S.L., Batley, G.E. (Eds.), Sediment Quality Assessment: A Practical Handbook. CSIRO Publishing, Canberra, Australia, pp. 157–193.

UNEP (United Nations Environment Programme), 2010. Harmful Substances and Hazardous Waste: Factsheet. Available at: www.unep.org/pdf/brochures/Harmful Substances.pdf. Accessed: October, 2015.

USEPA (US Environmental Protection Agency), 1993. Guidance Manual – Bedded Sediment Bioaccumulation Test. EPA-600-R-93–183. Office of Research and Development, Washington, DC, USA.

USEPA, 1994. Methods for Assessing the Toxicity of Sediment-associated Contaminants with Estuarine and Marine Amphipods. 600-R-94–025. Office of Research and Development, Washington, DC, USA.

USEPA, 1996. Marine Toxicity Identification Evaluation (TIE) Procedures Manual. Phase I Guidance Document. EPA-600-R-96–054. Office of Research and Development, Washington, DC, USA.

USEPA, 2000. Methods for Assessing the Toxicity and Bioaccumulation of Sediment-associated Contaminants with Freshwater Invertebrates. EPA-600-R-99-064, second ed. Office of Research and Development, Washington, DC, USA.

USEPA, 2001. Methods for Collection, Storage and Manipulation of Sediments for Chemical and Toxicological Analyses. Technical Manual EPA-823-B-01–002. Office of Water, Washington, DC, USA.

USEPA, 2003. Procedures for the Derivation of Equilibrium Partitioning Sediment Benchmarks (ESBs) for the Protection of Benthic Organisms: PAH Mixtures. EPA-600-R-02–013. Office of Research and Development, Washington, DC, USA.

USEPA, 2005. Procedures for the Derivation of Equilibrium Partitioning Sediment Benchmarks (ESBs) for the Protection of Benthic Organisms: Metal Mixtures (Cadmium, Copper, Lead, Nickel, Silver, and Zinc). EPA-600-R-02–011. Office of Research and Development, Washington, DC, USA.

USEPA, 2006a. Guidance on Systematic Planning Using the Data Quality Objectives Process. EPA-240-B-06–001. Office of Environmental Information, Washington, DC, USA.

USEPA, 2006b. Data Quality Assessment: Statistical Methods for Practitioners. EPA-240-B-06-003. Office of Environmental Information, Washington, DC, USA.

USEPA, 2007. Sediment Toxicity Identification Evaluation (TIE). Phases I, II, and III Guidance Document. EPA-600-R-07–080. Office of Research and Development, Washington, DC, USA.

USEPA, 2012. Equilibrium Partitioning Sediment Benchmarks (ESBs) for the Protection of Benthic Organisms: Procedures for the Determination of the Freely Dissolved Interstitial Water Concentrations of Nonionic Organics. EPA-600-R-02–012. Office of Research and Development, Washington, DC, USA.

USEPA, 2014. Screening for Dioxin-like Chemical Activity in Soils and Sediments Using the CALUX® Bioassay and TEQ Determinations. Method 4435–59. Test Methods for Evaluating Solid Waste. SW-846 On-line. US Environmental Protection Agency, Office of Solid Waste, Economic, Methods, and Risk Analysis Division, Washington, DC, USA. http://www.epa.gov/waste/hazard/testmethods/sw846/online/index.htm.

Van Geest, J.L., Burridge, L.E., Kidd, K.A., 2014a. The toxicity of the anti-sea lice pesticide AlphaMax® to the polychaete worm *Nereis virens*. Aquaculture 430, 98–106.

Van Geest, J.L., Burridge, L.E., Kidd, K.A., 2014b. Toxicity of two pyrethroid-based anti-sea lice pesticides, AlphaMax(R) and Excis(R), to a marine amphipod in aqueous and sediment exposures. Aquaculture 434, 233–240.

van Midwoud, P.M., Verpoorte, E., Groothius, G.M.M., 2011. Microfluidic device for in vitro studies on liver drug metabolism and toxicity. Integr. Biol. 3, 509–521.

Vandegehuchte, M.B., Nguyen, L.T.H., de Laender, F., Muyssen, B.T.A., Janssen, C.R., 2013a. Whole sediment toxicity tests for metal risk assessments: on the importance of equilibration and test design to increase ecological relevance. Environ. Toxicol. Chem. 32, 1048–1059.

Vangheluwe, M.L.U., Verdonck, F.A.M., Besser, J.M., Brumbaugh, W.G., Ingersoll, C.G., Schlekat, C.E., Garman, E.R., 2013b. Improving sediment-quality guidelines for nickel: development and application of predictive bioavailability models to assess chronic toxicity of nickel in freshwater sediments. Environ. Toxicol. Chem. 32, 2507–2519.

Villeneuve, D.L., Crump, D., García-Reyero, N., Hecker, M., Hutchinson, T.H., LaLone, C.A., Landesmann, B., Lettieri, T., Munn, S., Nepelska, M., Ottinger, M.A., Vergauwen, L., Whelan, M., 2014a. Adverse outcome pathways (AOP) development I: strategies and principles. Toxicol. Sci. 142, 312–320.

Villeneuve, D.L., Crump, D., García-Reyero, N., Hecker, M., Hutchinson, T.H., LaLone, C.A., Landesmann, B., Lettieri, T., Munn, S., Nepelska, M., Ottinger, M.A., Vergauwen, L., Whelan, M., 2014b. Adverse outcome pathways (AOP) development II: best practices. Toxicol. Sci. 142, 321–330.

Volkenborn, N., Polerecky, L., Wethey, D.S., Woodin, S.A., 2010. Oscillatory porewater bioadvection in marine sediments induced by hydraulic activities of *Arenicola marina*. Limnol. Oceanogr. 55, 1231–1247.

Von Bertalanffy, L., 1957. Quantitative laws in metabolism and growth. Q. Rev. Biol. 32, 217–231.

Wang, F.Y., Chapman, P.M., 1999. Biological implications of sulfide in sediment – a review focusing on sediment toxicity. Environ. Toxicol. Chem. 18, 2526–2532.

Ward, D.J., Simpson, S.L., Jolley, D.F., 2013a. Slow avoidance response to contaminated sediments elicits sub-lethal toxicity to benthic invertebrates. Environ. Sci. Technol. 47, 5947–5953.

Ward, D.J., Simpson, S.L., Jolley, D.F., 2013b. Avoidance of contaminated sediments by an amphipod (*Melita plumulosa*), a harpacticoid copepod (*Nitocra spinipes*) and a snail (*Phallomedusa solida*). Environ. Toxicol. Chem. 32, 644–652.

Williams, K., Green, D.W.J., Paseoe, D., Gower, D.E., 1986. The acute toxicity of cadmium to different larval stages of *Chironomus riparius* (Diptera: Chironomidae) and its ecological significance for pollution regulation. Oecologia 70, 362–366.

Word, J.Q., Gardiner, W.W., Moore, D.W., 2005. Influence of confounding factors on SQGs and their application to estuarine and marine sediment evaluations. In: Wenning, R.J., Batley, G.E., Ingersoll, C.G., Moore, D.W. (Eds.), Use of Sediment Quality Guidelines and Related Tools for the Assessment of Contaminated Sediments. Society of Environmental Toxicology and Chemistry. Pensacola, FL, USA, pp. 633–686.

Zhao, Y., Newman, M.C., 2007. The theory underlying dose-response models influences predictions for intermittent exposures. Environ. Toxicol. Chem. 26, 543–547.

Zheng, G., Wang, Y., Wang, Z., Zhong, W., Wang, H., Li, Y., 2013. An integrated microfluidic device in marine microalgae culture for toxicity screening application. Mar. Pollut. Bull. 72, 231–243.

Zheng, G., Li, Y., Liu, X., Wang, H., Yu, S., Wang, Y., 2014. Marine phytoplankton motility sensor integrated into a microfluidic chip for high-throughput pollutant toxicity assessment. Mar. Pollut. Bull. 84, 147–154.

CHAPTER

8

Mesocosm and Field Toxicity Testing in the Marine Context

A.C. Alexander[1], E. Luiker[1], M. Finley[2], J.M. Culp[1]

[1]University of New Brunswick, Fredericton, NB, Canada; [2]Government of Wisconsin, Madison, WI, United States

8.1 INTRODUCTION

8.1.1 What Is a Mesocosm?

Mesocosms are outdoor enclosure facilities (1 to >10,000 L) that control for biological, habitat, and chemical conditions (Fig. 8.1). They are used to simulate complex exposure dynamics under realistic field conditions (Culp and Baird, 2006). Mesocosms are a hybrid of field and laboratory techniques allowing for control of some parameters of interest (eg, habitats or species included) while permitting near natural environmental conditions (eg, diel temperature cycles) to create more realistic exposure scenarios than laboratory tests. Mesocosms can be short (<1 month), medium (1 month–1 year), or long (>1 year) in duration with the timeline selected often based on the lifecycle of the focal organism, population, or community under study.

Marine mesocosms have been used in a variety of applications (reviewed in Grice and Reeve, 1982; Clark and Noles, 1994; Oviatt, 1994; Peterson et al., 2009; Stewart et al., 2013). These systems are thought to be particularly useful for the increased understanding of the relationship between biodiversity and ecosystem function (Emmerson et al., 2001), anthropogenic and habitat effects (Renick et al., 2015), and more recently, the complex responses to climate change (Lejeusne et al., 2010; Stewart et al., 2013). Marine mesocosms pose unique challenges to experimentation in that equipment must be sturdily constructed to withstand saltwater corrosion and in the case of subtidal studies, robust enough to withstand wave action. Systems can also be sizeable (>1000 L per replicate cell), a scale that poses its own challenges and often precludes the creation of a mobile mesocosm facility.

Advantages of mesocosms include increased control and replication compared to field studies and more realistic conditions than laboratory bioassays (Table 8.1). Mesocosms can be expensive and logistically complex but can also help reduce uncertainty. This type of experiment tends to produce high-quality, reproducible data that is easier to collect and interpret than conventional field studies.

Marine Ecotoxicology
http://dx.doi.org/10.1016/B978-0-12-803371-5.00008-4

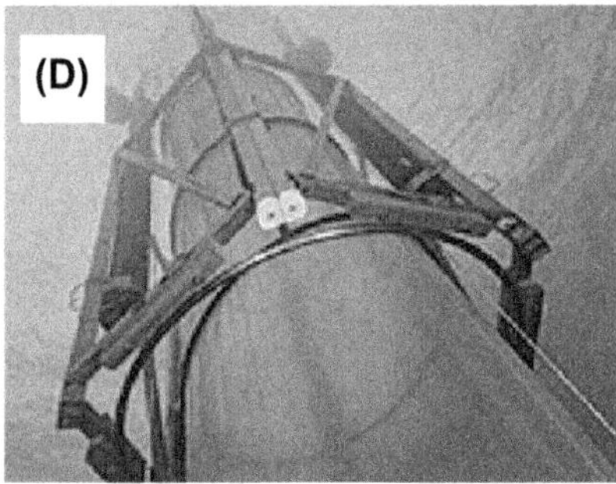

FIGURE 8.1 (A–D). Select contemporary marine mesocosm examples. Land-based marine mesocosm facilities (eg, A, B) or enclosed columns that separate test water from the surrounding water body (C, D). Facilities such as those in (A) Wageningen in the Netherlands and at the (B) University of Connecticut have been used to assess complex effluents in tidal systems and test contrasting management scenarios on species invasions in coastal communities. In situ techniques, such as the large floating polyurethane bags shown in (C) and (D) were deployed in a Swedish Fjord in 2013 to test rates of ocean acidification. *Photo credits (A) www.wageningenur.nl, (B) cfpub.epa.gov, (C, D) e360.yale.edu.*

TABLE 8.1 Strengths, Weakness, Opportunities, and Threats of Mesocosm Studies in the Marine Context

Strengths	**Weaknesses**
• ability to study two or more trophic levels over an extended period • ability to manipulate the environment by natural (eg, upwelling) or unnatural means such as the introduction of a pollutant • control and replication of treatments	• expensive, logistically demanding • loss of realism compared to field studies • confounding effects of enclosure (eg, wall effects, mixing)
Opportunities	**Threats**
• ability to extrapolate beyond historic and contemporary datasets • ability to bring together scientists from a variety of disciplines in a single study	• knowledge gaps, as in any experiment, limit the true understanding of the system under investigation

Despite criticisms that restrictions of spatial scale and duration limit their usefulness compared to whole-ecosystem studies (Carpenter, 1996, 1999; Carpenter et al., 1998), some of the most significant whole-ecosystem experiments were only possible after extensive mesoscale studies were conducted (Schindler, 1998). Additionally, unlike whole-ecosystem studies, mesocosms can provide the means to simulate processes without destroying the natural system of interest (Guckert, 1993; McIntire, 1993; Lawton, 1996; Boyle and Fairchild, 1997; Drenner and Mazumder, 1999; Clements et al., 2002).

Mesocosms are most useful when they simulate a process or cause–effect relationship (Lamberti and Steinman, 1993). According to de Lafontaine and Leggett (1987) four criteria should be fulfilled by an ideal mesocosm enclosure system: (1) good replication in time and space, (2) reproducibility of the natural environment, (3) establishment of a representative community, and (4) allow for observations of predator–prey relationships at natural densities. At their best, mesocosm experiments address fundamental questions or confounding issues clarifying the interpretation of field data. Further, these types of studies will continue to be essential for isolating the impacts of extreme events, species invasion, and habitat fragmentation where partial isolation from the surrounding landscape is required (Stewart et al., 2013). Mesocosms are useful experimental tools as they enable the examination of populations and ecosystems simultaneously to, according to Odum (1984), "reveal basic properties of the whole."

8.1.2 Linking Field Biomonitoring to Mesocosm Experiments

Field surveys cannot easily link cause and effect (Adams, 2003) because effluents from industrial and municipal activity tend to be diffuse, complex mixtures that can have stimulatory (eg, nutrient masking) or inhibitory effects (eg, contaminant toxicity). In situ experiments (eg, enclosures, limnocorrals) can be useful in this regard because the components of the mixture can be separated in the bioassay. For instance, the effect of natural stressors (eg, predation) can be separated from the effect of contaminants. Thus, the confounding factors of multiple interacting stressors can be teased apart and the cause of biological responses revealed.

In contrast to field biomonitoring or enclosure experiments, mesocosms are able to control more variables and thus isolate interacting conditions and contaminants that are influencing biological responses. Mesocosms incorporate greater complexity than is possible to include in laboratory toxicity tests and can generate important information about the sublethal and chronic effects of pollutants on aquatic communities. Integration of mesocosm studies with field biomonitoring has been particularly beneficial for ecological risk assessments (Norton et al., 2014) and has been used to generate weight-of-evidence risk assessments (Culp et al., 2000a). The specific objectives of the following chapter are to (1) illustrate the usefulness of mesocosm studies to investigate complex stressor and biological effects, (2) describe the basic methodology of different marine mesocosm designs, (3) discuss the needs of larger organisms in mesocosm experiments, and (4) outline issues specific to marine mesocosm studies.

8.2 MESOCOSMS AS A MANAGEMENT TOOL FOR THE PROTECTION OF MARINE ECOSYSTEMS

Environmental managers need tools to assess and manage environmental risks from anthropogenic stressor impacts. Mesocosms provide a unique experimental system allowing the researcher to control the receptor, hazard, and exposure pathway in realistic environmental conditions involving different food chain levels (Perceval et al., 2009). These mesocosm attributes can enhance the depth of information available to support ecological risk assessments and develop environmental standards and guidelines.

"Risk assessments are a process to evaluate the likelihood that adverse ecological impacts may occur or are occurring as a result of exposure to one or more stressors" (US EPA, 1992). Mesocosm experiments can support environmental risk assessments as they allow for control of the hazard–receptor interaction in semi-natural conditions, which is critical in the determination of risk. In artificial stream mesocosms developed by Environment Canada (Culp and Baird, 2006), the quantity and concentration of the hazard exposure can be controlled through dilution and flow rate of the effluent, and time of hazard exposure can be controlled from pulse to extended press periods. The receptor was a stream benthic community, and the exposure pathway/habitat was natural river substrate. In these experiments, the stressors have included municipal, pulp mill and mine tailings effluent, pesticides, nutrients, and sediments.

Mesocosm studies can also be used to establish cause and effect in multiple stressor scenarios. In their experiment investigating the cause and effect of effluents from two sources on a river in Alberta, Canada, Culp et al. (2003) used four treatments: control, 1% municipal sewage effluent (MSE), 3% pulp mill effluent (PME) (percentages were based on typical effluent concentrations in the river), and a mixture of both. With this experimental design, they were able to determine that both the MSE and PME were important sources of nutrients; MSE was a major source of nitrogen, and PME the dominant source for phosphorus. Combined, the two effluent sources created a synergistic effect for primary production creating eutrophic conditions downstream. This study provided information for watershed managers to look at the whole river system, and its effluent sources and manage accordingly. Similar approaches have been used by Clark and Clements (2006), looking at metal mine effects in an Arkansas river.

To establish a strong cause and effect case in multiple stressor/cumulative effects scenarios, a weight of evidence approach combining information from a variety of sources is required (Lowell et al., 2000; Culp and Baird, 2006). Mesocosm studies can bridge the gap of field results, where you can measure contaminants or other indicators and ecosystem components (however, level of exposure and ecological interactions are unknown), with laboratory-based bioassay studies (single species studies done at specific concentrations in sterile environments). Mesocosms can verify stressor effects under controlled conditions in relation to field data and also provide a steep exposure and response curve (Lowell et al., 2000).

Using similar methods in the marine environment, the National Oceanic and Atmospheric Administration has developed a modular saltwater mesocosm testing system to predict impacts from pollution, particularly petroleum hydrocarbons and chemical contaminants on salt marsh flora and fauna (Scott et al., 2013). These mesocosm systems, as in the freshwater example above, allow for the control of the hazard (in concentration, and combination with other stressors), and use saltwater marsh habitats and their natural biota for exposure. The results from this modular mesocosm system are used to assist in prespill contingency planning and predictive ecological risk assessments (Scott et al., 2013).

For new or other unevaluated substances, mesocosm studies can be very beneficial in establishing realistic NOEC (no observed effect concentration), or the amount of a substance that can be released without observable effects. This approach has been proven effective in both freshwater and marine ecosystems. Mesocosms have the benefit of testing the substance in a realistic environmental setting, involving a component of the ecosystem. Single species bioassay tests cannot replicate the effect of the substance released in the ecosystem, as the effect on nontarget/untested taxa can be missed. This was demonstrated in studies involving pesticide mixtures where different taxa and guilds of benthic invertebrates showed different sensitivities to treatment (Alexander and Culp, 2013). A similar approach was used by Foekema et al. (2015) to determine NOEC of dissolved copper on marine benthic and plankton communities using outdoor mesocosms, comparing species sensitivity distribution results from laboratory-based single species tests. Environmental fate and toxicity of silver nanoparticles was examined in a marine intertidal mesocosm study using two endobenthic species (ragworm, *Hediste diversicolor* and the bivalve mollusk, *Scrobicularia plana*) (Buffett et al., 2014). Unique to their approach was simulation of the tide by pumping water into and out of the mesocosm chambers, simulating a 6-h tide cycle. Endpoints included fate of Ag nanoparticles, Ag bioaccumulation, and biomarker responses (Buffett et al., 2014).

In addition, with mesocosm studies, the substance of concern can be tested on its own, or in combination with other substances/stressors. These studies could then identify potential additive, synergistic, or antagonistic effects. This was shown by Alexander et al. (2013) in studying the effects of single and multiple pesticides in combination with nutrients. Their results reveal that with exposure to low concentrations of pesticides, nutrients "masked" the pesticide effect on the benthic community with a stimulatory effect. Under higher concentrations of pesticide, nutrients had an antagonistic effect, with an overall increased toxicity to the benthic community (Alexander et al., 2013).

WHY DO A MESOCOSM? JURISDICTIONAL APPLICATIONS IN CANADA, THE UNITED STATES, AND THE EUROPEAN UNION

Because of the useful information derived from mesocosm studies, several jurisdictions around the world have incorporated mesocosms into their environmental assessment/risk regulatory process. In Canada, metal mining and pulp and paper environmental effects monitoring programs include mesocosms as an accepted alternative monitoring method for fish studies (Environment Canada, 2010, 2012). The European Union, in an effort to protect aquatic organisms in edge of field surface waters, includes mesocosms as acceptable Tier 3 (population and community level) experiments to determine the regulatory acceptable concentration for the ecological threshold for plant protection products (pesticides and their residues) (EFSA, 2013). The US EPA has incorporated mesocosms into their Environmental Sensitivity Index which is used to "assess, forecast, and mitigate oil spill impacts throughout the coastal regions of the United States" (Scott et al., 2013). The mesocosm studies provide information that allows for prespill contingency planning and predictive ecological risk assessments (Scott et al., 2013).

8.3 GENERAL DESIGN OF AQUATIC MESOCOSMS

8.3.1 Land-Based Mesocosms

Investigators have used various land-based mesocosm systems to evaluate the fate and effects of stressors and to quantify cause-and-effect relationships in a multispecies context (Fig. 8.1A and B). The choice of mesocosm method depends on the study question (eg, inclusion of fish predators), but the large investment of time and money required often directs researchers toward the use of smaller mesocosms. Fortunately, smaller systems can produce causal information similar to that generated in larger systems at substantial cost savings. For example, Howick et al. (1994) found that fiberglass tank experiments, despite being 80% less expensive, produced identical effects of insecticides on invertebrates as earthen pond experiments. Land-based mesocosms have a long history of use in assessing the ecological risk of pesticides. Land-based mesocosms tend to range in size from 100 to 1000 m^2 and have been used in higher tier-testing procedures required by the US EPA for pesticide registration (Graney et al., 1994).

Some land-based mesocosms are constructed ponds that are normally less than 3 m deep with side slopes approaching 2:1 (Christman et al., 1994; Howick et al., 1994; Johnson et al., 1994). Water is typically supplied from groundwater wells or nearby surface waters.

The most controlled land-based mesocosms are fabricated tanks where the sediment and water source are the same among replicates, and biological communities are carefully manipulated. Rand et al. (2000) provide a detailed example and methods for setting up this type of system. Tanks vary from 2000 to 20,000 L are large enough to be representative of simple food webs (often excluding fish), and have ambient environmental conditions of temperature, light, and wind (Graney et al., 1994). Primary advantages of tank mesocosms are lower cost and the ability to prevent contamination of the natural environment with study compounds. The small size of mesocosm tanks often limits the inclusion of larger predators and can result in undesirable wall effects. However, even some of the largest mesocosms ever tested (1300 m^3) report considerable wall effects in long-term (eg, >50 days) experiments (Grice et al., 1980). Of particular importance is the choice of sampling technique as destructive sampling (eg, sediment grab samples) can affect experimental results in these small systems (Graney et al., 1995).

MERL is a classic land-based marine mesocosm (Marine Ecosystems Research Laboratory in Narragansett, Rhode Island, USA). Established in 1976, MERL is based at the University of Rhode Island and was first funded by the US EPA (Grice and Reeve, 1982; Odum, 1984) (Fig. 8.2). The level of control compared to laboratory studies or enclosures is intermediate with temperature, mixing, exchange rate, and biological material all regulated. The experimental system contains 14 relatively small, rigid fiberglass tanks (2 m diameter, 6 m depth, 0.76 m^2 sediment surface, 13 m^3 total volume) and is seeded with field-collected planktonic and benthic communities scaled to match that of nearby Narragansett Bay. Each MERL tank contains a multitrophic benthic community suitable for long-term experimentation. A number of tests have shown that MERL mesocosms adequately replicated sufficient field variation for the tests conducted and, once stocked, remained similar to field communities (Oviatt et al., 1980; Pilson and Nixon, 1980). Longer-term experiments (eg, >1 year) have routinely been conducted. For example, recent experiments have investigated seasonal patterns in coastal acidification and how these can be altered by nutrient enrichment (Nixon et al., 2015).

8.3.2 Enclosure Studies

Mesocosm methods also include limnocorrals, littoral, and pelagic enclosures. Limnocorrals isolate replicate subsections of the aquatic environment using dividing sidewalls anchored to bottom sediments and filled with filtered (Forrest and Arnott, 2006) or unfiltered pelagic water (Thompson et al., 1994). Limnocorrals have similar advantages to other mesocosm approaches (eg, replicated design, standardized physicochemical environment, multispecies interactions). Limitations that must be considered are wall effects that result from restricted vertical and horizontal mixing. Wall effects can affect nutrient and chemical dynamics, physicochemical properties, and interspecies interactions (eg, predation) resulting in conditions that diverge with time relative to the surrounding water body (Graney et al., 1995). Littoral enclosures differ from limnocorrals in that they have a natural shoreline and three plastic walls embedded in the sediments (Graney et al., 1995). While they have comparable advantages to limnocorrals, their limitations include high replicate variation in top predator density and the possibility that fluctuations in water depth will compromise enclosure integrity.

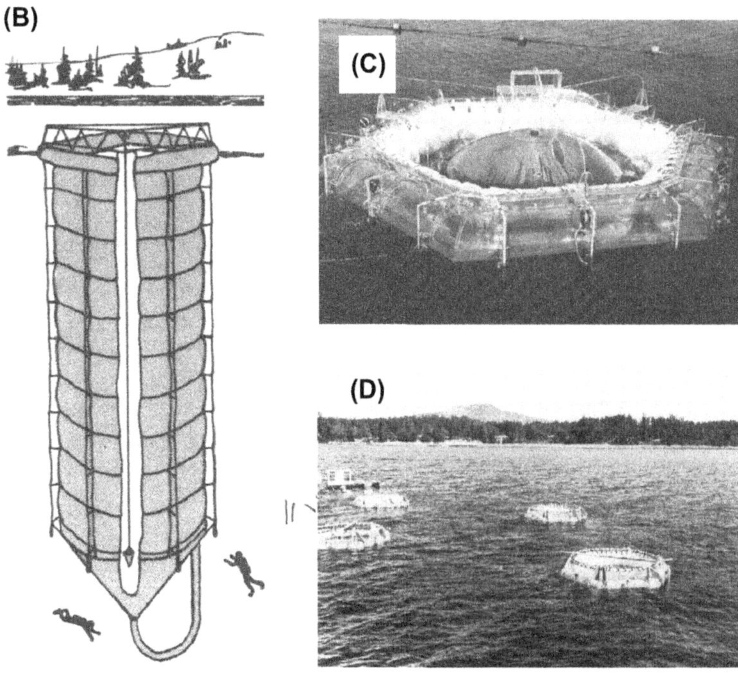

FIGURE 8.2 (A–D). Classic examples of (A) land-based and (B–D) pelagic enclosure experiments. Land-based experiments such as MERL (Marine Ecosystems Research Laboratory) in Narranganset Bay is operated by the University of Rhode Island and the US EPA. MERL contains 14 relatively small, rigid fiberglass tanks (2 m diameter, 6 m depth, 0.76 m^2 sediment surface, 13 m^3 total volume) and is seeded with field-collected planktonic and benthic communities scaled to match that of nearby Narragansett Bay. The pelagic enclosure experiment (B–D) shown is the CEPEX (Controlled Ecosystems Populations Experiment) in Saanich Inlet, British Columbia (Canada). The first CEPEX was composed of three flexible cone-shaped bags that were 68 m^3, 2.5 m diameter at the top, 0.2 m at the bottom, and 16 m in depth (see Parsons, 1978). (C) Photograph of earlier CEPEX design with acrylic dome for the collection of gases at the sea surface. Later versions (B, D) of CEPEX were larger (1300 m^3, 10 m diameter at the top, 0.2 m diameter bottom, and 24 m depth). *Images modified from Parsons, T.R., 1978. Controlled aquatic ecosystem experiments in ocean ecology research. Mar. Pollut. Bull. 9, 203–205 and Odum, E.P., 1984. The mesocosm. Bioscience 34, 558–562 with permission.*

Very few littoral enclosures have been constructed (but see the Marine Research Station in Solbergstrand, Norway, and the Baltic Sea Laboratory in Karlskrona, Sweden). At Solbergstrand in Norway, eight concrete littoral mesocosms (23 m^3) were built with wave bars to mimic the intertidal zone of the Oslofjord (Gray, 1987; Bokn et al., 2001). The original experiments using these hard-bottomed enclosures were seeded 3 years prior and benthic communities were allowed to develop before any experiments being conducted (Bakke, 1990). Littoral studies at Solbergstrand were among the first to suggest that care should be taken to ensure the effects of enclosing the system are documented as the recruitment of some species were strongly affected by the enclosures (notably mollusks) (Gray, 1987). Meticulous care has subsequently been taken in more recent experiments to describe how the enclosures were similar to the 50 species of macroscopic algae and animals typical of medium-sheltered shores of the Oslofjord (Bakke, 1990; Bokn et al., 2001).

A classic pelagic enclosure study was the CEPEX (Controlled Ecosystems Populations Experiment) but see also the Loch Ewe experiments in Scotland (Gamble et al., 1977). CEPEX is a moored pelagic enclosure deployed to examine plankton dynamics in Saanich Inlet, BC, Canada (Fig. 8.2). The initial experimental system was composed of three cone-shaped bags that were 68 m^3, 2.5 m diameter at the top, 0.2 m at the bottom, and 16 m in depth (Parsons, 1978). Later versions were larger (1300 m^3, 10 m diameter at the top, 0.2 m diameter bottom, and 24 m depth) and both the earlier and later versions captured static water columns but had no tidal exchange for dilution or recruitment (Grice et al., 1980; Menzel and Case, 1977). CEPEX was originally designed to examine natural unpolluted phytoplankton dynamics and offered, at the time, groundbreaking insight into effects of selective grazing and predation in addition to population recovery after exposure to copper (Grice et al., 1980; Menzel and Case, 1977). More recent pelagic enclosure studies in Norway (Riebesell et al., 2008) and Korea (Kim et al., 2008) have been used to examine ocean acidification, and considerable guidance is available from the European Commission as to their setup and deployment (Riebesell et al., 2009). The foundation of much of the recent pelagic work on ocean acidification is based on earlier mesocosm studies examining coral reefs (see 1996 special issue in *Ecological Engineering* v6: 1–225). Extensive reviews of these studies have also shown that given appropriate resources, mesocosm studies have the potential to improve model accuracy and are unique tools for the ecology and restoration of coral reefs globally (Luckett et al., 1996; Kleypas et al., 2006).

8.3.3 Artificial Stream Approaches

Insights gained from artificial stream approaches may be particularly useful in the marine context, as the means to overcome technical issues with mixing and water movement have been extensively investigated. Stream mesocosms have been used in freshwater ecological and ecotoxicological research for over 50 years (McIntire, 1993). They range in size from large, constructed channels (>100 m^3) with once-through flow, to smaller (<1 m^3) systems consisting of fabricated tanks with partial-recirculation or once-through flow (Swift et al., 1993; Culp and Baird, 2006). Large stream mesocosms (>50 m in length; >100 m^3) are a very rare resource (Swift et al., 1993). The high costs of construction and maintenance greatly limit the number of replicate streams and potential for variety in experimental design. Their operation has been largely restricted to government agencies or large consortia because of cost and logistical complexity. Although large experimental streams could facilitate the examination of long-term, multigenerational studies of recovery from stressor disturbance, few of these experiments exceed 12 months (Swift et al., 1993). Recognizing the need for larger, replicated systems which are ecologically realistic, Mohr et al. (2005) have developed a mesocosm facility that offers a highly flexible design which can be arranged as eight replicate streams (up to 106 m long each) or joined into a single stream approximately 850 m long.

In contrast, small stream mesocosms have been widely used by ecologists and ecotoxicologists (Lamberti and Steinman, 1993; Culp et al., 2000b,c; Dubé et al., 2002; Schulz et al., 2002; Crane et al., 2007). These systems range from flow-through flumes (Bothwell, 1993), to transportable streamside mesocosms (Culp and Baird, 2006), to greenhouse systems with naturally colonized substrates (Clark and Clements, 2006). Water sources include groundwater and surface water, with water velocity in the systems controlled by stirring mechanisms, water pumps, or the slope of the experimental unit. Normally, water temperature is not actively manipulated. Establishment of benthic communities varies and incorporates natural colonization, seeding of benthic communities from reference riffles (Alexander et al., 2008, 2013), and the use of trays of substrate that have been colonized in reference streams (Clark and Clements, 2006). Because of their small scale, researchers generally conduct shorter-term experiments (<30 days). Shorter studies decrease the possibility of wall effects and undesirable divergence from the reference stream composition as a result of the absence of biological processes such as invertebrate drift from upstream habitats. Overall, small-scale stream mesocosm experiments have proven to be particularly informative when combined with field and laboratory studies (Culp et al., 2000a; Clements et al., 2002; Clements, 2004).

8.4 THINKING BIGGER: DEVELOPING SYSTEMS FOR LARGE FISH

Large fish species present a unique challenge to experimental design. Their particular needs can determine the size, complexity, and cost of an experiment. However, the benefits of using a larger species, such as increased ecological realism, as well as the availability of larger tissue quantities for biochemical testing and the ability to perform repeat measurements, may outweigh these hardships. Over the last 20–30 years, the use of fish in research has substantially increased. Fish are hardy, adaptable, and have wide species diversity. Additionally, their morphological and physiological systems (Pohlenz and Gatlin, 2014; Bolon and Stoskopf, 1995; Sunyer, 2013), basic biochemical pathways, and responses are often analogous to those of mammals (Bolon and Stoskopf, 1995). Fish also exhibit similar behavioral, physiological, and hormonal stress responses to painful stimuli as mammals (Sneddon, 2009; Braithwaite and Huntingford, 2004; Baker et al., 2013). These similarities between some species of fish and other vertebrates make them suitable models for many research disciplines. Guidelines for proper care and use of fish as laboratory animals are available and are continually being updated, as our knowledge of their biology, behavior, and husbandry expands (eg, CCAC: Canadian Council on Animal Care, AVMA: American Veterinary Medical Association, AAALAC: Association for Assessment and Accreditation of Laboratory Animals International). Proper fish welfare ensures that their optimal health and well-being is provided for (food, shelter, ability to express normal behavior) and that they are not subjected to undue pain or distress.

Large fish species require a greater volume and supply of suitable water than many current mesocosm designs (eg, >10× more for some species). Water provides oxygen and removes toxic metabolic wastes, thus acting as a life support system for fish. Some species, like coldwater salmonids, are adapted to fast-flowing, oxygen-rich waters with a narrow temperature range and are more susceptible to small changes in water quality. Other species, like warmwater cyprinids, can tolerate a wider range of environmental conditions, such as water flow, dissolved oxygen, and temperature. Maintaining proper water chemistry with an adequate oxygen supply will maximize the number of fish that can be housed within a given aquarium and can even minimize the size of experimental housing required. Larger tanks with consistently high rates of water turnover are more resilient than small tanks, in terms of maintaining temperature, dissolved oxygen levels and water chemistry parameters, as well as minimizing the build-up of toxic metabolic wastes.

Traditional fish culture systems are static systems that typically use smaller tanks (eg, 1–300 L tanks) with minimal inputs of water and occasionally have small filtration units attached. These systems have lower carrying capacities and are highly susceptible to severe water quality changes (toxic build-ups, low dissolved oxygen). More sophisticated systems have been developed and are commonly used for smaller fish species. For large fish species, either flow-through or recirculation systems are recommended. In addition to being more resilient, these systems allow enough space for fish to exhibit more normal swimming behavior. Flow-through systems require a continuous supply of high-quality water. Recirculation systems require less water; however, they often require more expertise, time, and care and in water quality can be more variable. A common water quality problem in recirculation systems can occur when the biofilter capacity is overwhelmed by high fish densities and/or feeding rates that result in a build-up of toxic ammonia (Yanong, 2003). Other water quality issues, such as low dissolved oxygen, high carbon dioxide, decrease in pH, and increase in water temperature can also occur over time.

Tank design may also be dictated by the particular requirements of the chosen research species. More active fish (eg, salmonids) require higher flows and room for continuous swimming. Circular tanks provide high flows and have a higher carrying capacity with the extra benefit of being self-cleaning due to the

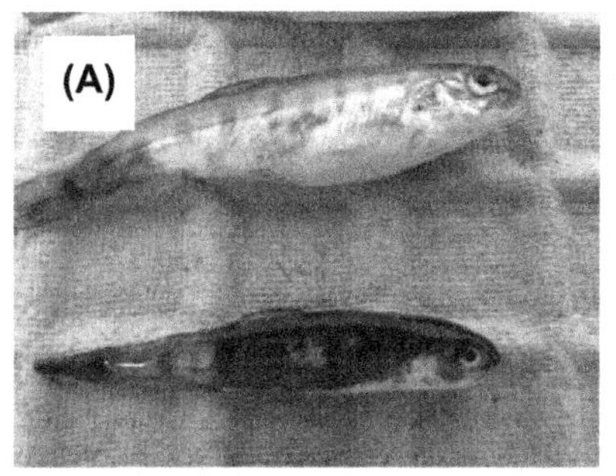

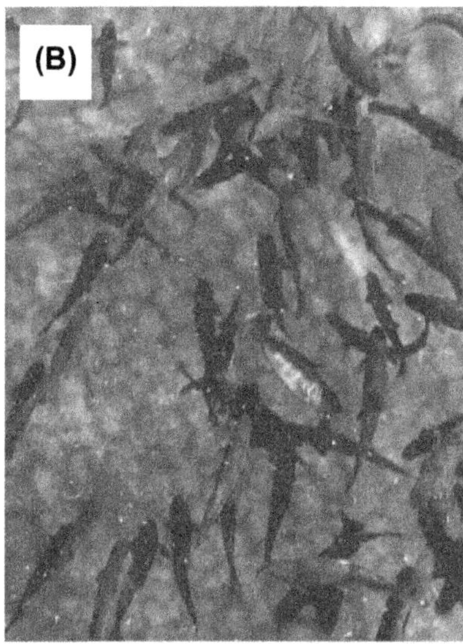

FIGURE 8.3 Fish condition deteriorates in raceways. (A) Steelhead trout (*Oncorhynchus mykiss*) with melanization as a result of sickness. Healthy fish (top). (B) Captive brook trout (*Salvelinus fontinalis*) exhibiting an abnormal whirling swimming behavior as a result of neurological damage due to skeletal deformation at the St. Croix and (C) Osceola State Fish Hatcheries, Wisconsin, USA. *Photos from Dr. M. Finley, with permission.*

ability of fish to continuously swim. This circular design works poorly for less active fish (eg, esocids) or small fish that become overwhelmed by higher water flows. Swedish ponds, which are square tanks with rounded edges, are thought to be best for fish that need more surface area and do not concentrate vertically within the water column (eg, Atlantic salmon (*Salmo salar*); Piper et al., 1982). The rounded edges allow for the same self-cleaning mechanism as found in the circular tank. Rectangular tanks or raceways do not reflect normal environmental conditions. They have poor water flow patterns with no flow refuges and cannot attain the high water velocities that allow for self-cleaning (Piper et al., 1982). Thus, overall water quality in raceways is poor with significant deterioration toward the outflow end of the raceway, often resulting in fish congregating near the inflow. Under these conditions, fish become stressed and disease incidence and transmission increases (Fig. 8.3). Rectangular designs are best suited for low densities of small fish. Earthen ponds can be used for hardier fish (eg, ictalurids) that do not require high water quality and/or intensive environmental control in the experimental design. Benefits to earthen ponds are their ability to produce natural feed (eg, zooplankton and aquatic invertebrates) and that they allow for more natural environments. However, ponds often acquire excessive vegetation, making these designs difficult to clean, distribute feed and collect fish. Ponds also tend to require aeration and reduced cleanliness can increase pathogen exposure.

MESOCOSMS: IS BIGGER BETTER?

The physical size of mesocosms ranges from artificially constructed systems (fabricated tanks, constructed ponds, ditches) to isolated subsections of the natural habitat (limnocorrals, littoral enclosures) (Boyle and Fairchild, 1997). Previous reviews of the mesocosm literature have found that studies were on average 49 days in duration and enclosures were 1.7 m^3 (Fig. 8.4) (Petersen et al., 1999, 2009). Although far shorter and smaller than the ecological processes under examination, both are concessions to the time and cost of running replicated and controlled mesocosms experiments. Reviews of the potential ecological effects and fate of pollutants suggest that perturbation studies combined with field data can create population and community data that is relevant and defensible at scales of months and tens of meters (Sanders, 1985). Determining the appropriate level of ecological realism is, however, dependent on the research question and the funds available to conduct the experiment. Larger systems, due to cost constraints, tend to sacrifice replication for large-scale, single-treatment mesocosms (Riebesell et al., 2008). Common throughout the mesocosm literature is a combined approach where the weight-of-evidence at different scales offers overlapping, unique perspectives on the same scientific narrative (mesocosm combined with field: Schindler, 1998; in situ: Culp and Baird, 2006; laboratory: Sanders, 1985; or modeling approaches: Stewart et al., 2013).

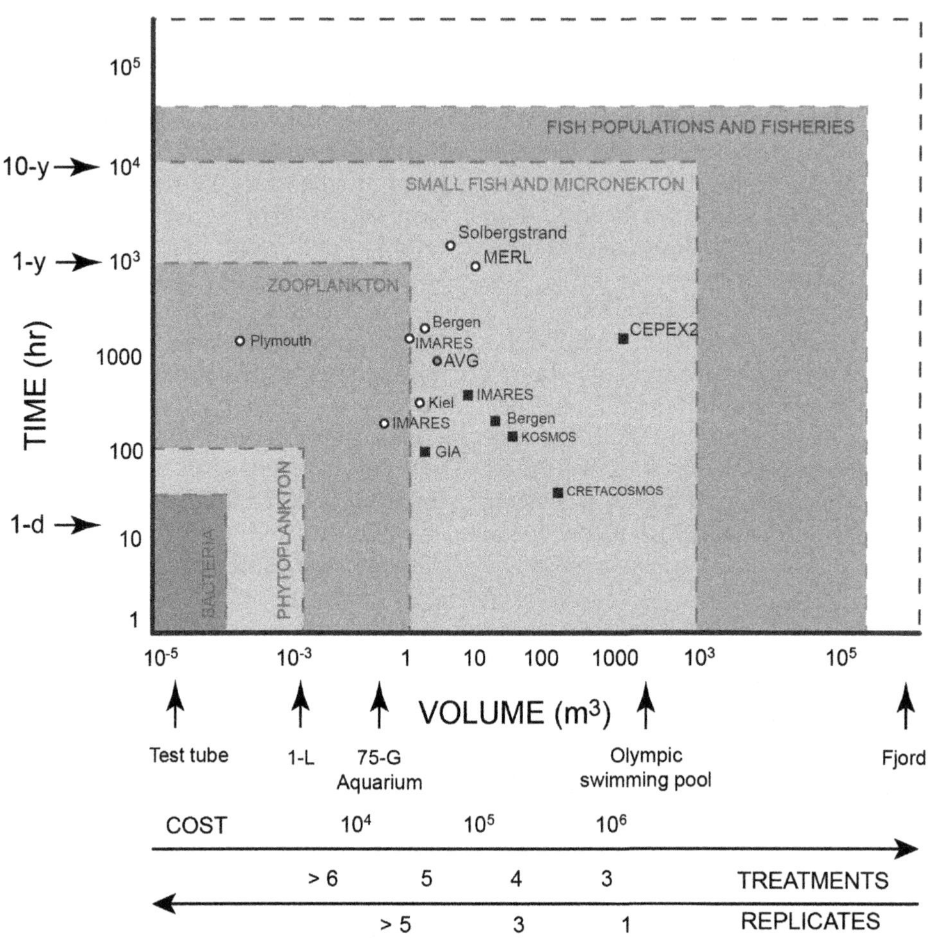

FIGURE 8.4 Previously proposed relationships between trophic levels of an enclosed littoral and pelagic community with respect to enclosure size and the duration of the experiment. *(Modified from Parsons, T.R., 1982. The future of controlled ecosystem enclosure experiments. In: Grice G.D., Reeve M.R. (Eds.), Marine Mesocosms Biological and Chemical Research in Experimental Ecosystems. Springer, New York, NY, USA, pp. 411–418 with permission.)* Cost estimates reflect minimum startup costs derived from benthic mesocosms (of which the authors are most familiar). Annual maintenance costs were not included but are easily 5% of the total value of the system and increase over time due to wear and tear. Data on treatment levels and replicates inferred from Petersen et al. (2009). Information on land-based (○) and enclosure (■) mesocosms were inferred from www.mesocosm.eu and compared to values reported in the scientific literature.

8.5 SOME SPECIFIC CONSIDERATIONS FOR THE DESIGN OF MARINE MESOCOSMS

Marine mesocosms first became popular in the 1970s. By the 1980s it was apparent that mesocosm methods would require intense study of the effects of scale to apply research findings to real systems (Petersen et al., 2009). It is well known that small enclosures and short experiments can limit the number of trophic levels included in an experiment (Fig. 8.4). The size of the enclosure can also create artifacts that can alter the nature of the exposure (eg, wall growth, animal movements, reduced mixing, and altered water exchange rates). All of the above have been linked to weak experimental designs—an issue that is not unique to mesocosm studies (discussed in: Kuiper, 1982; Bloesch et al., 1988).

The use of open systems has recently gained favor in freshwater research where populations of aquatic invertebrates are permitted to colonize or leave pond mesocosm habitats. In the marine context, open systems can flush the enclosed water permitting the inflow of nutrients and new recruits closer to that of natural systems. Closed systems have different benefits and are more appropriate when toxic substances are to be tested as a closed system prevents the release of chemicals or exposed organisms back into the environment. Another advantage of the closed system approach is that it mimics broad nonpoint source pollution providing a worse case scenario for management and modeling.

Enclosing a water body alters the mixing regime. Turbulence can affect species interactions and growth as well as particle movement. According to Sanford (1997) turbulent mixing is as important as light, temperature, salinity, and nutrients and should be directly measured and reported. Among the best-described mixing systems that can be found are the Multiscale Experimental Ecosystem Research Center benthic-pelagic mesocosms (Petersen et al., 2009) that extensively applied engineering principles to develop appropriate mixing regimes (Tatterson, 1991). In contrast, early flexible enclosures such as CEPEX were typically not mixed (Fig. 8.2b). Steele et al. (1977) estimated CEPEX's vertical turbulent diffusion at 0.1 cm^2/s as much as 10× smaller than the surrounding surface water. More recent studies have increased vertical mixing by bubbling and horizontal grids to create turbulence within the range of natural systems (Svensen et al., 2001; Nerheim et al., 2002).

8.6 FUTURE POSSIBILITIES

With the onset of emerging issues in the marine and aquatic environment, come new challenges and interesting opportunities for future mesocosm research. In relation to climate change, we anticipate an increase in marine and freshwater mesocosm experimentation. For these authors, our future possibilities tend toward studies of the cumulative impact of multiple stressors on riverine ecosystems using stream mesocosms, models, and field surveys. Investigations of cumulative effects are also relevant to marine mesocosm studies where the importance of natural (eg, seasonal) and anthropogenic (eg, urbanization, resource exploitation) gradients impact marine communities.

Another area of continued and expanding interest will include the rise in ocean temperatures and acidification. This is such a major issue and will affect the marine community in many different aspects. Some other future research questions might include investigations on the effect of climate change on Arctic ice melt or sea level rise. Generally, how can the changing timing and duration of the ice melt period influence the structure or function of aquatic communities? Will estuarine areas be exposed to different levels of salinity, what is the effect on the estuarine community?

Investigating invasive species is another area of interest that can be looked at using marine mesocosms. Why are species or diseases invading new habitats? Why are some areas more prone to invasion than others? Are the invaders altering interactions between ecosystems components (eg, predator–prey cycles, resource use, carbon flux)? Can native species coexist with lower-latitude invaders or will the food web be disrupted? Can conservation efforts protect regions sensitive to climate change?

Whatever the topic of investigation, noteworthy mesocosm studies tend to rely on innovation. The most influential mesocosm studies will be interdisciplinary efforts with innovative approaches to study questions, as opposed to simply conducting lab experiments outdoors. They will also incorporate relevant sources of natural variation (eg, temperature, mixing, colonization) to untangle important interactions between species and their environment. As with any experiment, understanding the limitations and appropriate scale of the study are of paramount importance.

8.7 CONCLUSIONS: UTILITY OF MESOCOSM TESTING IN MARINE ECOTOXICOLOGY

Mesocosms are effective tools for interdisciplinary research and have historically yielded new concepts of how seascapes function and react to perturbation. These types of studies tend to create high-quality data that can be particularly useful for teasing apart confounding factors to reveal underlying patterns in ecosystems. Mesocosms have also at times reminded researchers of the limitations of our understanding by unveiling both the complexity and the simplicity of relationships between organisms and ecosystems (*sensu* Benton et al., 2007). In a regulatory context such as Canada's Environmental Effects Monitoring program, mesocosms are an accepted alternative monitoring method for fish studies. Although large fish species present a challenge to experimental design, the benefits of using larger organisms may outweigh the difficulties. Increasingly, mesocosms explore very complex global issues including the effects of climate change (Davis et al., 1998; Petchey et al., 1999), biodiversity loss (Balvanera et al., 2006; Cardinale et al., 2006), pollution, and fishery overharvesting (Micheli, 1999) at times in very extreme habitats (Stark et al., 2014). Given the threat to the world's oceans posed by a changing climate, ocean acidification, increasing coastal development, and urbanization, future interdisciplinary studies will be needed and mesocosm studies can be a powerful tool in any research program.

References

Adams, S.M., 2003. Establishing causality between environmental stressors and effects on aquatic ecosystems. Hum. Ecol. Risk Assess. 9, 17–35.

Alexander, A.C., Culp, J.M., 2013. Predicting the effects of insecticide mixtures on non-target aquatic communities. (Chapter 3). In: Trdan, S. (Ed.), Insecticides – Development of Safer and More Effective Technologies. InTech, Rijeka, Croatia, pp. 83–101. Open Access. http://www.intechopen.com/books/insecticides-development-of-safer-and-more-effective-technologies/predicting-the-effects-of-insecticide-mixtures-on-non-target-aquatic-communities.

Alexander, A.C., Heard, K.S., Culp, J.M., 2008. Emergent body size of mayfly survivors. Freshwater Biol. 53, 171–180.

Alexander, A.C., Luis, A.T., Culp, J.M., Baird, D.J., Cessna, A.J., 2013. Can nutrients mask community responses to insecticide mixtures? Ecotoxicology 22, 1085–1100.

Baker, T.R., Baker, B.B., Johnson, S.M., Sladky, K.K., 2013. Comparative analgesic efficacy of morphine sulfate and butorphanol tartrate in koi (*Cyprinus carpio*) undergoing unilateral gonadectomy. J. Am. Vet. Med. Assoc. 243, 882–890.

Bakke, T., 1990. Benthic mesocosms: II. Basic research in hard-bottom Benthic mesocosms. In: Lalli, C.M. (Ed.), Coastal and Estuarine Studies. Springer, New York, NY, USA, pp. 122–135.

Balvanera, P., Pfisterer, A.B., Buchmann, N., He, J.S., 2006. Quantifying the evidence for biodiversity effects on ecosystem functioning and services. Ecol. Lett. 9, 1146–1156.

Benton, T.G., Solan, M., Travis, J.M.J., Sait, S.M., 2007. Microcosm experiments can inform global ecological problems. Trends Ecol. Evol. 22, 516–521.

Bloesch, J., Bossard, P., Buhrer, H., Burgi, H.R., Uehlinger, U., 1988. Can results from limnocorral experiments be transferred to in situ conditions? Hydrobiologia 159, 297–308.

Bokn, T.L., Hoell, E.E., Kersting, K., Moy, F.E., 2001. Methods applied in the large littoral mesocosms study of nutrient enrichment in rocky shore ecosystems—EULIT. Cont. Shelf Res. 21, 1925–1936.

Bolon, B., Stoskopf, M.K., 1995. Fish. In: Rollin, B.E. (Ed.), The Experimental Animal in Biomedical Research: Care, Husbandry, and Well-Being: An Overview by Species, vol. II. CRC Press, pp. 15–30.

Bothwell, M.L., 1993. Artificial streams in the study of algal/nutrient dynamics. In: Lamberti, G.A., Steinman, A.D. (Eds.), Research in Artificial Streams: Applications, Uses, and Abuses, J. N. Am. Benthol. Soc. 12, pp. 327–333, 313–384.

Boyle, T.P., Fairchild, J.F., 1997. The role of mesocosm studies in ecological risk analysis. Ecol. Appl. 7, 1099–1102.

Braithwaite, V.A., Huntingford, F.A., 2004. Fish and welfare: do fish have the capacity for pain perception and suffering. Anim. Welfare 13, S87–S92.

Buffet, P.-E., Zalouk-Vergnoux, A., Châtel, A., Berthet, B., Métais, I., Perrein-Ettajani, H., Poirier, L., Luna-Acosta, A., Thomas-Guyon, H., Risso-de-Faverney, C., 2014. A marine mesocosm study on the environmental fate of silver nanoparticles and toxicity effects on two endobenthic species: the ragworm *Hediste diversicolor* and the bivalve mollusk *Scrobicularia plana*. Sci. Total Environ. 470–471, 1151–1159.

Cardinale, B., Srivastava, D., Duffy, J., Wright, J., Downing, A., Sankaran, M., Jouseau, C., 2006. Effects of biodiversity on the functioning of trophic groups and ecosystems. Nature 443, 989–992.

Carpenter, S.R., Cole, J.J., Essington, T.E., Hodgson, J.R., Houser, J.N., Kitchell, J.F., Pace, M.L., 1998. Evaluating alternative explanations in ecosystem experiments. Ecosystems 1, 335–344.

Carpenter, S.R., 1996. Microcosm experiments have limited relevance for community and ecosystem ecology. Ecology 77, 677–680.

Carpenter, S.R., 1999. Microcosm experiments have limited relevance for community and ecosystem ecology: reply. Ecology 80, 1085–1088.

Christman, V.D., Voshell Jr., J.R., Jenkins, D.G., Rosenzweig, M.S., Layton, R.J., Buikema Jr., A.L., 1994. Ecological development and biometry of untreated pond mesocosms. In: Graney, R.L., Kennedy, J.H., Rodgers, J.H. (Eds.), Aquatic Mesocosm Studies in Ecological Risk Assessment. CRC Press, Boca Raton, FL, pp. 105–129.

Clark, J.L., Clements, W.H., 2006. The use of in situ and stream microcosm experiments to assess population- and community-level responses to metals. Environ. Toxicol. Chem. 25, 2306–2312.

Clark, J.R., Noles, J.L., 1994. Contaminant effects in marine/estuarine systems: field studies and scaled simulations. In: Graney, R.L., Kennedy, J.H., Rodgers, J.H. (Eds.), Aquatic Mesocosm Studies in Ecological Risk Assessment. CRC Press, Boca Raton, FL, USA, pp. 47–60.

Clements, W.H., Carlisle, D.M., Courtney, L.A., Harrahy, E.A., 2002. Integrating observational and experimental approaches to demonstrate causation in stream biomonitoring studies. Environ. Toxicol. Chem. 21 (6), 1138–1146.

Clements, W.H., 2004. Small-scale experiments support causal relationships between metal contamination and macroinvertebrate community responses. Ecol. Appl. 14, 954–967.

Crane, M., Burton, G.A., Culp, J.M., Greenberg, M.S., Munkittrick, K.R., Ribeiro, R., Salazar, M.H., 2007. Review of aquatic in situ approaches for stressor and effect diagnosis. Integr. Environ. Assess. Manage. 3, 234–245.

Culp, J.M., Baird, D.J., 2006. Establishing cause-effect relationships in multi-stressor environments. In: Hauer, Fr, Lamberti, G.A. (Eds.), Methods in Stream Ecology, second ed. Academic Press, Burlington, MA, USA, pp. 835–854.

Culp, J.M., Lowell, R.B., Cash, K.J., 2000a. Integrating in situ community experiments with field studies to generate weight-of-evidence risk assessment for large rivers. Environ. Toxicol. Chem. 19, 1167–1173.

Culp, J.M., Podemski, C.L., Cash, K.J., 2000b. Interactive effects of nutrients and contaminants from pulp mill effluents on riverine benthos. J. Aquat. Ecosyst. Stress Recovery 8 (1), 67–75.

Culp, J.M., Podemski, C.L., Cash, K.J., Lowell, R.B., 2000c. A research strategy for using stream microcosms in ecotoxicology: integrating experiments at different levels of biological organization with field data. J. Aquat. Ecosyst. Stress Recovery 7, 167–176.

Culp, J.M., Cash, K.J., Glozier, N.E., Brua, R.B., 2003. Effects of pulp mill effluent on benthic assemblages in mesocosms along the Saint John River, Canada. Environ. Toxicol. Chem. 22, 2916–2925.

Davis, A.J., Jenkinson, L.S., Lawton, J.H., Shorrocks, B., 1998. Making mistakes when predicting shifts in species range in response to global warming. Nature 391, 783–786.

de Lafontaine, Y., Leggett, W.C., 1987. Evaluation of in situ enclosures for larval fish studies. Can. J. Fish Aquat. Sci. 44, 54–65.

Drenner, R.W., Mazumder, A., 1999. Microcosm experiments have limited relevance for community and ecosystem ecology: comment. Ecology 80, 1081–1085.

Dubé, M.G., Culp, J.M., Cash, K.J., Glozier, N.E., MacLatchy, D.L., Podemski, C.L., Lowell, R.B., 2002. Artificial streams for environmental effects monitoring (EEM): development and application in Canada over the past decade. Water Qual. Res. J. Can. 37, 155–180.

EFSA PPR Panel (European Food Safety Authority Panel on Plant Protection Products and their Residues), 2013. Guidance on tiered risk assessment for plant protection products for aquatic organisms in edge-of-field surface waters. EFSA J. 11 (7), 3290. http://dx.doi.org/10.1016/B978-0-12-803371-5.00008-4, 268 pp. 2013.3290. http://www.efsa.europa.eu/sites/default/files/scientific_output/files/main_documents/3290.pdf.

Emmerson, M.C., Solan, M., Emes, C., Paterson, D.M., 2001. Consistent patterns and the idiosyncratic effects of biodiversity in marine ecosystems. Nature 411, 73–77.

Environment Canada, 2010. Pulp and Paper Environmental Effects Monitoring Technical Guidance Document—Chapter 8 Alternative Monitoring Methods. http://www.ec.gc.ca/esee-eem/3E389BD4-E48E-4301-A740-171C7A887EE9/PP_full_versionENGLISH[1]-FINAL-2.0.pdf.

Environment Canada, 2012. Metal Mining Technical Guidance for Environmental Effects Monitoring. Chapter 9 Alternative Monitoring Methods. https://ec.gc.ca/Publications/default.asp?lang=En&xml=D175537B-24E3-46E8-9BB4-C3B0D0DA806D.

Foekema, E.M., Kaag, N.H., Kramer, K.J.M., Long, K., 2015. Mesocosm validation of the marine No Effect Concentration of dissolved copper derived from a species sensitivity distribution. Sci. Total Environ. 521–522, 173–182.

Forrest, J., Arnott, S.E., 2006. Immigration and zooplankton community responses to nutrient enrichment: a mesocosm experiment. Oecologia 150, 119–131.

Gamble, J.C., Davies, J.M., Steele, J.H., 1977. Loch Ewe bag experiment, 1974. Bull. Mar. Sci. 27, 146–175.

Graney, R.L., Kennedy, J.H., Rodgers, J.H., 1994. Introduction. In: Graney, R.L., Kennedy, J.H., Rodgers, J.H. (Eds.), Aquatic Mesocosm Studies in Ecological Risk Assessment. CRC Press, Boca Raton, FL, USA, pp. 1–4.

Graney, R.L., Giesy, J.P., Clark, J.R., 1995. Field Studies. In: Rand, G.M. (Ed.), Fundamentals of Aquatic Toxicology: Effects, Environmental Fate, and Risk Assessment. Taylor & Francis, Washington, DC, USA, pp. 257–305.

Gray, J.S., 1987. Oil pollution studies of the Solbergstrand mesocosms. Philos. Trans. R. Soc. B 316, 641–654.

Grice, G.D., Reeve, M.R., 1982. Marine Mesocosms. Springer, New York, NY, USA.

Grice, G.D., Harris, R.P., Reeved, M.R., 1980. Large-scale enclosed water-column ecosystems an overview of foodweb I, the final CEPEX experiment. J. Mar. Biol. Assoc. UK 60, 401–414.

Guckert, J.B., 1993. Artificial streams in ecotoxicology. In: Lamberti, G.A., Steinman, A.D. (Eds.), Research in Artificial Streams: Applications, Uses, and Abuses, J. N. Am. Benthol. Soc. 12, pp. 350–356, 313–384.

Howick, G.L., deNoyelles Jr., F., Giddings, J.M., Graney, R.L., 1994. Earthen ponds vs. fiberglass tanks as venues for assessing the impact of pesticides on aquatic environments: a parallel study with Sulprofos. In: Graney, R.L., Kennedy, J.H., Rodgers, J.H. (Eds.), Aquatic Mesocosm Studies in Ecological Risk Assessment. CRC Press, Boca Raton, FL, USA, pp. 321–336.

Johnson, P.C., Kennedy, J.H., Morris, R.G., Hambleton, F.E., Graney, R.L., 1994. Fate and effects of Cyfluthrin (Pyrethroid insecticide) in pond mesocosms and concrete microcosms. In: Graney, R.L., Kennedy, J.H., Rodgers, J.H. (Eds.), Aquatic Mesocosm Studies in Ecological Risk Assessment. CRC Press, Boca Raton, FL, USA, pp. 337–371.

Kim, J.M., Shin, K., Lee, K., Park, B.K., 2008. In situ ecosystem-based carbon dioxide perturbation experiments: design and performance evaluation of a mesocosm facility. Limnol. Oceanogr.: Methods 6, 208–217.

Kleypas, J.A., Feely, R.A., Fabry, V.J., Langdon, C., Sabine, C.L., Robbins, L.L., 2006. Impacts of Ocean Acidification on Coral Reefs and Other Marine Calcifiers: A Guide for Future Research. Report of workshop held 18–20 April 2005, St. Petersburg, FL, sponsored by NSF, NOAA, and the US Geological Survey, 88 pp. http://www.ucar.edu/communications/Final_acidification.pdf.

Kuiper, J., 1982. Ecotoxicological experiments with marine plankton communities in plastic bags. In: Grice, G.D., Reeve, M.R. (Eds.), Marine Mesocosms. Springer, New York, NY, USA, pp. 181–193.

Lamberti, G.A., Steinman, A.D., 1993. Conclusions. In: Lamberti, G.A., Steinman, A.D. (Eds.), Research in Artificial Streams: Applications, Uses, and Abuses, J. N. Am. Benthol. Soc. 12, p. 370, 313–384.

Lawton, J.H., 1996. The Ecotron facility at Silwood Park: the value of "big bottle" experiments. Ecology 77, 665–669.

Lejeusne, C., Chevaldonné, P., Pergent-Martini, C., Boudouresque, C., Pérez, T., 2010. Climate change effects on a miniature ocean: the highly diverse, highly impacted Mediterranean Sea. Trends Ecol. Evol. 25, 250–260.

Lowell, R.B., Culp, J.M., Dubé, M.G., 2000. A weight-of-evidence approach for northern river risk assessment: integrating the effects of multiple stressors. Environ. Toxicol. Chem. 4 (2), 1182–1190.

Luckett, C., Adey, W.H., Morrissey, J., Spoon, D.M., 1996. Coral reef mesocosms and microcosms – successes, problems, and the future of laboratory models. Ecol. Eng. 6, 57–72.

McIntire, C.D., 1993. Historical and other perspectives of laboratory stream research. In: Lamberti, G.A., Steinman, A.D. (Eds.), Research in Artificial Streams: Applications, Uses, and Abuses, J. N. Am. Benthol. Soc. 12, pp. 318–323, 313–384.

Menzel, D.W., Case, J., 1977. Concept and design: controlled ecosystem pollution experiment. Bull. Mar. Sci. 27, 1–7.

Micheli, F., 1999. Eutrophication, fisheries, and consumer-resource dynamics in marine pelagic ecosystems. Science 285, 1396–1398.

Mohr, S., Feibicke, M., Ottenstroer, T., Meinecke, S., Berghahn, R., Schmidt, R., 2005. Enhanced experimental flexibility and control in ecotoxicological mesocosm experiments – a new outdoor and indoor pond and stream system. Environ. Sci. Pollut. Res. Int. 12, 5–7.

Nerheim, S., Stiansen, J.E., Svendsen, H., 2002. Grid-generated turbulence in a mesocosm experiment. Hydrobiologia 484, 61–73.

Nixon, S.W., Oczkowski, A.J., Pilson, M., Fields, L., 2015. On the response of pH to inorganic nutrient enrichment in well-mixed coastal marine waters. Estuaries Coasts 38, 232–241.

Norton, S.B., Cormier, S.M., Suter, G.W., 2014. Ecological Causal Assessment. CRC Press, Boca Raton, FL, USA.

Odum, E.P., 1984. The mesocosm. Bioscience 34, 558–562.

Oviatt, C.A., Walker, H., Pilson, M.E.Q., 1980. An exploratory analysis of microcosm and ecosystem behavior using multivariate techniques. Mar. Ecol. Progr. Ser. 2, 179–191.

Oviatt, C.A., 1994. Biological considerations in marine enclosure experiments: challenges and revelations. Oceanography 7, 45–51.

Parsons, T.R., 1978. Controlled aquatic ecosystem experiments in ocean ecology research. Mar. Pollut. Bull. 9, 203–205.

Parsons, T.R., 1982. The future of controlled ecosystem enclosure experiments. In: Grice, G.D., Reeve, M.R. (Eds.), Marine Mesocosms Biological and Chemical Research in Experimental Ecosystems. Springer, New York, NY, USA, pp. 411–418.

Perceval, O., Caquet, T., Lagadic, L., Bassères, A., Azam, D., Lacroix, G., Poulsen, V., October 2009. Mesocosms: their value as tools for managing the quality of aquatic environments. In: Recap Prepared From the Meeting of the Ecotoxicology Symposium in Le Croisic, France.

Petchey, O.L., McPhearson, P.T., Casey, T.M., Morin, P.J., 1999. Environmental warming alters food-web structure and ecosystem function. Nature 402, 69–72.

Petersen, J.E., Cornwell, J.C., Kemp, W.M., 1999. Implicit scaling in the design of experimental aquatic ecosystems. Oikos 85, 3–18.

Petersen, J.E., Kennedy, V.S., Dennison, W.C., Kemp, W.M., 2009. Enclosed Experimental Ecosystems and Scale: Tools for Understanding and Managing Coastal Ecosystems. Springer, New York, NY, USA.

Pilson, M.E.Q., Nixon, S.W., 1980. Annual nutrient cycles in a marine microcosm. In: Giesy, J.P. (Ed.), Microcosms in Ecological Research, DOE Symposium Series, 52. Springfield, VA, USA, pp. 753–778. CONF-781101.

Piper, R.G., McElwain, I.B., Orme, L.E., McCraren, J.P., Flower, L.G., Leonard, J.R., 1982. Fish Hatchery Management. United States Fish and Wildlife Service, Washington, DC.

Pohlenz, C., Gatlin III, D.M., 2014. Interrelationships between fish nutrition and health. Aquaculture 431, 111–117.

Rand, G.M., Clark, J.R., Holmes, C.M., 2000. Use of outdoor freshwater pond microcosms: II. Responses of biota to pyridaben. Environ. Toxicol. Chem. 19 (2), 396–404.

Renick, V.C., Anderson, T.W., Morgan, S.G., Cherr, G.N., 2015. Interactive effects of pesticide exposure and habitat structure on behavior and predation of a marine larval fish. Ecotoxicology 24, 391–400.

Riebesell, U., Lee, K., Nejstgaard, J.C., 2009. Pelagic mesocosms (Chapter 6). In: Riebesell, U., Fabry, V.J., Hansson, L. (Eds.), Guide to Best Practices for Ocean Acidification Research and Data Reporting. Report for the European Commission EUR 24328 EN.

Riebesell, U., Bellerby, R.G.J., Grossart, H.-P., Thingstad, F., 2008. Mesocosm CO_2 perturbation studies: from organism to community level. Biogeosciences 5, 1157–1164.

Sanders, F.S., 1985. Use of large enclosures for perturbation experiments in lentic ecosystems: a review. Environ. Monit. Assess. 5, 55–99.

Sanford, L.P., 1997. Turbulent mixing in experimental ecosystem studies. Mar. Ecol. Progr. Ser. 161, 265–293.

Schindler, D.W., 1998. Whole-ecosystem experiments: replication versus realism: the need for ecosystem-scale experiments. Ecosystems 1, 323–334.

Schulz, R., Thiere, G., Dabrowski, J.M., 2002. A combined microcosm and field approach to evaluate the aquatic toxicity of azinphosmethyl to stream communities. Environ. Toxicol. Chem. 21 (10), 2172–2178.

Scott, G.I., Fulton, M.H., DeLorenzo, M.E., Wirth, E.F., Key, P.B., Pennington, P.L., Kennedy, D.M., Porter, D., Chandler, G.T., Scott, C.H., Ferry, J.L., 2013. The environmental sensitivity index and oil and hazardous materials impact assessments: linking prespill contingency planning and ecological risk assessment. J. Coastal Res. 69, 100–113.

Sneddon, L.U., 2009. Pain perception in fish: indicators and endpoints. ILAR J. 50, 338–342.

Stark, J., Johnstone, G., Riddle, M., 2014. A sediment mesocosm experiment to determine if the remediation of a shoreline waste disposal site in Antarctica caused further environmental impacts. Mar. Pollut. Bull. 89, 284295.

Steele, J.H., Farmer, D.M., Hendersen, E.W., 1977. Circulation and temperature structure in large marine enclosures. J. Fish Res. Board Can. 34, 1095–1104.

Stewart, R.I., Dossena, M., Bohan, D.A., Jeppesen, E., Kordas, R.L., Ledger, M.E., Meerhoff, M., Moss, B., Mulder, C., Shurin, J.B., Suttle, B., Thompson, R., Trimmer, M., Woodward, G., 2013. Mesocosm experiments as a tool for ecological climate-change research (Chapter 2). Adv. Ecol. Res. 48, 71–181.

Sunyer, J.O., 2013. Fishing for mammalian paradigms in the teleost immune system. Nat. Immunol. 14, 320–326.

Svensen, C., Egge, J.K., Stiansen, J.E., 2001. Can silicate and turbulence regulate the vertical flux of biogenic matter? A mesocosm study. Mar. Ecol. Progr. Ser. 217, 67–80.

Swift, M., Troelstrup Jr., N., Detebeck, N., Foley, J., 1993. Large artificial streams in toxicological and ecological research. J. N. Am. Benthol. Soc. 12, 359–366.

Tatterson, G.B., 1991. Fluid Mixing and Gas Dispersion in Agitated Tanks. McGraw-Hill, New York, NY, USA.

Thompson, D.G., Holmes, S.B., Pitt, D.G., Solomon, K.R., Wainio-Keizer, K.L., 1994. Applying concentration-response theory to aquatic enclosure studies. In: Graney, R.L., Kennedy, J.H., Rodgers, J.H. (Eds.), Aquatic Mesocosm Studies in Ecological Risk Assessment. CRC Press, Boca Raton, FL, USA, pp. 129–156.

US EPA, 1992. Framework for Ecological Risk Assessment. US Environmental Protection Agency, Washington, DC, USA. Risk Assessment Forum Report N. EPA/630/P-04/068B. http://www.epa.gov/raf/publications/pdfs/FRMWRK_ERA.PDF.

Yanong, R.P.E., 2003. Fish Health Management Considerations in Recirculating Aquaculture Systems – Part 1: Introduction and General Principles. Circular FA-120. UF IFAS Cooperative Extension Service. http://edis.ifas.ufl.edu/fa099.

CHAPTER

9

Ecological Risk and Weight of Evidence Assessments

P.M. Chapman

Chapema Environmental Strategies Ltd, North Vancouver, BC, Canada

9.1 INTRODUCTION

Previous chapters have dealt with various aspects of marine ecotoxicology, ranging from contaminants and experimental design to modeling and monitoring, and including specialized components and applications (biomarkers, water and sediment testing, micro- and mesocosms). This chapter provides the framework for informed decision-making based on applied marine ecotoxicology.

Informed decision-making does not comprise perfect knowledge and total certainty. Rather, it comprises best-available knowledge coupled with best professional judgment and a reasonable reduction in uncertainty. Uncertainty is a reality to be recognized and embraced, particularly given the reality of global climate change (Landis et al., 2012; cf. Chapter 10 of this book); it can be reduced and bounded but not eliminated, as discussed herein.

Two mutually complementary decision-making frameworks are introduced in this chapter: ecological risk assessment (ERA) and weight of evidence (WoE) assessments. The roles of marine ecotoxicology (both toxicity tests and bioaccumulation assessments) within both these frameworks are then explored followed by future possibilities and probabilities related to those roles.

9.2 ECOLOGICAL RISK ASSESSMENT

9.2.1 Introduction to Ecological Risk Assessment

ERA comprises a framework for gathering data and evaluating their sufficiency for informed decision-making. It is based on evaluating and assigning probabilities for adverse effects of different activities or stressors on the environment. A risk refers to a probability; a hazard refers to a possibility. In other words, if there is a risk that a certain amount of copper in the aquatic environment could adversely affect fish, there is a probability of this occurring. If it is possible that a certain amount of copper in the aquatic environment could adversely affect fish, this is a possibility; but whether it is probable remains to be determined. The ERA framework recognizes, considers, and reports uncertainties in estimating adverse effects of stressors.

Marine Ecotoxicology
http://dx.doi.org/10.1016/B978-0-12-803371-5.00009-6

ERA and the framework it comprises do not provide 100% certainty; there is no such thing. However, a properly conducted ERA will provide two products: risk-related information and associated uncertainty for informed decision-making and the key uncertainties that should be addressed/resolved to improve the certainty in such decision-making.

ERA focuses on stressors, including contaminants that are biologically available (ie, they can affect biological systems). Exposure to stressors can, at the extreme, lead to three major adverse outcomes affecting both the structure and function of ecosystems:

- population-level alteration (including demographic bottlenecks and stressor-induced selection);
- changes in genetic diversity; and
- changes in evolutionary trajectories.

The above changes can be negative, neutral, or positive, depending on one's point of view. For instance, rainbow trout is an invasive species which, before the mid-1850s, was only found east of the Rocky Mountains in North America. It is now, thanks to humans, found across all continents except Antarctica. It is prized and protected as a sport and commercial fishery. Yet its introduction clearly changed the aquatic ecosystems it now lives in. From a human needs viewpoint, the introduction of rainbow trout was a positive change; however, this introduction changed those aquatic ecosystems with likely both positive and negative effects depending on the resident species and their role in the food chain.

Globally, the following are the five major environmental stressors causing environmental changes, in order of importance:

1. Global Climate Change;
2. Habitat Change;
3. Exotic Species Introductions/Invasions;
4. Eutrophication including Harmful Algae Blooms; and
5. Chemical Contamination.

ERA is conducted for the following reasons: to predict the ecological consequences of different stressors and activities, to evaluate the potential impact(s) from action(s), and to determine their significance now and in the future:

- *Retroactively assess risk*, to assess the significance of what has already occurred. This is the most common use of ERA, but not necessarily the most useful. It is better to predict and prevent environmental damage before it occurs than to try to remediate it after it occurs, as remediation itself can have risks (Moriarty, 2015; cf. Section 9.2.3).
- *Proactively predict risk*, to assess the significance of what may occur in future. This is a less common but increasingly important use of ERA.
- *Compare the risk posed by different actions or developments*, to assess and prioritize risks from different activities, including management actions such as risk reduction (cf. Section 9.2.3). Everything human beings do has negative environmental implications; human beings by their very existence change ecosystems (Chapman, 2013a). For instance, treatment of wastewater will reduce the contaminants in the wastewater by concentrating them in a semi-liquid residue, a sludge, which must be disposed of; the treatment and sludge disposal processes have energetic costs, generate greenhouse gases, and remove habitat from other uses (Chapman, 2013b).
- *Rank the risk posed by different stressors or actions*, similar to the point above but ranking rather than comparing, and including in that ranking the five major stressors listed above.
- *Develop protective (and sometimes site-specific) benchmarks* as necessary and appropriate. Numerical benchmarks can be developed and used to differentiate areas requiring management actions (eg, remediation) from those that do not pose significant environmental risks. This allows for focused

management actions with consequent reduced environmental damage from those actions (Chapman and Smith, 2012). For example, national governments typically develop environmental quality benchmarks for generic, conservative protection (eg, of aquatic and terrestrial environments); these national benchmarks can be used or modified site-specifically for that purpose.

A risk does not exist unless two conditions are satisfied. First, the stressor (ie, the chemical, biological, or physical entity) has the inherent ability to cause one or more adverse effects. Second, the stressor co-occurs with or contacts a receptor (ie, an ecological functional resource (individual, population, community, habitat) potentially affected by a stressor) with sufficient duration and magnitude to elicit the identified adverse effect.

There are four scientific components to an ERA, all of which have an ecotoxicological component as discussed in Section 9.2.2 (Fig. 9.1):

1. *Problem Formulation*: Define the problem, the assessment and measurement endpoints (ie, respectively, the environmental values to be protected such as commercial fisheries and its measurable aspect that can be extrapolated to represent the assessment endpoint (eg, the survival, growth, and reproduction of key fish species)); summarize available information; develop a conceptual model (Section 9.2.2).
2. *Exposure Assessment*: Identify exposure concentrations or levels, bioavailability (for chemicals), sensitive species/populations.
3. *Effects Assessment*: Identify effects that could occur to individuals directly or indirectly (eg, via biomarkers linked to effects at higher levels of biological organization), populations, and communities.
4. *Risk Characterization*: Estimate risk based on exposure compared to effects.

The nonscientific components of an ERA (Fig. 9.1) are the initial and final discussions with and inputs received from risk managers, which include all stakeholders—to provide initial input to and from the scientists and to provide final input to and from decision-makers regarding risk management. Risk management is a nonscientific process in which scientific evidence is only one consideration and not always the most important consideration; there are three other considerations: economic, social, and political.

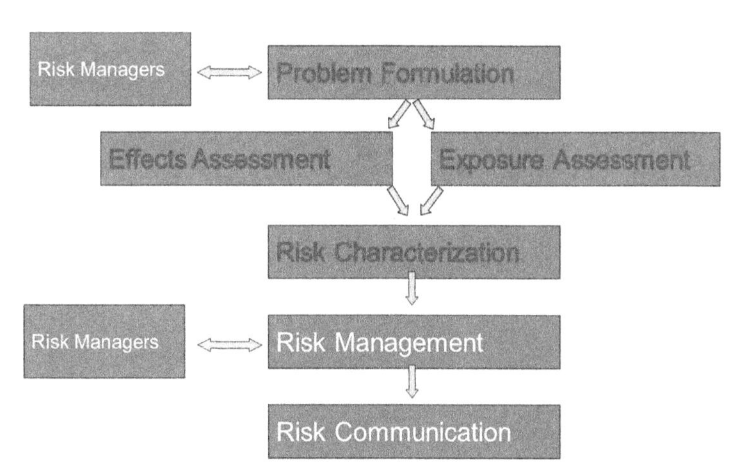

FIGURE 9.1 The scientific [*red* (gray in print versions)] and the nonscientific components (*white*) of ecological risk assessment. Additional detail is provided in USEPA (1992, 2007).

9.2.2 Role of Marine Ecotoxicology in Ecological Risk Assessment

Marine ecotoxicology has important roles in three of the four ERA components: problem formulation, effects assessment, and risk characterization. It does not have a role in exposure assessment, which is solely concerned with stressors, not with the potential effects of those stressors.

Problem formulation involves the following components where marine ecotoxicology has a role in terms of knowledge of potential effects, either generic or site-specific:

- identification and characterization of stressors (physical, chemical, biological);
- identification of receptors (biota potentially at risk);
- identification of potential ecological effects;
- determination of assessment and measurement endpoints;
- developing the conceptual model that illustrates schematically or in cartoon format the relationship between stressors and receptors (an example of a simplified conceptual model is shown in Fig. 9.2); and
- developing the risk hypotheses and the protection goals (human values, eg, ecosystem services) for the environment to determine what exactly will be assessed/tested in the next ERA components (exposure and effects assessments).

Effects Assessment describes the relationship between the stressor(s) and the receptor(s) and serves to determine any relationship between stressors and biological responses and/or ecological components. Toxicity tests and the calculation of a species sensitivity distribution (Fig. 9.3, see next paragraph for further explanation) are key contributors to the ERA component. Combined toxicity testing and bioaccumulation assessment are expected to be of

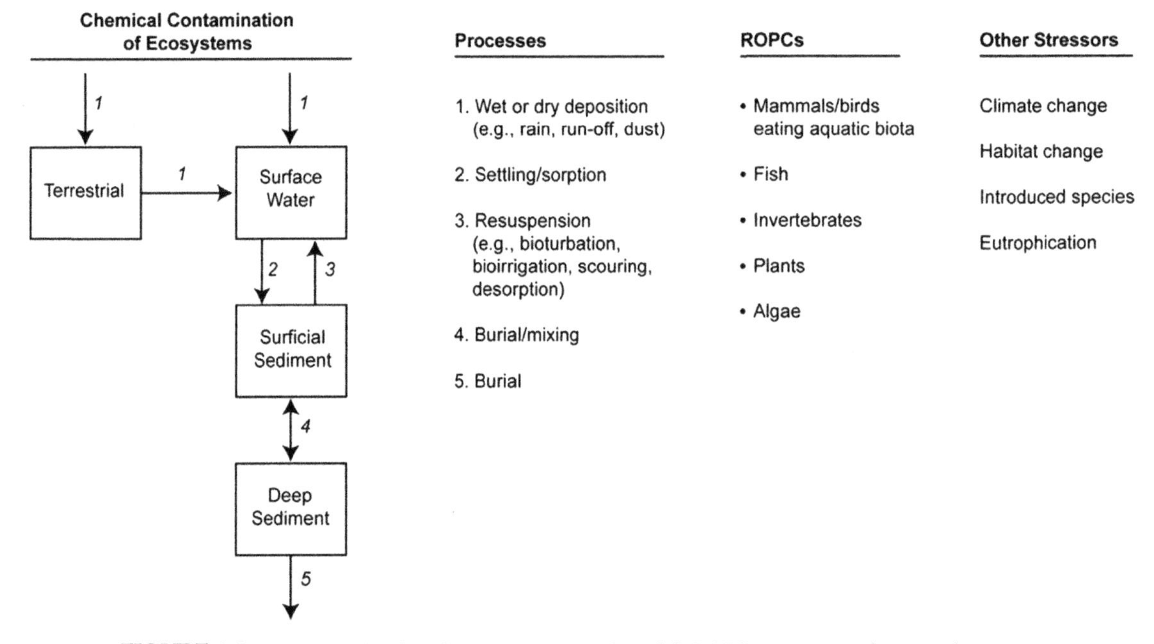

FIGURE 9.2 An example of a schematic conceptual model. *ROPCs*, receptors of potential concern.

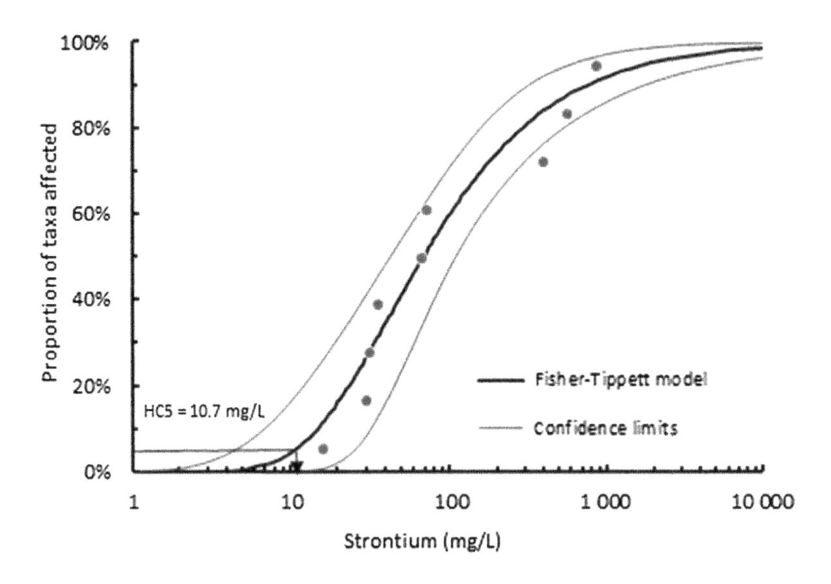

FIGURE 9.3 Example of a species sensitivity distribution. *From McPherson C., Lawrence G., Elphick J., Chapman P.M., 2014. Development of a strontium chronic effects benchmark for aquatic life in freshwater. Environ. Toxicol. Chem. 33, 2472–2478, with permission.*

increasing importance relative to determining benchmarks based on chemical contaminant concentrations in biota (Meador et al., 2014). Such benchmarks are needed not solely to understand the significance of contaminant doses in resident biota that are important for ecosystem function but also in toxicity test organisms16152, since it is the dose (ie, concentration and form) in the organism, not the concentration in the environment, which determines toxicity.

A species sensitivity distribution results when toxicity test results are plotted graphically against the percent of organisms tested. Specifically, concentrations of a stressor (in the example shown in Fig. 9.3, strontium, from McPherson et al., 2014) resulting in no or negligible effects (10–20% effect levels, eg, IC(inhibition concentration)10 to IC20) for each species tested are compared against the percent of organisms tested. This results in a curve from the most tolerant organisms tested to the least tolerant. A best-fit line is drawn and, where it intersects the 5% of species tested line, this is the point at which 95% of species will be unaffected. This point is termed the HC5 and is built upon Posthuma et al. (2002), who set the HC5 on the basis that (CCME, 2007, Part II, Section 3.1–5) "enough redundancy exists within aquatic communities to allow some loss." As CCME (2007, Part II, Section 3.1–5) notes regarding allowing for the possible impairment or loss of up to 5% of possible species, this issue "is less important when plotting low- or no-effect level data," hence the focus on no or negligible effects data (eg, 10–20% effect levels).

Risk Characterization is the final tier of an ERA; it integrates the other components of the ERA. Conclusions from individual lines of evidence (LoE) are considered relative to each other similar to a WoE determination (Section 9.3) in which concordance is sought among multiple LoE and different points of view help to determine possible mechanisms. The Risk Characterization estimates the magnitude and the probability of effects, either qualitatively or quantitatively.

9.2.3 Role of Marine Ecotoxicology in Assessing Risk Management

Marine ecotoxicology provides, as discussed in Section 9.2.2, information on potential adverse effects of stressors. This information provides for knowledge-based decision-making by nonscientists (Fig. 9.1) relative to risk management decisions.

Once risk management decisions have been made, their effectiveness needs to be assessed. Specifically, a determination is required as to whether the right decisions have been made to reduce or minimize potential adverse effects of stressors. In addition, because every action by human beings has potential negative effects, the risk management itself needs to be assessed to determine whether unexpected adverse effects are occurring which are of more concern than those being addressed, ie, whether the risk management is causing more harm than good (Moriarty, 2015).

Marine ecotoxicology thus has a continuing role following initial assessment. The exact role will be site- and situation-dependent, but will likely require some of the same tests and assessments conducted initially to determine the effectiveness of risk management. The possibility of unexpected adverse effects will require different tests and assessments, determined based on both new information and best professional judgment.

9.3 WEIGHT OF EVIDENCE

9.3.1 Introduction to Weight of Evidence

WoE refers to integration of data generated from multidisciplinary environmental studies involving multiple, independent LoE, which typically comprise both chemical and biological measurements. It is a determination related to possible ecological impacts based on multiple LoE.

WoE assessments provide three types of information: (1) relative certainty of adverse environmental effects due to stressors; (2) possible causation; and (3) key uncertainties that, if resolved, will improve management decision-making. Environment Canada (2012) defines WoE as "any process used to aggregate information from different lines of scientific evidence to render a conclusion regarding the probability and magnitude of harm." ECHA (2014, p. 59) states "The most defensible approach to measure and predict effects in risk assessments includes a combination of carefully designed observational studies and experiments in a WoE approach that are targeted to demonstrate key mechanisms and linkages."

WoE assessments provide an established and accepted method for integrating environmental assessment data (eg, Chapman and Anderson, 2005; McDonald et al., 2007; Environment Canada and Ontario Ministry of the Environment, 2008; Cormier et al., 2010; Suter et al., 2010; ECHA, 2010; Suter and Cormier, 2011; Chapman and Smith, 2012; Cormier and Suter, 2013; ECHA, 2014; Hope and Clarkson, 2014) as well as other data (eg, Borgert et al., 2011). Linkov et al. (2015) note significant efforts to formalize WoE methodologies particularly in quantitative, transparent, and objective terms; they state (p. 4) "integration of individual lines of evidence is an essential component of environmental assessments that should be standardized to establish consistency and comparability across similar efforts."

The basis for decision-making within a WoE analysis is a combination of one or more of statistical analyses, scoring systems, and qualitative judgments incorporated into a logic system. Qualitative WoE approaches typically involve classification of results (eg, low, medium, high), followed by integration of these based on best professional judgment to obtain an overall conclusion (eg, the Sediment Quality Triad: Chapman, 1990; Chapman et al., 2002; Chapman and McDonald, 2005). Semi-quantitative approaches begin in the same way, but then convert the classifications to numerical scores, to which numerical weighting factors are applied, to obtain an overall WoE "score" (eg, frameworks proposed by Menzie et al., 1996; Hope and Clarkson, 2014). Hybrid approaches have also been applied whereby numerical ratings and weightings are systematically applied, but then the overall WoE score is interpreted qualitatively rather than absolutely (eg, McDonald et al., 2007). The USEPA Causal Analysis/Diagnosis Decision Information System uses WoE together with environmental epidemiology principles to assess causation (CADDIS; http://www.epa.gov/caddis/).

Most WoE assessments share three common considerations: (1) the relative weighting of endpoints; (2) rating (qualitatively or quantitatively) the magnitude of response in individual endpoints; and (3) the concurrence responses of related endpoints (Menzie et al., 1996). A key requirement for technically defensible WoE assessments is transparency such that findings are reproducible (Linkov et al., 2009). To achieve this transparency and reproducibility, the quality of evidence provided by LoE is typically weighted a priori (ie, before results are known) to prevent potential bias associated with knowing responses before weightings are applied, with a posteriori (ie, after the fact) adjustments applied in some situations to account for concurrence (or lack thereof) in responses among related LoE. As exampled by Haake et al. (2010) and Wiseman et al. (2010) in case studies, structured, transparent WoE assessments can, even in the face of uncertain causation and substantial uncertainty, provide enough certainty to inform decision-making.

Weighting of different LoE has been suggested for WoE assessments (Menzie et al., 1996; McDonald et al., 2007; Suter and Cormier, 2011; Hope and Clarkson, 2014). Weighting considerations to address specific environmental issues typically fall into four categories: (1) strength of association between LoE and ecological values being protected; (2) representativeness (spatial, temporal, site-specificity); (3) study design and execution (for site-specific investigations); and (4) data quality. In the interest of maintaining clarity and transparency, the preferred approach is to standardize weightings according to these considerations a priori, such that they can be applied to LoE in a systematic and nonbiased manner. In practice, the nature of environmental assessment data is complex, and response combinations not foreseen a priori may occur; however, it is possible to allow for a posteriori judgments (ie, considering the nature and direction of actual LoE responses) to fine-tune WoE outcomes where warranted (Menzie et al., 1996; McDonald et al., 2007; SABCS, 2010). Weighting should include a priori considerations, consideration of the direction of change or response, and a posteriori considerations based on the nature, complexity, and certainty of the findings. Suter and Cormier (2011) and Hope and Clarkson (2014) caution that the weighting factors and scores in quantitative WoE are initiated using best professional judgment and therefore should be interpreted qualitatively.

9.3.2 Role of Marine Ecotoxicology in Weight of Evidence

WoE assessments comprise LoE for both exposure and effects. Marine ecotoxicology provides information on effects to supplement and assist in understanding information on the status of resident exposed communities. Although ecotoxicological studies are by definition not wholly realistic (levels of realism increase in the order: laboratory toxicity studies to microcosms to mesocosms to in situ exposures), they provide necessary information as to possible responses to stressors. Resident community assessments, although realistic, can be difficult to interpret due to stochasticity (natural variability) unless stressor effects are extreme.

The use of marine ecotoxicology in WoE is best developed based on a conceptual site-specific model developed to illustrate interactions of stressors of potential concern (SOPCs), exposure pathways, and receptors of potential concern (ROPCs) relative to effects (ie, change, which can be positive or negative). An example is provided in Fig. 9.4 for a mine discharge to a marine embayment. SOPCs in this case are potential toxicants and nutrients. ROPCs are epilithic algae (previously known as periphyton), phytoplankton, and zooplankton in the water column, benthic invertebrates in the sediment, and demersal and pelagic fish. Toxicity testing of the receiving environment would involve both water and sediment toxicity tests. Bioaccumulation assessments would more likely involve ROPCs that are large enough for relatively easy collection of sufficient biomass for chemical analyses, but depending on the substance(s) of potential concern and their pathway through the food web could involve smaller ROPCs such as benthic invertebrates and plankton.

Table 9.1 provides an example of how toxicity testing could be used, together with chemical analyses and resident community assessment, in a WoE assessment as to whether there are low, moderate, or high levels of potential concern that chemicals are adversely affecting resident biota. This example is clearly simplistic since it only involves three LoE; other LoE are possible as suggested by Chapman and Hollert (2006). For example, those few organic chemicals that biomagnify (eg, methyl mercury; 2,3,7,8-TCDD; DDT, PCBs) would be considered as a separate LoE. Critical body residues (ie, concentrations that cause adverse effects to the affected organisms) derived from bioaccumulation assessments can provide LoE in the form of

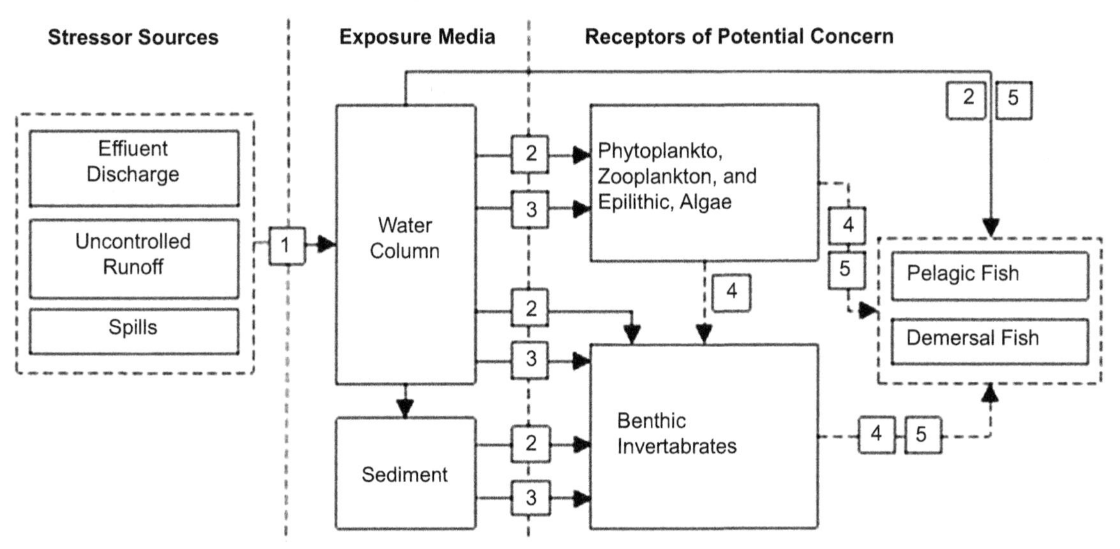

FIGURE 9.4 An example of a site-specific conceptual model for a mine discharge to a marine embayment. *ROPCs*, receptors of potential concern.

TABLE 9.1 Example of WoE Categorization for Chemical Stressors, Toxicity, and Resident Communities in Sediments.

	Categorization relative to level of concern		
Lines of evidence	**High**	**Moderate**	**Low**
Bulk chemistry (compared to SQG)	**Adverse effects likely:** One or more exceedances of SQG-high	**Adverse effects may or may not occur:** One or more exceedances of SQG-low	**Adverse effects unlikely:** All contaminant concentrations below SQG-low
Toxicity endpoints (relative to reference)	**Major:** Statistically significant reduction of more than 50% in one or more toxicological endpoints	**Minor:** Statistically significant reduction of more than 20% in one or more toxicological endpoints	**Negligible:** Reduction of 20% or less in all toxicological endpoints
Overall toxicity (based on more than one test)	**Significant**: Multiple tests/ endpoints exhibit major toxicological effects	**Potential**: Multiple tests/ endpoints exhibit minor toxicological effects and/or one test/endpoint exhibits major effect	**Negligible**: Minor toxicological effects observed in no more than one endpoint
Resident communities alteration (univariate and multivariate assessment)	"Different" or "very different" from reference stations	"Possibly different" from reference stations	"Equivalent" to reference stations
Overall WoE assessment	**Significant adverse effects**:	**Potential adverse effects**:	**No significant adverse effects:**
	Elevated chemistry	Elevated chemistry	Minor reduction in no more than one toxicological endpoint
	Greater than a 50% reduction in one or more toxicological endpoints	Greater than a 20% reduction in two or more toxicological endpoints	Resident community structure and function not different from reference
	Resident community structure and function different from reference	Resident community structure and function possibly different from reference	

SQG, sediment quality guideline; *WoE*, weight of evidence; SQG-low defines no effects and possible effects; SQG-high defines possible effects and probable effects. Note that the overall definition of "no significant adverse effects" is independent of elevated chemical stressors (ie, the findings of the biological LoE—toxicity and resident communities—overrule chemical LoE). Note further that not all possible LoE are shown (ie, biomagnification is not considered—see Environment Canada and Ontario Ministry of the Environment, 2008; Chapman and Hollert, 2006).

benchmarks below which effects are unlikely and above which effects are likely and can also be used to determine causation related to chemical stressors. Biomarkers (specific biological responses measured within individual organisms that range from molecular through cellular, metabolic, and physiological responses, to behavioral changes) of effects, not simply of exposure, can provide an additional LoE. Stressors other than chemical contamination, for example habitat changes, can also provide an additional LoE.

The key determinant of appropriate LoE depends on the SOPCs, ROPCs, and the questions to be answered (ie, the hypotheses to be tested). There are no cookbook formulae for such determinations which, again, will be site- and situation-specific. However, it is likely that in the case of chemically contaminated sediments the LoE of chemistry, toxicity, and

TABLE 9.2 Decision Matrix for Weight of Evidence Categorization for Chemical Stressors, Toxicity, and Resident Communities.

Scenario	Chemistry	Toxicity	Resident communities alteration	Assessment
1	Low	Low	Low	No further actions required
2	Moderate-high	Low	Low	No further actions required; chemical contaminants not bioavailable
3	Low	Low	Moderate-high	Determine reason(s) for resident communities alteration[a]
4	Low	Moderate-high	Low	Toxicity not related to elevated chemistry or resident communities alteration—may be laboratory artifact or unmeasured stressors; investigate
5	Moderate-high	Moderate-high	Low	Toxicity may be related to chemistry but no evidence of resident communities alteration—may be laboratory artifact or unmeasured stressors; investigate
6	Moderate-high	Low	Moderate-high	Determine reason(s) for resident communities alteration—lack of toxicity suggests no linkage between chemistry and alteration but investigate[a]
7	Low	Moderate-high	Moderate-high	Determine reason(s) for toxicity *and* resident communities alteration[a]
8	Moderate-high	Moderate-high	Moderate-high	Management actions required[b]

Separate endpoints can be included within each LoE (eg, metals, PAHs, PCBs for chemistry; survival, growth, reproduction for toxicity; abundance, diversity, dominance for resident communities).

[a] *Alteration of resident communities may be due to factors other than chemical stressors, either natural (eg, competition/predation, habitat differences), human-related (eg, habitat differences, introduced species), or both (eg, harmful algal blooms, invasive species).*

[b] *Definitive determination possible. Ideally elevated chemistry should be shown to in fact be linked to observed biological effects (ie, causal), to ensure management actions address the problem(s). Ensuring causality may require additional investigations such as toxicity identification evaluation and/or contaminant body residue analyses.*

resident communities that comprise the Sediment Quality Triad (Chapman, 1990; Chapman et al., 2002; Chapman and McDonald, 2005) will be appropriate and necessary. Table 9.2 provides a decision matrix for the three Sediment Quality Triad LoE.

9.4 MARINE ECOTOXICOLOGY WITHIN ECOLOGICAL RISK ASSESSMENT AND WEIGHT OF EVIDENCE

There are two primary roles for marine ecotoxicology within ERA and WoE: toxicity and bioaccumulation testing (collectively termed bioassay testing). Toxicity testing needs to be focused on determining no or negligible effect levels for both acute (mortality) and chronic responses (eg, growth, fecundity, behavior). While no effect (ie, 10% effect) levels are preferred, 20% effect levels are accepted by CCME (2007) as a "threshold level of negative effects." In other words, negative effects may occur above a 20% effect level. Neither CCME (2003) nor CCME (2007) state that reliance can only be placed on 10% effect levels. In fact, Canada's national water quality guideline for cobalt is based on a 25% effect level. USEPA (2002) whole effluent toxicity testing guidelines rely on the IC25 (a 25% effect level) to assess effluent toxicity. Bruce and Versteeg (1992) estimated a 20% effect concentration as an environmentally relevant concentration that would minimize adverse effects on a population as compared to natural variability. Barnthouse et al. (1987) and Norberg-King (1993) set this minimal adverse effect threshold at 25%. Multiple authors consider 10% and 20% effect thresholds as surrogates for no effects or minimal effects concentrations (eg, Dyer et al., 1997; Grist et al., 2003; Versteeg and Rawlings, 2003; Barnthouse et al., 2007; Oris et al., 2012). USEPA also accepts 20% effect levels for development of national water quality criteria (eg, USEPA Water Quality Criteria for ammonia (http://water.epa.gov/scitech/swguidance/standards/criteria/aqlife/ammonia/upload/AQUATIC-LIFE-AMBIENT-WATER-QUALITY-CRITERIA-FOR-AMMONIA-FRESHWATER-2013.pdf)).

With the exception of threatened and endangered species (eg, salmon along the west coast of North America), a 10–25% chronic effect to a population should not adversely affect that population. However, it is also important to understand the population dynamics and life history of a species and not solely rely on toxicity tests to predict effects (Stark et al., 2004; Stark, 2005; Hanson and Stark, 2011, 2012).

For example, in the case of the threatened North American west coast salmon, low levels of toxicity affecting growth of the young can reduce population viability. The rate of growth in rivers before the fish go to sea has a large impact on survivability and return; larger fish have greater survivability through to their return to spawn in their natal streams (Spromberg and Meador, 2005). Intermittent effects to one salmon run competing for habitat with other salmon runs can reduce overall population size (Spromberg and Scholz, 2011), essentially "action at a distance" (Spromberg et al., 1998). Thus, as in ERA and WoE, marine ecotoxicology is only one line of evidence for determining and avoiding adverse environmental effects and impacts.

9.5 FUTURE POSSIBILITIES AND PROBABILITIES

9.5.1 The Five Major Environmental Stressors

As previously noted (Section 9.2.1), there are five major environmental stressors causing environmental changes, of which chemical contamination is only one and, in relative terms, arguably the least important. Marine ecotoxicology within ERA and WoE will need to adapt and develop to provide useful component information related to all five stressors, not just individually, but in combination. Commonalities between species or higher levels of biological organization may be found (Sulmon et al., 2015).

With regard to climate change (discussed in more detail in Chapter 10), the direct effects of temperature and ocean acidification on marine fauna will need to be assessed in terms of both toxicity and bioaccumulation. Temperature and ocean acidification are also modifying factors for chemical toxicity (Chapter 10), thus resulting in indirect effects.

Habitat changes will alter spatiotemporal patterns of both water flow and sediment movement/loadings (Chapter 10), which will affect marine fauna. Again, both the direct and indirect effects of habitat changes will need to be assessed.

Similarly, direct and indirect effects of eutrophication, including harmful algal blooms, will need to be assessed. Species introductions and invasions will require ecotoxicology within ERA and WoE both to determine whether and how such species may impact resident ecosystems and to develop appropriate toxicity tests and bioaccumulation assessments with appropriate introduced/invasive species (ie, species that have the potential to enhance ecosystem services).

It is expected that all of these probable future roles for marine ecotoxicology will require consideration of acquired tolerance and the effects of the associated energetic requirements. Control for interpopulation tolerance variation will be required (Sun et al., 2015).

9.5.2 Good Ecosystem Status

The European Union, under its Marine Strategy Framework Directive (MSFD, 2008), integrates the concepts of environmental protection and sustainable use and requires that Member States attain Good Environmental Status (GES). GES is defined in Annex I of the MSFD (2008) by 11 qualitative descriptors, of which seven require ERA and/or WoE assessments that include marine ecotoxicology: (1) maintaining biological diversity; (2) invasive/introduced species do not adversely alter ecosystems; (3) commercially exploited fish and shellfish stocks remain healthy; (4) marine food webs are maintained; (5) sea-floor integrity does not adversely affect benthic or other ecosystems; (6) contaminants are not pollutants; and (7) fish and other seafood are safe to eat. The other four are (1) human-induced eutrophication is minimized including harmful algal blooms; (2) permanent alteration of hydrology does not adversely affect marine ecosystems; (3) marine litter is not harming the coastal and marine environments; and (4) energy including underwater noise does not adversely affect the marine environment.

The above qualitative descriptors do not explicitly state that ecosystems cannot change, for instance that species cannot be lost or replaced, only that ecosystems remain diverse, healthy, and sustainable, and that human access to edible seafood is maintained. The focus is on ecosystem function and ecosystem services and not on ecosystem structure. As such, truly determining whether GES has been attained/can be maintained requires changes in how we monitor and assess marine (and other) ecosystems.

Robust tools are required for determining GES; Lyons et al. (2010) and Robinson et al. (2012) emphasize the importance of combined monitoring of chemical contaminant concentrations and biological effect (eg, marine ecotoxicology) measurements. Combining chemistry and biology is particularly important as chemical contaminants, which traditionally included inorganic substances such as metals and organic chemicals, now include an increasing variety of emerging chemicals (eg, microplastics, pharmaceuticals and personal care products, fire retardants) whose mode of action and potential biological effects remain to be fully determined. Further, the combination of chemical contaminants and other stressors (eg, climate change and ocean acidification, habitat changes, invasive/introduced species, harmful algal blooms, eutrophication) increases the complexity of determining what effects and impacts could occur, and what they may mean to ecosystem function and, ultimately, to ecosystem services.

Assessment of GES should be based not on measuring the presence or absence of individual species (ie, not structure) but rather on what is essential to maintain ecosystem function. This can involve invasive/introduced species so long as they "are at levels that do not adversely alter the ecosystem" (MSFD, 2008).

Early warning is required of both unacceptable changes to ecosystem function and of contaminant or other stressor effects that could change that function. Early warning will require both biomarkers and bioindicators (eg, whole organism toxicity tests) but not necessarily as they are presently developed and used.

Biomarkers for key ecosystem functions are needed, which will provide early warning of chemical and/or other stressors that could eventually overwhelm specific ecosystem functions and thus adversely affect ecosystem services. It is uncertain what presently available biomarkers, primarily focused on exposure rather than effects, are truly suitable for this purpose. Omics techniques may help in developing necessary new effects-based biomarkers.

Presently, marine toxicity tests are generally conducted in the laboratory with naïve organisms and typically measuring the endpoints of survival, growth, and reproduction. Again the focus needs to change from protecting structure to protecting function. This may mean that toxicity testing is focused on specific functional groups and perhaps is conducted on organisms that are less sensitive to stressor effects than has presently been the case. For example, there may be no point in protecting a component of the food chain that, although sensitive, is not essential to ecosystem function. Some well-established toxicity tests may no longer be appropriate.

Benchmarks based on contaminant concentrations in biota (ie, critical body residues) need to be developed for inorganic, organic, and emerging contaminants (Meador et al., 2014). Such benchmarks are needed not just to understand the significance of contaminant doses in resident biota that are important for ecosystem function but also in toxicity test organisms.

All of this information needs to be integrated into multistressor, ecosystem-specific ERA/WoE assessments that do not rely on simplified indices (Green and Chapman, 2011). The end goal must be to determine and predict thresholds/tipping points for the function of specific ecosystems to contaminants and/or other stressors, and thus provide reliable information for decision-making relative to achieving and maintaining GES.

References

Barnthouse, L.W., Suter, G.W., Rosen, A.E., Beauchamp, J.J., 1987. Estimating responses of fish populations to toxic contaminants. Environ. Toxicol. Chem. 6, 811–824.

Barnthouse, L.W., Munns Jr., W.R., Sorensen, M.T., 2007. Population-level Ecological Risk Assessment. CRC Press, Boca Raton, FL, USA.

Borgert, C.J., Mihaich, E.M., Ortego, L.S., Bentley, K.S., Holmes, C.M., Levine, S.L., Becker, R.A., 2011. Hypothesis-driven weight of evidence framework for evaluating data within the US EPA's Endocrine Disruptor Screening Program. Regul. Toxicol. Pharmacol. 61, 185–191.

Bruce, R.D., Versteeg, D.J., 1992. A statistical procedure for modeling continuous toxicity data. Environ. Toxicol. Chem. 11, 1485–1494.

CCME (Canadian Council of Ministers of the Environment), 2003. Canadian Water Quality Guidelines for Protection of Aquatic Life – Guidance for Site-Specific Application of Water Quality Guidelines in Canada and Procedures for Deriving Numerical Water Quality Objectives. Winnipeg, MB, Canada.

CCME, 2007. A protocol for the derivation of water quality guidelines for the protection of aquatic life 2007. In: Canadian Environmental Quality Guidelines. Winnipeg, MB, Canada. Available at: http://ceqg-rcqe.ccme.ca/.

Chapman, P.M., Anderson, J., 2005. A decision-making framework for sediment contamination. Integr. Environ. Assess. Manag. 1, 163–173.

Chapman, P.M., Hollert, H., 2006. Should the sediment quality triad become a tetrad, a pentad, or possibly even a hexad? J. Soils Sed. 6, 4–8.

Chapman, P.M., McDonald, B.G., 2005. Using the sediment quality triad in ecological risk assessment. In: Blaise, C., Férard, J.-F. (Eds.), Small-Scale Freshwater Toxicity Investigations, Hazard Assessment Schemes, vol. 2. Kluwer Academic Press, Netherlands, pp. 305–330.

Chapman, P.M., Smith, M., 2012. Assessing, managing and monitoring contaminated aquatic sediments. Mar. Pollut. Bull. 64, 2000–2004.

Chapman, P.M., McDonald, B.G., Lawrence, G.S., 2002. Weight of evidence frameworks for sediment quality and other assessments. Hum. Ecol. Risk Assess. 8, 1489–1515.

Chapman, P.M., 1990. The sediment quality triad approach to determining pollution-induced degradation. Sci. Total Environ. 97-98, 815–825.

Chapman, P.M., 2013a. Polluting to pollute or "polluting" to protect? Mar. Pollut. Bull. 70, 1–2.

Chapman, P.M., 2013b. Treatment by any other name would be more environmentally friendly. Mar. Pollut. Bull. 77, 1–2.

Cormier, S.M., Suter II, G.W., 2013. A method for assessing causation of field exposure-response relationships. Environ. Toxicol. Chem. 32, 272–276.

Cormier, S.M., Suter II, G.W., Norton, S.B., 2010. Causal characteristics for ecoepidemiology. Hum. Ecol. Risk Assess. 16, 53–73.

Dyer, S., Lauth, J., Morrall, S., Herzog, R., Cherry, D., 1997. Development of a chronic toxicity structure-activity relationship for alkyl sulfates. Environ. Toxicol. Water Qual. 12, 295–303.

ECHA (European Chemicals Agency), 2010. Practical Guide 2: How to Report Weight of Evidence. Helsinki, Finland. http://echa.europa.eu/documents/10162/13655/pg_report_weight_of_evidence_en.pdf.

ECHA, 2014. Principles for Environmental Risk Assessment of the Sediment Compartment. In: Proceedings of the Topical Scientific Workshop, Helsinki, Finland, 1–8 March, 2013. http://echa.europa.eu/documents/10162/13639/environmental_risk_assessment_final_en.pdf.

Environment Canada and Ontario Ministry of the Environment, 2008. Canada-Ontario Decision-making Framework for Assessment of Great Lakes Contaminated Sediment. Ottawa, ON, Canada. Available at: http://publications.gc.ca/site/archivee-archived.html?url=http://publications.gc.ca/collections/collection_2010/ec/En164-14-2007-eng.pdf.

Environment Canada, 2012. Federal Contaminated Sites Action Plan (FCSAP) Ecological Risk Assessment Guidance. Prepared by Azimuth Consulting Group, Vancouver, BC, Canada. Available at: http://www.ec.gc.ca/Publications/default.asp?lang=En&xml=D86920CE-2DFE-40CB-985F-60A3EA6069A9.

Green, R., Chapman, P.M., 2011. The problem with indices. Mar. Pollut. Bull. 62, 1377–1380.

Grist, E.P., Wells, N.C., Whitehouse, P., Brighty, G., Crane, M., 2003. Estimating the effects of 17α-ethinylestardiol on populations of the fathead minnow *Pimephales promelas*: are conventional endpoints adequate? Environ. Sci. Technol. 37, 1609–1616.

Haake, D.M., Wilton, T., Krier, K., Stewart, A.J., Cormier, S.M., 2010. Causal assessment of biological impairment in the Little Floyd River, Iowa, USA. Hum. Ecol. Risk Assess. 16, 116–148.

Hanson, N., Stark, J.D., 2011. Utility of population models to reduce uncertainty and increase values relevance in ecological risk assessments of pesticides: an example based on acute mortality data for daphnids. Integr. Environ. Assess. Manag. 8, 262–270.

Hanson, N., Stark, J.D., 2012. Comparison of population level and individual level endpoints to evaluate ecological risk of chemicals. Environ. Sci. Technol. 46, 5590–5598.

Hope, B.K., Clarkson, J.R., 2014. A strategy for using weight-of-evidence methods in ecological risk assessments. Hum. Ecol. Risk Assess. 20, 290–315.

Landis, W.G., Durda, J.L., Brooks, M.L., Chapman, P.M., Menzie, C., Stahl Jr., R.G., Stauber, J.L., 2012. Ecological risk assessment in the context of global climate change. Environ. Toxicol. Chem. 32, 1–14.

Linkov, I., Loney, D., Cormier, S., Satterstrom, F.K., Bridges, T., 2009. Weight-of-evidence evaluation in environmental assessment: review of qualitative and quantitative approaches. Sci. Total Environ. 407, 5199–5205.

Linkov, I., Massey, O., Keisler, J., Rusyn, I., Hartung, T., 2015. From "weight of evidence" to quantitative data integration using multicriteria decision analysis and Bayesian methods. Altex 32, 3–8.

Lyons, B.P., Thain, J.E., Stentiford, G.D., Hylland, K., Davies, I.M., Vethaake, A.D., 2010. Using biological effects tools to define good environmental status under the European Union marine strategy framework directive. Mar. Pollut. Bull. 60, 1647–1651.

McDonald, B.G., de Bruyn, A.M.H., Wernick, B.G., Patterson, L., Pellerin, N., Chapman, P.M., 2007. Design and application of a transparent and scalable weight-of-evidence framework: an example from Wabamun Lake, Alberta, Canada. Integr. Environ. Assess. Manag. 3, 476–483.

McPherson, C., Lawrence, G., Elphick, J., Chapman, P.M., 2014. Development of a strontium chronic effects benchmark for aquatic life in freshwater. Environ. Toxicol. Chem. 33, 2472–2478.

Meador, J.P., Warne St, J.M., Chapman, P.M., Chan, K.M., Yu, S., Leung, K.M.Y., 2014. Tissue-based environmental quality standards. Environ. Sci. Pollut. Res. 21, 28–32.

Menzie, C., Henning, M.H., Cura, J., Finkelstein, K., Gentile, J., Maughan, J., Mitchell, D., Petron, S., Potocki, B., Svirski, S., Tyler, P., 1996. Special report of the Massachusetts weight-of-evidence work group: a weight-of evidence approach for evaluating ecological risks. Hum. Ecol. Risk Assess. 2, 277–304.

Moriarty, P., 2015. Reliance on technical solutions to environmental problems: caution is needed. Environ. Sci. Technol. 49, 5255–5256.

MSFD (Marine Strategy Framework Directive), 2008. Directive 2008/56/EC of the European Parliament and the Council of 17 June, 2008 Establishing a Framework for Community Action in the Field of Marine Environmental Policy. http://ec.europa.eu/environment/marine/eu-coast-and-marine-policy/marine-strategy-framework-directive/index_en.htm.

Norberg-King, T.J., 1993. A Linear Interpolation Method for Sublethal Toxicity: The Inhibition Concentration (ICp) Approach. National Effluent Toxicity Assessment Center Technical Report 39, pp. 3–93 (Duluth, MN, USA).

Oris, J.T., Belanger, S.E., Bailer, A.J., 2012. Baseline characteristics and statistical implications for the OECD 210 fish early life stage toxicity test. Environ. Toxicol. Chem. 31, 370–376.

Posthuma, L., Suter II, G.W., Traas, T. (Eds.), 2002. Species Sensitivity Distributions in Ecotoxicology. CRC Press, Boca Raton, FL, USA.

Robinson, C.D., Gubbins, M.J., Lyons, B.P., Bignell, J., Bean, T., MacNeish, K., Dymond, P., Dobson, J., Webster, L., Thain, J., 2012. Assessing Good Environmental Status for Descriptor 8—An Integrated

Assessment of Contaminants and Their Biological Effects across Multiple Matrices in the Firth of Forth, Scotland. ICES CM 2012/G,16. http://www.ices.dk/sites/pub/CMDoccuments/CM-2012/G/G1612.pdf.

SABCS (Science Advisory Board for Contaminated Sites in British Columbia), 2010. Guidance for a Weight of Evidence Approach in Conducted Detailed Ecological Risk Assessments (DERA) in British Columbia. Ministry of Environment, Victoria, BC, Canada.

Spromberg, J.A., Meador, J.P., 2005. Relating results of chronic toxicity responses to population-level effects: modeling effects on wild Chinook salmon populations. Integr. Environ. Assess. Manag. 1, 9–21.

Spromberg, J.A., Scholz, N.L., 2011. Estimating the future decline of wild Coho salmon populations resulting from early spawner die-offs in urbanizing watersheds of the Pacific Northwest, USA. Integr. Environ. Assess. Manag. 7, 648–656.

Spromberg, J.A., Johns, B.M., Landis, W.G., 1998. Metapopulation dynamics: indirect effects and multiple discrete outcomes in ecological risk assessment. Environ. Toxicol. Chem. 17, 1640–1649.

Stark, J.D., Banks, J.E., Vargas, R., 2004. How risky is risk assessment: the role that life history strategies play in susceptibility of species to stress. PNAS 101, 732–736.

Stark, J.D., 2005. How closely do acute lethal concentration estimates predict effects of toxicants on populations? Integr. Environ. Assess. Manag. 1, 109–113.

Sulmon, C., van Baaren, J., Cabello-Hurtado, F., Gouesbet, G., Hennion, F., Mony, C., Renault, D., Bormans, M., El Amrani, A., Wiegand, C., Gérard, C., 2015. Abiotic stressors and stress responses: what commonalities appear between species across biological organization levels? Environ. Pollut. 202, 66–77.

Sun, P.Y., Foley, H.B., Bao, V.W.W., Leung, K.M.Y., Edmands, S., 2015. Variation in tolerance to common marine pollutants among different populations in two species of the marine copepod *Tigriopus*. Environ. Sci. Pollut. Res. 22, 16143–16152.

Suter II, G.W., Cormier, S.M., 2011. Why and how to combine evidence in environmental assessments: weighing evidence and building cases. Sci. Total Environ. 409, 1406–1417.

Suter II, G.W., Norton, S.B., Cormier, S.M., 2010. The science and philosophy of a method for assessing environmental causes. Hum. Ecol. Risk Assess. 16, 19–34.

USEPA (US Environmental Protection Agency), 1992. Framework for Ecological Risk Assessment. EPA/630/R-92/001. Washington, DC, USA.

USEPA, 2002. Short-Term Methods for Estimating the Chronic Toxicity of Effluents and Receiving Water to Freshwater Organisms. EPA-821-r-02-13. Office of Water, Washington, DC, USA.

USEPA, 2007. Framework for Metals Risk Assessment. EPA/20/R-07/001. Washington, DC, USA.

Versteeg, D., Rawlings, J., 2003. Bioconcentration and toxicity of docecylbenzene sulfonate (C12LAS) to aquatic organisms exposed in experimental streams. Arch. Environ. Contam. Toxicol. 44, 237–246.

Wiseman, C.D., LeMoine, M., Cormier, S., 2010. Assessment of probable causes of reduced aquatic life in the Touchet River, Washington, USA. Hum. Ecol. Risk Assess. 16, 87–115.

CHAPTER

10

Global Change

J.L. Stauber[1], A. Chariton[2], S. Apte[1]

[1]CSIRO Land and Water, Kirrawee, NSW, Australia; [2]CSIRO Oceans and Atmosphere, Kirrawee, NSW, Australia

10.1 INTRODUCTION

Human activities are increasingly altering the composition and integrity of our coastal and marine ecosystems. Land use changes, increasing coastal urbanization and industrialization, population growth, altered water availability and quality, and climate change are already having a major impact on marine habitats, ecological processes and communities, and the livability of our coastal cities.

Global change may arise from anthropogenic pressures or from natural disruptive events. While natural events are unpredictable and cannot be managed, anthropogenic global change can be managed and predicted, albeit with great uncertainties (Duarte, 2015). This chapter focuses on anthropogenic global change, which is defined by Duarte (2015) as:

> The global-scale changes resulting from the impact of human activity on the major processes that regulate the functioning of the biosphere.

Only changes at the global scale are considered here, and these changes may be negative or positive.

The ocean is the largest biome on Earth and a key receptor of human pressures (Duarte, 2015). Halpern et al. (2008) developed a global map of cumulative impact for 17 anthropogenic drivers of ecosystem change in the world's oceans. They found that the marine ecosystems that had the highest predicted cumulative impact were hard and soft continental shelves, rocky reefs, and coral reefs. Global climate change was an important driver for high impacts, particularly for offshore systems.

Understanding and predicting the effects of chemicals on marine and estuarine biota, populations and communities is a primary focus of ecotoxicology globally. One of the particular challenges for ecotoxicologists is to predict how the joint effects of global change and contaminants on individual organisms will manifest at the population and community level. Contaminants are substances in the environment that are present at concentrations above natural background. They may be physical (eg, salinity), chemical (eg, metals), or biological (eg, microbial pathogens).

Globally, we have had limited success in considering multiple environmental stressors

Marine Ecotoxicology
http://dx.doi.org/10.1016/B978-0-12-803371-5.00010-2

and how direct and indirect stressors interact with contaminants and impact marine and estuarine ecosystems. Organisms may become more or less sensitive to chemical stressors, and the effects of these multiple stressors on ecosystem health are likely to become increasingly important in the future.

This chapter will discuss the drivers of global change including catchment-land use change, coastal development, including port activities and industrialization, and climate change. We consider multiple stressors and how they may interact to potentially impact marine and estuarine ecosystems in the Anthropocence (present era), from now and over the next 50–100 years.

10.2 CATCHMENT-LAND USE CHANGES

10.2.1 Population Growth and Landscape Changes

In 2012, the world's population reached 7 billion people (UNFPA, 2014). Even with an overall decline in the population growth rate, the world's population grew by 30% between 1990 and 2010 and is expected to reach 10 billion by 2083 (UNFPA, 2014). To support this growth and its increasing demand for resources, between one-third and one-half of the Earth's surfaces have been transformed, with 25% of the terrestrial environment now consisting of agricultural lands. Although coastal environments account for less than 5% of the Earth's land area, they are the epicenters for population growth, with 39% of the world's population living within 100 km of the coast. In countries such as Australia, where much of the landscape is incapable of supporting a high population density, the figure is much higher, with 85% of Australians residing within 50 km of the coast (ABS, 2003). Migration of people to coastal regions is also common in developing nations. Sixty percent of the world's 39 metropolises with a population of over 5 million are located within 100 km of the coast, including 12 of the world's 16 cities with populations greater than 10 million (Nicholls et al., 2007).

The combination of human population growth, particularly in coastal areas, and the rapid and ever-increasing changes to coastal land use, has resulted in pronounced losses of key marine environments. For example, the global extent of seagrass meadows is now less than 29% of pre-1880s estimates. In the last two decades of the 20th century, 20% of the world's coral reefs have been lost, with a similar proportion being degraded. During the same period, it is estimated that 35% of mangrove forests were also lost (Valiela et al., 2001). While the loss and degradation of these environments can be attributed to many causes—most notably habitat destruction—environmental contaminants have also directly or indirectly contributed to these trends. At the beginning of the 21st century, it was estimated that 2.7 million km^2 of marine habitat was being affected by nonpoint source organic and inorganic contaminants, with an additional 1.6 million km^2 affected by nutrient inputs (Halpern et al., 2008).

Coastal ecosystems are also under threat worldwide from chemical contaminants released through anthropogenic activities such as agriculture, aquaculture, mining, industrial development, dredging, large-scale manufacturing, and urbanization. Contamination of marine and coastal environments is predominantly a result of human activities. The United Nations Environment Programme estimates there are around 50,000 industrial, agricultural, and household chemicals that are commercially available (UNEP, 2013). These can enter waterways through diffuse source inputs such as stormwater runoff and atmospheric deposition, or by direct discharge of treated wastewaters from sewage treatment plants and industry. Such contaminants may comprise naturally occurring constituents (such as metals/metalloids), nutrients, or man-made chemicals (eg, industrial chemicals, pesticides, and pharmaceuticals).

Many of these pressures (ie, contaminant and noncontaminant stressors) co-occur and give rise to a complex mixture of stressors and effects which are intensifying as our activities in the coastal zone grow. Research focuses on understanding the risks to marine ecosystems, predicting impacts, and developing mitigation options.

The discharge of contaminants into the aquatic environment can originate from either point or diffuse sources (Fig. 10.1). Point sources are derived from specific fixed locations, eg, wastewater outlets associated with manufacturing sites, sewage treatment plants, and stormwater pipes. While the types of sources vary, in general there are often particular groups of contaminants associated with specific activities, for example: metals from mining operations; and nutrients, pharmaceuticals, and personal care products (PCPs) from sewage treatment plants. Consequently, in some cases, it is possible to use chemical signatures within the sediment to identify the likely source of the contaminant (Costanzo et al., 2001; Hajj-Mohamad et al., 2014). However, such forensic approaches are difficult in estuaries where inputs may be numerous and the dispersal of contaminants can be strongly modified by hydrodynamic conditions. Diffuse sources for contaminants are generally associated with

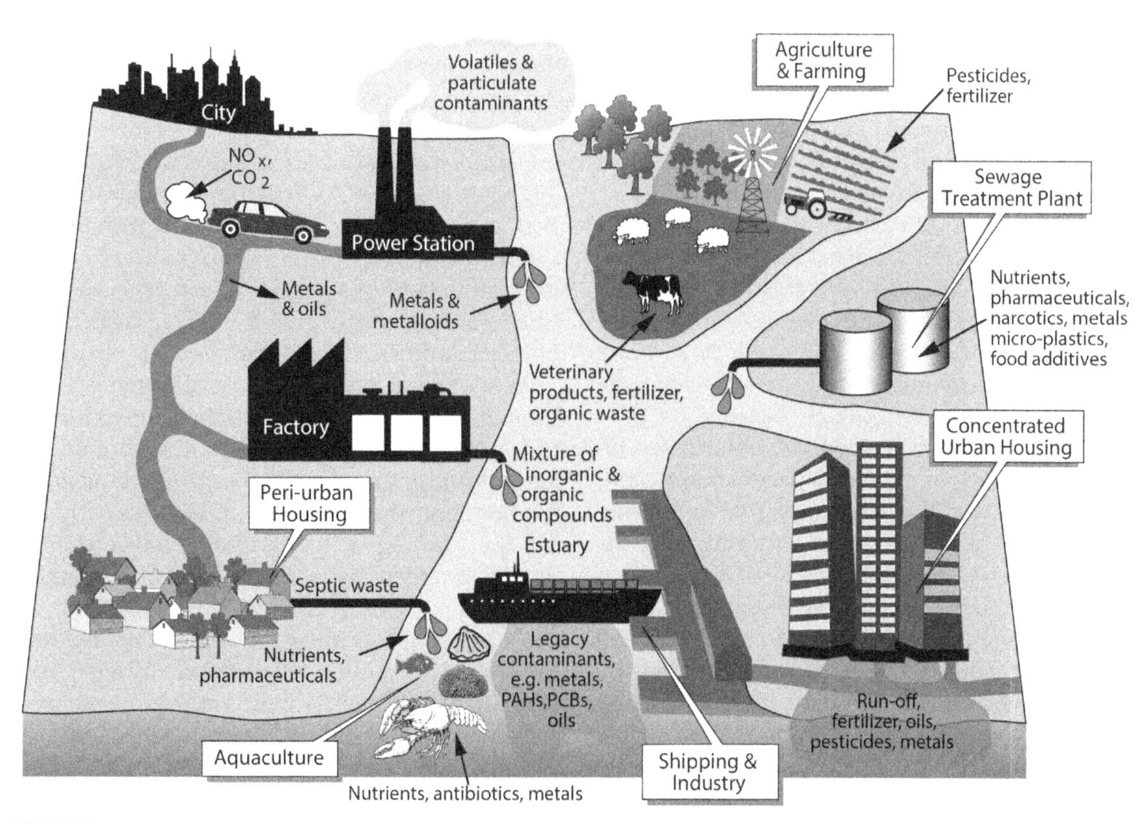

FIGURE 10.1 Conceptual model illustrating the relationships between land-use and the sources and types of contaminants which enter estuaries and marine environments.

catchment runoff, and loadings of these contaminants are often episodic, with the constituents reflecting the makeup of the surrounding landscape, eg, pesticides from agriculture and petroleum hydrocarbons from roads.

Additional diffuse sources include deposition from atmospheric-derived contaminants (eg, via smoke stacks, car emissions, and volatile contaminants) as well as legacy contaminants associated with former practices (eg, old industrial sites). Legacy contaminants can originate from both point and diffuse sources, and in many cases contain persistent organic pollutants (POPs) whose production has been phased out or dramatically reduced. For example, in Sydney Harbour, Australia, elevated concentrations of dioxins and furans (eg, TCDD) are still present in the sediments surrounding a former chemical industrial site that was used to manufacture a wide range of chemicals, including timber preservatives, herbicides, pesticides, and plastics. They are subsequently taken up by sediment-dwelling organisms and passed on via the food chain to fish, crustaceans, and marine reptiles/mammals. Given the significant health implications associated with dioxins and the high concentrations observed in several of the local fish species, fishing is currently banned within a significant proportion of the harbor some 30 years since chemical manufacturing ceased at the site (Manning and Ferrell, 2007). Because of their persistence, organochlorines, polychlorinated biphenyls (PCBs), and other legacy organic contaminants may still play a major role in altering the ecology of estuarine and marine systems (Chariton et al., 2010). Consequently, it is often important to consider the role these legacy contaminants may be having on an environment, even if the study is focused on examining the contaminants which are actively entering the system.

While a varied range of contaminants have been shown to have broad impacts in marine environments, here we focus on nutrients, pesticides, and the diversity of emerging contaminants of concern.

10.2.2 Nutrients and Eutrophication

Nutrient inputs into estuaries are naturally derived from coastal upwelling, the bacterial breakdown of organic material, and geological weathering. In particular, nitrogen and phosphorus are essential for maintaining the primary production and natural functioning of estuaries and marine environments, with both elements also providing the basic building blocks of life such as amino and nucleic acids. While nitrogen in the form of N_2 is the largest constituent of the Earth's atmosphere, elemental nitrogen must be converted (or fixed) to a reduced form to be used by organisms. Only 0.002% of the Earth's nitrogen is present in biological materials (eg, living organisms and detrital material), with the proportion in naturally occurring inorganic forms such as nitrate, nitrite, and ammonium being orders of magnitudes less (Howarth, 2008; Vitousek et al., 1997a). It is these inorganic forms that are pivotal for aquatic plant growth and consequently limit primary production in a majority of the world's temperate estuaries and marine environments (Enell and Fejes, 1995; Howarth and Marino, 2006).

While the global amount of nitrogen is fixed, human activities—most notably the production of synthetic nitrogen fertilizers, the adoption of agricultural practices which encourage nitrogen fixation, and the formation of reactive nitrogen via the burning of fossils fuels—have resulted in a disproportional shift toward more reactive forms of nitrogen (Galloway et al., 2004). At the turn of the century, it was estimated that the global formation of reactive nitrogen had increased by 33–55%, with anthropogenic processes now exceeding the natural formation of reactive nitrogen (Howarth and Marino, 2006). By 2030, annual production rates of reactive nitrogen are expected to exceed the 1990 level of 80 million tons by 1.7 times (Vitousek et al., 1997b).

In aquatic systems, phosphorus can exist in both dissolved and particulate forms. Particulate phosphorus can be transported into marine environments via absorption to sediment particles and organic material (eg, decaying material), or bound to proteins. The number of dissolved forms is numerous and includes inorganic orthophosphate, low molecular weight phosphate esters, pyrophosphate, and longer-chain polyphosphates. Within waters, phosphate compounds can be enzymatically or chemically hydrolyzed to orthophosphate, the form which is assimilated by plants, algae, and bacteria (Correll, 1998). Through a combination of assimilation and deposition, sediments can become a repository for phosphorus, and while they are relatively stable under oxic conditions, under anoxic conditions phosphorus can be re-released into the water column (Correll, 1998). While it is commonly viewed that phosphorus limits primary production in freshwater systems and nitrogen in marine systems, several studies have challenged this oversimplified view. For example, in the Peel-Harvey Estuary (Western Australia), the relationship between primary production limitation and nutrient availability has been shown to be seasonal, nitrogen-limiting in summer and phosphorus in winter (McComb and Davis, 1993).

The shift toward more urban, peri-urban (hybrid landscapes of fragmented urban and rural characteristics), and agriculturally intensive coastal catchments has resulted in marked increases of nutrient loadings into estuaries. It is estimated that more than 60 teragrams (Tg, $1\ \text{Tg} = 1 \times 10^9$ kg) of nitrogen flow annually into the world's oceans, doubling the estimated loadings from the mid-19th century (Boyer and Howarth, 2008; Galloway et al., 2004). Increases in nitrogen loads have been most pronounced in temperate regions in the Northern Hemisphere, where agricultural intensity is greatest. For example, in South Korea and northeastern USA, annual loadings of nitrogen into coastal oceans have increased from a baseline of 100 kg to 1700 and 1000 kg N/km^2, respectively (Howarth and Marino, 2006).

The trend for phosphorus is similar, with an estimated 600 million tons of phosphorus being applied to terrestrial environments between 1950 and 1995 (Brown et al., 1998). While a majority of phosphorus is artificially mineralized as fertilizer, in many less industrialized regions, phosphorus is primarily obtained in the form of manure (Carpenter et al., 1998). A major concern is that in many systems the phosphorus inputs greatly exceed the requirements of agriculture, and consequently large reserves of phosphorus accumulate in the soil, which eventually runs into catchments (Carpenter et al., 1998).

Marine waters can be classified by their inputs of growth-limiting nutrients (Fig. 10.2), with the term eutrophication commonly used

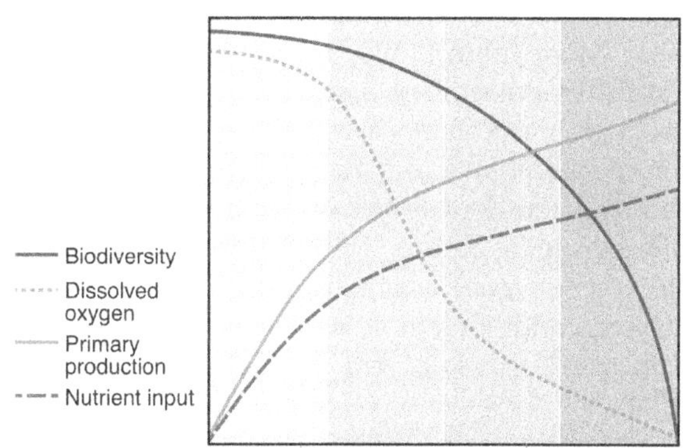

Nutrient loads and water quality characteristics	Oligotrophic	Mesotrophic	Eutrophic	Hypertrophic
TN (μg/L)	<260	260-350	350-400	>400
TP (μg/L)	<10	10-30	30-40	>40
Chlorophyll *a* (μg/L)	<1	1-350	3-5	>5
Secchi depth (m)	>6	3-6	1.5-3	<1.5

FIGURE 10.2 Classification and key changes in marine waters based on their nutrient loads. *TN*, total nitrogen, *TP*, total phosphorus. *Modified from Smith, V.H., Tilman, G.D., Nekola, J.C., 1999. Eutrophication: impacts of excess nutrient inputs on freshwater, marine, and terrestrial ecosystems. Environ. Pollut. 100, 179–196 and Correll, D.L., 1998. The role of phosphorus in the eutrophication of receiving waters: a review. J. Environ. Qual. 27, 261–266.*

to define an ecosystem's response to an oversupply of nutrients. In contrast to most contaminants where the ecotoxicological endpoints are direct acute or chronic toxicity, it is the indirect effects of eutrophication rather than actual toxicity of the nutrients which is of primary concern.

As both nitrogen and phosphorus limit plant growth, it is unsurprising that excess inputs of these nutrients can lead to marked increases in primary production, predominantly via increases in the biomass of marine phytoplankton and epiphytic algae (Smith, 1998). Such changes have far broader consequences than simply a decline in water clarity and other aesthetic values. The proliferation of primary producers increase shading and can lead to large-scale reductions in the quality of benthic habitats, including key ecosystems such as coral reefs and seagrass meadows (Bellwood et al., 2004). A reduction in the integrity of these environments can lead to further declines in biodiversity. More importantly, the overproduction of autochthonous plant biomass and the increase in microbial activity associated with breakdown of the additional organic material (ie, organic carbon) can deplete the dissolved oxygen from the overlying waters, resulting in hypoxic (dissolved oxygen concentrations <2 mg/L) or anoxic conditions (no dissolved oxygen), which can cause fish kills and mass die-offs in other biota.

Eutrophication can also lead to composition changes in the smallest constituents of an ecosystem (microbial eukaryotes and picophytoplankton) all the way up to macroalgae and fish (Brodie et al., 2005; Chariton et al., 2015a; Smith, 1998). In the case of estuarine macrobenthic invertebrates, eutrophication generally leads to communities which are dominated by a few highly abundant opportunistic, but small-bodied taxa, eg, capetillid and spionid polychates. Collectively, this is reflected as declines in diversity, evenness, and biomass (Warwick, 1986). In phytoplankton communities, the dominance of opportunistic taxa can have both severe and broad environmental implications, creating conditions suitable for the dominance of one of the estimated 60–80 toxic species (Smayda, 1997). One of the most significant occurrences of a toxic phytoplankton bloom occurred in the Baltic Sea in 1988, where anecdotal evidence suggests that eutrophication led to a 75,000 km^2 bloom of *Chrysochromulina polyepsis*, resulting in widespread and massive losses in fish, plants, and macroinvertebrates (Rosenberg et al., 1988). More recently, there has been growing evidence to suggest that eutrophication is contributing to outbreaks of the coral-eating Crown of Thorns starfish (COT), a species that is significantly contributing to the decline of Australia's Great Barrier Reef (Brodie et al., 2005; Box 10.1). However, the relationship between COT outbreaks and nutrients is complex, with some research indicating that nutrients may be indirectly driving COT populations by promoting larger phytoplankton species that are preferentially consumed by COT larvae (Brodie et al., 2005).

The ecological impacts of eutrophication are both widespread and often severe; however, in contrast to persistent contaminants (eg, organochlorines and polycyclic aromatic hydrocarbons (PAHs)) whose concentrations can remain stable long after inputs have ceased, there is growing evidence to suggest that eutrophication can be effectively counteracted by abatement programs which collectively reduce nutrient use, minimize input, and promote the transfer of materials to oceanic waters (Boesch et al., 2001). For example, in Tampa Bay (FL, USA), a 50% reduction of total nitrogen over a 20-year period has resulted in a 50% increase in water clarity and the recovery of 27 km^2 of seagrass meadows (Greening and Janicki, 2006). For more detailed information on the chemistry and ecotoxicology of nutrients, the reader is referred to the following articles: Smith (1998), Howarth and Marino (2006), and Kennish and de Jonge (2011).

BOX 10.1

GREAT BARRIER REEF CASE STUDY

The Great Barrier Reef (GBR), Queensland, Australia, is the largest living structure on earth, stretching 2300 km and covering an area of 344,400 km^2. It is a World Heritage site that supports 600 types of soft and hard corals, more than 100 species of jellyfish, 3000 varieties of mollusks, 500 worm species, 1625 types of fish, 133 varieties of sharks and rays, and over 30 species of whales and dolphins. Coral reef communities are both economically important and biologically diverse, providing essential ecosystem services including fisheries, coastal protection, tourism, and novel pharmacologically active compounds (Moberg, 1999). Anthropogenic climate change, together with other stressors such as sediments, nutrients, and contaminants, are having a major impact on the long-term resilience of these reef communities. These anthropogenic stressors interact with other large-scale disturbances, especially tropical storms and population outbreaks of the coral-eating crown-of-thorns starfish (COTS) (Fig. 10.3). De'ath et al. (2012) estimated that there has been a decline in coral cover in the GBR from 28 to 14% over the period 1985 to 2012. Tropical cyclones, coral predation by COTS, and coral bleaching accounted for 48%, 42%, and 10% of the losses, respectively. The GBR is increasingly being impacted by multiple stressors:

- sediment runoff from coastal catchments, which increases turbidity and reduces light penetration. Sedimentation kills corals through microbial processes triggered by the organic matter in the sediments. Microbial respiration results in anoxia and reduced pH, which initiates coral tissue degradation (Weber et al., 2012);
- contaminants such as diuron, a herbicide commonly used in agriculture (eg, cane farming) and in boat antifouling paints. Diuron is a potent inhibitor of photosynthesis and has potential impacts on coral zooxanthellae, crustose coralline algae, sea grasses, and phytoplankton in tropical systems (Haynes et al., 2000; Harrington et al., 2005). Concentrations of diuron of up to 10 μg/kg wet weight in subtidal sediments in the GBR have been found, and studies by Harrington et al. (2005) have shown that sedimentation stress in coralline algae is significantly enhanced in the presence of trace concentrations of diuron;
- nutrient inputs from catchments leading to potential eutrophication;
- oil spills from shipping activities;
- increasing water temperatures leading to coral bleaching; and
- ocean acidification.

GBR building corals such as *Porites* have already shown a reduction in growth (from reduced skeletal density and linear extension rate) of 21% over the last two decades due to reduced calcification as a likely consequence of climate change (Hoegh-Guldberg et al., 2007). Reduced coral skeletal density may also lead to increased coral erosion rates, increasing their vulnerability to grazing and storm damage, which consequently affects habitat structure, diversity, and ecosystem services such as coastal protection. The energetic cost of maintaining skeletal growth and density under unfavorable conditions may also impact coral reproduction.

In addition to impacts on corals, coralline algae, which are a key settlement substrate for corals, are sensitive to pH and require magnesium and calcium for exoskeleton formation.

Continued

BOX 10.1 (*cont'd*)

GREAT BARRIER REEF CASE STUDY

Thus, coral recruitment may be compromised if coralline algal abundance declines, providing an ideal habitat for macroalgae, which compete with them for space and light. Macroalgae form stable communities that impair the ability of corals to reestablish, thus affecting ecological resilience and tipping the ecosystem to an alternative state (Mumby et al., 2007).

Scenarios for predicting change from these multiple stressors, intended to provide a framework for future adaptive management, are well summarized in the review by Hoegh-Guldberg et al. (2007). Management of coral reefs should initially focus on local stressors such as water quality, which could potentially reduce COTS outbreaks. Management of fisheries near and even on coral reefs is also required. The objective of management interventions should be to assist coral communities to adapt to present and future multiple stressors including climate change. Mitigation of CO_2 emissions, to keep atmospheric CO_2 below 500 ppm, is considered vital for coral reefs and the human populations that depend on them now and into the future.

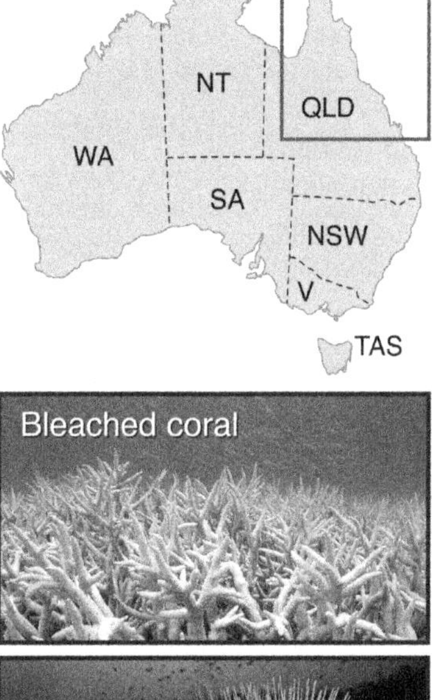

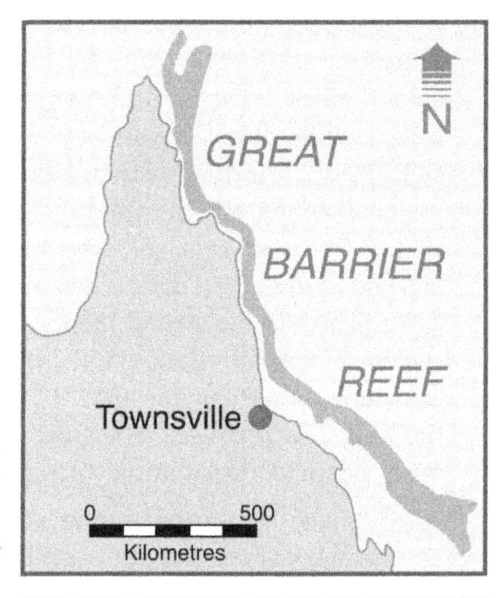

FIGURE 10.3 Selected stressors on the Great Barrier Reef, Australia.

Predicting the long-term ecological impact of excess nutrients on marine environments is difficult. While it is clear that aquaculture will have an increasing role in the eutrophication of marine and coastal systems, with the area allocated to aquaculture in 2090 forecast to be 1800 times greater than it was in 1990 (Duarte et al., 2009), predicting the loadings of agriculturally derived nitrogen and phosphorus is more uncertain. It may be logical to assume that agricultural runoff will continue to increase proportionally with population growth and the cultivation of coastal environments; however, the economic and environmental costs are increasingly being scrutinized. For example, while there was a sevenfold increase in the application of nitrogen on cereal crops between 1960 and 1995, crop yields only doubled, with the efficiency of nitrogen decreasing from 70 to 25 kg of grain per kg N (Keating et al., 2010). In Europe, the environmental cost of nitrogen losses, which includes abatement measures and the broader loss of resources to society, has been estimated at €70–€320 billion per year, outweighing the economic benefits of nitrogen application in agriculture (Brink et al., 2011). Although nitrogen is clearly being overused, because it is continually being fixed there is no prospect of it running out. However, the same is not true for phosphorus: like oil, phosphorus fertilizer is extracted from mineral deposits which have been formed over millions of years. While the demand for phosphate fertilizer will undoubtedly increase, its availability will be increasingly limited. Collectively, this suggests an overall trend toward the increasing role of aquaculture in the eutrophication of coastal and marine environments.

10.2.3 Pesticides

Pesticides are chemical mixtures that are applied to kill, repel, or mitigate pest species. Since their broad-scale use in the mid-1940s, the composition and attributes of pesticides have undergone several major changes. The first generation of synthetic pesticides was the organochlorines, which included aldrin, dieldrin, DDT, and endosulfan. Initially, these chemicals were proven to be highly efficient insecticides and were used extensively to reduce agricultural pests as well as vectors of diseases such as typhus and malaria. During its peak use (1950–80), more than 40,000 tons of DDT was applied annually (Geisz et al., 2008). However, by the 1960s, it was becoming clear that due to their persistence, overuse, and capacity to biomagnify through the food web, organochlorines were having a significant and highly visible impact on the environment. As early as 1968, Hungary had banned the use of DDT, with most organochlorines being outlawed in Sweden, Norway, and the United States by the early 1970s. Despite global bans, up to 4000 tons of DDT is still produced annually and is primarily used for disease vector control, although some agricultural use still occurs in India and possibly North Korea (van den Berg et al., 2012).

The second generation of pesticides is the organophosphates, including chemicals such as diazinon, chlorpyrifos, and malathion. In contrast to organochlorines, organophosphates generally only persist for a few days or weeks in the environment; however, they are far more toxic than organochlorines. Toxicity by organophosphates is caused by the inhibition of acetylcholinesterase, an enzyme that underpins neurological activity. This mechanism renders organophosphate exposure potentially toxic to a wide range of organisms, including humans, marine fish, and invertebrates (Bouchard et al., 2011; Janaki Devi et al., 2012; Canty et al., 2007). Given that only 1% of sprayed pesticides are effective, with the remaining 99% being applied to nontargeted environments such as soils, water bodies, and the atmosphere, the potential ecological impacts of highly toxic, nonspecific organophosphates are significant. However, because of their low environmental persistence, linking long-term alterations in marine systems to organophosphate use is challenging.

More recently developed insecticides include the pyrethroids, which use pyrethrin, a natural derivative of chrysanthemums. Newer herbicides include widely used chemicals such as glyphosate, commonly known as Roundup. Glyphosate contains a phosphonyl group that inhibits the synthesis of the amino acids tyrosine, tryptophan, and phenylalanine. Another group of commonly used herbicides, which is of increasing concern to the marine environment, are diuron and substituted ureas such as atrazine and trazine. These herbicides operate by disrupting photosynthesis. Importantly, a number of noninsecticide approaches for maximizing the resistance of plants to pests species are currently being developed and used, most notably, this includes the genetic modification of crops via a process called RNA interference (RNAi) (Castel and Martienssen, 2013).

10.2.3.1 Pesticides in Marine Environments

The legacy of organochlorines still continues. At the turn of the 21st century, relatively low concentrations of DDT and its metabolites were still detectable in the world's oceans. However, because organochlorines are hydrophobic, concentrations of DDT and its metabolites remain far greater in sediments and biota, and in many cases at concentrations that still warrant concern (Galanopoulou et al., 2005). There is, however, an overall decline in total DDT concentrations within marine biota (body burden), with the chemical having a half-life of between 10 and 14 years (Sericano et al., 2014). This trend is typified by a long-term survey in Sweden where researchers found total DDT concentrations in marine biota (guillemot eggs and herrings) declined by 96–99% between 1969 and 2012 (Nyberg et al., 2015). The researchers observed similar trends in polychlorinated biphenyls (PCBs), hexachlorocyclohexanes (HCHs), and hexachlorobenzene (HCBs).

Although increasing amounts of pesticides are being applied, there has been a global shift in the types of pesticides being used, with herbicides (47.5%) now representing the largest proportion of pesticide use, followed by insecticides (29.5%), fungicides (17.5%), and others (5.5%) (Pimentel, 2009). Even though herbicides may degrade relatively rapidly, their use in coastal catchments is both extensive and diverse (eg, agricultural and urban use), and continues to increase as coastal environments become more developed. Even in highly protected environments such as Australia's Kakadu World Heritage Area, diuron is detectable in estuarine sediments due to its continual application for the control of invasive floodplain weeds (Chariton unpublished).

As herbicides such as diuron and atrazine are designed to interfere with photosynthesis, runoff into coastal environments may have wide implications on nontarget marine photosynthesizing species such as algae and seagrass. The loss of such species may have a cascading effect; for example, seagrass meadows are both essential for habitat and as a food source for iconic species such as dugong, manatee, and green turtles (Reich and Worthy, 2006; Bjorndal, 1980; Bayliss and Freeland, 1989). Equally, concentrations in coastal waters adjacent to coral reefs have frequently been shown to be sufficient to induce sublethal effects, eg, growth and photosynthetic inhibition in marine microalgae (Magnusson et al., 2008). Given that microalgae form the basis of both benthic and pelagic food webs, the potential for broad-scale adverse effects to microalgae and other primary producers is of great concern. Growing evidence suggests that herbicides are contributing to the loss of coral reefs by affecting *Symbiodinium*, a dinoflagellate alga that provides nutrition to scleractinian corals via a mutualistic relationship (Jones, 2005; Veron, 1995). Interestingly, regardless of their mode of action, toxicological responses of herbicides are not limited to photosynthesizers, with herbicides also toxic to a range of estuarine invertebrates and other organisms (Macneale et al., 2010).

10.2.4 Emerging Contaminants of Concern

Marked changes to coastal landscapes and the socioeconomic makeup of coastal communities have led to an increasingly complex mixture of contaminants entering marine systems (Fig. 10.1). These include pharmaceuticals, PCPs, microplastics, nanomaterials, and narcotics. Collectively these are referred to as "emerging contaminants of concern." For many of these groups, the issue is by no means new, for example, concerns regarding antibiotics in the aquatic environment have been raised in the literature since the mid-1970s (eg, Hignite and Azarnoff, 1977). However, our understanding of the distribution and ecotoxicology of these contaminants has been previously constrained by our capacity to identify and quantify environmentally relevant concentrations, often requiring detection limits within the ng/L to μg/L range.

The mode of action and ecotoxicological effects of many emerging contaminants are quite different from those associated with pesticides and metals. For example, pharmaceuticals convey a particular physiological mode of action (eg, antiinflammatory and lipid regulators) used to target specific biochemical pathways or biological systems. In contrast, pesticides are generally designed to be lethal to a broad range of species within a particular taxonomic group. There are notable exceptions, eg, broad-spectrum antibiotics, which are designed to kill or inhibit broad taxonomic groups of bacteria.

In contrast to pesticides, whose type and application frequency is generally determined by the crop, pest, and season, marine environments may be continuously exposed to low concentrations of various mixtures of pharmaceuticals. The most prevalent of these include nonsteroidal antiinflammatory drugs, antibiotics, blood lipid lowering agents, sex hormones, and antiepileptics (Santos et al., 2010).

Many emerging contaminants of concern lack such specific modes of action associated with pharmaceuticals. For example, triclosan, an additive in a range of PCPs including toothpaste and liquid soaps, is an antimicrobial agent. Because of its wide taxonomic target (bacteria), persistence, and therefore capacity to bioaccumulate, triclosan may have more in common with traditional contaminants such as PCBs and metals. Furthermore, the toxicology of triclosan extends past its targeted taxonomic group, and it is also toxic to many meio- and macrofauna at high concentrations (Chariton et al., 2014). While triclosan is still used in many products, its use is being phased out in the EU, with increasing pressure for a similar trend in the USA.

The structural attributes of some emerging contaminants of concern, most notably plastics, also pose a significant ecotoxicological concern to marine environments. Global production of plastics commenced in the 1950s, and in 2010 the annual production was estimated at 265 million tons per year, increasing fivefold since the 1970s (Cózar et al., 2014). The ecological impact of plastics is strongly determined by size and material type. For example, larger floating pieces of plastic may cause structural damage (eg, choking and clogging of digestive pathways) to seabirds, turtles, and fish, while smaller pieces (eg, microplastics used as pellets in facial scrubs) can elicit a similar response in invertebrates. An extensive meta-analysis on the distribution of oceanic plastic debris demonstrated the strong relationships between plastic debris accumulation, oceanic currents, and on-land use (Cózar et al., 2014). For example, the North Pacific Ocean, which captures much of the east coast of Asia, contributed between

33% and 35% of the total 7000–35,000 tons of estimated plastic load within the world's oceans. The accumulation of oceanic plastic debris is most evident in a region dubbed the "Eastern Garbage Patch" in the Northern Pacific Ocean whose size, while difficult to quantify, has been estimated to range between 700,000 and 15,000,000 km^2, equating to 0.41–8.1% of the size of the Pacific Ocean, respectively (Moore, 2003).

Many plastics also contain potentially toxic chemicals such phthalates, flame retardants, and bisphenol-A, and consequently the accumulation of these chemicals, either from the breakdown of plastics or via consumption by marine organisms, including sea birds, also poses a concern (Tanaka et al., 2013). Even inert plastics can absorb hydrophobic contaminants, providing an additional pathway for accumulation. In a random survey on plastic debris from the East Garbage Patch, more than 50% of the samples contained pesticides, PCBs, and PAHs, with this being particularly evident in polyethylene plastics, which are highly resistant to degradation (Rios et al., 2010). Similar concerns have also been identified with microplastics (Teuten et al., 2007).

While the diversity and complexity of contaminant mixtures entering marine systems poses some huge challenges, environmental impacts can be effectively minimized with appropriate regulation. This has not only been demonstrated for organochlorine pesticides but also for a broader range of contaminants. For example, since the EU ban of the fire-retardant penta-PBDE in 2004, there has been an overall decline in the concentration of brominated diphenyl ethers in the blubber of Greenland ringed seals (Law, 2014). A similar trend has also being observed with the International Maritime Organization's 2008 banning of tributyltin antifoulant paints on large sea-going vessels (Law, 2014). However, for ecotoxicologists the challenge of emerging contaminants of concern is not only understanding the ecological impact of particular contaminants but also the interplay between numerous contaminants, including legacy contaminants, whose concentrations and mixture can vary greatly over time. The potential role of environmental genomics to address some of these challenges is addressed later in this chapter.

10.3 PORTS AND INDUSTRY-ASSOCIATED CHANGES

The oceans are crucial for global food security, human health, and regulation of climate. The livelihoods of over 3 billion people worldwide already depend upon services from marine and coastal biodiversity and it is inevitable that the demand for more food and resources from the seas will grow as populations increase (EASAC, 2015).

Utilization of the coast increased dramatically during the 20th century. Coastal population growth in many of the world's deltas, islands, and estuaries led to widespread conversion of natural coastal landscapes (including coastal forests, wetlands, coral reefs) to agriculture, aquaculture, industrial, and residential uses (Valiela, 2006). Centers of urbanization often occur near ecologically important coastal habitats. For instance, 58% of the world's major reefs occur within 50 km of major urban centers of 100,000 people or more, while 64% of all mangrove forests and 62% of all major estuaries occur near such centers (Agardy et al., 2005).

Understanding how our use of the marine environment and its resources will change over the next 50 years is of critical importance to the industrial sector and governments as this informs planning. In this section we look at the future global changes that may impact marine environments. Understanding future trajectories will also inform marine ecotoxicology in terms of understanding where most effort will be required.

10.3.1 Demands on Coastal Waters

There are multiple pressures on the coastal zone that can cause ecosystem degradation and impacts on coastal communities. Substantial degradation of coasts, bays, and estuaries occurred in the mid-19th and 20th centuries; the impacts principally arose from unregulated human activities in river catchments, urban and coastal developments, and fishing. The greatest threat to coastal systems is development-related loss of habitats and services. Many areas of the coast are degraded or altered, such that humans are facing increasing coastal erosion and flooding, declining water quality, and increasing health risks. Engineered structures, such as damming, channelization, and diversions of coastal waterways, change circulation patterns and alter freshwater, sediment, and nutrient delivery.

A significant challenge lies in improving our ability to predict change in coastal ecosystems. Predictions of ecosystem responses remain vague and largely qualitative and are therefore still of limited use (Duarte, 2015). The notion that changes in ecosystems in response to pressures are smooth, linear, and reversible is challenged by widespread evidence that complex natural systems, composed of multiple interacting elements, tend to show a nonlinear response to pressures where initially smooth, gradual responses to pressures are replaced by an abrupt shift in state once the pressure exceeds a limit, termed a threshold or tipping point (Andersen et al., 2009; Duarte et al., 2009, 2012a,b). An additional complicating factor is that the trajectory of recovery of the system following reduction of a pressure typically follows a different pathway, often to a different equilibrium state.

10.3.2 Coastal Industries

Globally, the geographic footprint of established industries is changing with a shift of traditional heavy industries and chemical industries to developing countries. Materials production has relocated, with Organization for Economic Cooperation and Development (OECD) countries increasingly relying on imports from countries such as China and India (World Ocean Review, 2010). It is predicted that this change will continue as the global demand for industrial products is expected to more than double by 2050. Currently, China is the largest producer of ammonia, cement, iron and steel, and methanol. Production in China will follow OECD trends, likely leveling out by 2050, as its economy shifts toward the services industry (World Ocean Review, 2010). By comparison, industrial activity in India, Africa, and the Middle East is expected to increase significantly by 2050. Established maritime industries will be undergoing significant change in the coming decades through the adoption of enabling technologies such as 3D printing, advanced robotics, lightweight materials, nanotechnology, and marine biotechnology. Predicting the impact of disruptive technologies is very difficult and may lead to dramatic departures from predictions based on simple extrapolations of current trajectories.

Oil refineries are located predominantly in coastal locations. The majority of the world's 10 largest refineries are situated in the Asia Pacific region, with India hosting the world's largest refinery complex, followed by Venezuela and South Korea. Owing to a global rationalization of the industry and a growing reliance on large-scale refineries, one in five oil refineries are expected to cease operations over the next 5 years.

The petrochemical industry converts feedstocks such as naphtha and natural gas components such as butane, ethane, and propane through steam cracking or catalytic cracking into petrochemical building blocks such as ethylene, propylene, benzene, and xylene. These chemicals are further processed to yield final products such as paints, tires, detergents, agrochemicals, and plastic products. Around 80% of petrochemicals' manufacturing costs are related to energy and to oil and gas as feedstock. Currently, the petrochemical industry is concentrated in North America, Western Europe, and Asia. However, in the coming years, the Middle East is likely to emerge as a significant producer owing to abundant natural gas reserves. European petrochemical production is in decline owing to old plant and high production costs. By contrast, Brazil is leading the world in terms of bioderived chemicals and fuels.

10.3.3 Oil and Gas Exploration

Since industrial oil extraction began in the mid-19th century, 147 billion tons of oil have been extracted from reserves around the world—over half of it since 1990 (World Ocean Review, 2010). Over the next 20 years there will be a decline in production from the world's older oil fields and greater reliance on offshore oil and gas sources (Figs. 10.4 and 10.5).

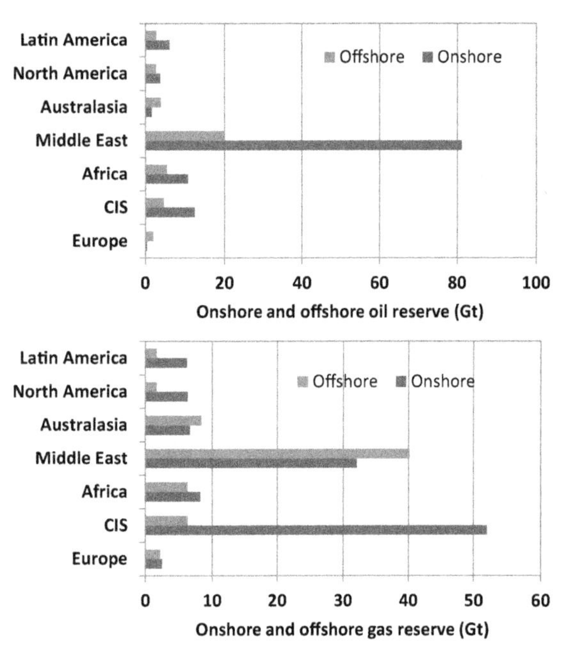

FIGURE 10.4 Geographic distribution of oil and gas reserves by region. *Adapted from World Ocean Review, 2010. Living with the Oceans, vol. 1. Maribus, Hamburg, Germany.*

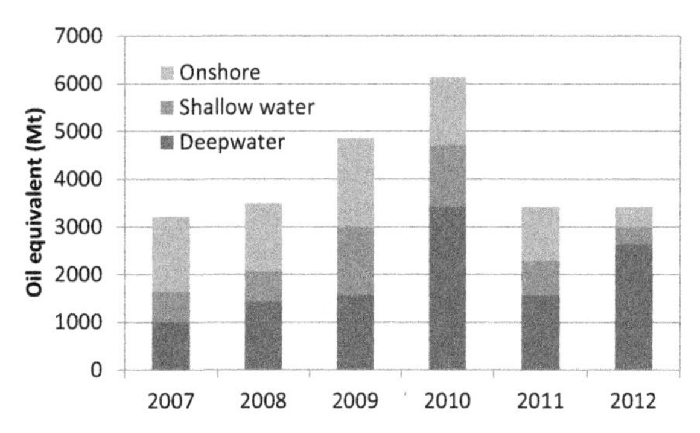

FIGURE 10.5 Oil and gas fields discovered between 2007 and 2012. *Adapted from World Ocean Review, 2014. Marine Resources – Opportunities and Risks, vol. 3. Maribus, Hamburg, Germany.*

The most productive offshore areas are currently the North Sea and the Gulf of Mexico, the Atlantic Ocean off Brazil and West Africa, the Arabian Gulf and the seas off South East Asia. Almost half of the remaining recoverable conventional oil is estimated to be in offshore fields and a quarter of that in deep water (IEA, 2012). Offshore oil extraction currently accounts for 37% of global production. As the depletion of shallow water (<400 m depth) offshore hydrocarbon reserves continues, the focus is shifting increasingly toward exploration and exploitation of oil and gas reserves in deep (500–1500 m) and ultra-deep (depths greater than 1500 m) water.

Interest in the Arctic is growing, as it is estimated that about 30% of the world's undiscovered gas and 13% of its undiscovered oil may be found in the marine areas north of the Arctic Circle (USGS, 2008). While most of the drilling would be offshore in less than 500 m of water, the conditions in the Arctic are extremely hostile. As the Arctic sea ice melts as a result of climate change, tapping the oil and natural gas deposits in the northern Polar Regions will become increasingly feasible. The oil and gas reserves of Antarctica have not as yet been characterized.

Oil slicks resulting from the unintended release of crude oil at sea often drift toward the coasts and kill seabirds and marine mammals. However, oil tanker disasters account for only around 10% of global marine oil pollution. Most of the oil enters the seas along less obvious pathways, making it difficult to precisely estimate global oil inputs into the marine environment. Around 5% comes from natural sources, and approximately 35% comes from tanker traffic and other shipping operations, including illegal discharges and tank cleaning (World Ocean Review, 2010).

Offshore gas production of 65 trillion cubic meters currently accounts for a third of the worldwide total. At present, 28% of global gas production takes place offshore. The North Sea is currently the most important gas-producing area, but will be overtaken by other regions such as the Middle East in the near future, as well as off India and Bangladesh, Indonesia, and Malaysia (World Ocean Review, 2010).

Unconventional gas, which includes coal seam, shale, and tight gas, is a relatively new source of gas that has dramatically lowered the price of gas on the international market. Unconventional gas is abundant and present on every continent. The volume of shale gas and coalbed methane resources is currently estimated to represent about 51% of global gas resources. The availability of cheap shale gas has changed the face of the US petrochemicals industry. Ethane crackers are now being built and US-produced ethylene products will be exported.

Liquefied natural gas (LNG) plays a crucial role in the gas industry as it is cheaper to ship cooled and liquefied natural gas across the oceans in large tankers than through pipelines. LNG already accounts for a quarter of today's global trade in gas. In future, natural gas is more likely to be moved by ship than via pipelines.

10.3.4 Shipping and Ports

Shipping accounts for 90% of global commercial trade and is increasing by around 10% per year (IMO, 2012). Ocean shipping can be divided into two submarkets: liquid cargo such as oil, petroleum products, LNG; and dry cargo. Dry cargo is made up of bulk goods; the five most important being iron ore, coal, grain, phosphates, and bauxite. The single most significant type of cargo worldwide is crude oil, which alone accounts for around 25% of all goods transported by sea. The world's main shipping lanes therefore stretch from oil-producing hubs such as the Arabian Gulf around the Cape of Good Hope or through the Suez Canal, and from Africa northward and westward to Europe and North America. Other shipping lanes connect the Arabian Gulf to East Asia and the Caribbean to the Gulf Coast of the United States.

In terms of quantity, iron ore and coal are significant dry-bulk goods. Iron ore is transported over long distances in very large ships, mainly from Brazil to Western Europe and Japan, and from Australia to Japan. The most important coal routes are from the major export countries of Australia and South Africa to Western Europe and Japan and also from Colombia and the East Coast of the United States to Western Europe, as well as from Indonesia and the West Coast of the United States to Japan. Shipping lanes traverse some of the most ecologically sensitive marine areas, and regular groundings and accidents at sea place additional pressure on the marine environment. A number of environmental impacts are associated with shipping. These include oil spills and transfer of invasive species who "hitchhike" in ballast waters or on the hulls of ships. Also of concern is the frequency of ship strikes on marine mammals. Antifouling paints (mainly copper-based biocides) may also degrade water quality in harbors or marinas where there is a high density of shipping. Also, emissions from the propulsion systems of commercial vessels in marine waters constitute a significant proportion of total worldwide emissions of air pollutants and greenhouse gases which eventually reach the oceans (Blasco et al., 2014).

Even with only modest assumptions of economic growth, port cargo volumes are expected to rise by 57% by 2030. The average size of ships has increased substantially and port authorities must respond to increasing vessel sizes by expanding port infrastructure and improving port access (eg, by deepening navigation channels). By 2060, it is envisaged that ports will be located offshore on artificial islands where layouts can be optimized. These ports will be supported by floating feeder/river terminals that can be moved around in line with changing demands.

Container shipping was first introduced in the USA during the 1960s and is considered to be one of the key transport revolutions of the 20th century. The use of standardized containers saves costs, as the goods are packed only once and can be transported over long distances using various modes of transport—truck, rail, or ship. Since 1985, global container shipping has increased by about 10% annually to 1.3 billion tons (2008) and is set to increase, with volumes tripling by 2030 (World Ocean Review, 2010). In the future it is envisaged that the container will remain in use, based on the same compact, standardized format and may have inbuilt intelligence to communicate destination, contents, and journey details; the next generation of ultra-large vessels will carry 18,000 containers (McKinnon et al., 2015).

10.3.5 Dredging

Dredging involves the excavation or removal of sediment and/or rock from the seabed and is a routine part of port operations and of coastal and marine infrastructure developments. There are two major types of dredging operations (McCook et al., 2015): capital dredging is carried out to open up new developments such as marina or port basins or widen existing channels. Maintenance dredging keeps previously dredged areas at the required depth. Maintenance dredging campaigns are undertaken at regular intervals (eg, annually), are typically of short duration (days to weeks), and generally remove sediments with a higher proportion of finer particles. The sediment removed by dredging can be used for reclamation, or disposed on land. However, most of the material dredged in harbors, estuaries, and at sea is dumped at sea and only minor amounts of this dredged material are beneficially used. In most developed countries, there are strict controls on the disposal of potentially toxic dredged sediments which require treatment before disposal and/or placement in a secure landfill.

Dredging operations will almost always resuspend sediments and increase turbidity, but the level of resuspension and associated impacts depends on the physical and chemical characteristics of the sediment, as well as the site conditions, type of equipment, and dredging method. The sediment released into the water column can directly affect marine organisms such as corals, sponges, and shellfish and can cause membrane irritation and gill abrasion in fin fish (OSPAR, 2004, 2008b). Elevated suspended sediment levels can absorb and scatter or reduce light levels that are fundamentally important to the many photosynthetic benthic organisms including hard and soft corals, seagrasses, mangroves, macroalgae (including seaweeds), and algae. Sediments can also settle out of suspension and can potentially smother bottom-dwelling organisms. Additionally, release of nutrients from sediments may result in increased eutrophication and consumption of oxygen (OSPAR, 2004, 2008a,b).

10.3.6 Land Reclamation

Land reclamation is the process of creating new land from the sea. The simplest method of land reclamation involves simply filling the area with large amounts of heavy rock and/or cement, then filling with clay and soil until the desired height is reached. Draining of submerged wetlands is often used to reclaim land for agricultural use.

The first major land reclamations were carried out in the 1970s, when the Port of Rotterdam in the Netherlands was extended (OSPAR, 2008a). This was the start of the modern era of land reclamation, which rapidly spread around the world. In 1975 the government of Singapore commenced the construction of a new airport on the eastern tip of Singapore. Changi airport was built with over 40 million cubic meters of sand reclaimed from the seabed. Notable examples of coastal land reclamation include Hong Kong, Singapore, the Netherlands (OSPAR, 2008a,b; Hilton and Manning, 1995) and much of the coastline of mainland China (An et al., 2007). Artificial islands are an example of land reclamation. The Flevopolder (970 km^2) in the Netherlands, reclaimed from the IJsselmeer, is the largest reclaimed artificial island in the world. Kansai International Airport (in Osaka, Japan) and Hong Kong International Airport are also examples.

Marine habitats are permanently lost where land is reclaimed from the sea. It is estimated that nearly 51% of coastal wetlands in China have been lost due to land reclamation (An et al., 2007). Land reclamation may also influence habitat types of coastal and terrestrial origin such as sand dunes or freshwater bodies. Subsidence can be an issue, both from soil compaction on filled land and also when wetlands are enclosed by levees and drained to create polders (ie, low-lying land reclaimed from the sea or a river and protected by dikes) (Hoeksema, 2007).

10.3.7 Beach Restoration

Beach rebuilding is the process of repairing beaches using materials such as sand or mud from inland or offshore. This can be used to build up beaches suffering from beach starvation or erosion from longshore drift (Nordstrom, 2000; Hamm and Stive, 2002). Although it is not a long-lasting solution, it is cheap compared to other types of coastal defenses. Beach nourishment has long been seen as a necessity for coastal protection but is also a form of extending living and recreational possibilities. These improve the quality of life for millions of people. For instance, Australia's coastline and sandy beaches are an essential recreational and tourist resource. Currumbin-Tugin Beach on the Gold Coast of Australia was severely eroded before reclamation took place. The same applies to Spain's Mediterranean and Atlantic coasts and many other coastal areas. The east and west coasts of the United States, Netherlands, and Belgium are also replenished annually.

10.3.8 Acid Sulfate Soils

In many coastal areas, acid sulfate soils have long been recognized as a problem for landholders and the environment. The soils, which are rich in iron sulfides, are benign while covered with water, but when they dry out, oxygen combines with the sulfide to produce sulfuric acid that acidifies the soil. After rain, the acid washes into waterways causing acidification and reductions in oxygen. The acid can dissolve metals, such as aluminum, and if discharged to rivers and estuaries, the combination of metals and acidity can kill plants and animals. The polluting effects of acid sulfate soils were realized when fish kills due to hypoxia, dissolved aluminum toxicity, and fish disease (eg, red spot ulceration) were observed in estuarine waters (Sammut et al., 1995).

Geographically, the majority of acid sulfate soils occur in coastal areas, developing from recent or semi-recent sediments. They are usually restricted to areas relatively close to the sea, where they have formed marine and estuarine deposits. Acid-forming soils can be found at many coastal locations and are particularly prevalent in many parts of coastal Australia. The cost of managing acid sulfate soils, including the replacement of damaged infrastructure, can be significant. In Queensland alone, the cost is approximately

$189 million per year not including direct losses to fisheries and agriculture (Ozcoasts, 2010).

Drainage of coastal land affects the water table and can trigger problems with acid sulfate soils. Acid sulfate soil disturbance is often associated with dredging, excavation, and dewatering activities during canal, housing, and marina developments. Droughts can also result in acid sulfate soil exposure and acidification. Preventative management actions include maintaining high water tables to prevent the soils drying out, either through filling in drains or holding water in drains. Another technique is to flush out the acidic drain water with tidal flows because alkaline seawater can neutralize the acid.

10.3.9 Fishing and Aquaculture

Worldwide demand for fish and fishery products is expected to increase in the coming years across all continents. Wild fish stocks are under great pressure. Based on global data collected in the year 2009, around a third of global fish stocks were found to be overexploited, depleted, or recovering from depletion, and over half were considered fully exploited (FAO, 2012). It is therefore highly unlikely that wild capture fisheries will be able to produce higher yields in future. Most of the growth to meet demand will need to come from aquaculture (Fig. 10.6).

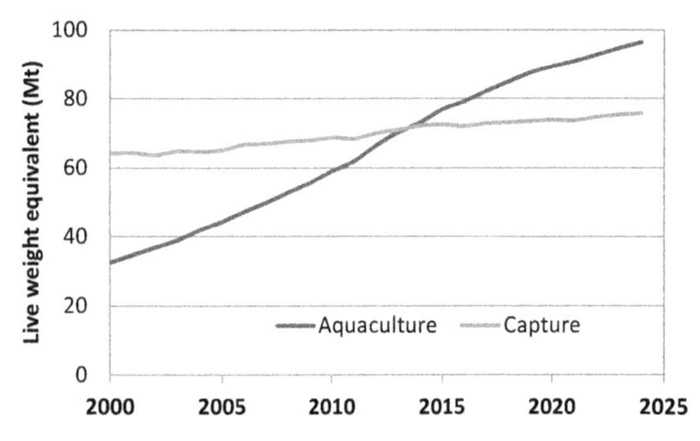

FIGURE 10.6 The growth of aquaculture versus conventional commercial fishing (ie, capture). *Adapted from World Ocean Review, 2013. Living with the Oceans. The Future of Fish – the Fisheries of the Future, vol. 2. Maribus, Hamburg, Germany.*

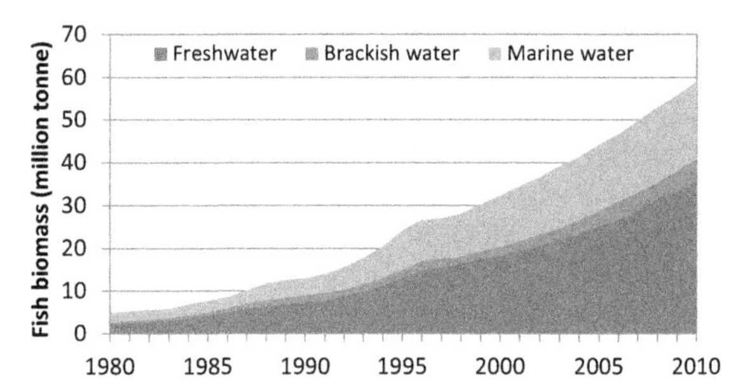

FIGURE 10.7 Growth of the global aquaculture industry since 1980. *Adapted from World Ocean Review, 2013. Living with the Oceans. The Future of Fish – the Fisheries of the Future, vol. 2. Maribus, Hamburg, Germany.*

Aquaculture already provides more than 40% of the global consumption of fish and shellfish. No other food production sector has grown as fast over the past 20 years (Fig. 10.7). Today about 60 million tons of fish, mussels, crab, and other aquatic organisms are farmed around the world each year. Asia, particularly China, is the most important aquaculture region, currently supplying 89% of global production. Coastal aquaculture has been growing in a number of industrialized and developing countries, particularly in parts of Southeast and East Asia and parts of Latin America, and there is considerable scope for development and expansion in other regions (FAO, 2012).

There are many constraints affecting aquaculture. These include the growing scarcity of suitable water, limited opportunities for sites for new operations along increasingly crowded, multiple-user coastal areas, limited carrying capacity of the environment for nutrients and pollution, and more stringent environmental regulations. Most of the future expansion in aquaculture production capacity will probably occur in the ocean, with some of it moving increasingly offshore to escape the constraints of coastal waters.

The sustainable intensification and exploitation of aquaculture is a major challenge for global seafood security. Many fish species raised in the aquaculture sector are predatory fish, which rely on a supply of other fish for food. Although the amounts vary considerably according to species, it takes an average of around 5 kg of fish meal and fish oil to produce 1 kg of farmed fish. Nevertheless, an advantage of aquaculture is that much fewer feedstuffs are needed to farm fish and seafood than domestic animals such as beef cattle and pigs. It takes 15 times as much feed to produce 1 kg of beef as to produce 1 kg of fish (World Ocean Review, 2010). Nevertheless, it is problematic that aquaculture still requires large amounts of wild fish, which is processed into fishmeal and fish oil and used as feed. Aquaculture can thus still be a contributor to the problem of overfishing.

There are a number of environmental issues that relate to the treatment and impacts of wastes and impacts of biocides and veterinary chemicals, which are used widely in the industry. Food, fecal, and metabolic wastes from intensive fish farms can lead to the eutrophication of water. Many fish farms are more environmentally friendly than, for instance, the intensive farming of pigs or cattle. The latter emit large quantities of nitrogen and phosphorus from the slurry and manure used to fertilize the land. Aquaculture produces far lower emissions of nitrogen and phosphorus and can be compared with those from the farming of poultry (World Ocean Review, 2013).

Fish farmed in intensive systems to provide maximum yields are more susceptible to disease than their relatives in the wild. For this reason antibiotics and other drugs are widely used, especially in South East Asia. Already there are indications that these are no longer effective. The antibiotics used in aquaculture have, in recent years, led to the spread of multiresistant pathogens, against which most established antibiotics are ineffective. Another concern is disease transmission from cultured fish to wild populations. Aquaculture in coastal waters has resulted in major disease outbreaks that have affected the ecology of native species (State of Environment, 2011).

To meet future needs, sustainable intensification of aquaculture will be required. Sustainable intensification in terrestrial systems has been defined as a form of production wherein "yields are increased without adverse environmental impact and without the cultivation of more land" (Garnett and Godfray, 2012). It is a response to the challenges of increasing demand for food from a growing global population, in a world where land, water, energy, and other inputs are in limited supply and used unsustainably.

10.3.10 Mining and Mine Waste Disposal

10.3.10.1 Seabed Mining

Seabed mining is in its infancy. However, some successful mining has already occurred in relatively shallow waters. In the 1960s, the Marine Diamond Corporation recovered over 1 million carats from the coast of Namibia in waters of less than 20 m depth. Today, the world's leading diamond company, De Beers obtains a significant portion of its total diamond production from the continental shelf of southern Africa, in water shallower than 300 m.

Interest in deep-sea mining for minerals, and especially metals, has increased in recent years. Commercial interest is particularly strong in manganese nodules and seafloor massive sulfides (SMS). SMS are base-metal sulfur-rich mineral deposits that precipitate from the hydrothermal fluids as these interact with the cooler ambient sea water at hydrothermal vent sites. The sulfide deposits that are created contain valuable metals such as silver, gold, copper, manganese, cobalt, and zinc. Massive sulfides and sulfide muds form in areas of volcanic activity that occur near plate boundaries, at depths of 500–4000 m (World Ocean Review, 2010).

Cobalt crusts are found along the edges of undersea mountain ranges (between 1000 and 3000 m). Manganese nodules are usually located at depths below 4000 m and are composed primarily of manganese and iron and elements of economic interest, including cobalt, copper, and nickel and make up a total of around 3% by weight (Margolis and Burns, 1976). In addition, there are traces of other significant elements such as platinum or tellurium. Covering large areas of the deep sea with masses of up to 75 kg/m^2, manganese nodules range in size from a potato to a soccer ball. The greatest densities of nodules occur off the west coast of Mexico, in the Peru Basin, and the Indian Ocean. The actual mining process does not present any major technological problems because the nodules can be collected fairly easily from the surface of the seafloor.

Deposits can be mined using either hydraulic pumps or bucket systems that take ore to the surface to be processed. A deep-sea mining operation might consist of: a mining support platform or vessel; a launch and recovery system; a crawler with a mining head, centrifugal pump, and vertical transport system; and electrical, control, instrumentation, and visualization systems. The mining industry has been developing specialized dredgers, pumps, crawlers, drills, platforms, cutters, and corers, many of them robotic designed to work in the harsh conditions of the deep ocean. Submarine vehicles are being developed that can operate down to 5000 m depth.

The international regulations on deep-sea mining are contained in the United Nations Conventions on the Law of the Sea, which came into force in 1994 (Glasby, 2000). The convention set up the International Seabed Authority (ISA), which regulates deep-sea mining ventures outside each nation's exclusive economic zone (a 200-nautical-mile area surrounding coastal nations). To date, the ISA has entered into 17 15-year contracts for exploration for polymetallic nodules and polymetallic sulfides in the deep seabed with 13 contractors. Eleven of these contracts are for exploration for polymetallic nodules in the Clarion Clipperton Fracture Zone in the Pacific, with two contracts for exploration for polymetallic sulfides in the Southwest Indian Ridge and the Mid Atlantic Ridge.

Nautilus Minerals Inc. was granted the first mining lease for polymetallic SMS deposits at the prospect known as Solwara 1, in the territorial waters of Papua New Guinea. Located in the Bismarck Sea, the Solwara 1 project will be the world's first deep-seabed mining project where it is aiming to extract copper, gold, and silver (Batker and Schmidt, 2015). The Solwara 1 site is a small area of 11 ha at a depth of 1600 m, with a total maximum predicted disturbance of 14 ha. It is located 30 km off the shore of Papua New Guinea near an area known as the "Coral Triangle." This area occupies approximately 2% of the Earth's seafloor but contains 76% of the world's corals and 37% of the world's coral fish population.

The potential impacts of deep-sea mining are likely to be associated with disturbance and suspension of sediments. Turbid plumes may be caused when the tailings from mining (usually fine particles) are deposited back into the ocean. The plumes could impact zooplankton and light penetration, in turn affecting the food web of the area (Ahnert and Borowski, 2000; Nath and Sharma, 2000). Removal of parts of the seafloor will result in significant disturbances to the habitats of benthic organisms (Ahnert and Borowski, 2000). In some cases, waste will represent most (90%) of the volume of materials pumped to the surface, and thus, seabed operations will deposit massive amounts of waste on the seafloor. Deep-sea communities are very poorly characterized and mapped and it is not yet known how, or even whether, recovery of the excavated areas would occur. Research is needed to understand ecological restoration and recovery following the impacts of deep-sea mining.

10.3.10.2 Methane Hydrates

Methane hydrates are white, icelike solids that consist of methane and water. They are an untapped potential future energy source. The methane molecules are enclosed in microscopic cages composed of water molecules. Methane gas is primarily formed by microorganisms that live in the deep sediment layers and slowly convert organic substances to methane. Methane hydrates are only stable under pressures in excess of 35 bar and at the low temperatures of the bottom waters of the oceans and the deep seabed, which almost uniformly range from 0 to 4°C. Below a water depth of about 350 m, the pressure is sufficient to stabilize the hydrates. Methane hydrates therefore occur mainly near the continental margins at water depths between 350 and 5000 m. At the bottom of the expansive ocean basins, scarcely any hydrates are found because there is insufficient organic matter embedded in the deep-sea sediments.

10.3.10.3 Mine Waste Disposal

About 2500 industrial-sized mines are operating around the world (Vogt, 2012). Almost all of them dispose of their tailings on land, usually in tailings impoundments (also known as tailings dams). However, in some locations, land-based disposal may not be the most technically feasible option. For instance, in Indonesia and Papua New Guinea, the challenges of mountainous terrain, high rainfall, and earthquakes combine to make the development of effective tailings impoundment very difficult; in countries such as Norway, lack of available land is a problem (Vogt, 2012).

An alternative waste management strategy, suitable for mines that are relatively close to coastal locations or have access by pipeline, is marine disposal. Early operations were largely unplanned and tailings plus other waste materials were directly discharged into the sea. Over time, design modifications to both the tailings outfalls and the final tailings deposition basins have located them at progressively greater depths (Ellis and Ellis, 1994; Ellis et al., 1995). Currently, there are at least 15 mining and mineral processing operations around the world using engineered submarine tailings disposal in the marine environment (Vogt, 2012).

Marine tailings disposal generally refers to tailings disposal in the shallow marine environment (surface discharge); submarine tailings disposal or placement is where the deposited tailings are intended to settle at depths of 100–1000 m; deep-sea tailings placement (DSTP), a more recent practice, is where tailings are intended to settle at depths greater than 1000 m. A conceptual diagram showing the operational components and potential environmental impacts of DSTP is given in Fig. 10.8. Tailings are discharged from a submerged pipeline at depth to avoid tailings resuspension, particularly when this affects the biologically productive euphotic zone.

The fundamental premise of DSTP is that tailings can be discharged at a depth below the euphotic zone in the form of a stable plume that descends to the ocean floor. The tailings solids eventually deposit on the ocean floor as a footprint with the local ocean floor topography determining the shape of the deposit. DSTP requires that pipeline discharge of tailings occurs at depths greater than the maximum depth of the surface mixed layer, euphotic zone, and the upwelling zone. Placing tailings below these zones maximizes their stable deposition on the seafloor. Both the tailings liquids and solids behave as stressors to the marine environment, potentially affecting pelagic and benthic organisms. The processes that could lead to contamination of the productive surface waters of the euphotic zone from DSTP are depicted in Fig. 10.8. Strong currents on the seafloor may remobilize deposited tailings which, in the event of upwelling, may be brought to the surface and made available to the biota in the euphotic zone. Similarly, some organisms (eg, zooplankton and micronektonic fish) migrate up and down the

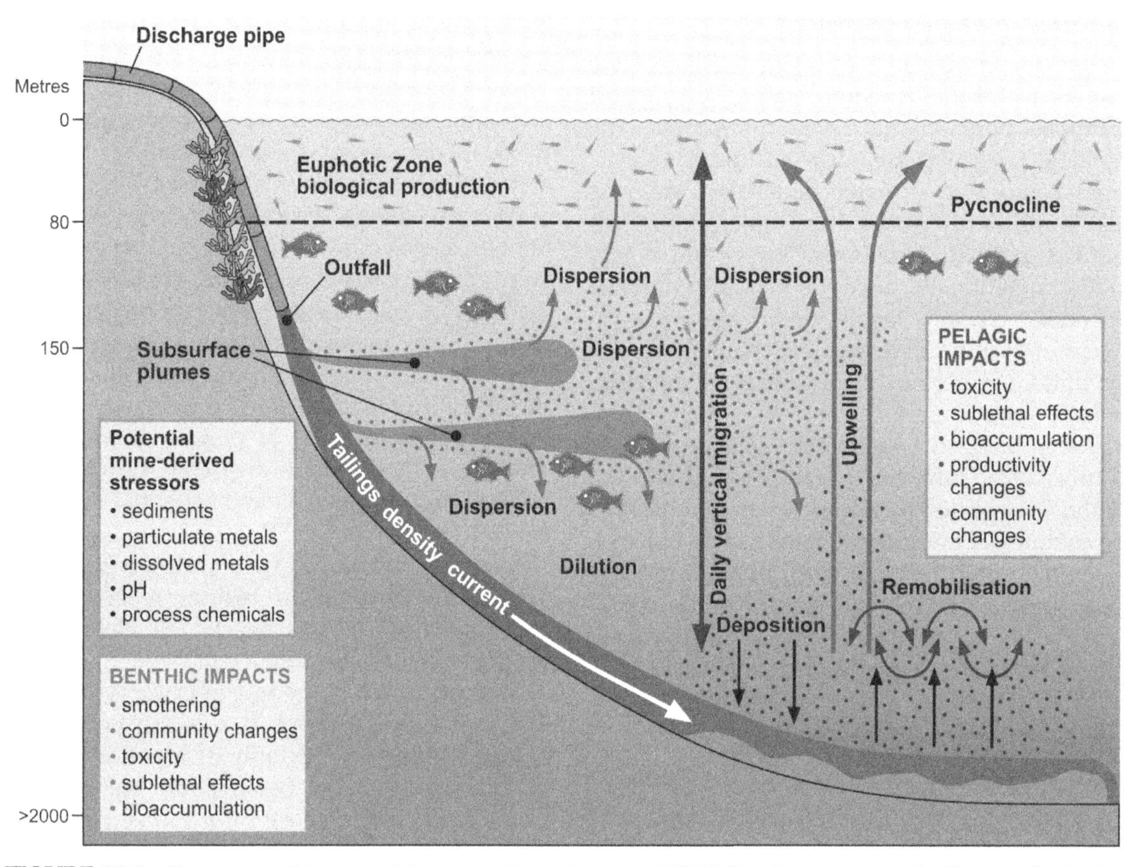

FIGURE 10.8 Environmental impacts of deep-sea tailings placement (DSTP). Depths are not to scale. Biomagnification is a rare phenomenon, restricted to a few organic chemicals and defined as increasing contaminant concentrations from food alone up three or more trophic levels. *CSIRO.*

water column daily and may act as carriers of potentially toxic contaminants to their predators in the surface layer. Suitable sites are restricted to some oceanic islands and archipelagos where very deep water occurs close to shore. Submarine canyons and naturally excised channels beyond fringing coral reefs are regarded as being particularly suitable sites.

10.4 CLIMATE CHANGE

Climate change is an increasingly urgent problem with potentially far-reaching consequences for life on earth. Climate change refers to the long-term trends in climate over many decades and differs from climate variability, which refers to year-to-year variations (Mapstone, 2011). In its 2014 annual report on global risks, the World Economic Forum (WEF, 2014) ranked climate change as fifth among the top 31 global risks, where global risk is defined as *"an occurrence that causes significant negative impacts for several countries and industries over a time frame of up to 10 years."* Potential environmental risks associated with climate change include increased temperature, changing rainfall patterns, sea-level rise, ocean acidification, increasing numbers of extreme weather events such as floods and droughts, and increasing numbers and magnitudes of natural catastrophes such as cyclones, fires, and landslides, with resulting major shifts in the biogeographical distribution of species (including pests) and major losses of biodiversity.

The Intergovernmental Panel on Climate Change in its 5th Assessment Report (IPCC, 2014) concluded that warming of the climate system is unequivocal, and many of the observed changes are unprecedented. The globally averaged, combined land and ocean surface temperature has increased by nearly 1°C since 1880. In the Northern Hemisphere, the period from 1983 to 2012 was the warmest 30-year period of the last 1400 years. Australia's mean temperature has warmed 0.9°C since 1910, with the frequency of very warm months increasing fivefold in the last 15 years, compared to the period 1951–80 (State of the Climate, 2014).

It is well recognized that the rates and trajectory of climate change and its effects will be highly variable at regional and local scales. As well as affecting the inputs, transport, and fate of contaminants, climate change may also affect the structure and functioning of ecosystems and their vulnerability to chemical contaminants and other stressors.

In the following sections, individual climate change stressors relevant to marine ecosystems are first described, and then the likely interactions between these stressors and chemical contaminants is explored. Assessing the impact of these multiple stressors necessitates changes to ecological risk assessment approaches and adaptive management.

10.4.1 Environmental Variables Affected by Climate Change

10.4.1.1 Temperature

Model projections of climate change suggest that global mean surface air temperatures will increase by between 1 and 5°C by 2100, compared to the reference period 1986–2005, but that this will be very dependent on location (IPCC, 2014). Best estimates of ocean warming in the top 100 m of the water column are about 0.6–2.0°C during this same time period, with the biggest increase likely in tropical and Northern Hemisphere subtropical regions. At greater depths, the warming will be most pronounced in the Southern Ocean (IPCC, 2014).

Most marine species are ectotherms, making temperature an important variable controlling physiological processes. Many species have adapted to cope with daily and seasonal temperature fluctuations. However, for a combination of stressors, the resilience of ecosystems may be exceeded. While genetic adaptation to temperature stress may allow populations to persist under relatively strong selection pressures, a drawback is that this usually also causes a reduction in genetic diversity, so adapting to one set of environmental stressors has a fitness cost and will likely increase susceptibility to other stressors (Moe et al., 2013).

Tropical species generally have less tolerance to temperature variation than their temperate counterparts, and symbiosis in tropical species may be less stable during thermal fluctuations than in temperate species (Przeslawski et al., 2008). The sensitivity of corals and their dinoflagellate endosymbionts to rising ocean temperatures has been well documented (Hoegh-Guldberg, 1999). A 1°C warming has been predicted to bleach corals of the Great Barrier Reef in Australia by 65% (Hennessy, 2011). When temperatures exceed summer maxima of a few degrees for extended periods (3–4 weeks), the zooxanthellae are expelled leading to coral bleaching and potentially coral mortality. While corals may recover after mild thermal stress, they typically show reduced growth, calcification, and fecundity and may experience greater disease (Hoegh-Guldberg et al., 2007).

Water temperature also affects the timing of reproduction for many marine invertebrates. Rising sea temperatures may cause changes in temporal patterns of spawning in temperate regions (Przeslawski et al., 2008). Many invertebrates (echinoderms, molluscs, corals, and polychaetes) spawn en masse annually in response to lunar cues and annual temperature ranges. Thermal change is likely to have direct or indirect impacts on these spawning events (Przeslawski et al., 2008). For example, if spawning does not coincide with phytoplankton availability as a food source, larval survival and settlement may be inhibited. In addition to affecting the timing of reproduction, warming temperatures may also decrease or increase fecundity of tropical invertebrates such as sponges (Ettinger-Epstein et al., 2007).

10.4.1.2 Global Water Cycles and Salinity

The effect of climate change on global hydrological cycles and the consequences for marine and estuarine systems, eg, changes in salinity, are predicted to be large in coastal areas. On a global scale, a warming climate leads to increased evaporation, increased water vapor in the atmosphere, and increased precipitation; however, this is predicted to be highly variable regionally. Recent detection of increasing trends in extreme precipitation and discharge in some catchments implies greater risks of flooding and storm surges in coastal environments.

In some areas, such as southeastern Australia, there has been decreased coastal rainfall, decreased cloud cover, increased evaporation, and decreased water levels leading to a net moisture deficit, eg, the decade-long drought in the Murray River and Estuary. Prolonged drought due to the southward movement of the subtropical ridge atmospheric pressure system as a direct result of climate change, compounded by overallocation of water upstream, resulted in very low water flows from 2005 to 2010 (Stauber et al., 2008). Consequently, the Murray River and Estuary and adjacent wetlands were seriously impacted by a combination of low water levels and the presence of sulfuric materials in acid sulfate soils and sediments.

Observations of changes in ocean surface salinity also provide indirect evidence for changes in the global water cycle over the ocean. It is very likely that regions of high salinity, where evaporation dominates, have become more saline, while regions of low salinity, where precipitation dominates, have become more fresh since the 1950s (IPCC, 2014).

Echinoderms, in particular sea urchins, have very limited tolerance to decreases in salinity and many cases of mortality of adults caused by freshwater runoff have been reported (Lawrence, 1996). Species with larvae that are intolerant to freshwater are particularly vulnerable because a whole season's recruitment can be lost. Timing of reduced salinity events is critical to predict population response.

10.4.1.3 Hypoxia

Hypoxia, ie, low dissolved oxygen, is another climate change stressor that may increasingly impact vulnerable species. Hypoxia has been reported in large coastal areas such as the Baltic Sea, the Gulf of Mexico, and Chesapeake Bay, and this is expected to worsen as a result of climate change (Hooper et al., 2013). Elevated water temperatures reduce oxygen availability which, in combination with increased precipitation, can bring nutrient-rich warm waters to sensitive areas, leading to eutrophication and increased organic matter loads. Hypoxia can decrease an organism's ability to detoxify contaminants such as PAHs and dioxins, disrupting endocrine systems and reproduction (Wu, 2002). Exposure to these chemical classes may also hinder the ability of species to respond to increased hypoxia under climate change.

Of the invertebrates, crustaceans are thought to be most susceptible to hypoxia and organic matter loading (Gray et al., 2002). For example, there has been a large decline in populations of the amphipod *Monoporeia affinis* in the Baltic Sea over the last 40 years, with abundances of less than 10% now compared to the 1970s (Eriksson-Wiklund and Sundelin, 2004). *Monoporeia affinis* is extremely sensitive to oxygen deficiency and temperature during oogenesis, which occurs in August to November in the Baltic Sea, at a time when oxygen is low and surface temperatures are at a maximum. Thus this species may be trapped in a dwindling habitat between oxygen-depleted deeper waters and too warm shallow waters. In its place, more tolerant species such as polychetes are invading, and as they are good bioturbators in sediments, this may lead to more contaminant resuspension and flow-on effects to other sensitive species (Hedman, 2006).

10.4.1.4 Sea-Level Rise and Storm Surges

Increased sea-surface temperatures and melting of Arctic and Antarctic ice sheets have caused global mean sea-level rises of 0.19 m (0.17–0.21 m) over the period 1901 to 2010. The rate of sea-level rise since the mid-19th century has been larger than the mean rate during the previous two millennia (IPCC, 2014). In the Arctic, sea-ice extent has decreased in every season and in every successive decade since 1979 at a rate of about 3.5–4.1% per decade. There are strong regional differences in Antarctica, with ice extent increasing in some regions and decreasing in others (IPCC, 2014).

Global mean sea level is projected to rise between 0.26 and 0.82 m by 2100, relative to 1986–2005 (IPCC, 2014), with large local differences (resulting from tides, wind and atmospheric pressure patterns, changes in ocean circulation, vertical movements of continents) in the relative sea-level rises (IPCC, 2014). The impacts of sea-level rise are therefore expected to be localized. Under the worst scenario, the majority of the people who would be affected live in China (72 million), Bangladesh (13 million people and loss of 16% of national rice production), and Egypt (6 million people and 12–15% of agricultural land lost) (Nicholls and Leatherman, 1995). Low-lying Pacific coral atolls such as Kiribati and the Marshall Islands are under severe threat. Even more significant than the direct loss of land caused by the sea rising are associated indirect factors, including erosion patterns and damage to coastal infrastructure, salinization of wells, suboptimal functioning of the sewage systems of coastal cities (with resulting health impacts), loss of littoral ecosystems, and loss of biotic resources.

If the rate of sea-level rise is high or if man-made barriers such as breakwaters restrict expansion of mangrove communities, invertebrate community composition may be substantially altered due to loss of intertidal habitats. This will depend on tidal ranges and local geomorphology. Sea-level changes may also alter larval dispersal patterns (Przeslawski et al., 2008).

Storm events, predicted to increase in frequency and severity due to climate change, can cause major physical disturbance to marine habitats. Fabricius et al. (2008) found inshore reefs were more vulnerable to storm damage than offshore reefs, particularly when pollution from coastal runoff was found. Sessile invertebrates that are detached from their substrate by wave action are particularly at risk if there is a poor supply of larval recruits or lack of sufficient intact settlement substrate.

10.4.1.5 Ocean Acidification

Since the beginning of the industrial era, oceanic uptake of CO_2 has resulted in acidification of the ocean. The pH of ocean surface water has decreased by 0.1 pH unit, corresponding to a 26% increase in acidity, measured as hydrogen ion concentration (IPCC, 2014). A global decrease in ocean pH of between 0.06 and 0.32 is predicted, depending on the modeled scenario, by the end of the 21st century and reduction in carbonate saturation levels below those required to sustain coral reef accretion by 2050 (IPCC, 2014).

Approximately 25% of global anthropogenic CO_2 emissions enter the oceans and react with water to produce carbonic acid, which dissociates to bicarbonate and H^+ (Fig. 10.9). The proton then combines with more carbonate ions to

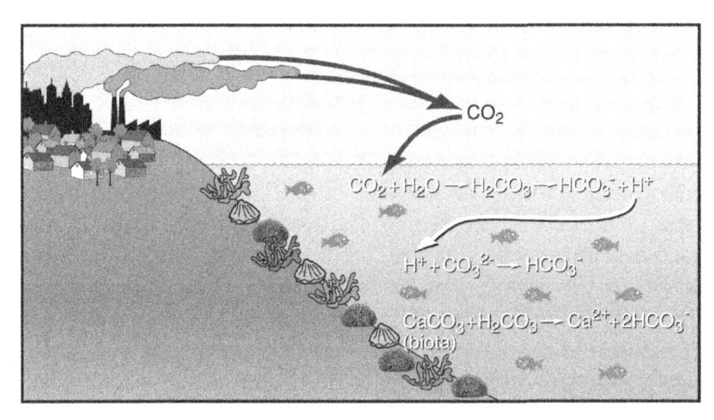

FIGURE 10.9 Ocean acidification process. *Modified from Hoegh-Guldberg, O., Mumby, P.J., Hooten, A.J., Steneck, R.S., Greenfiled, P., Gomez, E., Harvell, C.D., Sale, P.F., Edwards, A.J., Caldeira, K., Knowlton, N., Eakin, C.M., Iglesias-Prieto, R., Muthiga, N., Bradbury, R.H., Dubi, A., Hatziolos, M.E., 2007. Coral reefs under rapid climate change and ocean acidification. Science 318, 1737–1742.*

produce more bicarbonate, which further reduces carbonate concentrations making it unavailable for marine biota that form calcium carbonate shells such as corals, oysters, sea urchins, and foraminifera (Fig. 10.9). The carbonic acid also reacts with calcium carbonate in the shells, resulting in shell dissolution. Weakening shell formation will compromise survivorship of both planktonic and benthic life stages of coral reef invertebrates by providing less protection from predators, physical damage, and desiccation (Przeslawski et al., 2008). Increased carbon dioxide in surface waters may also lower the metabolic rate of invertebrates due to acidosis, which could then impact their feeding, growth, and reproduction. However, the effects of gradually increasing ocean acidification on invertebrates is poorly understood as most studies are laboratory based and have used large CO_2 changes (Przeslawski et al., 2008).

Changes in pH also affect biogeochemical processes such as alteration of metal speciation, which can have substantial biological effects. Ocean acidification may also affect ion and nutrient assimilation of algae either directly by altering ion channels, or indirectly by changes in nutrient availability (Li et al., 2013). These authors suggested that the combined effects of ocean acidification and nitrogen limitation could act synergistically to affect marine diatoms and potentially marine food webs.

10.4.2 Multiple Stressors: Interactions of Climate Change and Contaminants

While global climate change is increasingly accepted within scientific and regulatory communities and by the informed public, to date the debate has not included the role of contaminants as additional stressors in the environment.

Noyes et al. (2009) published one of the first descriptions of the potential interactions between climate change and chemical contaminants. Environmental variables altered by climate change can affect the environmental fate and behavior of contaminants as well as the toxicokinetics of chemical adsorption, distribution and metabolism, and toxicodynamic interactions between chemicals and target molecules and receptors in biota.

Multiple stressors can interact in two ways: climate change stressors can increase or decrease the toxicity of contaminants to biota; or the contaminants themselves can alter the ability of organisms to respond to climate change stressors. This can ultimately lead to ecological thresholds or tipping points, ie, abrupt changes in community structure or function in response to small perturbations. Populations living at the edge of their physiological tolerance range may be more vulnerable to the multiple stressors of increased temperature, decreased food supply, and contaminant exposure (Heugens et al., 2001).

Estimating the effects of multiple stressors on biota is complex as effects may be direct (such as decreased reproduction), indirect (such as altered predator–prey relationships), or induced (associated with physical or ecological changes not directly attributable to a chemical stressor). There is increasing evidence that multiple stressors will affect survival, growth, reproduction, metabolism, behavior, and recruitment of biota, especially early life stages. Water temperature may affect the timing of reproduction such as spawning time and planktonic duration. Biota will show species-specific responses to these stressors, and to predict effects, both the magnitude and duration of exposure to each stressor will be important (Przeslawski et al., 2008). Timing of the exposure is also important; if the stressor such as a pulse event coincides with a sensitive life stage such as spawning or maturation, then it may have a greater effect.

The interactions between climate change stressors, contaminants, and ecosystems are shown in Fig. 10.10. The consequences of climate change such as increased temperature, changes in rainfall patterns, ocean acidification, hypoxia, and increased extreme events can directly impact contaminant processes such as transport, transformation, bioaccumulation, and ultimately their bioavailability and toxicity to biota, ie, climate effects are co-stressors. Climate change can also have direct impacts on biota, eg, their thermal and salinity tolerances may be exceeded, leading to a change in their biogeographical distribution and invasion by more tolerant nonnative species. Contaminants can also act directly or indirectly on biota, rendering the biota more

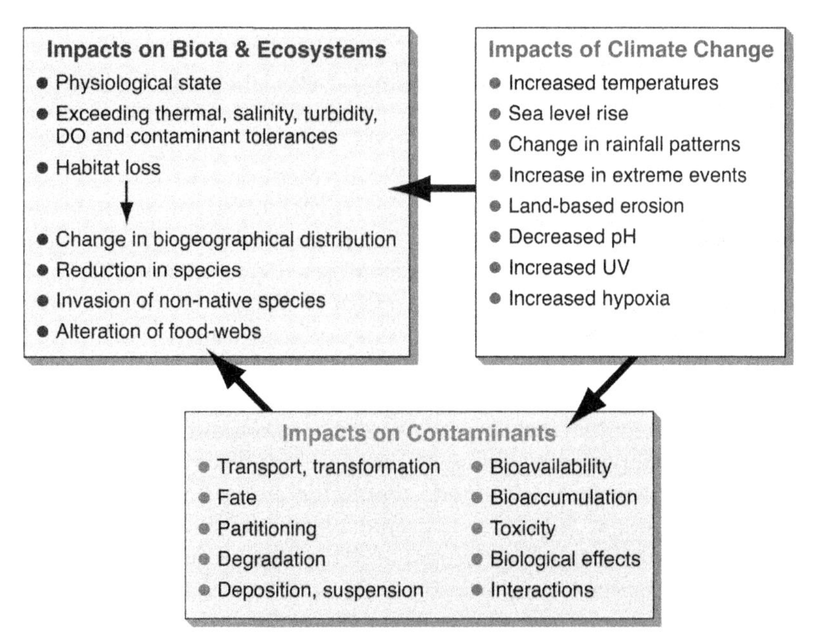

FIGURE 10.10 Impacts of climate change stressors and contaminants on ecosystems. *Modified from Schiedek, D., Sundelin, B., Readman, J.W., Macdonald, R.W., 2007. Interactions between climate change and contaminants. Mar. Pollut. Bull. 54, 1845–1856.*

sensitive to climate change impacts, and it is these effects which have been poorly studied to date.

Since it is impractical to collect empirical data for all potential interactions between environmental variables affected by climate change and contaminants of concern, it will be necessary to develop predictive approaches, incorporating mechanistic data, into the risk assessment process (Hooper et al., 2013). An additional challenge is that contaminants rarely occur alone, so that there is the unknown impact of multiple contaminants which may be synergistic or antagonistic. For simplicity, it is generally assumed that toxic effects are additive; however, climate co-stressors might affect different contaminants differently.

10.4.2.1 *Effects on Contaminant Fate, Transport, and Bioaccumulation*

Climate change, along with related land-use changes, will likely impact chemical usage patterns in the future. Increases in pests and disease vectors will increase the frequency and timing of use of pesticides, biocides, and pharmaceuticals, with potential increased discharge into coastal marine environments. Climate change will have an effect on the environmental fate and behavior of contaminants by altering physical, chemical, and biological drivers of partitioning between atmosphere, water, soil, and biota, including reaction rates. For example, increasing temperatures and subsequent melting of snow and ice can remobilize POPs into water and the atmosphere, with long range transport to higher latitudes thousands of kilometers from their source (Noyes et al., 2009). For legacy chemicals such as mercury in sediments, increased temperatures may accelerate mercury methylation and volatilization leading to remobilization and increased emissions (Bogdal and Scheringer, 2011).

Pulse releases of contaminants, together with diffuse sources, will be particularly important in extreme events such as cyclones and floods, frequencies of which are predicted to increase with climate change. Tropical marine and estuarine ecosystems may be more at risk of contaminant inputs from surface runoff and seasonal high rainfall events than those in temperate regions. However, tropical systems may also recover more quickly from disturbance.

Climate-related changes to food webs can also alter contaminant bioaccumulation and mobilization. Bioaccumulated contaminants stored in fat reserves, eg, in whales and polar bears, may be released in periods of stress. For example, when ice flows in the Arctic melt, seals, the food supply for polar bears, become scarce and bears have to use their stored fats as energy reserves to prevent starvation. This can lead in turn to release of POPs from energy stores; the combined effects of POPs and starvation have been shown to affect thyroid function and consequently behavioral and cognitive function, reducing bear hunting ability (Stirling et al., 1999). Thus small changes in food webs have potential consequences higher up the food chain.

Tools available to investigate climate change effects on contaminant bioaccumulation largely derive from the combined application of bioaccumulation models (that describe the uptake and distribution of contaminants in biotic systems) with temperature-dependent chemical fate and bioenergetics models (Gouin et al., 2013). Most data to date have focused on chemical distribution in a single region, with few data on indirect effects such as food web changes. Gouin et al. (2013) showed that the effect of climate change on long-term chemical fate and transport was estimated to be relatively small, with predicted changes less than a factor of 2 different from baseline for a limited number of organic chemicals. The direct climate change-induced effects on bioaccumulation potential in an aquatic food chain varied substantially depending on partitioning properties and biotransformation rate constants. Neutral organic chemicals, eg, POPs, usually considered problematic with respect to bioaccumulation, were also within a factor of 2 of the baseline scenario (Goin et al., 2013). However, uncertainties relating to the physicochemical properties of chemicals may lead to greater variability in exposure than predicted from these fate and bioaccumulation models. Changes in chemical usage and emissions were predicted to have a greater impact on contaminant exposure than climate effects on contaminant partitioning and bioaccumulation. It was concluded that both modeling and monitoring data were required, particularly in the Southern Hemisphere, to enable better prediction of climate change on contaminant exposure.

10.4.2.2 Effects on Contaminant Toxicity

Temperature: Climate change parameters, such as increases in sea-surface temperature, not only alter the environmental distribution of contaminants but also their toxicity. Generally higher temperatures lead to higher contaminant toxicity, but this is very species- and contaminant-specific. Higher water temperatures can increase the uptake and bioaccumulation of contaminants, eg, increased gill ventilation rates and increased metabolism can lead to high tissue concentrations of contaminants. Depuration and detoxification of contaminants can also increase with increasing temperature, as was found for the estuarine fish *Fundulus heterclitus*, which eliminated toxaphene congeners at double the rate at 25°C compared to 15°C (Maruya et al., 2005). However, this often comes at an energetic cost to the organism. Higher temperatures can also increase bioaccumulation of metals in marine biota, such as reported for crustaceans, echinoderms, and mollusks (Marques et al., 2010). Higher temperatures can also increase biotransformation to more toxic compounds, eg, PCBs to toxicologically active hydroxylated PCB metabolites. Unfortunately, many of the studies investigating the effect of temperature on contaminant toxicity have used very high contaminant concentrations and large temperature changes that are unlikely to occur in natural systems. Rather, biota may adapt to the gradual temperature changes expected due to global warming, such that the real issue is not temperature increase per se but how this interacts with other stressors, especially contaminants.

Negri (unpublished data) found that the percentage metamorphosis in the larval coral *Acropora millepora* was inhibited by both copper and temperature. As the temperature was increased from 28 to 34°C in 1° increments in the presence of copper (0.5–75 μg/L), metamorphosis was reduced, with EC50 values (ie, the effective concentration to inhibit metamorphosis by 50%) decreasing from ~35 μg Cu/L at 28°C to ~10 μg Cu/L at 32°C. At 34°C, metamorphosis was completely inhibited even in the absence of copper.

Exposure to contaminants can also alter an organism's thermal tolerance, usually measured as the critical thermal maximum (CTM). The CTM is the measure of the upper limit of thermal tolerance of aquatic animals, which may be modified as a result of exposure to toxicants. Ectotherms such as fish, reptiles, and amphibians may be particularly vulnerable as they are unable to regulate their own body temperatures. Tropical species, such as corals, which are already living close to their upper thermal tolerance, may be even more vulnerable than temperate species (Heugens et al., 2001). The physiological condition of an organism may also be modified by changing temperature, eg, induction of heat-stress proteins, which may influence the organism's sensitivity to toxicants.

Adams et al. (unpublished) investigated the combined acute effects of temperature and copper on immobilization of the tropical copepod *Acartia sinjiensis* (Fig. 10.11). They found that copper toxicity increased with increasing temperature from 30 to 34°C, with 48-h EC50 values decreasing from 44 to 10 μg Cu/L. They also investigated the effect of copper on the thermal tolerance of the copepod, which was found to decrease several degrees with increasing copper concentrations over the range 0–38 μg Cu/L (Fig. 10.11).

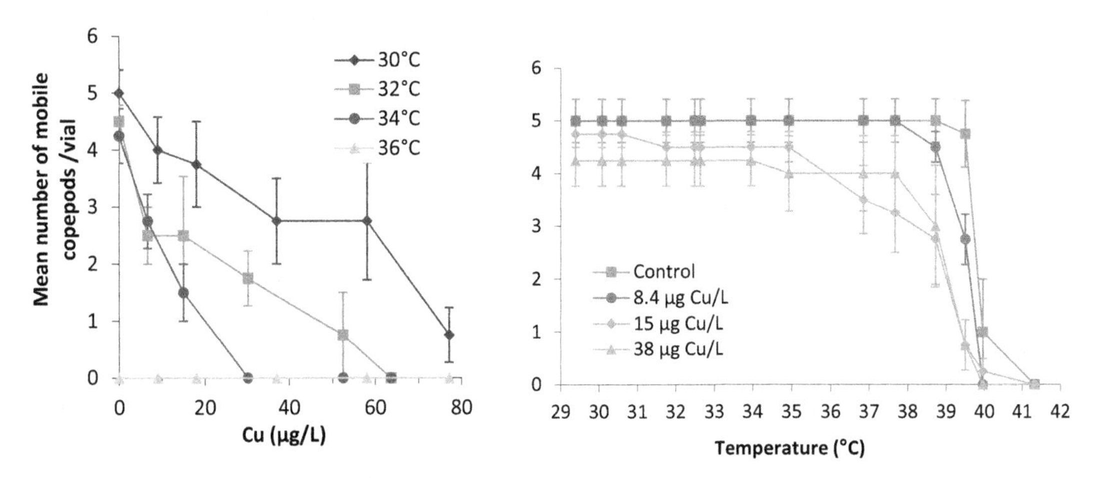

FIGURE 10.11 Combined effects of copper and temperature on survival of the marine copepod *Acartia sinjiensis* over 48 h. *Adams et al. (unpublished).*

Salinity: In estuaries, organisms are exposed to contaminants as well as fluctuating salinity. Interactions between salinity and contaminants are complex because salinity can influence both the chemical speciation of the contaminant as well as physiological processes and hence toxicity. For metals, increased salinity often decreases bioavailability and toxicity due to increasing metal complexation, but this depends on the metal's uptake route and mode of action. For example, Wildgust and Jones (1998) found that the estuarine mysid *Neomysis integer* was most tolerant to cadmium at an optimal salinity of 20‰, with greater mortalities at higher and lower salinities. Differences in cadmium toxicity were not completely explained by differences in cadmium speciation, with increases in osmotic stress contributing to reduced cadmium tolerance.

In contrast, the toxicity of many organic compounds, such as organophosphates, generally increases with increasing salinity due in part to decreased solubility, higher persistence, and increased bioaccumulation. The brine shrimp *Artemia* sp., when exposed to the organophosphate pesticide dimethoate, showed increased mortality at three to four times its iso-osmotic salinity (Song and Brown, 1998). Contaminants can also alter osmoregulation in biota, eg, fish larvae normally tolerant to a wide range of salinities showed increased mortality at high and low salinities when exposed to 5 μg/L atrazine (Fortin et al., 2008).

Reduced salinity from flood events, coupled with nutrients and contaminants such as pesticides has been shown to adversely impact corals, making them more susceptible to fungal infections, colonization by algae and barnacles, and causing increased mortality (Smith et al., 1996).

UV: Increasing UV radiation can enhance the toxicity of PAHs by up to 100 times in marine systems. Laboratory studies have shown that the toxicity of anthracene, pyrene, and fluoranthene to marine invertebrate larvae and embryos was significantly increased in the presence of UVA (320–400 nm) radiation, compared to embryos exposed to these PAHs alone, at concentrations previously thought to have no effect (Pelletier et al., 1997). The mechanism of toxicity is via the formation of reactive single oxygen species which interact with various macromolecules to cause cell damage and lethality in acute exposures (Hooper et al., 2013).

10.4.3 Implications for Environmental Risk Assessment

The complexity, uncertainty, and variability of climate drivers pose major challenges for the prediction of effects and implementation of environmental management programs. Interactions between contaminant impacts and climate change may occur on vastly different spatial and temporal scales. Changes in the types and quantities of chemicals used and released; their transport, fate, and accumulation in the environment; and their effects on biota all need to be considered in risk assessments that incorporate the effects of multiple stressors including climate change (Stauber et al., 2012).

Current ecological risk assessment (ERA) frameworks were developed to examine risks from particular stressors (usually chemical) acting on particular receptors within small geographic boundaries and largely ignored other noncontaminant stressors (physical and biological). Traditionally, ERAs evaluate whether there is a change in ecosystem services relative to a reference site or condition. Ecosystem services are the products of ecological functions or processes that directly or indirectly contribute to human well-being. In effect, they are the benefits of nature to households, communities, and economies (Costanza and Daly, 1992). They fall into four general categories: provisioning (food, water, energy), regulating (eg, flood control, erosion prevention), cultural (eg, recreational), and supporting (eg, nutrients, oxygen). Endpoints for assessing some components of ecosystem services and models at the regional scale have only recently been developed and have not been examined under conditions of multiple stressors. In addition, ERAs rarely include mechanistic aspects of biological effects of contaminants such as altered gene expression or histopathology (Hooper et al., 2013).

Because ecological conditions will change unpredictably with global climate change, simplistic assumptions of static conditions and unidirectional change will no longer apply. Global climate change brings with it the need to consider interactions between both contaminant and noncontaminant stressors, which may lead to either negative or positive impacts. Depending on the focus of the ERA, substantial effort is required at the problem formulation stage to ensure that all climate change stressor interactions, relevant to marine species or habitats of concern, are available for the assessment. Because there is considerable uncertainty associated with predicting risk associated with global climate change and with identifying appropriate management actions, an adaptive management approach will be essential (Landis et al., 2013).

Landis et al. (2013) recommend seven critical changes to ecological risk assessment to take account of climate change:

1. Consider whether climate change magnitude, rate, and scale will be an important factor in the particular ERA and whether the consequences will be long term.
2. Express assessment endpoints in terms of ecosystem services to provide a common understandable terminology between stakeholders.
3. Develop cause–effect conceptual models for climate change and the stressors of interest.
4. Consider that responses to climate change may be nonlinear, and both adverse and advantageous to different receptors.
5. Consider a regional and multiple stressor approach to ERA, eg, using the relative risk model.
6. Determine the major drivers of uncertainty spatially and temporally.
7. Plan for adaptive management to account for changing environmental conditions.

Multiple stressors may interact additively, synergistically, or antagonistically. From the prospective risk assessment perspective, contaminant effects will be of greater consequence under climate change in the case of synergistic interactions; this would require more stringent environmental quality standards for chemical contaminants. From an ecological restoration perspective, removing one stressor may result in greater benefit than expected in case of a synergistic interaction or less than expected in the case of antagonistic interactions. In a review of multiple stressor effects from climate change and contaminants, Moe et al. (2013) suggested that interactions, whether additive, synergistic, or antagonistic, vary with the specific stressor combination, the species, the trophic level, and the response level (population, community). Multiple stressors are therefore likely to give ecological surprises in real ecosystems, hindering our ability to predict impacts in ERAs.

Stressor responses can be nonlinear, with tipping points, so tolerance of biota (both physiological acclimation and genetic adaptation) also needs to be considered. The resilience of ecosystems is likely to be exceeded by a combination of stressors. Since organisms can adapt rapidly to stressors, understanding evolutionary responses to stressors is also an important component for ERAs. Adaption can occur rapidly and within a few generations of the exposure, thus within the timeline of ERAs (Kimberley and Salice, 2012). Responses may be tolerance at first, then acclimation, avoidance, and finally heritable variation (adaptive potential) and appearance of novel traits. However, there is evidence that when exposed to multiple co-occurring stressors, populations have decreased ability to adapt to novel environmental stressors with flow-on effects to population responses.

10.5 FUTURE RESEARCH NEEDS AND NEW TOOLS FOR ASSESSING IMPACTS IN MARINE ECOSYSTEMS

A challenge for ecotoxicology is to predict how joint effects of global change and contaminants at the individual level, such as decreased survival, reproduction, or growth, will manifest at the population level (eg, abundance) and community level (eg, biodiversity and food webs) (Stahl et al., 2013). More data at multiple levels of organization are required to understand and predict the effects of, eg, climate change: at the organism level—physiology, toxicity, and genetics; at the population level—reproduction dispersal and recruitment; at the community level—species interactions and habitat; and at the ecosystem level—global processes (Przeslawski et al., 2008). Both modeling and monitoring approaches will be required to address this data gap.

Our understanding of marine ecotoxicology from a multiple stressor perspective has benefited from the development and application of a range of new tools for assessing ecosystem health. Epigenetics, omics, and modeling approaches are just some of the new tools that can assist in assessing responses to global change and these are discussed in the following sections.

10.5.1 Epigenetics

Epigenetics is an emerging field that is rapidly being incorporated into ecotoxicological studies. It investigates the alterations in gene function or cell phenotype, without changes in DNA sequences, that may result from methylation or histone modification (Connon et al., 2012). Environmental exposure to, eg, metals, POPs, or endocrine-disrupting chemicals has been shown to modulate epigenetic markers in environmentally relevant species such as fish or water fleas (Vandegehuchte and Janssen, 2011). Epigenetic changes can in some cases be transferred to subsequent generations, even when these generations are no longer exposed to the external factor which induced the epigenetic change (Vandegehuchte and Janssen, 2011).

Currently, epigenetic mechanisms are not considered in chemical risk assessment or used in the monitoring of the exposure and effects of chemicals and environmental change. Epigenetic profiling of organisms could identify classes of chemical contaminants to which they have been exposed throughout their lifetime (Mirbahai and Chipman, 2014). The potential implications of epigenetics in an ecotoxicological context, which require further investigation, are the possibility of trans-generationally inherited, chemical stress-induced epigenetic changes and epigenetically induced adaptation to stress upon long-term chemical exposure. Epigenetic changes following exposure to multiple stressors is another area for further research (Vandegehuchte and Janssen, 2014).

10.5.2 Environmental Genomics in Marine Ecotoxicology

The need to understand marine ecotoxicology from a multiple stressor perspective (eg, the interplay between multiple contaminants and climate change) is highlighted throughout this chapter. To date, however, ecotoxicology studies have been heavily skewed toward examining responses to single or occasional binary stressors, under controlled laboratory conditions. While such tests are critical and underpin our understanding of stressors, there are many limitations with regards to expanding this knowledge to understanding the broader implications of multiple stressors on marine environments: (1) only a relatively few number of organisms are routinely used in ecotoxicological assays, and these are disproportionally biased toward temperate species; (2) laboratory bioassays cannot accurately mimic the complex interaction of multiple contaminants and their variations in exposure across space and time; and (3) laboratory studies provide an oversimplistic view of how contaminants may alter biotic interactions, including indirect effects, eg, how alterations to primary producers and microbial processes may affect higher trophic organisms.

While more realistic environmental scenarios may be carried out using manipulative experiments, eg, mesocosm and transplant studies, the number of scenarios that can be measured is heavily constrained by experimental design and logistics (eg, number of replicates). Field studies, which examine communities in natural scenarios, also have their limitations. For example, a benthic survey may identify a strong relationship between benthic community structure and metals; however, it is not possible to determine that metals were causing the response, as the trend is correlative and other factors, eg, unmeasured contaminants and natural changes across space, may be driving community change. Another important and overlooked limitation of benthic surveys is that macrobenthic invertebrate communities only capture a minute fraction of a system's true diversity. This can lead to the naive assumption that changes at the macrobenthic level reflect the overall response of the ecosystem.

While by no means a panacea, the field of environmental genomics is opening up new and exciting opportunities for exploring the ecological effects of multiple stressors across multiple levels of biological organization, from the subcellular to the community level (Chariton et al., 2015b). Environmental genomics can be broadly defined as the study of genetic material (both DNA and RNA) derived from environmental samples, with the aim of understanding biological structure, function, and response. At the lower level of biological organization, transcriptomics, which produces gene expression profiles of RNA transcripts from organisms exposed to various scenarios, is showing great promise as an early indicator of stress and has the potential to provide unique signatures based on the stressor's mode of action (eg, metals versus endocrine disruptors). Other additional complementary "omic" techniques at the organism level include proteomics and metabolomics, whose endpoints are protein and metabolite profiles, respectively.

As previously noted, one of the greatest challenges is understanding how communities, all organisms and not just those traditionally sampled, respond to a complex array of natural (eg, salinity) and anthropogenic stressors (eg, metals and nutrients). By using metabarcoding, a high-throughput sequencing approach which targets taxonomic informative regions of DNA, researchers are now able to obtain biodiversity profiles that potentially capture all of life, providing a previously unattainable view of biodiversity (eg, diatoms, microbes, fungi, invertebrates). While not yet widely adopted into the ecotoxicological community, this approach has proven highly suited for examining the effects of anthropogenic activities on coastal benthic systems. For example, using metabarcoding, Chariton et al. (2015a) were not only able to distinguish estuaries subjected to varying levels of anthropogenic activity but also able to identify how key environmental stressors (eg, salinity and nutrients) altered the individual distributions of hundreds of taxa and communities as a whole.

As many key biogeochemical cycles are driven by microbial processes, alterations at the compositional level can also affect ecosystem function. Using an approach similar to metabarcoding, profiles of microbial organisms and genes associated with particular steps in key biogeochemical pathways can also be analyzed, for example, nifH and NifK genes are commonly used as molecular markers for nitrogen-fixing bacteria. In addition to metabarcoding, which examines amplified genes of interest, another approach to microbial community analysis is shot-gun sequencing, which examines random sequence fragments from the total DNA (metagenomics) or RNA (metatranscriptomics). The fragments (reads) can then be examined or assembled to provide composition profiles as well as functional profiles by identifying protein coding reads and comparing these to databases with sequences of known biological function.

While these "omic" approaches currently provide a diverse range of informative data at varying levels of biological organization, it is yet to be determined whether an expression at the lower levels of organization (eg, transcriptomes and metabolomes) translates to alterations in community function and structure. Given the broad nature of this topic and the complexity of the approaches, interested readers are directed toward the following publications for a more detailed understanding of the topic: Chariton et al. (2014), Hook (2010), Paulsen and Holmes (2014), Piña and Barata (2011), Sheehan (2013), van Straalen and Feder (2012), and van Straalen and Roelofs (2011).

10.5.3 Ecological Modeling

Since it is impractical to collect empirical data concerning all the potential interactions between environmental variables affected by global change and contaminants of concern, it is necessary to develop predictive approaches to help assess where, when, and how these interactions might affect potential risk (Hooper et al., 2013). To support predictive approaches, mechanistic data need to be included in the assessment process; this is being pursued through the Adverse Outcome Pathway (AOP) framework (Ankley et al., 2010). AOP depict linkages between molecular initiating events (interaction of chemicals with biological targets) and the subsequent cascade of responses across individuals, populations, and communities. Hooper et al. (2013) provide examples of how climate change can impact exposures and bioavailability of chemicals and their interactions with toxicodynamics and toxicokinetics in organisms.

Ecological models are increasingly being used to predict global change impacts on population dynamics, species distributions, and biodiversity. Ecological models integrate physical and biological processes and incorporate mechanistic linkages (van de Brink et al., 2015). Modeling can be particularly useful for scenario testing to identify the most promising options (and eliminate less effective ones) and to identify key knowledge gaps. Recent advances include the development of "whole of system" models, which attempt to include interactions amongst multiple stressors, alternative uses, and increasing pressures on ecosystems—expanding populations, new industries, climate change, and climate variability (van den Brink et al., 2015).

There needs to be better integration of these models with ecotoxicology to consider and predict nonadditive interactions between contaminants and climatic stressors, adaptation, and recovery. For example, the trait-based framework, in which population vulnerability is defined by external exposure, intrinsic sensitivity, and population sustainability, could be expanded to assess population vulnerability to contaminants and climate stressors (Moe et al., 2013).

10.6 CONCLUSIONS

Global change is leading to major shifts in the concentrations and types of contaminants that are reaching our estuarine and marine ecosystems. In addition, additional complexity arises from the changes in ecosystem physicochemistry that result from climate change pressures. Prediction of the impacts of single contaminants in field situations is difficult enough and becomes even more difficult dealing with multiple contaminants and new stressors. Further, while a system may be able to recover from a single event, recovery from multiple stressors or recurrent events may be compromised. It is the pace, frequency, and magnitude of changes that are the major threats (Przeslawski et al., 2008). While some species may be especially vulnerable to climate change, the additional stress of not only contaminants, but also pathogens, invasive species, overharvesting, and habitat destruction magnifies the impacts (Noyes et al., 2009).

While climate change models provide insight into future climate change scenarios, they do not predict what altered climate will mean for marine systems. Species must adapt or move, otherwise populations will be vulnerable to extinction because of their inability to respond to the rate and magnitude of climate change. However, until we are able to understand the causes of such declines, our ability to predict the vulnerability of populations to environmental change based on their geographic range will be limited. Ecotoxicology can assist by providing further information on individual organism physiology, mobility, and habitat requirements to predict species redistribution as a result of climate change.

The absence of consideration of contaminants in IPCC reports to date suggests that better communication is needed between climate change scientists and ecotoxicologists, both in the research and policy sphere. It has been suggested that one immediate management action should be to monitor baseline contaminant concentrations (time series) in water and sediments and to reduce exposure to contaminants (one stressor) where possible, as this may be more easily tackled initially than removing ongoing stressors due to climate change. However, in cases where there are antagonistic interactions between stressors, local interventions may lead to ineffective, costly management actions, and wasted management effort. Careful assessment is required on a case-by-case basis.

Clearly, greater effort is required to understand multiple stressors and how to manage them in the broader context of earth systems science as we move forward into a period of unprecedented change. National and international legal and management frameworks for environmental regulation of contaminants do not yet incorporate global change in their assessment framework. Policy makers and industry around the world therefore need to begin to understand the implication of future changes on the risks of chemicals in the marine environment to ecosystem and human health so that they can begin to implement necessary adaptation and mitigation strategies.

Acknowledgments

The authors would like to thank Drs. Graeme Batley and Stuart Simpson (CSIRO) for reviewing the chapter, Merrin Adams for providing the data for Fig. 10.11, and Greg Rinder for assistance with the figures.

References

ABS (Australian Bureau of Statistics), 2003. Regional Population Growth, Australia and New Zealand, 2001–02. Catalogue number 3218.0. Australian Bureau of Statistics, Canberra, ACT, Australia.

Agardy, T., Alder, J., Dayton, P., Curran, S., Kitchingman, A., Wilson, M., Catenazzi, A., Restrepo, J., Birkeland, C., Blaber, S., Saifullah, S., Branch, G., Boersma, D., Nixon, S., Dugan, P., Davidson, N., Vorosmarty, C., 2005. Coastal systems. In: Reid, W.V. (Ed.), Millennium Ecosystem Assessment: Ecosystems and Human Well-Being, Current State and Trends, vol. 1. Island Press, Washington, DC, USA, pp. 513–549.

Ahnert, A., Borowski, C., 2000. Environmental risk assessment of anthropogenic activity in the deep-sea. J. Aquat. Ecosys. Stress Recov. 7, 299–315.

An, S., Li, H., Guan, B., Zhou, C., Wang, Z., Deng, Z., Zhi, Y., Liu, Y., Xu, C., Fang, S., Jiang, J., Li, H., 2007. China's natural wetlands: past problems, current status, and future challenges. Ambio 4, 335–342.

Andersen, T., Carstensen, J., Hernandez-Garcia, E., Duarte, C.M., 2009. Ecological thresholds and regime shifts: approaches to identification. Trends Ecol. Evol. 24, 49–57.

Ankley, G.T., Bennett, R.S., Erickson, R.J., Hoff, D.J., Hornung, M.W., Johnson, R.D., Mount, N.J.W., Russom, C.L., Schmieder, P.K., Serrrano, J.A., Tietge, J.E., Villeneuve, D.L., 2010. Adverse outcome pathways: a conceptual framework to support ecotoxicology research and risk assessment. Environ. Toxicol. Chem. 29, 730–741.

Batker, D., Schmidt, R., 2015. Environmental and Social Benchmarking Analysis of the Nautilus Minerals Inc. Solwara 1 Project. Earth Economics Report. Available at: http://www.nautilusminerals.com/irm/content/pdf/eartheconomics-reports/earth-economics-may-2015.pdf.

Bayliss, P., Freeland, W., 1989. Seasonal distribution and abundance of dugongs in the Western-Gulf-of-Carpentaria. Wildl. Res. 16, 141–149.

Bellwood, D.R., Hughes, T.P., Folke, C., Nystrom, M., 2004. Confronting the coral reef crisis. Nature 429, 827–833.

Bjorndal, K.A., 1980. Nutrition and grazing behavior of the green turtle *Chelonia mydas*. Mar. Biol. 56, 147–154.

Blasco, J., Durán-Grados, V., Hampel, M., Moreno-Gutiérrez, J., 2014. Towards an integrated environmental risk assessment of emissions from operating ships. Environ. Int. 66, 44–47.

Boesch, D.F., Burroughs, R.H., Baker, J.E., Mason, R.P., Rowe, C.L., Siefert, R.L.M., 2001. Marine Pollution in the United States. Technical Report. Prepared for the Pew Oceans Commission, Arlington, VA, USA.

Bogdal, C., Scheringer, M., 2011. Release of POPs to the Environment. In Climate Change and POPS: Predicting the Impacts. UNEP/AMAP Expert Group Report, Geneva, Switzerland.

Bouchard, M.F., Chevrier, J., Harley, K.G., Kogut, K., Vedar, M., Calderon, N., Trujillo, C., Johnson, C., Bradman, A., Boyd Barr, D., 2011. Prenatal exposure to organophosphate pesticides and IQ in 7-year old children. Environ. Health Perspect. 119, 1189–1195.

Boyer, E.W., Howarth, R.W., 2008. Nitrogen fluxes from rivers to the coastal oceans. In: Capone, D., Carpenter, E.J. (Eds.), Nitrogen in the Marine Environment, second ed. Academic Press, San Diego, CA, USA, pp. 1565–1587.

Brink, C., van Grinsven, H., Jacobsen, B.H., Velthof, G., 2011. Costs and Benefits of Nitrogen in the Environment. The European Nitrogen Assessment: Sources, Effects and Policy Perspectives. Cambridge University Press, Cambridge, UK, pp. 513–540.

Brodie, J., Fabricius, K., De'ath, G., Okaji, K., 2005. Are increased nutrient inputs responsible for more outbreaks of crown-of-thorns starfish? An appraisal of the evidence. Mar. Pollut. Bull. 51, 266–278.

Brown, L.R., Renner, M., Flavin, C., Starke, L., 1998. Vital Signs 1998: The Environmental Trends That Are Shaping Our Future. Worldwatch Institute, W.W. Norton and Company, New York, NY, USA.

van den Berg, H., Zaim, M., Yadav, R.S., Soares, A., Ameneshewa, B., Mnzava, A., Hii, J., Dash, A.P., Ejov, M., 2012. Global trends in the use of insecticides to control vector-borne diseases. Environ. Health Perspect. 120, 577–582.

Canty, M.N., Hagger, J.A., Moore, R.T.B., Cooper, L., Galloway, T.S., 2007. Sublethal impact of short term exposure to the organophosphate pesticide azamethiphos in the marine mollusc *Mytilus edulis*. Mar. Pollut. Bull. 54, 396–402.

Carpenter, S.R., Caraco, N.F., Correll, D.L., Howarth, R.W., Sharpley, A.N., Smith, V.H., 1998. Nonpoint pollution of surface waters with phosphorus and nitrogen. Ecol. Appl. 8, 559–568.

Castel, S.E., Martienssen, R.A., 2013. RNA interference in the nucleus: roles for small RNAs in transcription, epigenetics and beyond. Nat. Rev. Genet. 14, 100–112.

Chariton, A.A., Roach, A.C., Simpson, S.L., Batley, G.E., 2010. Influence of the choice of physical and chemistry variables on interpreting patterns of sediment contaminants and their relationships with estuarine macrobenthic communities. Mar. Freshwater Res. 61, 1109–1122.

Chariton, A.A., Ho, K.T., Proestou, D., Bik, H., Simpson, S.L., Portis, L.M., Cantwell, M.G., Baguley, J.G., Burgess, R.M., Pelletier, M.M., Perron, M., Gunsch, C., Matthews, R.A., 2014. A molecular-based approach for examining responses of eukaryotes in microcosms to contaminant-spiked estuarine sediments. Environ. Toxicol. Chem. 33, 359–369.

Chariton, A.A., Stephenson, S., Morgan, M.J., Steven, A.D.L., Colloff, M.J., Court, L.N., Hardy, C.M., 2015a. Metabarcoding of benthic eukaryote communities predicts the ecological condition of estuaries. Environ. Pollut. 203, 165–174.

Chariton, A.A., Sun, M.Y., Gibson, J., Webb, J.A., Leung, K.M.Y., Hickey, C.W., Hose, G.C., 2015b. Emergent technologies and analytical approaches for understanding the effects of multiple stressors in aquatic environments. Mar. Freshwater Res. http://dx.doi.org/10.1071/MF15190.

Connon, R.E., Geist, J., Werner, I., 2012. Effect-based tools for monitoring and predicting the ecotoxicological effects of chemicals in the aquatic environment. Sensors 12, 12741–12771.

Correll, D.L., 1998. The role of phosphorus in the eutrophication of receiving waters: a review. J. Environ. Qual. 27, 261–266.

Costanza, R., Daly, H.E., 1992. Natural capital and sustainable development. Conserv. Biol. 6, 37–46.

Costanzo, S.D., O'Donohue, M.J., Dennison, W.C., Loneragan, N.R., Thomas, M., 2001. A new approach for detecting and mapping sewage impacts. Mar. Pollut. Bull. 42, 149–156.

Cózar, A., Echevarría, F., González-Gordillo, J.I., Irigoien, X., Úbeda, B., Hernández-León, S., Palma, Á.T., Navarro, S., García-de-Lomas, J., Ruiz, A., Fernández-de-Puelles, M.L., Duarte, C.M., 2014. Plastic debris in the open ocean. Proc. Natl. Acad. Sci. U.S.A. 111, 10239–10244.

De'ath, G., Fabricius, K.E., Sweatman, H., Puotinen, M., 2012. The 27-year decline of coral cover on the Great Barrier Reef and its causes. Proc. Natl. Acad. Sci. U.S.A. 109, 17995–17999.

Duarte, C.M., Holmer, M., Olsen, Y., Soto, D., Marbà, N., Guiu, J., Black, K., Karakassis, I., 2009. Will the oceans help feed humanity? Bioscience 59, 967–976.

Duarte, C.M., Lenton, T.M., Wadhams, P., Wassmann, P., 2012a. Abrupt climate change in the Arctic. Nat. Clim. Change 2, 60–62.

Duarte, C.M., Agustí, S., Wassmann, P., Arrieta, J.M., Alcaraz, M., Coello, A., Marbà, N., Hendriks, I.E., Holding, J., García-Zarandona, I., Kritzberg, E., Vaqué, D., 2012b. Tipping elements in the Arctic marine ecosystem. Ambio 41, 44–55.

Duarte, C.M., 2015. Global change and the future ocean: a grand challenge for marine sciences. Front. Mar. Sci. 1, 1–16.

EASAC (European Academies Science Advisory Council), 2015. Marine Sustainability in an Age of Changing Oceans and Seas. Available at: http://www.academies.fi/wp-content/uploads/2015/06/EASAC-JRC-Marine-Sustainability_Summary.pdf.

Ellis, D., Ellis, K., 1994. Very deep STD. Mar. Pollut. Bull. 28, 472–476.

Ellis, D.V., Poling, G.W., Baer, R.L., 1995. Submarine tailings disposal (STD) for mines – an introduction. Mar. Geores. Geotechnol. 13, 3–18.

Enell, M., Fejes, J., 1995. The nitrogen load to the Baltic Sea – Present situation, acceptable future load and suggested source reduction. Water Air Soil Pollut. 85, 877–882.

Eriksson Wiklund, A.K., Sundelin, B., 2004. Sensitivity to temperature and hypoxia – a seven year field study. Mar. Ecol. Progr. Ser. 274, 209–214.

Ettinger-Epstein, P., Whalan, S.W., Battershill, C.N., de Nys, R., 2007. Temperature cues gametogensis and larval release in a tropical sponge. Mar. Biol. 153, 171–178.

Fabricius, K.E., De'ath, G., Puotinen, M.L., Done, T., Cooper, T.F., Burgess, S.C., 2008. Disturbance gradients on inshore and offshore coral reefs caused by severe tropical cyclone. Limnol. Oceanogr. 53, 690–704.

FAO (Food and Agriculture Organisation of the United Nations), 2012. The State of the World Fisheries and Aquaculture 2012. Rome, Italy.

Fortin, M.G., Couillard, C.M., Pellerin, J., Lebeuf, M., 2008. Effects of salinity on sublethal toxicity of atrazine to mummichog (*Funduluus heteroclitus*) larvae. Mar. Environ. Res. 65, 158–170.

Galanopoulou, S., Vgenopoulos, A., Conispoliatis, N., 2005. DDTs and other chlorinated organic pesticides and polychlorinated biphenyls pollution in the surface sediments of Keratsini Harbour, Saronikos Gulf, Greece. Mar. Pollut. Bull. 50, 520–525.

Galloway, J.N., Dentener, F.J., Capone, D.G., Boyer, E.W., Howarth, R.W., Seitzinger, S.P., Asner, G.P., Cleveland, C., Green, P., Holland, E., 2004. Nitrogen cycles: past, present, and future. Biogeochemistry 70, 153–226.

Garnett, T., Godfray, C., 2012. Sustainable Intensification in Agriculture. University of Oxford, Oxford, UK.

Geisz, H.N., Dickhut, R.M., Cochran, M.A., Fraser, W.R., Ducklow, H.W., 2008. Melting glaciers: a probable source of DDT to the Antarctic marine ecosystem. Environ. Sci. Technol. 42, 3958–3962.

Glasby, G.P., 2000. Lessons learned from deep-sea mining. Science 289, 551–553.

Gouin, T., Armitage, J., Cousins, I., Muir, D.C.G., Ng, C.A., Reid, L., Tao, S., 2013. Influence of global climate change

on chemical fate and bioaccumulation: the role of multimedia models. Environ. Toxicol. Chem. 32, 20–31.
Gray, J.S., Wu, R.S.S., Or, Y.Y., 2002. Effects of hypoxia and organic enrichment on the coastal marine environment. Mar. Ecol. Progr. Ser. 238, 249–279.
Greening, H., Janicki, A., 2006. Toward reversal of eutrophic conditions in a subtropical estuary: water quality and seagrass response to nitrogen loading reductions in Tampa Bay, Florida, USA. Environ. Manag. 38, 163–178.
Hajj-Mohamad, M., Aboulfadl, K., Darwano, H., Madoux-Humery, A.S., Guérineau, H., Sauvé, S., Prévost, M., Dorner, S., 2014. Wastewater micropollutants as tracers of sewage contamination: analysis of combined sewer overflow and stream sediments. Environ. Sci. Process Impacts 16, 2442–2450.
Halpern, B.S., Walbridge, S., Selkoe, K.A., Kappel, C.V., Micheli, F., D'Agrosa, C., Bruno, J.F., Casey, K.S., Ebert, C., Fox, H.E., Fujita, R., Heinemann, D., Lenihan, H.S., Madin, E.M.P., Perry, M.T., Selig, E.R., Spalding, M., Steneck, R., Watson, R., 2008. A global map of human impact on marine ecosystems. Science 319, 948–952.
Hamm, L., Stive, M.J.F., 2002. Shore nourishment in Europe. Coast. Eng. 47, 79–263.
Harrington, L., Fabricus, K., Raglesham, G., Negri, A., 2005. Synergistic effects of diuron and sedimentation on photosynthesis and survival of crustose coralline algae. Mar. Pollut. Bull. 51, 415–427.
Haynes, D., Ralph, P., Prange, J., Dennison, B., 2000. The impact of the herbicide diuron on photosynthesis in three species of tropical seagrass. Mar. Pollut. Bull. 41, 288–293.
Hedman, J., 2006. Fate of Contaminants in Baltic Sea Sediment Ecosystems: Experimental Studies on the Effects of Macrofaunal Bioturbation (Licenthiate thesis). ISSN: 1401-4106. Department of Systems Ecology, Stockholm University, Stockholm, Sweden.
Hennessy, K., 2011. Climate change impacts. In: Cleugh, H., Stafford Smith, M., Battaglia, M., Graham, P. (Eds.), Climate Change. CSIRO Publishing, Collingwood, VIC, Australia, pp. 45–57.
Heugens, E.H.W., Henrick, A.J., Dekker, T., Van Stralen, N.M., Admiraal, W., 2001. A review on the effects of multiple stressors on aquatic organisms and analysis of uncertainty factors for use in risk assessment. Crit. Rev. Toxicol. 31, 247–284.
Hignite, C., Azarnoff, D.L., 1977. Drugs and drug metabolites as environmental contaminants: chlorophenoxyisobutyrate and salicylic acid in sewage water effluent. Life Sci. 20, 337–341.
Hilton, M.J., Manning, S.S., 1995. Conversion of coastal habitats in Singapore: indications of unsustainable development. Environ. Conserv. 22, 307–322.
Hoegh-Guldberg, O., Mumby, P.J., Hooten, A.J., Steneck, R.S., Greenfiled, P., Gomez, E., Harvell, C.D., Sale, P.F., Edwards, A.J., Caldeira, K., Knowlton, N., Eakin, C.M., Iglesias-Prieto, R., Muthiga, N., Bradbury, R.H., Dubi, A., Hatziolos, M.E., 2007. Coral reefs under rapid climate change and ocean acidification. Science 318, 1737–1742.
Hoegh-Guldberg, O., 1999. Climate change, coral bleaching and the future of the world's coral reefs. Mar. Freshwater Res. 50, 839–866.
Hoeksema, R.J., 2007. Three stages in the history of land reclamation in the Netherlands. Irrig. Drain. 56, 113–126.
Hook, S., 2010. Promise and progress in environmental genomics: a status report on the applications of gene expression-based microarray studies in ecologically relevant fish species. J. Fish Biol. 77, 1999–2022.
Hooper, M.J., Ankley, G.T., Cristol, D.A., Maryoung, L.A., Noyes, P.D., Pinkerton, K.E., 2013. Interactions between chemical and climate stressors: a role for mechanistic toxicology in assessing climate change risks. Environ. Toxicol. Chem. 32, 32–48.
Howarth, R.W., Marino, R., 2006. Nitrogen as the limiting nutrient for eutrophication in coastal marine ecosystems: evolving views over three decades. Limnol. Oceanogr. 51, 364–376.
Howarth, R.W., 2008. Coastal nitrogen pollution: a review of sources and trends globally and regionally. Harmful Algae 8, 14–20.
IEA (International Energy Agency), 2012. World Energy Outlook 2012. Paris, France.
IMO (International Maritime Organisation), 2012. International Shipping Facts and Figures – Information Resources on Trade, Safety, Security, Environment. Available at: http://www.imo.org/en/KnowledgeCentre/ShipsAndShippingFactsAndFigures/TheRoleandImportanceofInternationalShipping/Documents/International%20Shipping%20-%20Facts%20and%20Figures.pdf.
IPCC (International Panel on Climate Change), 2014. Climate Change 2014: Synthesis Report. Contribution of Working Groups I, II and III to the Fifth Assessment Report of the Intergovernmental Panel on Climate Change. Geneva, Switzerland, 151 pp.
Janaki Devi, V., Nagarani, N., Yokesh Babu, M., Vijayalakshimi, N., Kumaraguru, A.K., 2012. Genotoxic effects of profenofos on the marine fish, *Therapon jarbua*. Toxicol. Mech. Methods 22, 111–117.
Jones, R., 2005. The ecotoxicological effects of photosystem II herbicides on corals. Mar. Pollut. Bull. 51, 495–506.
Keating, B.A., Carberry, P.S., Bindraban, P.S., Asseng, S., Meinke, H., Dixon, J., 2010. Eco-efficient agriculture: concepts, challenges, and opportunities. Crop Sci. 50, S109–S119.
Kennish, M., de Jonge, V., 2011. Chemical introductions to the systems: diffuse and nonpoint source pollution from chemicals (nutrients: eutrophication). In: Wolanski, E., McLusky, D. (Eds.), Treatise on Estuarine and Coastal Science, vol. 8. Elsevier Science, UK, pp. 113–148.

Kimberly, D.A., Salice, C.J., 2012. Understanding interactive effects of climate change and toxicants: importance of evolutionary processes. Integr. Environ. Assess. Manag. 8, 385–386.

Landis, W.G., Durda, J.L., Brooks, M.L., Chapman, P.M., Menzie, C.A., Stahl, R.G., Stauber, J.L., 2013. Ecological risk assessment in the context of global climate change. Environ. Toxicol. Chem. 32, 1–14.

Law, R.J., 2014. An overview of time trends in organic contaminant concentrations in marine mammals: going up or down? Mar. Pollut. Bull. 82, 7–10.

Lawrence, J.M., 1996. Mass mortality of echinoderms from abiotic factors. In: Jangoux, M., Lawrence, J.M. (Eds.), Echinoderm Studies, vol. 5. Balkema, Rotterdam, Netherlands, pp. 103–137.

Li, W., Gao, K., Beardall, J., 2013. Interactive effects of ocean acidification and nitrogen limitation on the diatom *Phaeodactylum tricornutum*. PLoS One 7, e51590.

Macneale, K.H., Kiffney, P.M., Scholz, N.L., 2010. Pesticides, aquatic food webs, and the conservation of Pacific salmon. Front. Ecol. Environ. 8, 475–482.

Magnusson, M., Heimann, K., Negri, A.P., 2008. Comparative effects of herbicides on photosynthesis and growth of tropical estuarine microalgae. Mar. Pollut. Bull. 56, 1545–1552.

Manning, T., Ferrell, D., 2007. Dioxins in fish and other seafood from Sydney Harbour, Australia. Organohalogen Comp. 69, 343–346.

Mapstone, B., 2011. Introduction. In: Cleugh, H., Stafford Smith, M., Battaglia, M., Graham, P. (Eds.), Climate Change. CSIRO Publishing, Collingwood, VIC, Australia, pp. ix–xii.

Margolis, S.V., Burns, R.G., 1976. Pacific deep-sea manganese nodules: their distribution, composition, and origin. Annu. Rev. Earth Planet Sci. 4, 229–263.

Marques, A., Nunes, M.L., Moore, S.K., Strom, M.S., 2010. Climate change and seafood safety: human health implications. Food Res. Int. 43, 1766–1779.

Maruya, K.A., Smalling, K.L., Vetter, W., 2005. Temperature and congener structure affect the enantioselectivity of toxaphene elimination by fish. Environ. Sci. Technol. 39, 3999–4004.

McComb, A., Davis, J., 1993. Eutrophic waters of southwestern Australia. Fertil. Res. 36, 105–114.

McCook, L.J., Schaffelke, B., Apte, S.C., Brinkman, R., Brodie, J., Erftemeijer, P., Eyre, B., Hoogerwerf, F., Irvine, I., Jones, R., King, B., Marsh, H., Masini, R., Morton, R., Pitcher, R., Rasheed, M., Sheaves, M., Symonds, A., Warne, M., St Warne, J., 2015. Synthesis of Current Knowledge of the Biophysical Impacts of Dredging and Disposal on the Great Barrier Reef: Report of an Independent Panel of Experts. Great Barrier Reef Marine Park Authority, Townsville, Australia.

McKinnon, A., Allen, J., Woodburn, A., 2015. Development of greener vehicles, aircraft and ships. Chapter 8. In: McKinnon, A., Browne, M., Whiteing, A., Piecyk, M. (Eds.), Green Logistics: Improving the Environmental Sustainability of Logistics, third ed. Kogan Page Ltd, London, UK, pp. 165–193.

Mirbahai, L., Chipman, J.K., 2014. Epigenetic memory of environmental organisms: a reflection of lifetime stressor exposures. Mutat. Res. 764, 10–17.

Moberg, F., 1999. Ecological goods and services of coral reef ecosystems. Ecol. Econ. 29, 215–233.

Moe, S.J., de Schamphelaere, K., Clements, W.H., Sorensen, M.T., Van den Brink, P.J., Liess, M., 2013. Combined and interactive effects of global climate change and toxicants on populations and communities. Environ. Toxicol. Chem. 32, 49–61.

Moore, C., 2003. Across the Pacific Ocean, plastics, plastics, everywhere. Nat. Hist. Mag. New York, NY, USA. Available at: http://www.naturalhistorymag.com/htmlsite/master.html?http://www.naturalhistorymag.com/htmlsite/1103/1103_feature.html.

Mumby, P.J., Hastings, A., Edwards, H.J., 2007. Thresholds and the resilience of Caribbean coral reefs. Nature 450, 98–101.

Nath, B.N., Sharma, R., 2000. Environment and deep-sea mining: a perspective. Mar. Geores. Geotechnol. 18, 285–294.

Nicholls, R.J., Leatherman, S.P., 1995. The implications of accelerated sea-level rise for developing countries: a discussion. J. Coast. Res. 14, 303–323.

Nicholls, R.J., Wong, P.P., Burkett, V.R., Codignotto, J.O., Hay, J.E., McLean, R.F., Ragoonaden, S., Woodroffe, C.D., 2007. Coastal systems and low-lying areas. Contribution of working group II to the fourth assessment report of the intergovernmental panel on climate change. In: Parry, M.L., Canziani, O.F., Palutikof, J.P., van der Linden, P.J., Hanson, C.E. (Eds.), Climate Change 2007: Impacts, Adaptation and Vulnerability. Cambridge University Press, Cambridge, UK, pp. 315–356.

Nordstrom, K.F., 2000. Beaches and Dunes of Developed Coasts. Cambridge University Press, Cambridge, UK.

Noyes, P.D., McElwee, M.K., Miller, H.D., Clark, B.W., Van Tiem, L.A., Walcott, K.C., Erwin, K.N., Levin, E.D., 2009. The toxicology of climate change: environmental contaminants in a warming world. Environ. Int. 35, 971–986.

Nyberg, E., Faxneld, S., Danielsson, S., Eriksson, U., Miller, A., Bignert, A., 2015. Temporal and spatial trends of PCBs, DDTs, HCHs, and HCB in Swedish marine biota 1969–2012. Ambio 44, 484–497.

OSPAR (Oslo/Paris Convention (for the Protection of the Marine Environment of the North-East Atlantic)), 2004. Environmental Impacts to Marine Species and Habitats

of Dredging for Navigational Purposes. OSPAR Commission, Publication No. 208/2004, London, UK.
OSPAR, 2008a. Assessment of Environmental Impact of Land Reclamation. OSPAR Commission, London, UK.
OSPAR, 2008b. Literature Review on the Impacts of Dredged Sediment Disposed at Sea. OSPAR Commission Publication No. 362/2008, London, UK.
Ozcoasts, 2010. Economic Consequences of Acid Sulfate Soils. OzCoasts Australian Online Coastal Information. Australian Government, Geoscience Australia. Available at: http://www.ozcoasts.gov.au/indicators/econ_cons_acid_sulfate_soils.jsp.
Paulsen, I.T., Holmes, A.J., 2014. Environmental Microbiology: Methods and Protocols. Humana Press, Totowa, NJ, USA.
Pelletier, M.C., Burgess, R.M., Hi, K.T., Kuhn, A., McKinney, R.A., Ryba, S.A., 1997. Phototoxicity of individual polycyclic aromatic hydrocarbons and petroleum to marine invertebrate larvae and juveniles. Environ. Toxicol. Chem. 16, 2190–2199.
Pimentel, D., 2009. Pesticides and pest control. In: Peshin, R., Dhawan, A. (Eds.), Integrated Pest Management: Innovation-Development Process. Springer, Dordrecht, Netherlands, pp. 83–87.
Piña, B., Barata, C., 2011. A genomic and ecotoxicological perspective of DNA array studies in aquatic environmental risk assessment. Aquat. Toxicol. 105, 40–49.
Przeslawski, R., Ahong, S., Bryne, M., Worheides, G., Hutchings, P., 2008. Beyond corals and fish: the effects of climate change on noncoral benthic invertebrates of tropical reefs. Glob. Change Biol. 14, 2773–2795.
Reich, K.J., Worthy, G.A., 2006. An isotopic assessment of the feeding habits of free-ranging manatees. Mar. Ecol. Progr. Ser. 322, 303–309.
Rios, L.M., Jones, P.R., Moore, C., Narayan, U.V., 2010. Quantitation of persistent organic pollutants adsorbed on plastic debris from the Northern Pacific Gyre's "eastern garbage patch". J. Environ. Monit. 12, 2226–2236.
Rosenberg, R., Lindahl, O., Blanck, H., 1988. Silent spring in the sea. Ambio 289–290.
Sammut, J., Melville, M.D., Callinan, R.B., Fraser, G.C., 1995. Estuarine acidification: impacts on aquatic biota of draining acid sulphate soils. Aust. Geogr. Stud. 33, 89–100.
Santos, L.H., Araújo, A.N., Fachini, A., Pena, A., Delerue-Matos, C., Montenegro, M.C., 2010. Ecotoxicological aspects related to the presence of pharmaceuticals in the aquatic environment. J. Hazard. Mater. 175, 45–95.
Schiedek, D., Sundelin, B., Readman, J.W., Macdonald, R.W., 2007. Interactions between climate change and contaminants. Mar. Pollut. Bull. 54, 1845–1856.
Sericano, J.L., Wade, T.L., Sweet, S.T., Ramirez, J., Lauenstein, G.G., 2014. Temporal trends and spatial distribution of DDT in bivalves from the coastal marine environments of the continental United States, 1986–2009. Mar. Pollut. Bull. 81, 303–316.
Sheehan, D., 2013. Next-generation genome sequencing makes non-model organisms increasingly accessible for proteomic studies: some implications for ecotoxicology. J. Proteom. Bioinform. 6, 10000e21. http://dx.doi.org/10.4172/jpb.10000e21.
Smayda, T.J., 1997. Harmful algal blooms: their ecophysiology and general relevance to phytoplankton blooms in the sea. Limnol. Oceanogr. 42, 1137–1153.
Smith, G., Ives, L.D., Nagelkerken, I.A., Ritchie, K.B., 1996. Caribbean sea fan mortalities. Nature 383, 487.
Smith, V.H., Tilman, G.D., Nekola, J.C., 1999. Eutrophication: impacts of excess nutrient inputs on freshwater, marine, and terrestrial ecosystems. Environ. Pollut. 100, 179–196.
Smith, V.H., 1998. Cultural eutrophication of inland, estuarine, and coastal waters. In: Pace, M.L., Groffman, P.M. (Eds.), Successes, Limitations and Frontiers in Ecosystem Science. Springer, New York, NY, USA, pp. 7–49.
Song, M.Y., Brown, J.J., 1998. Osmotic effects as a factor modifying insecticide toxicity on *Aedes* and *Artemia*. Ecotoxol. Environ. Saf. 41, 195–202.
Stahl, R.G., Hooper, M.L., Balbus, J.M., Clements, W., Fritz, A., Gouin, T., Helm, R., Hickey, C., Landis, W.G., Moe, S.J., 2013. The influence of global climate change on the scientific foundations and applications of environmental toxicology and chemistry: introduction to a SETAC international workshop. Environ. Toxicol. Chem. 32, 13–19.
State of the Climate, 2014. Bureau of Meteorology and Commonwealth Scientific and Industrial Research Organisation, Canberra, Australia.
State of the Environment, 2011. Australia State of the Environment 2011. Independent Report to the Australian Government Minister for Sustainability, Environment, Water, Population and Communities. Australian Government, Canberra, Australia.
Stauber, J.L., Chariton, A., Binet, M., Simpson, S., Batley, G., Durr, M., Fitzpatrick, R., Shand, P., 2008. Water Quality Screening Risk Assessment of Acid Sulfate Soil Impacts in the Lower Murray, SA. CSIRO Land and Water Science Report 45/08, Sydney, Australia, 127 pp.
Stauber, J.L., Kookana, R.S., Boxall, A.B.A., 2012. Contaminants and climate change: multiple stressors in a changing world. In: Proceedings of Water and Climate: Policy Implementation Challenges National Conference, Canberra, ACT, Australia, 1–3 May, 2012, 10 pp.
Stirling, I., Lunn, N.J., Iacozza, J., 1999. Long term trends in the population ecology of polar bears in Western Hudson Bay in relation to climate change. Arctic 52, 294–306.
van Straalen, N.M., Feder, M.E., 2012. Ecological and evolutionary functional genomics–how can it contribute to

the risk assessment of chemicals? Environ. Sci. Technol. 46, 3–9.

van Straalen, N.M., Roelofs, D., 2011. An Introduction to Ecological Genomics. Oxford University Press, Oxford, UK.

Tanaka, K., Takada, H., Yamashita, R., Mizukawa, K., Fukuwaka, M.A., Watanuki, Y., 2013. Accumulation of plastic-derived chemicals in tissues of seabirds ingesting marine plastics. Mar. Pollut. Bull. 69, 219–222.

Teuten, E.L., Rowland, S.J., Galloway, T.S., Thompson, R.C., 2007. Potential for plastics to transport hydrophobic contaminants. Environ. Sci. Technol. 41, 7759–7764.

UNEP (United Nations Environment Programme), 2013. Global Chemicals Outlook - Towards Sound Management of Chemicals. New York, NY, USA.

UNFPA (United Nations Population Fund), 2014. UNFPA State of the World Population 2014. United Nations Population Fund. New York, NY, USA.

USGS (United States Geological Survey), 2008. Circum-Arctic Resource Appraisal: Estimates of Undiscovered Oil and Gas North of the Arctic Circle. Available at: http://pubs.usgs.gov/fs/2008/3049/fs2008-3049.pdf.

Valiela, I., Bowen, J.L., York, J.K., 2001. Mangrove forests: one of the world's threatened major tropical environments. Bioscience 51, 807–815.

Valiela, I., 2006. Global Coastal Change. Wiley-Blackwell, Oxford, UK.

Van den Brink, P., Choung, C.B., Landis, W., Mayer-Pinto, M., Pettigrove, V., Scanes, P., Smith, R., Stauber, J., 2015. New approaches to the ecological risk assessment of multiple stressors. Mar. Freshwater Res 67, 429–439.

Vandegehuchte, M.B., Janssen, C.R., 2011. Epigenetics and its implications for ecotoxicology. Ecotoxicology 20, 607–624.

Vandegehuchte, M.B., Janssen, C.R., 2014. Epigenetics in an ecotoxicological context. Mutat. Res. 764–765, 36–45.

Veron, J.E.N., 1995. Corals in Space and Time: The Biogeography and Evolution of the Scleractinia. Cornell University Press, Ithaca, NY, USA.

Vitousek, P.M., Aber, J.D., Howarth, R.W., Likens, G.E., Matson, P.A., Schindler, D.W., Schlesinger, W.H., Tilman, D.G., 1997a. Human alterations of the global nitrogen cycle: sources and consequences. Ecol. Appl. 7, 737–750.

Vitousek, P.M., Aber, J., Howarth, R.W., Likens, G.E., Matson, P.A., Schindler, D.W., Schlesinger, W.H., Tilman, G., 1997b. Human Alteration of the Global Nitrogen Cycle: Causes and Consequences. Ecological Society of America, Washington, DC, USA.

Vogt, C., 2012. International Assessment of Marine and Riverine Disposal of Mine Tailings. Secretariat, London Convention/London Protocol, International Maritime Organization, London, UK & United Nations Environment Programme – Global Program of Action.

Warwick, R.M., 1986. A new method for detecting pollution effects on marine macrobenthic communities. Mar. Biol. 92, 557–562.

Weber, M., de Beer, D., Lott, C., Polerecky, L., Kohls, K., Abed, R.M.M., Ferdelman, T.G., Fabricius, K.E., 2012. Mechanisms of damage to corals exposed to sedimentation. Proc. Natl. Acad. Sci. U.S.A. 109, 1558–1567.

WEF (World Economic Forum), 2014. Global Risk, ninth ed. Geneva, Switzerland.

Wildgust, M.A., Jones, M.B., 1998. Salinity change and the toxicity of the free cadmium ion ($Cd(aq)^{2+}$) to *Neomysis integer* (Crustacea: Mysidacea). Aquat. Toxicol. 41, 187–192.

World Ocean Review, 2010. Living with the Oceans, vol. 1. Maribus, Hamburg, Germany.

World Ocean Review, 2013. Living with the Oceans. The Future of Fish – the Fisheries of the Future, vol. 2. Maribus, Hamburg, Germany.

World Ocean Review, 2014. Marine Resources – Opportunities and Risks, vol. 3. Maribus, Hamburg, Germany.

Wu, R.S.S., 2002. Hypoxia: from molecular responses to ecosystem responses. Mar. Pollut. Bull. 45, 35–45.

Index

'*Note*: Page numbers followed by "f" indicate figures, "t" indicate tables and "b" indicate boxes.'

T

Printed in the United States
By Bookmasters